Periodic Table of the Elements with the Gmelin System Numbers

Each cell is given as: *atomic number · symbol · Gmelin System number*

1	2	3	4	5	6	7	8	9	10	11	12	13	14	15	16	17	18
1 H 2																	2 He 1
3 Li 20	4 Be 26											5 B 13	6 C 14	7 N 4	8 O 3	9 F 5	10 Ne 1
11 Na 21	12 Mg 27											13 Al 35	14 Si 15	15 P 16	16 S 9	17 Cl 6	18 Ar 1
19 K *22	20 Ca 28	21 Sc 39	22 Ti 41	23 V 48	24 Cr 52	25 Mn 56	26 Fe 59	27 Co 58	28 Ni 57	29 Cu 60	30 Zn 32	31 Ga 36	32 Ge 45	33 As 17	34 Se 10	35 Br 7	36 Kr 1
37 Rb 24	38 Sr 29	39 Y 39	40 Zr 42	41 Nb 49	42 Mo 53	43 Tc 69	44 Ru 63	45 Rh 65	46 Pd 64	47 Ag 61	48 Cd 33	49 In 37	50 Sn 46	51 Sb 18	52 Te 11	53 I 8	54 Xe 1
55 Cs 25	56 Ba 30	57** La 39	72 Hf 43	73 Ta 50	74 W 54	75 Re 70	76 Os 70	77 Ir 67	78 Pt 68	79 Au 62	80 Hg 34	81 Tl 38	82 Pb 47	83 Bi 19	84 Po 12	85 At 8a	86 Rn 1
87 Fr 25a	88 Ra 31	89*** Ac 40	104 71	105 71													

Lanthanides (39)

58 Ce	59 Pr	60 Nd	61 Pm	62 Sm	63 Eu	64 Gd	65 Tb	66 Dy	67 Ho	68 Er	69 Tm	70 Yb	71 Lu

***Actinides**

90 Th 44	91 Pa 51	92 U 55	93 Np 71	94 Pu 71	95 Am 71	96 Cm 71	97 Bk 71	98 Cf 71	99 Es 71	100 Fm 71	101 Md 71	102 No 71	103 Lr 71

* NH₄ 23

A Key to the Gmelin System is given on the Inside Back Cover

Gmelin Handbook of Inorganic and Organometallic Chemistry

8th Edition

Gmelin Handbook of Inorganic and Organometallic Chemistry

8th Edition

Gmelin Handbuch der Anorganischen Chemie

Achte, völlig neu bearbeitete Auflage

PREPARED
AND ISSUED BY

Gmelin-Institut für Anorganische Chemie
der Max-Planck-Gesellschaft
zur Förderung der Wissenschaften

Director: Ekkehard Fluck

FOUNDED BY

Leopold Gmelin

8TH EDITION

8th Edition begun under the auspices of the
Deutsche Chemische Gesellschaft by R. J. Meyer

CONTINUED BY

E. H. E. Pietsch and A. Kotowski, and by
Margot Becke-Goehring

Springer-Verlag Berlin Heidelberg GmbH 1993

Gmelin-Institut für Anorganische Chemie
der Max-Planck-Gesellschaft zur Förderung der Wissenschaften

Volumes published on "Beryllium" (Syst. No. 26)

Main Volume: Element and Compounds – 1930

Supplement Volume
A 1: The Element. Production, Atom, Molecules, Chemical Behavior, Toxicology – 1986
A 2: The Element. Physical Properties – 1991
A 3: The Element. Physical Properties (continued) and Electrochemical Behavior
     (**present volume**) – 1993
Organoberyllium Compounds 1 – 1987

Gmelin Handbook of Inorganic and Organometallic Chemistry

8th Edition

Be
Beryllium

Supplement Volume A 3

The Element. Physical Properties (continued)
and Electrochemical Behavior

With 68 illustrations

AUTHORS   Gudrun Bär, Lieselotte Berg, Gerhard Czack, Dieter Gras, Vera Haase

EDITORS   Gudrun Bär, Lieselotte Berg, Gerhard Czack, Dieter Gras,
                 Elisabeth Koch-Bienemann

CHIEF EDITOR   Lieselotte Berg

System Number 26

Springer-Verlag Berlin Heidelberg GmbH 1993

LITERATURE CLOSING DATE: PARTLY 1992

Library of Congress Catalog Card Number: Agr 25-1383

ISBN 978-3-662-10322-7   ISBN 978-3-662-10320-3 (eBook)
DOI 10.1007/978-3-662-10320-3

© by Springer-Verlag Berlin Heidelberg 1993
Originally published by Springer-Verlag, Berlin · Heidelberg · New York · London · Paris · Tokyo · Hong Kong · Barcelona in 1993
Softcover reprint of the hardcover 8th edition 1993

Preface

The present Supplement Volume Beryllium A 3 continues and completes the description of the physical properties of the element, begun in Supplement Volume A 2, 1991, and also treats the electrochemical behavior of the metal. The unique combination of the Be properties, which was pointed out in Supplement Volume A 2, is also demonstrated in the following chapters of this Volume A 3:

13. Electrical Properties
14. Electronic Properties
15. Optical Properties. Emission and Impact Phenomena
16. Electrochemical Behavior

Starting with the electrical properties, Be is a rather good electrical conductor in contrast to what might be expected. Superconductivity was studied, especially on films. Quantum effects, which are more pronounced in Be than in most other metals, are the reason for numerous investigations of the magnetoresistance and the magnetic-breakdown effect.

The basis for many of the characteristic properties is the unique nature of bonding in Be as a consequence of its peculiar electronic structure and the special shape of its Fermi surface which also gave rise to further numerous studies. Detailed cluster calculations were performed to better understand the bonding in the metal.

Regarding the optical properties, the high reflectivity of Be, particularly in the infrared region, makes it attractive for the fabrication of precision optical surfaces (mirrors); it is also useful for solar-collector surfaces in spacecraft applications. Emission and electron- and ion-impact phenomena as well as neutron optics are also discussed.

In the concluding chapter on the electrochemical behavior of Be as a cathode or anode, the large section on electrodeposition contains the studies of the fundamental processes important for the technical production of pure Be and for the refining of the crude metal.

Frankfurt/Main, September 1993   Lieselotte Berg

Table of Contents

13 Electrical Properties

13.1 General Remarks

Beryllium should be an insulator due to its electronic properties (see pp. 48 ff.); its two valence electrons nearly fill the first Brillouin zone. However, Be is a good electrical conductor. Its conductivity at room temperature is only four times lower than that of Cu under the same conditions. Two effects contribute to this high conductivity: first, band structure calculations indicate that there is an overlap of the electrons from the first into the second and higher Brillouin zones; second, the amplitude of atomic vibration and, hence, electron-phonon scattering is rather low, because of the relatively strong elastic forces binding the light Be atoms. The temperature dependence of the resistivity is difficult to interpret because of the high Debye temperature (see "Beryllium" Suppl. Vol. A 2, 1991, p. 252) and the anisotropy of the resistivity [1]. At low temperatures, between 200 and 40 K, the resistivity of Be is considerably lower, i.e., its conductivity is higher than that of Al or Cu; see figure in [2].

It should be born in mind that differences in both purity and thermal history of a sample create substantial differences in resistivity values. Therefore it seems reasonable to attribute at least some of the scatter in reported data to variations in these parameters [3]. Large differences among reported values for electrical properties of Be due to impurities, as shown by measurements of the residual resistance in materials of different degrees of purity, are also mentioned in the review [4]. The variation of the electrical resistivity of Be with the purity and preparation method of the samples is shown in a figure and discussed in the monograph [5].

Superconductivity is discussed especially for films as a function of the preparation, film thickness, and heat treatment.

Because of its high Debye temperature and its low density of carrier states at the Fermi level, a number of quantum effects are more pronounced in Be than in most other metals. This feature renders Be a useful material for investigating its magnetoresistance with respect to Kohler's rule and such effects as magnetic breakdown. The study of factors that influence magnetic breakdown is a topic of special interest for obtaining information on the electron dynamics of metals.

Compared with any other metal, Be single crystals show a strong anisotropy of the Hall effect and extremely large absolute values of the Hall coefficient.

The data available on the Be thermoelectric power vary considerably depending on the different sample characteristics and experimental conditions.

References:

[1] Mitchell, M. A. (J. Appl. Phys. **46** [1975] 4742/6).
[2] Lupton, D.; Aldinger, F. (Radex Rundschau **1/2** [1983] 43/51, 48), Aldinger, F.; Petzow, G. (Radex Rundschau **3/4** [1972] 275/83, 277).
[3] Stonehouse, A. J.; Carrabine, J. A.; Beaver, W. W. (in: Hausner, H. H.; BeRYLLIUM , Its Metallurgy and Properties, Univ. Calif. Press, Berkeley 1965, pp. 191/205, 199).
[4] Petzow, G.; Aldinger, F.; Jönsson, S.; Preuss, O. (Beryllium and Beryllium Compounds, Ullmann's Encycl. Ind. Chem. 5th Ed. A **4** [1985] 11/33, 12).
[5] Talbot, S. (Monographies sur les Metaux de Haute Purete **2** [1977] 88/112, 96/8).

13.2 Electrical Resistivity

13.2.1 Temperature Dependence

Recommended values for the electrical resistivity of Be and other alkaline earth elements (Mg, Ca, Sr, and Ba), available from the literature up to 1979, were compiled over a wide range of temperature, critically evaluated, analyzed, and synthesized. For Be 80 sets of experimental data were available. From these data provisional values of the total resistivity, ρ, for 99.9 + % Be and the intrinsic resistivity, ρ_i ($= \rho - \rho_0$, where ρ_0 is the residual resistivity at low temperatures), were deduced as functions of temperature, and from the single-crystal data resistivity values for polycrystalline samples were calculated from 10 to 1200 K. Above 1200 K the values follow experimental data from the literature. The provisional ρ values below 100 K are applicable to specimens with residual resistivity ratios ($\rho_{300}/\rho_{4.2}$) of 0.00718 $\mu\Omega \cdot$ cm ($\perp$c-axis), 0.00426 $\mu\Omega \cdot$ cm ($\|$c-axis), and 0.0332 $\mu\Omega \cdot$ cm (polycrystalline metal). The uncertainty of the ρ values was believed to be within 8% below 1000 K, and within $\pm$10% from 1000 to 1500 K. Above 40 K, the uncertainty of ρ_i is a little higher than that of ρ; below 40 K the uncertainty is so large that ρ_i values are not given. Selected ρ and ρ_i values (both in $\mu\Omega \cdot$ cm) according to Chi [1] are given in the following table:

T in K	$\perp$c-axis ρ	$\perp$c-axis ρ_i	$\|$c-axis ρ	$\|$c-axis ρ_i	polycrystalline ρ	polycrystalline ρ_i
1	0.0072		0.0043		0.0332	
10	0.0072		0.0043		0.0332	
20	0.0076		0.0046		0.0336	
30	0.0086		0.0054		0.0345	
40	0.0109	0.0037	0.0074	0.0031	0.0367	0.0035
50	0.0150	0.0078	0.0112	0.0069	0.0407	0.0075
70	0.0325	0.0253	0.0293	0.0250	0.0584	0.0252
100	0.103	0.0954	0.111	0.107	0.133	0.0993
150	0.447	0.440	0.550	0.546	0.510	0.447
200	1.16	1.15	1.48	1.48	1.29	1.26
273.15	2.72	2.71	3.54	3.54	3.02	2.99
293	3.21	3.20	4.19	4.19	3.56	3.53
300	3.38	3.38	4.43	4.43	3.76	3.73
400	6.08	6.07	8.07	8.07	6.76	6.73
500	8.91	8.90	12.0	12.0	9.94	9.91
600	11.8	11.8	16.0	16.0	13.2	13.2
700	14.8	14.8	20.2	20.2	16.5	16.5
800	17.9	17.9	24.5	24.5	20.0	20.0
900	21.1	21.1	28.9	28.9	23.7	23.7
1000	24.4	24.4	33.5	33.5	27.5	27.5
1100	27.8	27.8	38.3	38.3	31.5	31.5
1200	31.5	31.5	43.3	43.3	35.7	35.7
1300					40.1	40.1
1400					44.8	44.8
1500					49.9	49.9

Previously measured resistivity values over extended temperature ranges for Be samples of different purity and preparation methods are compiled and discussed in [2, 3].

Measurements of the electrical resistivity of very pure single crystals ($\rho_{273}/\rho_{4.2}$=10.0 and 15.5 for the sample axis ∥ and ⊥ to the c-axis, respectively) in the range 4.2 to 900 K gave the results shown in **Fig. 13-1** from Mitchell [4] in comparison with the results of Powell [2] obtained on well-annealed polycrystalline Be between 293 and ~1000 K. The anisotropy of the temperature-dependent part of the resistivity was about 35% at room temperature (in contrast to 12% found in earlier studies [5]). The anisotropy was approximately constant at temperatures above 273 K and decreased rapidly below [4]. Conversely, in the range 4.2 to 273 K, the anisotropy $\rho_{\|}/\rho_{\perp}$ of the Be resistance increased rapidly with the temperature and was practically independent of the impurity content (measurements on 99.99 and 99.6 wt% pure samples) [6].

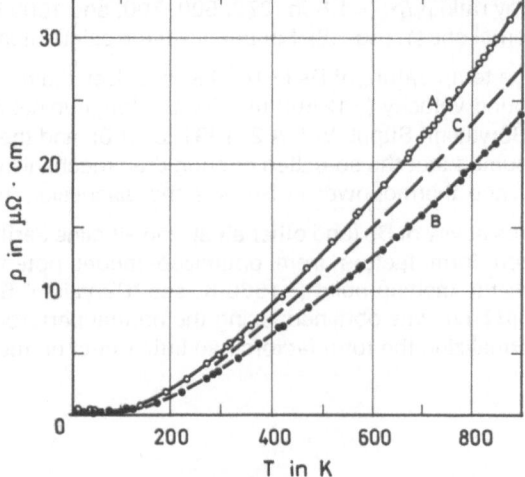

Fig. 13-1. Total electrical resistivity, ρ, of two Be single crystals.

A: c-axis ∥ sample axis; $\rho_{0\|}$=0.41 $\mu\Omega\cdot$cm [4].
B: c-axis ⊥ sample axis; $\rho_{0\perp}$=0.19 $\mu\Omega\cdot$cm [4].
C: polycrystalline sample, from Powell [2].

Theoretical ρ(T) curves for the range ~30 to 300 K are represented in a figure in the paper [7] together with experimental values selected by Meaden [8] from earlier measurements [2, 5]. An empirical pseudopotential fitted to Fermi surface measurements and a Heine-Animalu pseudopotential were used to calculate the electron-phonon coupling. Good agreement was obtained between the experimental results and the curve calculated with the empirical pseudopotential at temperatures above 150 K, whereas the Heine-Animalu pseudo-potential gave better agreement at lower temperatures [7].

The contribution of electron-electron collisions to the resistance of Be at low temperatures has been determined theoretically to be proportional to T^6 in a wide temperature range and not to T^2 as predicted by the Landau-Pomeranchuk theory. This results from the fact that the effective interaction between electrons is appreciably decreased due to compensation of their Coulomb repulsion and their attraction due to virtual phonon exchange [9].

A quantitative calculation of the electrical resistivity at lowest temperatures requires the application of quantum mechanics to electron waves. A qualitative estimation can be obtained by considering the electrons as particles which move freely through the metal except when they are scattered by impurities, other defects, quantized lattice vibrations (phonons), or other electrons. The impurity and defect scattering causes a temperature-independent contribution (residual resistivity), but electron-phonon scattering becomes most important in a pure metal at temperatures of a few K. The resistivity due to electron-electron scattering in metals is usually expected to vary with T^2 [10].

The phonon-limited electrical resistance of solid Be was calculated for the range 90 to 1000 K in the directions parallel ($\rho_{\parallel}$) and perpendicular ($\rho_{\perp}$) to the hexagonal c-axis using a modified single-site approximation, a realistic model for the shape and size of the Fermi surface (see p. 58), and also a Born-von Karman force-constant model to calculate the spectrum of the lattice vibrations. The resulting values of the mean resistivity were in good agreement with experimental data over a wide temperature range (see the figure in the paper). The calculated anisotropy ratio $\rho_{\parallel}/\rho_{\perp}$ (= 1.8 for 273, 500, 700, and 1000 K to 1.9 for 90 K) was larger than the experimental one (1.1 to 1.2). For details of the calculations, see the paper [11].

Due to the high Debye temperature of Be (~1440 K, see "Beryllium" Suppl. Vol. A 2, 1991, p. 252) and the high sound velocity (~12800 m/s for the longitudinal and 8890 m/s for the transverse wave, see "Beryllium" Suppl. Vol. A 2, 1991, p. 160), and the very small electronic specific heat, it was assumed that the so-called phonon drag mechanism will be important for the electrical resistivity and thermopower in Be; see the discussion in [12].

When the electrical resistivity of Be (and other alkali and alkaline earth metals) is calculated using mass-renormalized form factors from optimized model potentials and a modified perturbation theory near the melting point (~1560 K, see "Beryllium" Suppl. Vol. A 2, 1991, p. 242) a value of 68.7 $\mu\Omega \cdot$ cm was obtained; using the normal perturbation theory the value was 75.1 $\mu\Omega \cdot$ cm. Renormalizing the form factors had little effect on the calculated resistivity [13].

References:

[1] Chi, T. C. (J. Phys. Chem. Ref. Data **8** [1979] 439/97, 445/57, 493).
[2] Powell, R. W. (Phil. Mag. [7] **44** [1953] 645/63, 648, 653).
[3] Darwin, G. E.; Buddery, J. H. (Beryllium, Metallurgy of the Rarer Metals, Vol. 7, Butterworth, London 1960, p. 172).
[4] Mitchell, M. A. (J. Appl. Phys. **46** [1975] 4742/6).
[5] Grüneisen, E.; Erfling, H. D. (Ann. Physik [5] **38** [1940] 399/420).
[6] Pletenetskii, G. E.; Koshkarev, G. S. (Izv. Akad. Nauk SSSR Met. **1985** No. 2, pp. 60/1; Russ. Metall. **1985** No. 2, pp. 57/8).
[7] Borchi, E.; De Gennaro, S.; Tasselli, P. L. (Phys. Status Solidi B **46** [1971] 489/94).
[8] Meaden, G. T. (Electrical Resistance of Metals, Plenum, New York 1965, pp. 22/3).
[9] Akhiezer, A. I.; Akhiezer, I. A.; Bar'yakhtar, V. G. (Zh. Eksperim. Teor. Fiz. **65** [1973] 342/5; Soviet Phys.-JETP **38** [1974] 167/8).
[10] Bass, J. (Adv. Cryog. Eng. **30** [1984] 441/52).

[11] Sano, H. (J. Phys. Soc. Japan **39** [1975] 1268/76).
[12] Yamaguchi, M.; Takahashi, Y.; Takasaki, Y. (Bull. Fac. Eng. Yokohama Natl. Univ. **23** Pt. 2 [1974] 175/8).
[13] Brown, C.; Jarzynski, J. (Phil. Mag. [8] **30** [1974] 21/32).

13.2.2 Measurements on Single Crystals

At low temperatures the specific resistivity, ρ, of 99.99 wt% pure Be crystals was found to be virtually constant below 20 K, the values for $\rho_{\|}$ being greater than for $\rho_{\perp}$. Selected values from [1] are given:

T in K	4.2	20.4	50	110	190	273	293
$\rho_{\|}$ in $\mu\Omega\cdot$cm	0.0266	0.0268	0.0365	0.183	1.25	3.67	4.31
$\rho_{\perp}$ in $\mu\Omega\cdot$cm	0.016	0.017	0.0245	0.151	1.03	2.72	3.22

The values obtained between 4.2 and 80 K [1] differ basically from those of [2] where $\rho_{\perp}$ is greater than $\rho_{\|}$. The latter values [2] were obtained through generalization of the data determined on crystals of varying purity. For 273 and 293 K the ρ values of [2] agree well with the results of [1]. In the temperature range 60 to 130 K, the thermal constituent of the resistivity of Be is satisfactorily described by a $\sim T^5$ relation (simplified Grüneisen-Bloch equation). With rising temperature (T>130 K) the relation between ρ and T becomes complex. The deviation of the measured $\rho_i(T)$ values (where $\rho_i(T) = \rho_T - \rho_{4.2}$) from the calculated data (see figure in the paper) at temperatures below 60 K was considered to be due to impurity effects which alter the phonon-electron and electron-electron constituents of Be at low temperatures [1].

A single crystal sample, 2 cm in length and 2.3 mm thick, with the sample axis parallel to the c-axis and a residual resistivity ratio $\rho_{273}/\rho_{4.2} = 1750$ had a resistivity $\rho_4 = (1.86 \pm 0.02)$ n$\Omega \cdot$cm at 4 K. Values up to 100 K and older data are compiled in the paper [3]; cf. p. 2.

Measurements of the temperature dependence of the resistivity $\rho_{\|}$ and $\rho_{\perp}$ of Be crystals ($\rho_{296}/\rho_{4.2} \approx 300$) between 4.2 and 300 K can be described by $\rho(T) = \rho_{4.2} + A \cdot T^n$. In the range 60 to 130 K, n = 4 in both crystal directions and $A_{\|} = 11.8 \times 10^{-18}$ Ωm/K for the current j$\|$c-axis with $\rho_{4.2\|} = 1.98 \times 10^{-10}$ Ωm and $A_{\perp} = 9.47 \times 10^{-18}$ Ωm/K for j$\perp$c-axis with $\rho_{4.2\perp} = 1.17 \times 10^{-10}$ Ωm [4]. The results [4] obtained for j$\|$c-axis agree well with those of [5]. The Bloch law ($\Delta\rho \approx T^5$) is not satisfied. Apparently the resistivity has been measured at the "dirty limit" [6].

At temperatures below 60 K a minimum was observed in the temperature dependence of ρ at about 18 K, see **Fig. 13-2**, p. 6, from [7]. The value and the position of the minimum agree with theoretical estimates of Kozlov and Flerov [8]. The sample used for the measurements had a residual resistivity ratio $\rho_{300}/\rho_{4.2} = 11.5$ and was cut from a single crystal in the basal plane in the form of a serpentine. From the theoretical considerations a minimum in $\rho(T)$ was expected because at low temperatures, when the characteristic wavelength of the thermal phonons becomes larger than the electron path length (due to scattering by nonmagnetic impurities), the phonon contribution to the resistivity will be negative; see the discussion in [7].

Above room temperature, i.e., between 300 and 900 K, the electrical resistivity of two pure Be crystals ($\rho_{273}/\rho_{4.2} = 10.0$ and 15.5 for $\rho_{\|}$ and $\rho_{\perp}$ to the c-axis, respectively) increased from $\rho_{\|} = 5.10$ to 33.20 $\mu\Omega \cdot$cm and from $\rho_{\perp} = 3.67$ to 23.90 $\mu\Omega \cdot$cm [9], cf. Fig. 13-1 on p. 3; see also the selected provisional data in the table on p. 2 from [2]. Above 300 K the single crystal data are consistent within $\sim 3\%$ with the polycrystalline results ($\bar{\rho}$) of Powell [10] given by the relation $\bar{\rho} = (2\rho_{\|} + \rho_{\perp})/3$. In this temperature range the anisotropy $(\rho_{i\|} - \rho_{i\perp})/\rho_{i\perp}$ is fairly constant [9].

For the range 293 to 1323 K measurements on Be single crystals of 99.99 wt% purity showed that ρ is most fully represented by a fourth-power polynominal $\rho_{\perp} = -7.19 + 3.64 \times 10^{-2} T - 5.48 \times 10^{-6} T^2 - 6.375 \times 10^{-9} T^3 + 9.8 \times 10^{-12} T^4$ and $\rho_{\|} = -5.25 + 2.69 \times 10^{-2} T + 2.6 \times 10^{-5} T^2 - 2.4 \times 10^{-8} T^3 + 1.38 \times 10^{-11} T^4$ (in $\mu\Omega \cdot$cm). Values of $\rho_{\perp}$ and $\rho_{\|}$ at 50 K intervals are given in a table in the paper. At temperatures above 1323 K the values $\rho_{\perp,\|}$ increase virtually linearly up to the transition temperature of α-Be $\rightarrow$ β-Be (see "Beryllium" Suppl. Vol. A 2, 1991, p. 4) where the specific resistivity abruptly increased, reaching a value of $\rho_{\parallel}=142.0\ \mu\Omega\cdot cm$ (1553 to 1573 K) in the cubic β-Be phase. On further raising the temperature, ρ decreased to 137 $\mu\Omega\cdot cm$ at 1691 to 1707 K; i.e., it remained constant in molten Be [1]; cf. "Beryllium" Suppl. Vol. A 2, 1991, pp. 136/40.

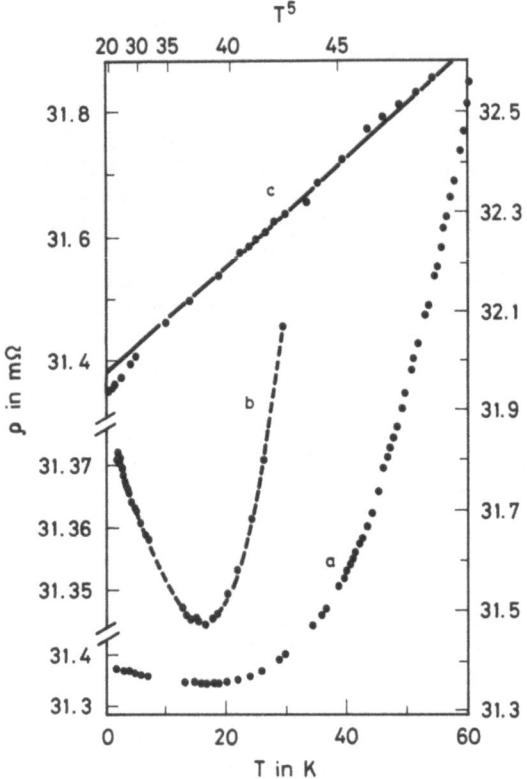

Fig. 13-2.

a) Temperature dependence of the resistivity of Be ($\rho_{300}/\rho_{4.2}=11.5$);
b) the same dependence in a larger scale;
c) the same dependence as a function of T^5.

The lower temperature scale is for a and b, the upper scale for c.

The temperature dependence of the specific resistivity anisotropy of Be has been a matter for discussion. The figure (in the paper [11]) showing this dependence from measurements on three Be crystals of slightly differing purity can be subdivided into three regions; the cryogenic temperatures (as mentioned before), the region between 273 and 1023 K, and the region above 1023 K up to the polymorphic transformation. In the first region, the anisotropy of the background component of the specific resistivity increased rapidly with the temperature, but was virtually independent of an impurity content. In the second region, above room tempera-ture, the temperature dependence of $\rho_{i\parallel}/\rho_{i\perp}$ is complex with a minimum at about 523 K in the case of high-purity (99.99 wt%) Be. Impurities slightly increase the anisotropy and smooth out the minimum (in the case of a Be crystal containing 1.34 at% Cu this minimum is absent). After a slight increase of the anisotropy at about 1023 K, the decrease at higher temperatures (from $\rho_{i\parallel}/\rho_{i\perp}=1.38$ to 1.28) was attributed to a variation in the state of the impurities in the metal. At the transition to β-Be the anisotropy is 1.29 for 99.6% and 1.28 for 99.99% pure Be [11].

At 273 and 293 K the resistivity of (~99.99 wt%) pure Be single crystals is practically independent of both the structural state and the orientation in the basal plane. But the relative resistance ρ_{77}/ρ_{293} depends on the orientation in the basal plane, as shown in **Fig.** 13-3, with a periodicity of 60° due to inhomogeneous impurity distribution. Resistivity values for 77, 273, and 293 K and orientations of 0°, 9.5°, and 30° related to $\langle 10\bar{1}0 \rangle$ are given in the paper [12].

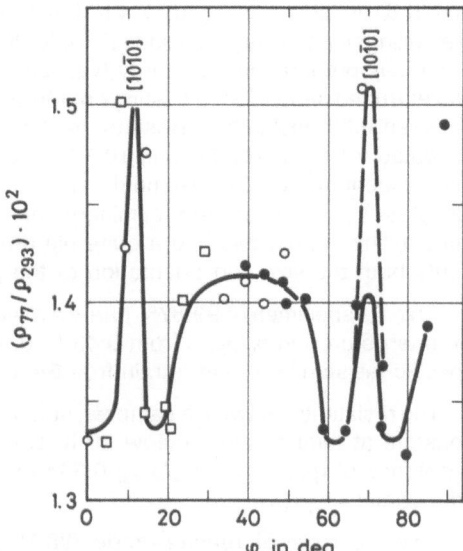

Fig. 13-3. Relative resistance of Be single crystals as a function of crystallographic directions in the basal plane: o plate 1, □ plate 2, ● plate 3.

References:

[1] Pletenetskii, G. E. (Izv. Akad. Nauk SSSR Met. **1984** No. 6, pp. 25/30; Russ. Metall. **1984** No. 6, pp. 24/9).

[2] Chi, T. C. (J. Phys. Chem. Ref. Data **8** [1979] 439/97, 445/57).

[3] Nelson, W. E. (Diss. Univ. Massachusetts 1977, pp. 1/207, 109/10, 120; Diss. Abstr. Intern. B **38** [1978] 3757).

[4] Morgun, V. N. (Fiz. Metal. Metalloved. **51** [1981] 756/61; Phys. Metals Metallog. [USSR] **51** No. 4 [1981] 68/73).

[5] Radebaugh, R. (J. Low Temp. Phys. **27** [1977] 91/105).

[6] Kagan, Yu.; Flerov, V. N. (Zh. Eksperim. Teor. Fiz. **66** [1974] 1374/86; Soviet Phys.-JETP **39** [1974] 673/9).

[7] Egorov, V. S.; Varyukhin, S. V. (Pis'ma Zh. Eksperim. Teor. Fiz. **25** No. 1 [1977] 58/61; JETP Letters **25** No. 1 [1977] 52/4).

[8] Kozlov, A. N.; Flerov, V. N. (Tezisy NT-19, Minsk 1976, p. 200 from [7]).

[9] Mitchell, M. A. (J. Appl. Phys. **46** [1975] 4742/6).

[10] Powell, R. W. (Phil. Mag. [7] **44** [1953] 645/63).

[11] Pletenetskii, G. E.; Koshkarev, G. S. (Izv. Akad. Nauk SSSR Met. **1985** No. 2, pp. 60/1; Russ. Metall. **1985** No. 2, pp. 57/8).

[12] Pletenetskii, G. E.; Tikhinskii, G. F.; Papirov, I. I. (Fiz. Metal. Metalloved. **53** [1982] 277/80; Phys. Metals Metallog. [USSR] **53** No. 2 [1982] 64/7).

13.2.3 Measurements on Polycrystalline Metal

The residual electrical resistivity, used as a measure of the purity and perfection of metal samples at low temperatures, was determined on Be of variable purity. Preliminary studies in the range 1.4 to 77.4 K revealed that the resistivity of a very pure Be specimen remained unchanged up to 20.4 K with the relative residual resistivity ratio $\delta = \rho_{1.4}/\rho_{rt} = 4.1 \times 10^{-3}$ (where $\rho_{1.4}$ is the residual resistivity at 1.4 K and ρ_{rt} is the resistivity at room temperature). Above

20.4 K δ rose to 1.3×10^{-2} at 77.4 K. For commercial-grade Be, δ hardly increased at all up to the boiling point of liquid nitrogen (77.4 K). In further studies $\delta_{4.2}$ was measured on Be samples of various preparation and purity (vacuum-distilled, foil produced by condensation, pressed and extruded metal, rolled castings of Mg-thermal and distilled Be with total impurity contents between 0.02 and 0.9%). Results are tabulated in the paper. All specimens were vacuum annealed for 1 h at 1000°C and 10^{-5} Torr to relieve stresses. The δ values were found to be independent of the grain size (in the range 20 to 200 μm) for commercial-grade (99 to 99.5%) samples. Variations in the preliminary annealing conditions led to a variation in the impurity state in the matrix; this had a quite big effect upon the residual resistivity. With increasing purity both the size and orientation of the grains began to affect δ [1].

On a Be specimen of 99.87% purity a residual resistivity of 0.034 $\mu\Omega \cdot$ cm was found [2]. For Be a large drop in $\rho_0/\rho_{273.2}$ from 0.43 to 0.06 resulted after a heat treatment at 700°C. The specific resistance of pure strain-free Be at 0°C was estimated to be $\rho_{273.2} \approx 3.2$ $\mu\Omega \cdot$ cm [3].

The resistance of two Be samples, stated to be of comparatively high purity, was effectively constant at temperatures below 20 K. Both specimens displayed a relatively high residual resistance of ($\rho_{20.4}/\rho_{273} = \rho_{4.2}/\rho_{273}$) 0.384 and 0.276, respectively. The results are compared with earlier data [4].

Two Be rods of reactor-grade (98.7%) purity with the sample axis parallel ($\parallel$) and perpendicular ($\perp$) to the pressing axis (without heat treatment) showed residual resistivities $\rho_4 = 0.64$ ($\parallel$) and 1.01 ($\perp$) $\mu\Omega \cdot$ cm and ratios $\rho_{273}/\rho_4 = 6.01$ ($\parallel$) and 4.22 ($\perp$). Resistivity values in the range 4 to >200 K for both directions (in comparison with isotropic Mo and W) are shown in a figure in the paper [5].

The sharp decrease of the resistivity of Be wires of 99.0 to 99.8% purity below room temperature levels off at ~20.3 K. This lower plateau corresponds approximately to the residual resistance since it was not possible to detect any further variation at 4.2 or even at 2 K (accuracy about 1%). The ρ values at 20.3 K and below, given in the paper, depend on the purity and to some degree on the preparation conditions of the samples (extrusion, annealing) [6]. The resistance ratio ρ_{293}/ρ_{77} decreased with increasing degree of wire drawing of 99.96 and 99.99% pure Be. Annealing at 620°C increased this ratio again; see figures and tables in the paper [7].

At 293 K the provisional values (in $\mu\Omega \cdot$ cm) of the total resistivity $\rho = 3.56$ and of the intrinsic resistivity $\rho_i = 3.53$ are recommended (as calculated from single-crystal data for 99.9 +% Be); at 273.15 K the corresponding values are $\rho = 3.02$ and $\rho_i = 2.99$ [8]; cf. table on p. 2. The data of Reich et al. [9] and Powell [10], which agree fairly well, were taken into account. Not considered in the generation of the recommended values were the data reported by Berteaux [11] whose graph gave the lower value $\rho = 3.0$ at 300 K [8].

Above 273 K the temperature dependence of the electrical resistivity of high-purity (99.97%) Be shows a monotonic resistivity increase (curve 1) with the temperature up to ~1000°C, whereas a material of lower purity (98%) shows an abnormal increase between 200 and 800°C, as represented in **Fig. 13-4** for the temperature course of the relative resistivity ρ_t-ρ_0/ρ_0. This anomaly (curve 2) depends upon the amount of impurities, their nature, and the experimental conditions. The abnormal course of the curve may be caused by residual stresses which are difficult to eliminate (because of the high recrystallization temperature of Be, see "Beryllium" Suppl. Vol. A 2, 1991, pp. 109 ff.). Annealing for 90 min at 900°C was not sufficient to attain stress relief in pure Be deformed 10% by rolling (curve 3) [12]. With technical-grade (Brush grade QM-V) Be the monotonic increase of resistivity from room temperature to ~1000°C was found to be different in a purified He atmosphere and in vacuum. It increased in vacuum (10^{-5} Torr) more sharply; see figure in the paper [13].

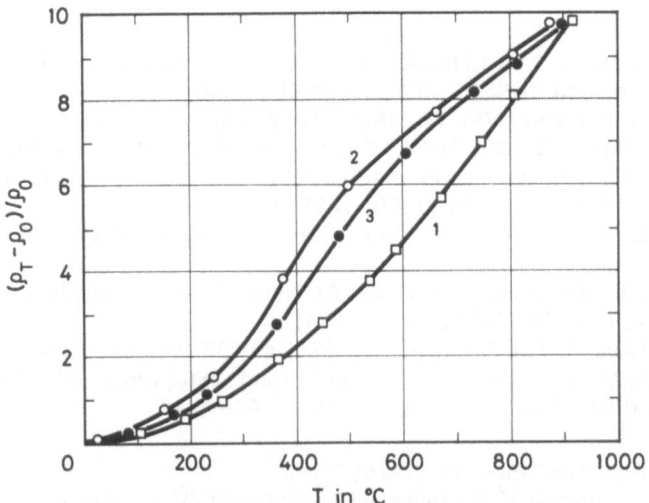

Fig. 13-4.   Temperature dependence of the relative electrical resistivity of Be with the following
purities: 1=99.97%; 2=~98%; 3=99.97% after
10% deformation and annealing for 1.5 h at 900°C.

With further rising temperature the electrical resistivity of 99.97% pure Be started to increase sharply at the (hcp→bcc) phase transition temperature at 1254±5°C, see **Fig.** 13-5, reaching in the course of 15 K values which are twice to three times higher than at 1250°C. This behavior could only be attributed to the polymorphic phase transition to (cubic) β-Be [12]. It agrees with the results of band structure calculations for bcc Be at the Hartree and exact exchange Hartree-Fock levels. For this phase there is no gap immediately above the valence band [14].

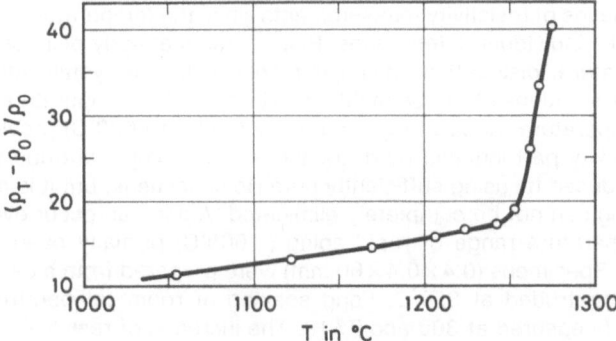

Fig. 13-5.   Temperature dependence of the relative electrical resistivity of 99.97% pure Be between 1000 and 1280°C.

When the electrical resistivity of Be (and other alkali and alkaline earth metals) is calculated using mass-renormalized form factors from optimized model potentials and a modified perturbation theory at the melting point (actually at 1280°C, i.e., not quite in the liquid phase) a value of $\rho=68.7$ $\mu\Omega\cdot$cm is obtained; $\rho=75.1$ $\mu\Omega\cdot$cm results from using the normal perturbation theory. Renormalizing the form factor had little effect on the calculated resistivity [15].

References:

[1] D'yakov, I. G.; Papirov, I. I.; Tikhinskii, G. F. (Fiz. Metal. Metalloved. **19** [1965] 768/90; Phys. Metals Metallog. [USSR] **19** No. 5 [1965] 135/6).
[2] Reich, R. (Mem. Sci. Rev. Met. **63** [1966] 21/58, 46).
[3] Denton, W. G. (AERE-GR-101 [1947] 24 pp. from N.S.A. **19** [1965] No. 20 543).
[4] MacDonald, D. K. C.; Mendelssohn, K. (Proc. Roy. Soc. [London] A **202** [1950] 523/33).
[5] Powell, R. L. (J. Appl. Phys. **31** [1960] 1221/4).
[6] Logerot, J. M.; Adenis, D. (Met. Soc. Conf. **33** [1964] 445/66; C.A. **66** [1967] No. 107 555).
[7] Pletenetskii, G. E.; Tikhinskii, G. F. (Vopr. At. Nauki Tekh. Ser. Obshch. Yad. Fiz. **1978** No. 6, pp. 70/6; C.A. **91** [1979] No. 115 871).
[8] Chi, T. C. (J. Phys. Chem. Ref. Data **8** [1979] 439/97, 445/57, 493).
[9] Reich, R.; Kinh, V. Q.; Bonmarin, J. (Compt. Rend. **256** [1963] 5558/61).
[10] Powell, R. W. (Phil. Mag. [7] **44** [1953] 645/63, 648/54).

[11] Berteaux, F. (Rev. Gen. Elec. **79** [1970] 7/14).
[12] Amonenko, V. M.; Ivanov, V. E.; Tikhinskii, G. F.; Finkel', V. A.; Shpagin, I. V. (Fiz. Metal. Metalloved. **12** [1961] 865/72; Phys. Metals Metallog. [USSR] **12** No. 6 [1961] 77/83).
[13] Lampson, F. K. (PB-140 790 [1951] 1/9; C.A. **1960** 20 758).
[14] Monkhorst, H. J.; Pack, J. D. (Solid State Commun. **29** [1979] 675/9).
[15] Brown, C.; Jarzynski, J. (Phil. Mag. [8] **30** [1974] 21/32).

13.2.4 Effects of Lattice Defects

Point defects (vacancies, interstitials) and line defects (dislocations) give rise to a considerable increase of the electrical resistivity. The removal of these defects by annealing (recovery) is often studied by means of resistivity measurements since the temperature ranges in which ρ decreases indicate the individual defect types. Point defects already disappear on annealing below room temperature; dislocations disappear only at the recrystallization temperature (rearrangement of dislocations starts below the recrystallization temperature). Therefore the annealing in the temperature range of rapid recovery ($\sim$420 to 620°C) provides a means of estimating the resistivity part induced by dislocations. The component due to scattering on impurities may be reduced by using sufficiently pure Be specimens, but it is emphasized that the influence of aging can not be completely eliminated. Aging can occur even in the purest metal, and the temperature range of rapid aging ($\sim$600°C) partially overlaps with that of recovery processes. Specimens (0.4 × 0.4 × 60 mm) were prepared from a cast of 99.9% pure Be which was hydroextruded at 350°C. Long soaking at room temperature produced no change in resistivity (measured at 300 and 77 K). The increase of resistivity depends on the degree of deformation ε (reduction of cross section). It is, for example:

ε in %	40	80	92
ρ in $\mu\Omega\cdot$cm	0.07	0.085	0.255

There is a particularly large increase at higher levels of deformation. The relative resistivity decrease due to annealing is shown in **Fig.** 13-6 where $\Delta\rho_1$ is the total increment due to dislocations after hydroextrusion and $\Delta\rho_2$ is the residue of the increment after each annealing treatment. The fact that the resistivity had not completely recovered by the moment of recrystallization was thought to be connected with the considerable grain refinement [1].

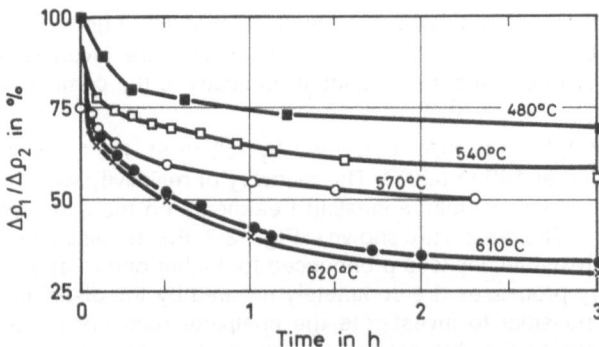

Fig. 13-6. Resistivity recovery after isothermal annealing of hydroextruded Be specimens (350°C, $\varepsilon = 92\%$) in the range of 0.05 to 3 h.

The influence of aging at 200 to 1000°C on compacted samples (extruded rods, 0.5 in in diameter) from commercial Be powder was investigated by measuring the electrical resistivity. On aging at 600°C the resistivity decreased rather quickly by more than 30%, depending on the aging time (up to about 600 h), see the figure in the paper [2]. This change in resistivity was suggested to be due to a precipitation process in the commercially pure Be. This process may occur also, but rather slowly, in the range 200 to 400°C. On aging at 800°C the process is already complete after 2 h. However, in this case the resistivity decrease is less (~15%), cf. figure in the paper. This result was explained by the fact that on cooling the material passes from a single- to a two-phase field between 800 and 1000°C. Aging at 1000°C followed by quenching produced only minor changes in the resistivity, regardless of the aging time [2]; cf. [3]. For the electrical resistance of specimens pressed from Be powders after varying duration of powder storage in air, see [4].

Measurements of the electrical conductivity of Be wires of 99.96 to 99.99% purity and 0.09 to 1.3 mm in diameter (for production conditions and applications, see the paper) at 77 K showed that tensile stresses up to 42 kg/mm² produced a resistivity increase of about 28 to 55%, depending on the purity. The resistivity of wires of commercial purity (99.9%) increased only slightly (~1%) with stresses up to mechanical failure (70 kg/mm²). The resistivity of the wire of highest purity (99.99%) changed reversibly (~2%) at stresses up to 20 kg/mm², due to elastic deformation. At higher stresses the change became irreversible, but the stress at which it happened was lower than the yield strength. A resistivity increase of 2% was attained with a 99.96% pure Be wire at a stress of 25 kg/mm² [5]. When Be wires of 99.96 to 99.99% purity (main impurities in 10^{-3} wt%: Al 3.5 to 6.5, Fe 1.8 to 7.5, Cr 1.0 to 1.5, and Ni 1.4 to 4.0) were drawn from extruded rods to 2.2 or 0.7 mm diameter, corresponding to 80 and 98% deformation, respectively, annealing at 600 to 620°C restored the original conductivity in 30 h for 80% deformation, but for 98% deformation a complete return to the initial value at 77 K was not reached. This was suggested to be due to the size effect (see p. 20). The oxygen content, converted to BeO, was 0.03%. For the resistance of Be wires of 99.0 to 99.8% purity below room temperature down to 2 K, see the paper [6]; cf. p. 3.

Investigations of the electrical resistance recovery in deformed Be involve difficulties. Due to the low solubility of impurities in Be (see "Beryllium" Suppl. Vol. A 2, 1991, pp. 29/33), the metal is a supersaturated solid solution at practically any level of purity up to 99.99%. Precipitates begin to dissolve above 600°C, and the resistance of the annealed metal grows. Because the processes of recovery and impurity dissolution at this temperature are simulta-

neous, the kinetic of resistance recovery can only be investigated up to annealing temperatures of 500 to 550°C. For the same reasons, difficulties are involved in the accurate
investigation of vacancy quenching in Be; another difficulty is the complex and nonuniform
substructure [7].

Casts of distilled (99.9% pure) Be, deformed by rolling at 400°C to 30, 70, and 90%
reduction, were annealed at 100 to 650°C. The recovery of resistivity was measured at 77 K
after successive isochronous (30 min) annealing treatments on the same specimen at ever
increasing temperatures. The results are shown in **Fig.** 13-7. For annealing temperatures up to
600°C the recovery of resistance is less pronounced for higher deformation levels. At 600°C
and above the recovery processes are completely masked by the dissolution of impurities.
Therefore it was not possible to investigate the complete recovery of resistance due to
deformation or to determine the dislocation component. The resistance recovery curve for
hydroextruded Be after a 92% reduction at 350°C, given in Fig. 13-7 for comparison, shows
that this specimen recovers almost 200 K lower [7].

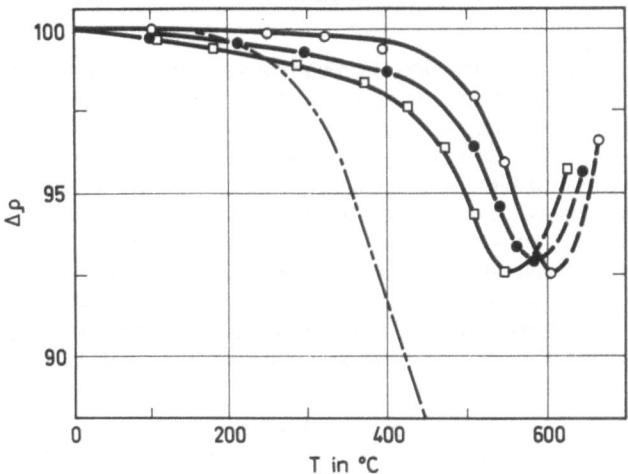

Fig. 13-7. Temperature curves of the resistance
recovery, $\Delta\rho = \rho_{annealed}/\rho_{initial}$, of Be specimens rolled
with 30 (□), 70 (●), and 90% (○) reductions; the - — -
curve refers to hydroextruded Be after 92% reduction at 350°C.

After hot rolling to 90% reduction with a gradual temperature decrease from 750 to 350°C,
recovery of the electroresistance was investigated after 2 to 300 min annealing at 300 to
700°C. The recovery of the electroresistance and microdeformation proceeded at lower
temperatures than the recovery of the mechanical properties, see figures in [8].

The activation energy, E_a, of the recovery process was estimated from the curves of the
resistivity change, $\Delta\rho(T)$, during annealing at 758, 773, and 788 K for times up to 240 min to be
$E_a = (17 \pm 3)$ kcal/mol for specimens rolled to 90% reduction as compared with $E_a = (18 \pm 3)$
kcal/mol for hydroextruded Be with 92% reduction at 350°C [7].

The variation of the resistivity of polycrystalline Be wires due to low-temperature thermal
cycling was measured at 77 K, as a result of thermal stresses which intensified the plastic
deformation by twinning and changed the initial grain size of the material. The wires of 99.96%
purity (12 cm in length and 0.045 cm in diameter) with a resistance ratio $\rho_{300}/\rho_{77} = 52$ showed

a maximum resistance increase to $\Delta\rho = 9\ \mu\Omega$ after 20 thermal cycles (carried out by cooling in liquid nitrogen, holding at 77 K for 1230 s, followed by 10 min heating in air at 300 K). After 30 thermal cycles the resistivity decreased to only $\Delta\rho = 3\ \mu\Omega$, then it rose again to about $\Delta\rho = 6\ \mu\Omega$ after 50 cycles (initial value $\Delta\rho = 1.5\ \mu\Omega$ after 3 cycles). X-ray investigations of the initial wire specimens with a mean grain size of 60 μm showed a nonuniform grain arrangement and a stressed state. After 30 thermal cycles the grain size was 45 μm and grew to the initial value of 60 μm after 60 cycles. Details are discussed in [9].

The introduction of lattice defects by **irradiation** with neutrons or electrons and their vanishing during recovery are frequently determined from changes of the electrical resistivity. For the recovery stages occurring in Be from low to high temperatures, determined by measuring the resistance changes, see "Beryllium" Suppl. Vol. A 2, 1991, pp. 106/9 (the resistance recovery after laser irradiation on heating in the range 77 to 300 K is shown in Fig. 9-26 on p. 107).

As was already mentioned in "Beryllium" Suppl. Vol. A 2, 1991, p. 106, the defect production in Be wires (of 99.9% nominal purity, $\rho_{300}/\rho_{4.2} \approx 70$, 0.5 mm in diameter, 10 cm long) by neutron irradiation at 4.2 K to total fluxes of 5×10^{16} n/cm^2 and 1.2×10^{18} n/cm^2 (E >0.1 MeV) and the subsequent isochronal recovery up to 350 K (fractional and derivative) of the electrical resistivity, shown in a figure in the paper [10], agreed well with the results of earlier studies in which high-purity Be was irradiated with fast neutrons at 19.5 K to 4×10^{17} n/cm^2, giving a resistivity increase of 0.173 $\mu\Omega \cdot$cm. Recovery started at about 30 K, reaching about 40% after 3 min at 50 K. The isochronal annealing data in the range 20 to 300 K (see figure in the paper) indicated that recovery becomes complete below room temperature [11]. In agreement with the radiation damage studies up to 280 K of Nicoud et al. [12] are the results of the flux dependence of ρ found by Williams et al. [13] on cyclindrical Be samples by measuring the electrical resistivity in situ in the longitudinal direction of the samples after irradiation with fast neutrons ($\sim 10^{19}$ n/cm^2, E >1 MeV) at ~ 110 K and subsequent annealing at successively increasing temperatures. The data of ρ as functions of the flux (corrected for a hypothetical measuring temperature of 90 K) show an increase from 0.773 to 1.892 $\mu\Omega \cdot$cm with increasing neutron flux from 0 to 2.2×10^{18} n/cm^2. The temperature dependence of ρ, measured before and after irradiation and after annealing at increasing temperature, is shown in a figure in the paper. The results indicated two stages of recovery, one extending from the lowest temperatures to 180 K and the second (larger) stage from there to about 280 K; the most pronounced recovery was around 240 K. Complete recovery was found by room temperature annealing [13].

Irradiation to 10^{19} n/cm^2 has an equivalent effect on the electrical resistivity of Be in the range 4 to 100 K as a few tenths of a percent impurities. Therefore not much improvement is gained by using high-purity Be in a radiation field at low temperatures. An exposure of 10^{19} n/cm^2 gave a resistivity increment of 4.33 $\mu\Omega \cdot$cm. Warming to 250 K resulted in rapid and complete annealing of the neutron-induced resistivity change [14].

No change of the electrical resistivity was found, within the accuracy of measurement (0.5%), after reactor irradiation of Be samples as forged, cast, and extruded [15]; for doses applied and resistance values measured, see [16].

A perfectly linear resistivity increase (slope not specified) was reported for hot-extruded Be (from powder at 1050°C; for impurity contents, see table in the paper) up to irradiation doses of 6×10^{20} n/cm^2 and temperatures in the range 60 to 80°C. The resistivity change was interpreted in terms of the distribution of He (produced by transmutation reactions) in the Be matrix; it was estimated to be of the order of 12 $\mu\Omega \cdot$cm per 1% He. Damage due to defect production is negligible for irradiation temperatures of $\gtrsim 75$°C. With annealed Be (before the experiments) the resistivity increased on irradiation and recovered in a steady manner on heating in the range 300 to 900°C. Recovery occurred over a wide spectrum of activation energies. Without annealing before irradiation, recovery started at 500°C and was complete at 800°C. The recovery processes are discussed in the paper [17]. For the effects of He formation by transmutation reactions in Be during neutron irradiation, see "Beryllium" Suppl. Vol. A 2, 1991, p. 64.

The electrical conductivity of Be samples, measured (with a special apparatus) under irradiation in a water-cooled, water-moderated reactor at an integral flux of fast neutrons of 1.5×10^{19} n/cm² together with a thermal neutron flux of 4×10^{19} n/cm² in the range 120 to 170°C, remained almost unchanged under these conditions [18].

The electrical conductivity, κ, of **porous** Be, prepared from electrolytic powder with a porosity of ~13 to 62%, was investigated in the range 20 to 650°C (tables and figures are given in the paper). The temperature dependence of the conductivity obeyed the equation $\kappa = A \cdot \exp(-bT)$; the constants A and b were determined over wide ranges of temperatures and porosity. The porosity dependence of κ agreed completely with theoretical calculations [19].

Calculations of the electrical resistivity increase due to impurity atoms or point lattice defects in metals as provided in the literature for monovalent metals were not applicable for multivalent metals. A procedure proposed for multivalent metals gave a value of $\rho_v = 0.652$ $\mu\Omega \cdot$ cm per 1% vacancies for the vacancy resistivity of Be. Details of the three-step procedure, especially for the appropriate interpreting of the temperature coefficient of the electronic specific heat, are given in the paper [20].

The dc resistivity due to electron scattering from a vacancy, an octahedral interstitial, and a tetrahedral interstitial in Be was computed by application of the diffraction model concept of pseudopotentials in the theory of metals. The resistivity values (in $\mu\Omega \cdot$ cm) were 0.296 for vacancies, 0.283 for octahedral interstitials, and 0.249 for tetrahedral interstitials, corresponding to 1 at% concentration of each defect type [21].

References:

[1] Papirov, I. I.; Stoev, P. I.; Taranenko, I. A. (Fiz. Metal. Metalloved. **34** [1972] 1022/6; Phys. Metals Metallog. [USSR] **34** No. 5 [1972] 114/8).

[2] Blainey, A.; Johnston, T. L.; Jones, J. W. S. (Rev. Met. [Paris] **52** [1955] 735/49, 744/5).

[3] Gelles, S. H.; Pickett, J. J.; Wolff, A. (J. Metals **12** [1960] 789/92).

[4] Gelles, S. H.; Wolff, A. (NMI-1238 [1961] 74 pp.; C.A. **1961** 25668).

[5] Koshkarev, G. S.; Pletenetskii, G. E.; Pikalov, A. I.; Tikhinskii, G. F. (Metalloved. Term. Obrab. Met. **22** [1980] No. 8, pp. 57/9; Metal. Sci. Heat Treat. Metals [USSR] **22** [1980] 610/2).

[6] Pletenetskii, G. E.; Koshkarev, G. S.; Papirov, I. I.; Tikhinskii, G. F. (Izv. Akad. Nauk SSSR Met. **1979** 207/9; Russ. Met. **1979** 180/4).

[7] Papirov, I. I.; Stoev, P. I.; Taranenko, I. A. (Fiz. Metal. Metalloved. **35** [1973] 1241/7; Phys. Metals Metallog. [USSR] **35** No. 6 [1973] 112/7).

[8] Papirov, I. I.; Volokita, G. I.; Kapcherin, A. S.; Stoev, P. I. (Izv. Vyssh. Uchebn. Zaved. Tsvetn. Metal. **17** [1974] 100/5; Soviet Non-Ferrous Metals Res. **17** [1974] 357/8; C. A. **83** [1975] No. 119685).

[9] Braude, I. S.; Lavrent'ev, F. F.; Nikiforenko, V. N.; Salita, O. P. (Fiz. Metal. Metalloved. **65** [1988] 1145/8; Phys. Metals Metallog. [USSR] **65** No. 6 [1988] 96/100).

[10] Delaplace, J.; Nicoud, J. C.; Schumacher, D.; Vogl, G. (Phys. Status Solidi **29** [1968] 819/24).

[11] Blewitt, T. H.; Coltman, R. R.; Klabunde, C. E.; Holmes, D. K.; Redman, J. K. (ORNL-2614 [1958] 1/176, 65/9; N.S.A. **13** [1959] No. 2275).

[12] Nicoud, J. C.; Delaplace, J.; Hillairet, J.; Schumacher, D.; Adda, Y. (J. Nucl. Mater. **27** [1968] 147/65, 151, 158; Colloq. Metall. Commis. Energ. At. [France] No. 10 [1966] 27/39 from C.A. **68** [1968] No. 5606).

[13] Williams, J. M.; Hinkle, N. E.; Eatherly, W. P. (ORNL-TM-3914 [1972] 1/145, 23/4; C.A. **78** [1973] No. 51334).

[14] Dienes, G. J.; Damask, A. C. (J. Nucl. Mater. **3** [1961] 16/20).

[15] Templeton, L. C. (ORNL-864 (Del.) [1951] 1/23, 6; N.S.A. **11** [1957] No. 14031).

[16] Templeton, L. C. (in: Billington, D. S.; Howe, J. T.; AECD-4027 [1956] 49/51; N.S.A. **12** [1958] No. 9226).

[17] Hickman, B. S.; Stevens, G. T. (AAEC/E-109 [1963] 1/42, 8; N.S.A. **17** [1963] No. 41480).

[18] Kharlamov, A. G.; Zakharova, N. P.; Batalov, A. A.; Zaikin, Yu. I.; Kolyadin, V. I. (At. Energ. [USSR] **29** [1970] 289/91; Soviet At. Energy **29** [1970] 1016/8; C.A. **74** [1971] No. 48772).

[19] Bykovskii, N. A.; Dubinin, V. A.; Nichkov, I. F.; Rezaev, M. D. (Deposited Doc. VINITI-4442-76 [1976] 1/10; C.A. **90** [1979] No. 96144).

[20] Reale, C. (Phys. Letters **2** [1962] 268/70).

[21] van Torne, L. I. (Phys. Status Solidi **18** [1966] 737/42).

13.2.5 Effect of Impurities

Impurities have a profound effect on the conductivity characteristics even when their solubility in Be is small. Studies performed with zone-refined single crystals as well as with polycrystalline metal revealed the marked effect of purity on the electrical resistance at room temperature and on the temperature coefficient of the resistance in the range 25 to 1000°C. At room temperature, the specific resistance was determined for two crystals prepared from zone-refined bars (about 6 in long, 0.5 in in diameter), which were divided into three 2 in long sections of pure, central, and impure portions of material (analyses of the specimens with impurity contents of Mg, Al, Si, Fe, and Mn at the pure and impure sections and pole figures of the crystals are given in the paper). The results revealed that the effect of impurity level on the specific resistance depended markedly on the crystal orientation. The lowest resistivity value recorded was 3.60 $\mu\Omega \cdot$ cm in a distance of 2 cm from the pure end of the triply refined crystal, rising to 4.25 $\mu\Omega \cdot$ cm in a distance of 2 cm from the impure end, compared with the corresponding values of 4.38 and 7.05 $\mu\Omega \cdot$ cm for the doubly refined crystal. The resistance vs. temperature curve between 25 and 1000°C showed inflections in the range 300 to 700°C which were affected by the impurity content and prior history of the material. The inflections and hysteresis effects of the resistance on heating and cooling are probably due to solid solution changes of the impurities; details and figures are given in the paper [1]. Previous data revealed little correlation with nonmetallic or total impurity content, verifying that the solubility limits of the nonmetallics were exceeded and they were present as second-phase particles such as BeO. The resistivity varied clearly as a function of metallic impurity content. For example, in hot-pressed block material at 4 K the resistivity increased from 0.5 to 5.5 $\mu\Omega \cdot$ cm for an increase of metallic impurities from 0.1 to 0.8 wt%; similarly, at 300 K, the resistivity increased from 4.5 to 10 $\mu\Omega \cdot$ cm, respectively. Between 100 and 700°C the resistivity of specimens from different lots of Be varied by an amount of 20 $\mu\Omega \cdot$ cm [2]. This variation is shown in **Fig.** 13-**8**, p. 16, from [3]; it may depend on the degree of precipitation of impurities according to prior thermal histories [2].

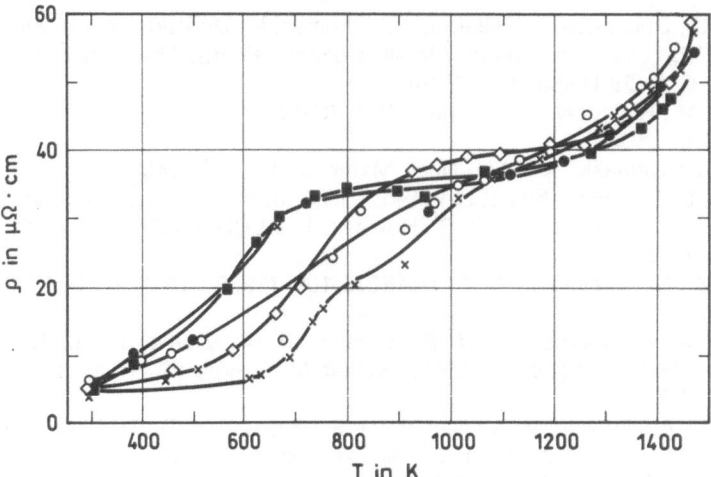

Fig. 13-8.   Variation in the electrical resistivity in different
lots of Brush (NRL) Be specimens as a function of temperature with the following impurity contents (in %):

	×	○	◇	●	■
BeO	2.13	0.85	0.83	1.20	1.43
metallic	0.272	0.217	0.358	0.371	0.266

Studies on two Be crystals of 99.49 and 98.89% purity (containing mainly BeO, C, Al, and Fe as definite impurities, see table in the paper) revealed a marked maximum near 850°C in the variation of the resistivity coefficient with the temperature up to ~1000°C, as well as a strong rise in the relative resistance at the same temperature for the sample of the lower purity. These anomalies were attributed to the dissolution of an impurity precipitate (of the type XBe_5, where X is possibly Fe with or without Al) [4]. A content of 0.5% metallic impurities caused an increase of the electrical resistivity of about 2.4 $\mu\Omega \cdot$ cm at 0°C and an increase in the resistance ratio $\rho_0/\rho_{273.2}$ by about 0.20; for pure strain-free Be at 0°C $\rho \approx 3.2$ $\mu\Omega \cdot$ cm was estimated [5]. Resistivity measurements on unalloyed Be after extrusion and heat treatment did not reveal any systematic relationship between the resistivity and the amounts of various impurities; see table in the paper [6].

The influence of a heavy substitutional impurity (1.34 at% Cu) on the resistivity of pure Be crystals parallel and perpendicular ($\parallel$ and $\perp$) to the c-axis was measured in the range 4.2 to 300 K. The maximum at T$\approx$80 K found with pure Be ($\rho_{296}/\rho_{4.2} \approx 300$) for the anisotropy of the temperature-dependent part of ρ, $\alpha = \rho_\parallel/\rho_\perp$, was not observed in the Be-Cu alloy ($\rho_{296}/\rho_{4.2} \approx 2$), see **Fig.** 13-**9** from [7]. At room temperature the resistivity anisotropy of the alloyed Be is somewhat higher ($\alpha \approx 1.44$) than for pure Be ($\alpha \approx 1.23$), in agreement with the measurements of Grüneisen [8] on Be with $\rho_{296}/\rho_{4.2} \approx 800$ and of Mitchell [9] on Be with $\rho_{296}/\rho_{4.2} \approx 10$. With the substitutional impurity (Cu) in Be the deviation from the Matthiessen standard exhibits a broadening of its peaks for the anisotropic crystal due to splitting of quasi-local oscillations. Electrons being translated parallel with the principal hexagonal plane are apparently more strongly scattered by the Cu atoms than those traveling perpendicular to the principal plane [7].

The resistance of four Be samples (three single crystals and one polycrystalline sample), differing with respect to magnetic impurity concentrations and to the residual mean free path of the electrons, was measured in the range 2 to 70 K (in addition also in a magnetic field). The impurities were (in 10^{-3} at%) Fe 1.3 to 14, Ni 0.65 to 9, Mn 0.1 to 1.3, in part Ti and V, and traces of Cr and Co (see table in the paper), and $\alpha = (\rho_{300} - \rho_{4.2})/\rho_{4.2}$ was in the range 1.1 to 61. All samples exhibited an increase in the resistance due to cooling with a minimum between 10 and 17 K. The dependences could be represented conveniently by $\rho(T) = \rho_0 + A \cdot T^{3.7} + \rho_{im}(T)$, where $\rho_{im}(T)$ is the correction to the resistance due to the presence of impurities. The observed anomalies were largely due to the influence of magnetic impurities. Figures and discussion are given in the paper [10].

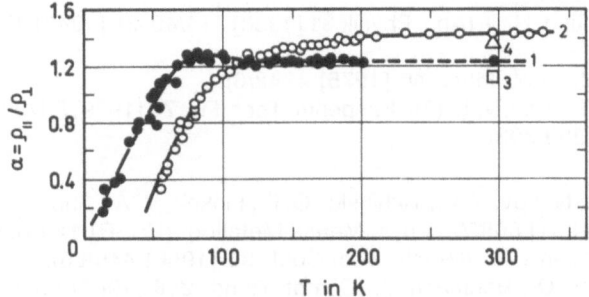

Fig. 13-9. Anisotropy of the temperature-dependent part of the resistivity of pure Be (1, $\rho_{296}/\rho_{4.2} \approx 250$), and a Be–1.34 at% Cu alloy (2, $\rho_{296}/\rho_{4.2} \approx 2$); results of Grüneisen [8] (3, $\rho_{296}/\rho_{4.2} \approx 800$) and Mitchell [9] (4, $\rho_{296}/\rho_{4.2}$ ≈ 10) are given for comparison.

Beryllium of low (98%) purity exhibits an abnormal temperature dependence of the resistance between 200 and 800°C (see Fig. 13-4 on p. 9). This anomaly was ascribed to the impurities and the pretreatment of the sample. Residual stresses are difficult to eliminate [11]; cf. p. 8.

The resistivity of Be wires (of 99.0 to 99.8% purity, 0.01 to 0.08 in diameter, produced from polycrystalline rods) was investigated simultaneously with the fabrication development (by casting, hot extrusion, swaging at 700 to 800°C to at least 0.04 in diameter and/or warm drawing after heating to 425°C down to <0.05 in diameter). The effects of thermal treatment on the electrical resistivity of Be were found to increase with the amount of impurities. The wires contained 0.1 to 1.2% BeO and 0.02 to 0.18% total metallic impurities. Grain sizes were in the ranges 10 to 80 or 30 to 200 μm. Details and tables are given in the paper [12].

Resistivity measurements were performed from 2 to 500 K on Be wires of varying purity (for the impurity contents of BeO, Si, Al, Cr, Mn, Ni, and Fe, see table in the paper), all with a fiber structure close to [10$\overline{1}$0], and grain sizes in the ranges 10 to 80 and 30 to 200 μm. The resistivity values (in the range 0.03 to 1.05 μΩ·cm) determined at 4.2 K on the samples without annealing and after vacuum (10^{-5} Torr) annealing for 1 h and 150 h at 800°C represent the residual resistivity because the same values were found at 2 K. Small differences in the values after 1 h and 150 h annealing of a sample were explained by a competition between recrystallization and the formation of a solid solution of the impurities. The measurements showed that the ideal resistivity as well as the residual resistivity depended on the sample purity at low temperatures, which is in contrast to the Matthiessen rule. Details such as figures and discussion are given in the paper [13].

References:

[1]   Martin, A. J.; Bunce, J. E. J.; Tilbury, P. D. (J. Less-Common Metals **4** [1962] 191/8).

[2]   Matthews, C. O. (J. Metals **12** [1960] 780/5).

[3]   Wright, E. S.; Ho, J. (LMSD-288218 [1960] from [2]).

[4]   Pointu, P.; Espagno, L.; Azou, P.; Bastien, P. (Compt. Rend. **250** [1960] 2365/7).

[5]   Denton, W. H. (AERE-GR-101 [1947] 1/24 from N.S.A. **19** [1965] No. 20543).

[6]   Gordon, P. (MDDC-1370 [1947] 1/39; Metall. Abstr. [2] **18** [1951] 689).

[7]   Morgun, V. N. (Fiz. Metal. Metalloved. **51** [1981] 756/61; Phys. Metals Metallog. [USSR] **51** No. 4 [1981] 68/73).

[8]   Grüneisen, E.; Erfling, H. D. (Ann. Physik **31** [1938] 714/40, **38** [1940] 399/421, **41** [1942] 89/99).

[9]   Mitchell, M. A. (J. Appl. Phys. **46** [1975] 4742/6).

[10] Varyukhin, S. V.; Egorov, V. S. (Zh. Eksperim. Teor. Fiz. **76** [1979] 597/606; Soviet Phys.-JETP **49** [1979] 299/303).

[11] Amonenko, V. M.; Ivanov, V. E.; Tikhinskii, G. F.; Finkel', V. A.; Shpagin, I. V. (Fiz. Metal. Metalloved. **12** [1961] 865/72; Phys. Metals Metallog. [USSR] **12** No. 6 [1961] 77/83).

[12] Logerot, J. M.; Adenis, D. (Metall. Soc. Conf. **33** [1966] 445/66).

[13] Reich, R.; Kinh, V. Q.; Bonmarin, J. (Compt. Rend. **256** [1963] 5558/61).

13.2.6 Pressure Dependence

The electrical resistance of commercially pure Be (Brush, wire of 0.5 mm diameter) was measured at room temperature under pressure up to 105 kbar. The results in terms of the ratio ρ/ρ_0 (where ρ_0 is the resistance at 25°C and 5 kbar) as a function of the pressure, p, are represented in **Fig.** 13-**10**. The resistance gradually decreased up to 55 kbar at which pressure a small deflection occurred which indicated a structural change (in the c/a ratio of the hcp structure or a phase transformation). This change was not identified, but the discontinuity at 55 kbar on loading occurred also at 50 kbar on releasing the pressure. When the pressure was further increased, the resistance again dropped gradually up to 93 kbar. At this pressure a discontinuity in the resistance curve corresponds to a sharp resistance drop of approximately 45% which was accompanied by drifting of the resistance with time. This drop is characteristic of a first-order phase transition. On releasing the pressure, a hysteresis was observed and complete transformation to the original structure was not achieved until 70 kbar was reached [1]; cf. "Beryllium" Suppl. Vol. A 2, 1991, pp. 2/3.

Previously, the effect of pressure on the resistance of Be was studied by Bridgman [2] who found a gradual resistance decrease up to 98 kbar without a discontinuity or inflections. But corrections of the pressure calibration [2] by other authors [3, 4] showed that resistance had been measured really up to 70 to 80 kbar only. The corrected Bridgman [2] data are also contained in Fig. 13-10 from [1].

According to measurements on 99.7% pure Be with an improved method, the ρ/ρ_0 values decreased monotonically from 1.0 to $\sim$0.9 with the pressure increasing from 0 to $\sim$140 kbar. In the range 0 to $\sim$20 kbar the relative resistance diminished with a comparatively steep slope (corresponding to that reported by Bridgman [2]). At higher pressures, the slope gradually became smaller, see figure in the paper [5]. More recently, the resistance measurements on Be were extended to 40 GPa (400 kbar) in a diamond-anvil cell. The data obtained were referred to the resistance at 10 GPa and 25°C. (The resistance determination at the high pressure was possible with the four-probe technique using insulating gaskets made of mica sheets and MgO powder). The following results were obtained with Be wires of 99% purity, annealed for 30 min at 900°C and etched in acid solution down to 20 μm diameter: a slight (~5%) decrease of the resistance was observed up to ~20 GPa, starting at ~2 GPa, and then a leveling off or a very slight increase up to ~40 GPa. A small jump appeared to exist at 25 GPa (see figure in the paper [6]), probably because of the scatter in the data sets. The discontinuity in the resistivity discovered at 93 kbar and room temperature by Marder [1] and interpreted as the transformation of the hcp structure to a bcc phase was not observed [6]. Agreement within ~8% exists with the data obtained by Stromberg and Stephens [5], which were collected to a maximum pressure of 13 GPa [6].

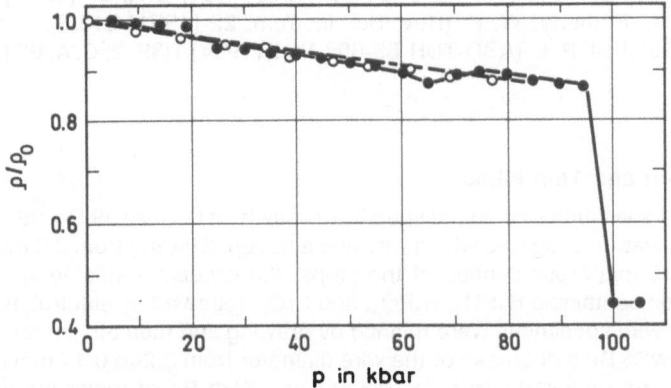

Fig. 13-10.   Resistance of Be in terms of the ratio ρ/ρ_0 (where ρ_0 is the resistance at 25°C and 5 kbar) as a function of pressure, p, at room temperature (●); superimposed are corrected data of Bridgman (○).

An attempt to reproduce the discontinuity at 93 kbar and room temperature observed by Marder [1] for Be of 97.5% purity was unsuccessful when using specimens of 99.85% purity. The measurements showed no such effect up to 100 kbar. This discrepancy was attributed to the difference in sample purity [7]. In the same investigation thermal cycles were performed at constant pressures up to about 76 kbar. An increase in temperature led to a slight change in the slope of the electrical resistivity and a fairly well-marked plateau occurred at temperatures lower than that of the hcp → cubic phase transformation. The initial temperature of the plateau varied inversely with the pressure. Qualitatively the same behavior of the resistivity was found at atmospheric pressure. On cooling, the final resistivity value was always slightly lower than the initial value. This different behavior of the electrical resistivity of Be on heating and cooling at a given pressure was explained by recrystallization following severe deformation brought about by applying the pressure in the solid state [7].

With an experimental arrangement as used by Donoghue and Eatherly [8] to measure the resistivity and Hall constant of semiconductors and metals under uniaxial compression, no change in the resistivity was observed for Be single crystals even at stresses large enough to cause plastic deformation [9].

References:

[1] Marder, A. R. (Science **142** [1963] 664).

[2] Bridgman, P. W. (Proc. Am. Acad. Arts Sci. **81** [1952] 165/251, 194/6, **79** [1951] 149/79, 152/3).

[3] Bundy, F. P.; Strong, H. M. (Solid State Phys. **13** [1962] 81/146, 115/6).

[4] Kennedy, G. C.; La Mori, P. N. (in: Bundy, F. P.; Hibbard, W. R., Jr.; Strong, H. M.; Progress in Very High Pressure Research, Wiley, New York 1961, p. 304 from [1]).

[5] Stromberg, H. D.; Stephens, D. R. (UCRL-7902-Rev. 1 [1964] 1/12; C.A. **62** [1965] 13967).

[6] Reichlin, R. L. (Rev. Sci. Instrum. **54** [1983] 1674/7).

[7] Francois, D.; Contre, M. (UCRL-trans-1299-L [1965] 1/31; C.A. **67** [1967] No. 76787).

[8] Donoghue, J. J.; Eatherly, W. P. (Rev. Sci. Instrum. **22** [1951] 513/6).

[9] Gross, G. E.; Gutshall, P. L. (ASD-TDR-63-605-Pt. II [1964] 1/35, 2; C.A. **62** [1965] 1423).

13.2.7 Size Effect and Thin Films

The influence of specimen size on the specific resistivity of Be with 99.94, 99.96, and 99.99 wt% purity at 77 K was investigated with a wire and a ribbon. Starting from 2.2 mm diameter of the wire and 0.5×5 mm cross section of the strips, the vacuum-annealed specimens were thinned in a solution containing H_2SO_4, H_3PO_4, and CrO_3, followed by electrolytic polishing of the surface. Other wire specimens were thinned by drawing and then etched and annealed. In the case of 99.99 wt% Be a decrease of the wire diameter from 2.2 to 0.14 mm increased the resistivity at 77 K from 4.8×10^{-2} to 6.2×10^{-2} $\mu\Omega \cdot cm$. With Be of lower purity and for the ribbons the increase was less pronounced. The effect results from the free path length of the conduction electrons, which at 77 K was estimated to be 45 and 32 μm for Be of 99.99 and 99.96% purity, respectively [1]; cf. [2]. Utilization of Be with a purity greater than 99.99% for cryoconductors is not recommended because of the substantial losses in conductivity by the size effect [2].

The resistivity of radiofrequency-sputtered thin Be films is ~ 18 $\mu\Omega \cdot cm$, i.e., more than about four times higher than the bulk value of about 4 $\mu\Omega \cdot cm$. Increasing the film thickness from 0.07 to 1.15 μm caused the resistivity to decrease from 50 to 16 $\mu\Omega \cdot cm$ (see figure in the paper). Both impurities (mainly oxygen and carbon) and grain size (~ 20 nm) contribute to the high resistivity of the films. Details are discussed in [3].

The change of the electrical resistivity of Be foils of 99.9% purity, obtained by evaporation and condensation in vacuum, after irradiation at 77 and 300 K by a pulsed laser (pulse duration $\sim 10^{-3}$ s, radiation energy 40 J, average radiation density $\sim 10^7$ erg/cm^2) was investigated. At the irradiated region of 2.5 mm diameter (in the central part of 10 mm long, 2 mm wide, and 0.2 mm thick strips) the rate of heating was about 10^6 K/s and the rate of cooling about 10^5 K/s. Under the above irradiation conditions the change of the electrical resistivity of the strips due to the shape distortion by melting was comparable with the sensitivity of the measurements and the loss of material did not exceed several percent. The variation of the resistance under irradiation is shown in a figure in the paper. After irradiation at 77 K the resistance was higher by $\Delta\rho \approx 0.6 \times 10^{-2}$ $\mu\Omega \cdot cm$ and then increased further for 30 to 40 s to $\Delta\rho \approx 1.2 \times 10^{-2}$ $\mu\Omega \cdot cm$ reaching a value 9% higher than the initial value. Annealing for 10 min at temperatures up to 300 K revealed a recovery stage at about 200 K in which about 20% of the initial (lower) resistance is restored. On further holding for 17 h at 300 K about 45% recovered. It was assumed that at least 50% of the electrical resistance increase due to irradiation can be attributed to point defects [4].

Thin films of various thicknesses of Be (and other metals) were prepared in vacuum (5×10^{-9} Torr) and subsequently heated for 1 to 1.5 h at 400 to 650 K. Their electrical resistance was measured in vacuum or in the presence of gases (H_2, O_2, CO) at low pressures. The resistance change under the influence of the gases depended on the metal, film thickness, temperature of mounting, kind and pressure of the gas, and the duration of exposure to the gas. With increasing oxygen pressure from 1×10^{-8} to 5×10^{-4} Torr the mean relative resistance change was 5 to 30%. The resistance of films of 100 to 200 Å thickness were most influenced by the gases, i.e., at thicknesses where the sharp increase of resistivity and the decrease of its temperature coefficient usually begin [5].

References:

[1] Pletenetskii, G. E.; Koshkarev, G. S.; Tikhinskii, G. F. (Fiz. Metal. Metalloved. **48** [1979] 889/91; Phys. Metals Metallog. [USSR] **48** No. 4 [1979] 192/4).
[2] Pletenetskii, G. E.; Tikhinskii, G. F. (Vopr. At. Nauki Tekh. Ser. Obshch. Yad. Fiz. **1978** No. 6, pp. 70/6; C.A. **91** [1979] No. 115871).
[3] Paulson, W. M.; Lorigan, R. P. (J. Vac. Sci. Technol. **14** [1977] 210/8).
[4] Avotin, S. S.; Krivchikova, E. P.; Papirov, I. I.; Stoev, P. I.; Tereshin, V. I. (Zh. Eksperim. Teor. Fiz. **62** [1972] 288/93; Soviet Phys.-JETP **35** [1972] 155/7).
[5] Nakhodkin, N. G.; Zykov, G. A.; Shaldervan, A. I. (Fiz. Metal. Plenok **1968** 165/7; C.A. **71** [1969] No. 75378).

13.3 Superconductivity

13.3.1 Bulk Metal

Bulk superconductivity with a zero field transition temperature $T_c = 0.026$ K was found in a pure (α-)Be sample containing the following impurities (in ppm): 5 N, 10 Si, 15 Al, 5 Mn, 1 Ni, and 3 Fe. The initial slope of the critical magnetic field curve was (68 ± 2) Oe/K [1]. This T_c value of 0.026 K is usually reported in the literature. The transition temperature of pure bulk Be with impurity traces as given in [1] was determined to be between 0.021 and 0.024 K. The typical transition width was 0.2 mK and the typical reproducibility upon thermal cycling 0.10 mK (average of the standard deviation of several samples) [2].

The transition temperature to superconductivity can be used as a temperature reference standard. A cryogenic temperature scale for the range below 0.5 K was obtained by developing a compact device for temperature calibration. This device provides five reference temperatures from 0.015 to 0.21 K of five different materials (W, Be, Ir, $AuAl_2$, and $AuIn_2$). The narrow and highly reproducible superconducting phase transitions of these materials were assigned to temperature values by means of fundamental thermometers used at NBS [2]. Measurements of the superconducting transition points of the materials mentioned in [2], using the developed device, are realized in [3]. For intercomparison of temperature scales with the five superconductors given above, see [4]. The superconducting transition temperatures of Be as measured with the Helsinki temperature scale (based on platinum NMR) agree well with those measured with the NBS noise and nuclear orientation temperature scale. The superconducting transition of Be occurred in a temperature interval of about 0.5 mK (above ~23 mK), but the width of the main part of the transition was less than 0.1 mK [5].

In early studies two "high-temperature allotropes" of Be were stabilized at room temperature by alloying with Co, Ni, or Cu (bcc phase) and Au (hcp phase); all showed superconductivity below 3 K (see "Beryllium" Suppl. Vol. A 2, 1991, p. 2). In addition, superconducting intermetallic compounds of Be (with Ni, Cu, Au, Ru, Os, and Re) with T_c between 0.44 and 9.5 K were prepared as well as superconducting alloys (with Co, Fe, Ag, Rh, Ir, and Pt) having T_c < 0.45 K [6]. A superconducting compound $Be_{13}W$ with T_c = 4.1 K (in form of sintered bars with tetragonal structure, a = 10.14, c = 4.23 Å) is reported [7].

A discussion regarding the upper limit of the critical temperature of superconducting materials led to the conclusion that a metal that does not have a high Debye temperature can not become a superconductor with high T_c. Because Be has a Debye temperature of nearly 1500 K (see "Beryllium" Suppl. Vol. A 2, 1991, p. 252) attempts were made to enhance the T_c of Be by quench condensing with various elements, by coevaporation with Li, and by treatment in an alloyed form ($Be_{22}Re$). Probably for metallurgical reasons the predicted maximum T_c could not be approached [8]. No superconducting transition was found for Be in its normal bulk structure down to 1 K according to [9].

A calculation of superconducting parameters for 18 nontransition metals gave T_c = 1.6 K for Be (experimental value 0.026 K). The calculation used formulas obtained by solving the gap equations of Eliashberg, taking into account the Umklapp process of the electron-phonon interaction. The calculated influence of pressure is discussed in [10]. Together with the electron-phonon mass-enhancement parameter, T_c = 0.15 K was determined for Be. The electron-phonon coupling for Be (and 18 other metals) was calculated in a one-orthogonalized-plane-wave (1-OPW) using empirical pseudopotentials and phonon spectra [11].

Using the electron-phonon coupling parameter which was determined from the measured phonon spectrum and an empirical nonlocal pseudopotential fit to the measured Fermi surface, T_c = 0.11 K was calculated for Be [12]. A linearized form for the screened form factors of electron-ion interaction was proposed and applied to predict the superconducting state parameters. On this basis T_c = 0.665 K was calculated for Be [13]. Previous calculations based on the electron-phonon interaction and the mass-enhancement factor gave no results (T_c = 0.0 K) for Be [14].

An ab initio calculation of the electron-phonon interaction in simple metallic systems including Be was performed and the electron mass-enhancement factor for the evaluation of T_c was determined. Within the framework of the rigid muffin-tin approximation, Hartree-Fock total energy cluster calculations could be combined with traditional band structure results to calculate the mass-enhancement parameter [15]. An empirical intercorrelation was found between the superconducting transition temperature T_c and the thermal expansion of Be and of other superconducting metals with two to five valence electrons [16].

An early approximate criterion for superconductivity, considering the interaction of electrons with lattice vibrations, was tested on a series of metals. According to this criterion Be belonged to nonsuperconducting metals, but had a very special position near to superconducting metals [17]; see also [18]. Another possible criterion for superconductivity, based on the concentration of holes in the lattice, the interatomic distance, and the absence of ferromagnetism and strong paramagnetism, revealed also a special (limiting) position of Be among the nonsuperconductors [19]. In agreement with previous results [20], a pseudopotential calculation of the superconducting state parameters (to estimate T_c and the isotope effect) in nontransition metals led to the conclusion that Be is nonsuperconducting [21].

References:

[1] Falge, R. L. (Phys. Letters A **24** [1967] 579/80).

[2] Soulen, R. J.; Dove, R. B. (NBS Spec. Publ. [U.S.] **260/262** [1979] 1/37; C. A. **91** [1979], No. 41241).

[3] Cao, L.; Xia, J.; Lin, P.; Zhang, Q.; Mao, Y. (Diwen Wuli Xuebao **10** [1988] 75/9 from C. A. **109** [1988] No. 84330).

[4] Utton, D. B.; Soulen, R. J.; Marshak, G. (Proc. 14th Intern. Conf. Low Temp. Phys., Otaniemi, Finland, 1975, Vol. 4, pp. 76/9; C. A. **85** [1976] No. 126 327).

[5] Lhota, E.; Manninen, M. T.; Pekola, J. P.; Soinne, A. T.; Soulen, J. R. (Physica B + C **107** [1981] 337/8).

[6] Olsen, C. E.; Matthias, B. T.; Hill, H. H. (Z. Physik **200** [1967] 7/12).

[7] Alekseevskii, N. E.; Mikhailov, N. N. (Zh. Eksperim. Teor. Fiz. **43** [1962] 2110/3; Soviet Phys.-JETP **16** [1963] 1493/5).

[8] Klein, J.; Léger, A.; de Cheveigné, S.; MacBride, D.; Guinet, C.; Belin, M.; Defourneau, D. (Solid State Commun. **33** [1980] 1091/5).

[9] Curzon, A. E.; Mascall, A. J. (J. Phys. C **2** [1969] 382/5).

[10] Kakitani, T. (Progr. Theor. Phys. Kyoto **42** [1969] 1238/64).

[11] Allen, P. B.; Cohen, M. L. (Phys. Rev. [2] **187** [1969] 525/38).

[12] Allen, P. B.; Cohen, M. L. (Solid State Commun. **7** [1969] 677/80).

[13] Sharma, R.; Sharma, K. S.; Dass, L. (Phys. Status Solidi B **133** [1986] 701/6).

[14] Papaconstantopoulos, D. A.; Boyer, L. L.; Klein, B. M.; Williams, A. R.; Morruzzi, V. L.; Janak, J. F. (Phys. Rev. [3] B **15** [1977] 4221/6).

[15] Zdetsis, A. D. (Proc. 18th Intern. Conf. Low Temp. Phys., Kyoto 1987; Japan. J. Appl. Phys. **26** Suppl. 3 [1987] 2135/6; C. A. **108** [1988] No. 210 420).

[16] Keppler, U. (Z. Metallk. **77** [1986] 519/21).

[17] Bardeen, J. (Phys. Rev. [2] **80** [1950] 567/74).

[18] Band, W. (Phys. Rev. [2] **79** [1950] 1005).

[19] Chapnik, I. M. (Dokl. Akad. Nauk SSSR **141** [1961] 70/3; Soviet Phys.-Dokl. **6** [1961/62] 988/91).

[20] Rajput, J. S.; Gupta, A. K. (Proc. 14th Nucl. Phys. Solid State Phys. Symp., Roorkee, India, 1969 [1970], Vol. 3, pp. 16/9; C. A. **75** [1971] No. 123 848; Phys. Rev. [2] **181** [1969] 743/52), Rajput, J. S. (Phys. Status Solidi B **45** [1971] 287/92).

[21] Jain, S. C.; Kachhava, C. M. (Indian J. Phys. A **55** [1981] 89/95).

13.3.2 Films

With regard to superconductivity and other properties, Be films depend strongly on their preparation conditions such as evaporation and ion-beam sputtering (IBS) as well as on their thickness and heat treatment. Impurities or additions are another important factor to influence the film behavior. Superconductivity of thin films was preliminarily reported in the discussion of a possible low-temperature polymorphism of Be, see "Beryllium" Suppl. Vol. A 2, 1991, pp. 3/4.

13.3.2.1 Films from Evaporation and Ion-Beam Sputtering

In early studies the presence of metallic impurities was thought to be a necessary condition for Be to become superconducting above 0.07 K. Therefore it was concluded that superconductivity of thin Be films reported in the literature was also caused by metal (W) impurities [1]. Later studies showed that many superconducting metals when quenched by evaporation on a low-temperature substrate exhibited a raised transition temperature, T_c, which could be sometimes further increased by coevaporation with another material. The effect of this treatment is to render the metal amorphous, to soften the phonon modes, and consequently enhance T_c. In the case of Be, **quench-condensing** produced the so-called γ-phase with an enhanced $T_c = 9.4$ K (as compared with 0.026 K for ordinary α-Be, cf. p. 21) and with a micro-crystalline rather than an amorphous structure [2].

Freshly prepared Be films of approximately 10^{-6} cm thickness displayed superconductivity at 4.2 K. An exact determination of T_c was not made, but it was roughly extrapolated (from the temperature dependence of the critical current) to be ~ 8 K. The films, obtained by evaporation of Be metal from a W helix and condensation on a backing cooled with liquid He, were stable only when thinner than 10^{-5} cm; increasing the thickness caused cracking of the films. Heating even to a temperature of liquid hydrogen (~ 11 K) transferred the superconductivity of the films into the liquid-He temperature range (~ 2.45 K) [3]. The presence of a magnetic field destroyed the superconductivity of the films [4].

In connection with studies on the existence of low-temperature modifications of Be, all films (from evaporation as described) with thicknesses between 400 and 2500 Å became superconducting in the region between 7 and 9 K. They showed, in common, a temperature range up to 20 to 30 K in which the superconducting phase existed. On heating to this temperature, the films became superconducting again when being subsequently cooled. Heating above these temperatures led to incomplete superconducting transition on cooling again (some resistivity remained which increased with increasing heating temperature). On heating above 60 K, superconductivity was completely lost; figures are given in the paper [5]. Two superconducting low-temperature modifications were found to exist in Be films condensed onto cold substrates under differing conditions. One of these modifications became superconducting at about 8 to 9 K and existed up to about 30 K (on heating to 60 K it disappeared completely). The other phase became superconducting at about 6 K and was stable up to at least 130 K (it survived 1 to 2 h heating to room temperature). The first modification occurred in films from rapid evaporation (~ 10 s, charge temperature ~ 1500°C), condensed onto a substrate cooled by liquid He (or liquid hydrogen). The second modification was obtained by slow evaporation from the solid state (~ 1000 s, charge temperature ~ 900°C) and condensation on a substrate cooled by liquid nitrogen. Evaporation from the solid state of Be onto a substrate cooled by liquid He gave a mixture of the two superconducting phases, and before annealing the first modification (with higher T_c) apparently shunted the second phase (with lower T_c). The different behavior of the films was ascribed to the vapor condensation in different molecular states (atomic vs. diatomic) [6]; see also [7].

A thin Be film with amorphous structure (from evaporation in an electron diffraction camera and condensation on a Cu substrate at liquid He temperature) was superconducting below ~ 7 K. After annealing at room temperature the film showed the usual hcp diffraction pattern and was not superconducting down to 4.2 K [8].

Using **ion-beam sputtering** (IBS), thin Be films were grown on room-temperature substrates. In contrast to the evaporation-deposited films on cooled substrate, the films from IBS with a transition temperature up to nearly 6 K showed such a high thermal stability that T_c exhibited no change by room-temperature annealing for three months. The IBS deposition was carried out on oxidized silicon wafers as substrate, maintained at room temperature during deposition under vacuum ($\sim 5 \times 10^{-8}$ Torr). Pure Ar gas (99.9995%) was introduced into an ion source, and the ion beam from the source bombarded a target of 99.9% pure Be [9]. In further studies T_c of IBS films was unchanged after being stored for more than one year at room temperature. However, annealing at 400 to 470 K led to a reduction of T_c and hcp crystals appeared in the highly disordered, metastable microstructure of the films [10]. In the case of a 20 nm thick Be film (from IBS), T_c was about 6.4 K (see figure in the paper) and did not decrease on annealing up to 480 K. On annealing at 520 K, T_c finally decreased below 1.6 K [11]. A method for fabricating thin superconducting Be films by irradiation with an ion beam or a neutral particle beam is described in [12].

The superconducting properties, especially T_c, of Be films from IBS depend on the ejection angle measured relative to the surface of the Be target (T_c tends to drecrease with increasing angle α, see figure in the paper). For explanation, the role of incorporated oxygen in the films and the contribution of highly energetic particles to the formation of amorphous Be are discussed in [13].

Although the mechanism for T_c enhancement in Be films from IBS was not clear, the films were suggested to be in a heavily disordered state considering the relatively high resistivity, the low residual resistance ratio, and the result of reflective electron diffraction [9]. In order to investigate the origin of the T_c enhancement in IBS films, the electronic density of states at the Fermi level, N(0), was estimated from the temperature dependence of the critical magnetic field and Hall coefficient measurements. This study showed that the T_c enhancement of Be was caused not only by N(0) increment but also by other factors [14]. For measurements of the upper critical magnetic field for a 20 nm thick Be film, see the figure in the paper [11].

Critical current densities and pinning-force densities were examined for thin amorphous Be films from IBS with $T_c \approx 6$ K, and the films were proved to be useful in an Abrikosov vortex memory [15]. The Be films from IBS were considered to be advantageous for applications as electrode material and (when oxidized) as good tunnel barriers [11].

The occurrence of superconductivity in thin (at 4.2 K vapor-quenched) Be films was attributed to phonon-induced attractive electron-electron interaction enhancement caused by phonon field perturbation due to gradually annealable, thickness-dependent lattice distortions and discontinuities frozen in during the deposition process, which are relatively more numerous the thinner the film [16].

References:

[1] Olsen, C. E.; Matthias, B. T.; Hill, H. H. (Z. Physik **200** [1967] 7/12).

[2] Klein, J.; Léger, A.; de Cheveigné, S.; MacBride, D.; Guinet, C.; Belin, M.; Defourneau, D. (Solid State Commun. **33** [1980] 1091/5).

[3] Lazarev, B. G.; Sudovtsov, A. I.; Smirnov, A. P. (Zh. Eksperim. Teor. Fiz. **33** [1957] 1059/60; Soviet Phys.-JETP **6** [1958] 816/7).

[4] Lazarev, B. G.; Semenenko, E. E.; Sudovtsov, A. I. (Zh. Eksperim. Teor. Fiz. **45** [1963] 391/2; Soviet Phys.-JETP **18** [1964] 270/1).

[5] Lazarev, B. G.; Sudovtsov, A. I.; Semenenko, E. E. (Zh. Eksperim. Teor. Fiz. **37** [1959] 1461/3; Soviet Phys.-JETP **10** [1960] 1035/6).

[6] Lazarev, B. G.; Semenenko, E. E.; Sudovtsov, A. I. (Zh. Eksperim. Teor. Fiz. **40** [1961] 105/8; Soviet Phys.-JETP **13** [1961] 75/7).

[7] Semenenko, E. E.; Sudovtsov, A. I. (Fiz. Metal. Plenok Akad. Nauk Ukr.SSR Resp. Mezhvedomstv. Sb. **1965** 97/109 from C.A. **65** [1966] 3145).

[8] Curzon, A. E.; Mascall, A. J. (J. Phys. C **2** [1969] 383/5).

[9] Takei, K.; Nakamura, K.; Maeda, Y. (J. Appl. Phys. **57** [1985] 5093/4).

[10] Takei, K.; Okamoto, M.; Nakamura, K.; Maeda, Y. (Japan. J. Appl. Phys. **26** Pt. 1 [1987] 386/90).

[11] Maeda, Y.; Takei, K.; Okamoto, M.; Nakamura, K.; Igarashi, M. (IEEE Trans. Magn. **23** [1987] 1022/5; C.A. **107** [1987] No. 188738).

[12] Takei, K.; Okamoto, M.; Nakamura, T.; Maeda, Y. (Japan. Kokai Tokkyo Koho 87-208678 [1986/87] 1/10 from C.A. **108** [1988] No. 47623).

[13] Takei, K.; Maeda, Y.; Beag, Y. W.; Shimizu, R. (Japan. J. Appl. Phys. **29** Pt. 1 [1990] 500/6).

[14] Okamoto, M.; Takei, K.; Maeda, Y. (Japan. J. Appl. Phys. **26** Suppl. 26-3 [1987] 1323/4; Proc. 18th Intern. Conf. Low Temp. Phys., Kyoto 1987).

[15] Okamoto, M.; Takei, K.; Kubo, S.; Mukaida, M.; Miyahara, K. (J. Appl. Phys. **62** [1987] 212/5).

[16] Reale, C. (Phys. Letters A **51** [1975] 353/4).

13.3.2.2 Effects of Film Thickness and Annealing

Ultrathin (~10 Å thick) Be films were quench-condensed in the usual manner at liquid He temperatures for investigation of the coexistence of superconductivity and localization of electron states. The degree of disorder in such films increased with decreasing thickness. As a consequence, strong electron localization, which completely suppresses superconductivity, was achieved when the films were freshly deposited (i.e., not annealed). The influence of annealing on the superconductivity of such thin films was studied by the temperature dependence of the electrical resistance per unit area, $\rho_\square$. The results of the measurements, shown in **Fig. 13-11**, indicate that a film not annealed after deposition at liquid He temperature has a reversible resistance in the temperature range 1.6 to 5 K with an exponential temperature (T) dependence $\sim\exp(1/T)^{1/4}$; no superconductivity was observed. From the high resistance values of the films strong localization of the electron states was assumed. Annealing the film at only 50 K resulted in a broad peak of resistance below 4 K, whereas on annealing at 100 K total superconductivity could be obtained at rather low temperatures. The T_c values increased with increasing annealing temperature until the maximum value, $T_c = 5.2$ K, was obtained on annealing at 340 K. All resistance vs. temperature curves were reversible below the annealing temperature for these films of ~10 Å thickness [1]. The dependence of resistance on annealing of a freshly deposited ultrathin film has the shape of a smooth curve over the entire investigated temperature range (1.6 to 340 K), see Fig. 13-11, without any plateau characteristic of thicker layers (>40 Å). For thicker layers the plateau extends from 40 to 50 K; this determines the temperature region beyond which heating of the layers results in the loss of superconductivity. When the annealing of ultrathin films is intensified, an increase in

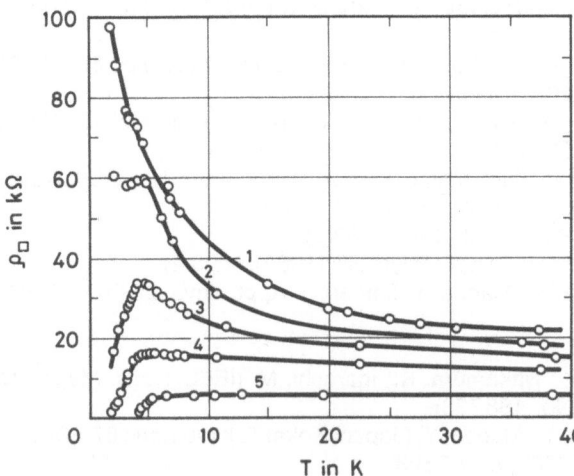

Fig. 13-11. Temperature dependence of the electrical resistance per unit area, $\rho_\square$, of a Be film of ~10 Å thickness: 1) for a freshly deposited layer (reversible behavior of the ρ(T) dependence in the range 5 to 1.6 K); 2 to 5) after annealing of the Be film at 50, 100, 180, and 290 K, respectively.

the T_c values and in the temperature range of the manifestation of superconducting fluctuations above T_c as well as a decrease in the width of the superconducting transition were observed. Ultrathin films seem to be homogeneous. It should be remarked that the described changes, the appearance of total superconductivity following a semiconducting behavior of the resistance vs. temperature, occur without any change in composition and size, only as a result of annealing. The phase state of the films (polycrystalline or amorphous) remains unchanged as a result of heating up to 340 K. From the statements above it was concluded that an increase in T_c with a decrease in the degree of disorder is mainly due to delocalization of electron states as a result of annealing of quench-condensed ultrathin highly deformed films. Superconductivity and a strong localization of electron states seem to coexist (T_c=1.7 K and $\rho_\square$=34 kΩ) [1].

Studies of the dependence of T_c on the film thickness (30 to 500 Å) showed that extremely thin films must possess an amorphous lattice order, i.e., a state where additional disorder gained by going to small dimensions is of no significance to the phonon distribution. If the films were microcrystalline, T_c would be enhanced by the increased disorder when the extremely thin films become still thinner. This mechanism competes with, and is finally overtaken by, the depression of T_c due to the nonsuperconducting surface layer. Thus, for microcrystalline films, a peaked T_c vs. thickness relation was expected. Other experiments indicated that in pure Be not only an amorphous state is present but also traces of a crystalline phase. The role of crystallinity in vapor-quenched Be films could not be definitely determined. Figures are discussed in the paper [2]. A model was proposed to calculate the superconducting transition temperature of very thin films. This model considers the proximity effect between an extremely thin nonsuperconducting surface sheath and the remaining superconducting part of the film. Since T_c is very sensitive to changes of the electron density of states at the Fermi level, it was assumed that superconductivity is lost because of the drop of the electron density in very thin layers at the film surface. The metallic contact of the resulting bilayer structure implies a pair breaking effect and can be treated in well-developed concepts of the proximity effect [3]. Pair breaking due to a surface sheath can explain not only the observed depression of T_c with decreasing film thickness, but also the detailed shape of the resistivity transition into the superconducting state. The agreement was good for $Be_{0.7}Al_{0.3}$ films whereas for pure Be films it was poor close to T_c^* (a parameter which represents the onset of finite conductivity). This discrepancy may be explained by the homogeneous state of the amorphous BeAl films whereas pure Be films show inhomogeneities (traces of a crystalline phase) [2].

When superconductivity of Be films up to 500 Å thickness was studied (thicker films cracked during condensation) by investigating T_c of the films after evaporation in stages of equal length of time, the T_c peak was observed at a film thickness of ~90 Å; this is called the critical film thickness. In films thicker than 90 Å, hexagonal Be appears along with the amorphous phase. The lowering of T_c for films thicker than 90 Å was associated with a certain "ordering" of the residues of the amorphous phase. When films of the critical thickness were annealed in the range 49 to 110 K, Be also transformed from the amorphous to the crystalline state and T_c dropped (T_c is taken as equal to the temperature at which the resistivity of a film on heating becomes equal to half its resistivity in the normal state). As **Fig. 13-12**, p. 28, shows, this temperature is highest when the film is completely in the amorphous state. With annealing at increasing temperatures, T_c decreases but superconductivity remains complete in spite of the partial transition to the crystalline state, probably because there are unbroken superconductivity paths from the amorphous phase. When the amount of the amorphous phase is reduced so much that breaks appear in the superconductivity paths, some residual resistivity is observed after the transition, see **Fig. 13-13**, p. 28. Finally, after complete transformation of amorphous to crystalline Be, all signs of superconductivity disappear [4]. From galvanomagnetic studies (see p. 42) followed that beginning at a thickness of about 20 Å the Be films consisted of a mixture of amorphous and crystalline phases and that the content of the latter increased with increasing thickness [5]. For previous studies of the influence of annealing at room temperature and the formation of the hexagonal phase, see [6].

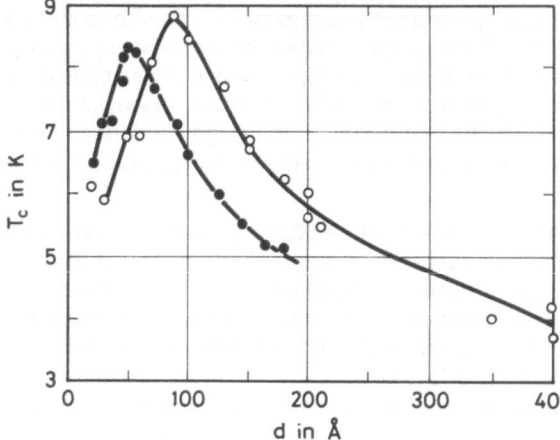

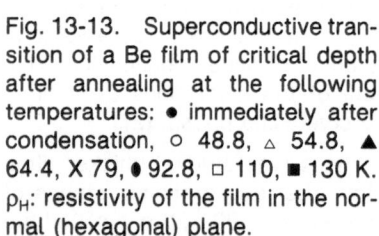

Fig. 13-12. Superconductive transition point, T_c, as a function of the depth, d, of freshly deposited Be films: ● data of Alekseevskii and Tsebro [8]; ○ data of Lazarev et al. [4].

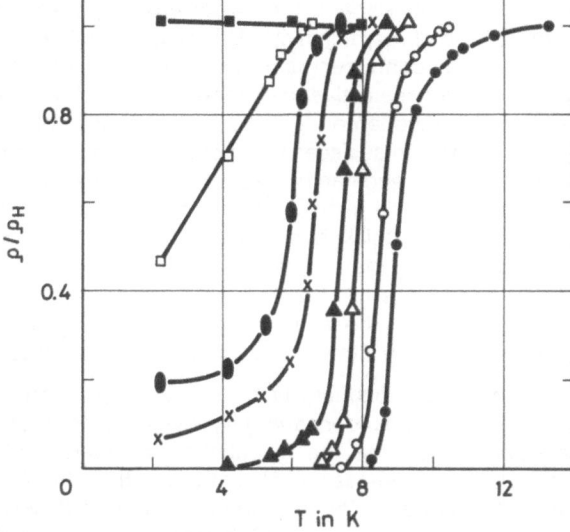

Fig. 13-13. Superconductive transition of a Be film of critical depth after annealing at the following temperatures: ● immediately after condensation, ○ 48.8, △ 54.8, ▲ 64.4, X 79, ● 92.8, □ 110, ■ 130 K. ρ_H: resistivity of the film in the normal (hexagonal) plane.

Quench-condensed, relatively thick (325 Å) Be films are inhomogeneous. It was assumed that both thin pure Be films as well as (thicker) films stabilized by Ge additions are not amorphous but have a disordered crystalline structure (different from the usual room temperature phase), possibly of a metastable high-temperature phase which transforms to the room temperature phase at about 60 K [7].

Previously, the dependence of T_c on the thickness of cold-deposited Be films was studied for the range 25 to 1000 Å. For films thicker than ~200 Å no superconductivity was observed down to 1.3 K; for thinner films T_c passed through a maximum of 8.2 K at about 60 Å [8]. The results [8] are shown in Fig. 13-12 together with the data [4] for newly deposited films which were assumed to consist of amorphous Be formed by condensation on a substrate cooled by liquid He. Qualitatively the shape of the curve agrees with the data [8]; the quantitative difference was ascribed to different experimental conditions [4]. The lowering of T_c on the left side of the peak was assumed to be due to fluctuations which, in the case of a monoatomic two-dimensional metal film, return T_c to zero (see [9]), the lowering of T_c on the right side of the peak as due to dimensional quantization [8, 10, 11].

Films prepared by different methods (by evaporating Be from a BeO crucible or by electron bombardment of the free end of a Be wire) were condensed on crystalline quartz and on glass substrates cooled by liquid He. Independent of the preparation method, films with a thickness up to 600 Å showed good agreement with respect to both the superconducting transition curves and the behavior after annealing. The superconducting transition occurred at 9.6 ± 0.1 K on cooling, but the shape of the curves indicated that the films contained material having a spread of T_c values with an average transition temperature between 8 and 9 K. The normal resistance of the films decreased irreversibly after annealing between 40 and 60 K, and at the same time the well-developed superconductivity transition disappeared. Films thicker than ~600 Å tended to crack during condensation or annealing [12].

It has been pointed out that the condensation rate strongly influences the structure of the Be films freshly deposited at liquid He temperature. At a thickness increase of less than 7 Å/s the layers were amorphous according to electron diffraction, whereas with a condensation rate of ~9 Å/s even at small film thicknesses the diffraction pattern showed a crystalline state corresponding to the usual hcp Be. The critical film thickness above which the crystalline Be pattern appeared (and T_c diminished) varied also with the condensation rate. At a rate of ~4 Å/s the critical thickness was 90 Å. It could not be excluded that Be crystallites in the films were separated by amorphous interlayers due to which superconductivity with a high T_c may be maintained [13]. For earlier electron diffraction studies, see [14, 15].

The evaporation procedure, i.e., evaporation from the Be melt at about 1300°C or from solid Be held at, e.g., 900°C, strongly influences the dependence of T_c on the film thickness (see figure in the paper [16]) for 900, 1100, 1200, and 1300°C. With films obtained by evaporation from the melt, T_c increases up to 9.75 K with the thickness increasing to 400 Å. At greater thickness the films cracked. Otherwise all films produced by one-step evaporation show a smoothly increasing T_c with increasing film thickness (up to ~2000 Å). The superconducting form of Be is retained under strictly controlled conditions even for larger film thickness. Details are described in [16]; cf. [17].

Some experimental results indicated that the Be films start to grow in a strongly disordered phase due to the influence of the substrate or residual gases; after that they grow in the usual hcp manner. Measurements of the mechanical stresses produced by the films onto the substrates can reveal this rather than diffraction experiments. Vapor deposition of Be onto crystalline quartz plates causes the latter to bend. The bending was determined by a multiple beam interference device which allows the simultaneous observation of mechanical stresses and the electrical resistivity of the same film. Films condensed at 4.2 K showed a superconducting transition temperature $T_c = 9.6$ K and a drop of the electrical resistivity at about 55 K for a film thickness of 21 nm, and at about 70 K for a thickness of 16 nm (see figure in the paper). The bending of the quartz substrate changed at practically the same annealing temperatures. The increasing negative values of the reciprocal bending radius, r^{-1}, indicated increasing tensile stress. This behavior was concluded to indicate that the strongly disordered Be phase immediately formed on the substrate is transformed into crystalline (hcp) Be which, because of its crystallite boundary mismatch, produces tensile stresses. It is pointed out that even films with $T_c = 9.6$ K produce such stresses and consequently can not be completely amorphous. The metastable state of the condensed films may be stabilized up to 120 K by means of residual gases (poor vacuum) [18].

The annealing behavior of so-called "inhomogeneous" films differs somewhat from that of "homogeneous" films. All these films show smaller drops in the electrical resistivity on heating and correspondingly smaller changes of the mechanical stresses at the transformation temperature. Further annealing reduced the stresses continuously. The stresses on the substrate during condensation of the inhomogeneous films at thicknesses below 9 nm are the same as with the homogeneous films, but above this thickness value the curves for r^{-1} bend abruptly downwards with altered slope. It was concluded that the films consist of two layers: the layer near the substrate corresponding to the structure of the homogeneous films, and the layer on top of it being a pure hcp Be phase [18].

Measurements of the volume dependence of T_c of quench-condensed Be films (by bending the substrate) revealed that all films investigated could be classified into one of two groups: homogeneous and inhomogeneous films (this classification, however, does not regard the microstructure of the films). The first group showed a temperature dependence of the paraconductivity according to the theory of Aslamazov and Larkin [19]; films belonging to the second group showed a quite different behavior. Their transition curves break up abruptly and can not be fitted by the theory. By coevaporation with small amounts of a noble metal it was possible to stabilize homogeneous films up to a thickness of ~30 nm. Other differences between the two groups of films are discussed in [20].

The Be phase (designated as γ-Be), which is obtained by cryogenic temperature condensation from the vapor and which is superconducting at 9.5 K, appears to be a weak-coupling superconductor. This resulted from studies of the phonon structure in the tunneling characteristics [21].

Any distortion in the lattice of a superconducting metal leads to a rise in the critical magnetic field, H_0. For a severe lattice distortion, the range of the conduction electrons decreases to a value of the order of the lattice constant. Consequently, it follows from the specific electrical resistance of Be condensed at liquid He temperature that the range of the electrons is ~10^{-7} cm, i.e., a small multiple of the lattice parameter. Since any distortion in the lattice of a superconducting metal raises H_0, large values of H_0 were expected for these Be films. At fields up to 110 kOe and in the temperature range of 2 to ~4 K the $H_0(T)$ straight lines for different film thicknesses were found to be parallel, i.e., they do not converge to a single critical temperature. This suggests that a number of Be forms may exist with T_c ranging from 6.5 to ~10 K. The critical magnetic fields determined for the Be films with their high degrees of lattice distortion equal $\geqq 2 \times 10^4$ T_c, and, found by linear extrapolation, the magnetic field eliminating superconductivity for $T \rightarrow 0$ K equals ~180 kOe [22].

The variation of the transition temperature, T_c, in zero field with the annealing temperature, T_a, was investigated for fine-grained, high-purity, continuous Be films of 100 to 2000 Å thickness. The films were vapor-quenched at 10^{-10} to 10^{-9} Torr on very smooth alkali-zinc borosilicate substrates held at 4.2 K inside an He cryostat. Results for a 100 Å thick film, represented in **Fig.** 13-**14**, indicate that the superconductivity is irreversibly destroyed by annealing at ~132 K. On annealing at lower temperatures, T_c decreases without steps with increasing T_a and/or increasing film thickness up to 2000 Å (see figure in the paper). Extrapolation led to $T_c > 0$ K for film thickness $\rightarrow \infty$. For 100 Å thick films the critical magnetic field at 0 K was determined to be $H_0 = 9.49$ kOe [23].

In the case of ion-beam sputtered Be films the superconducting transition temperature, T_c, as studied on films of 9 to 31 nm thickness, depends on the film thickness as shown in **Fig.** 13-**15** from [24]. The highest value of $T_c = 5.8$ K was found in a film of 21 nm thickness [24]. For the dependence of T_c on film thickness up to ~120 nm, see figure in the paper [25].

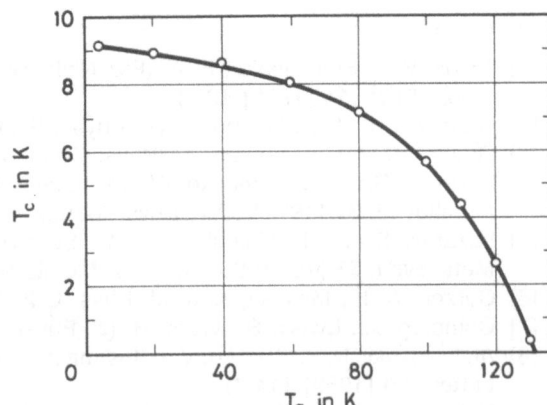

Fig. 13-14. Superconducting transition temperature, T_c, in zero field vs. annealing temperature, T_a, for a 100 Å thick high-purity Be film.

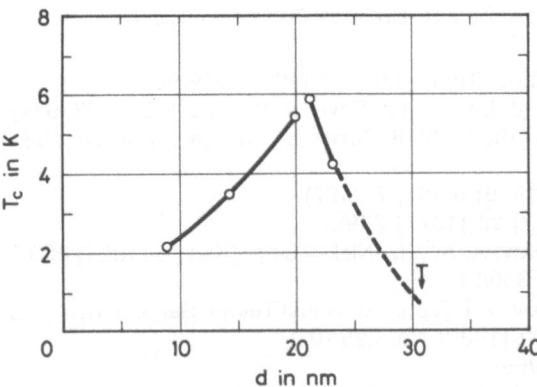

Fig. 13-15. Relation between the transition temperature (midpoint), T_c, and the film thickness, d, for ion-beam sputtered Be.

The superconducting critical temperature of Be films from IBS increases up to 7 K with increasing film thickness up to 20 nm (T_c being defined by the temperature corresponding to a 50% reduction of the film resistance). Films of this thickness gave almost featureless transition electron micrographs and diffraction rings were very diffuse. This indicates that the films were in an amorphous or very fine microcrystalline structure. In contrast, the micrograph of a 40 nm thick film showed densely distributed hcp Be crystal particles (see figures in the paper). The critical thickness at which hcp crystallites begin to grow is in the range 30 to 40 nm. The T_c values of thicker films ($\geqq$ 40 nm) were also smaller than those of 20 nm thick films and decreased further with increasing film thickness. At the same time the transition width was considerably increased, and in the thickness range from ~20 to ~40 nm an abrupt drop of the normal film resistance occurred (see figures in the paper). This was proposed to be due to the growth of low-resistive hcp crystallites. The growth of hcp crystals, which appeared even in thinner films after annealing at 400 to 470 K, could be suppressed in thicker films by incorporating oxygen impurity atoms into the film. For those thicker films (thickness between 40 and 60 nm) values of T_c and transition widths are displayed as a function of film resistivity, which was increased by decreasing the deposition rate (see figure in the paper). The highly resistive samples (film resistivity $\geqq$ 200 $\mu\Omega \cdot$ cm) exhibited a relatively sharp transition at about T_c = 5 K. These films seemed to be homogeneous [26].

References:

[1] Tutov, V. I.; Semenenko, E. E. (Fiz. Nizk. Temp. [Kiev] **14** [1988] 23/6; Soviet J. Low Temp. Phys. **14** [1988] 12/3).

[2] Granqvist, C. G.; Claeson, T. (Z. Physik B **20** [1975] 241/5).

[3] Granqvist, C. G.; Claeson, T. (Phys. Condens. Matter **18** [1974] 79/97).

[4] Lazarev, B. G.; Kuz'menko, V. M.; Sudovtsov, A. I.; Mel'nikov, V. I. (Fiz. Metal. Metalloved. **32** [1971] 52/7; Phys. Metals Metallog. [USSR] **32** No. 1 [1971] 49/54).

[5] Lazarev, B. G.; Kuz'menko, V. M.; Sudovtsov, A. I.; Mel'nikov, V. I. (Fiz. Metal. Metalloved. **33** [1972] 984/9; Phys. Metals Metallog. [USSR] **33** No. 5 [1972] 78/83).

[6] Curzon, A. E.; Mascall, A. J. (J. Phys. C **2** [1969] 383/5).

[7] Comberg, A.; Ewert, S.; Wühl, H. (Z. Physik B **20** [1975] 165/8).

[8] Alekseevskii, N. E.; Tsebro, V. I. (Pis'ma Zh. Eksperim. Teor. Fiz. **10** [1969] 181/4; JETP Letters **10** [1969] 114/7).

[9] Hohenberg, P. C. (Phys. Rev. [2] **158** [1967] 383/6).

[10] Alekseevskii, N. E.; Tsebro, V. I. (Wiss. Z. T.U. Dresden **20** [1971] 475/7).

[11] Alekseevskii, N. E.; Tsebro, V. I. (J. Low Temp. Phys. **4** [1971] 679/96).

[12] Glover, R. E.; Moser, S.; Baumann, F. (J. Low Temp. Phys. **5** [1971] 519/36, 528, 534).

[13] Yatsuk, L. A. (Fiz. Nizk. Temp. [Kiev] **8** [1982] 765/8; Soviet J. Low Temp. Phys. **8** [1982] 384/5).

[14] Fujime, S. (Japan. J. Appl. Phys. **5** [1966] 59/67, 778/87).

[15] Glover, R. E. (Bull. Am. Phys. Soc. [2] **22** [1977] 289).

[16] Tutov, V. I.; Semenenko, E. E.; Chupikov, A. A. (Fiz. Nizk. Temp. [Kiev] **8** [1982] 683/7; Soviet J. Low Temp. Phys. **8** [1982] 340/2).

[17] Lazarev, B. G.; Semenenko, E. E.; Tutov, V. I. (Vopr. At. Nauki Tekhn. Ser. Obshch. Yad. Fiz. **1979** No. 9, pp. 3/5 from C. A. **92** [1980] No. 225102).

[18] Buck, V. (Z. Physik B **33** [1979] 349/55).

[19] Aslamazov, L. G.; Larkin, A. I. (Phys. Letters A **26** [1968] 238/9).

[20] Müller, W. H.-G. (Z. Physik B **40** [1980/81] 203/7).

[21] Klein, J.; Léger, A.; de Cheveigné, S. (Proc. 14th Intern. Conf. Low Temp. Phys., Otaniemi, Finland, 1975, Vol. 2, pp. 71/4; C. A. **84** [1976] No. 188278).

[22] Lazarev, B. G.; Lazareva, L. S.; Semenenko, E. E.; Tutov, V. I.; Goridov, S. I. (Dokl. Akad. Nauk SSSR **196** [1971] 1063/4; Soviet Phys.-Dokl. **16** [1971] 147/8).

[23] Reale, C. (Phys. Letters A **51** [1975] 353/4).

[24] Takei, K.; Nakamura, K.; Maeda, Y. (J. Appl. Phys. **57** [1985] 5093/4).

[25] Maeda, Y.; Takei, K.; Okamoto, M. (IEEE Trans. Magn. **23** [1987] 1022/5; C. A. **107** [1987] No. 188738).

[26] Takei, K.; Okamoto, M.; Nakamura, K.; Maeda, Y. (Japan. J. Appl. Phys. **26** Pt. 1 [1987] 386/90).

13.3.2.3 Effects of Impurities and Additions

Coevaporation of Be with B, C, La, Ge, W, or Pd in various concentrations at liquid He temperature resulted in a decrease of T_c (see figure in the paper). It was thought that the metallurgical treatment applied did not result in a modification of the Be lattice, perhaps because the structure of γ-Be obtained at 4.2 K is more amorphous than generally believed [1].

When Be is coevaporated with Li in various concentrations at room temperature and the films were held at room temperature for 24 h before measuring T_c, a maximum for T_c of nearly 6 K was observed on coevaporation with $\sim$ 20 at% Li. This effect was explained by the fact that the low T_c value ($\sim$ 0.03 K) of ordinary hcp Be is favored by the position of the Fermi energy close to a minimum of the electronic density of states, and that the Fermi density will be enhanced by adding electrons or holes. When T_c is deduced from a rigid-band model as a function of band filling, $T_c = 3$ K can be obtained with 5 at% Li. Experimentally this T_c value is obtained with 10 at% Li. The discrepancy can be explained by the assumption that not all the Li atoms are in substitutional positions (this may explain why a day's annealing at room temperature is necessary to get superconductivity in the films) [1].

Theoretical considerations based on an electron-phonon interaction led to the conclusion that the upper limit of the superconducting transition temperature, T_c, is about 1/10 of the Debye temperature. Therefore the upper limit of T_c is 140 K for Be. Since Be is a weak coupling superconductor, T_c will be sensitive to small modifications of its coupling parameter λ. Especially this should be the case with the so-called amorphous (but possibly microcrystalline [1, 2]) γ-phase ($T_c = 9.5$ K), obtained by quench condensing of Be vapor. The value $\lambda = 0.6$ obtained from the phonon structure of the tunneling characteristics for the upper limit of the coupling parameter of this phase supports the assumption that it is a weak coupling superconductor. Although experimental attempts failed (to enhance T_c by codeposition of Be with BeO, SiO, Ge), enhancement of T_c should be possible if λ could be increased [3].

When Be films were produced by coevaporation with Bi, Li, Al, Ga, and Pb the superconducting transition temperatures (determined as the onset of resistivity) decreased with increasing content of the added element; at the same time the transition became sharper (some 10^{-3} K in the alloys compared to a few 10^{-1} K in pure Be films). The pure Be films seemed to be crystalline (from electron diffraction studies) while for intermediate concentrations of the added elements the structure of the alloy films was certainly amorphous. In other alloy films (Pb-Be, Pb-Bi, Pb-Ga) the transition crystalline–amorphous took place in a rather narrow concentration range; similar results were absent in Be-rich alloys. For explanation it was assumed that pure Be films were already structurally inhomogeneous, i.e., crystalline and amorphous [4]. For Be-Al alloy films, see [5].

Coevaporation of Be with Al was carried out from two simultaneously working vaporizers onto a glass or carbon substrate cooled by liquid He to give $\sim$ 150 Å thick layers containing 70% Be and 30% Al. The condensation rate was very low ($\sim$ 3 Å/s). The superconductive transition temperature was $T_c \approx 8.8$ K, even when the films were kept at room temperature for 1 h. After annealing at room temperature for 20 to 30 days was $T_c = 8.6$ K, but a "tail" appeared in the transition curves (see figure in the paper). Electron diffraction patterns indicated that the layers of $\sim$ 1000 Å thickness consisted of a mixture of Be and Al crystallites with average dimensions of $\sim$ 90 Å for Be and 60 to 80 Å for Al. While the size of the Be crystallites did not change on annealing at room temperature, the size of the Al crystallites increased from $\sim$ 60 to $\sim$ 80 Å. The tail in the transition curves was thought to indicate the formation of a discontinuity of the Be superconducting path. Details are discussed in [6], see also previous studies [7]. In contrast to the behavior of the Be-Al deposits, films consisting of Be and Mg in about equal proportions showed sharp transitions even after annealing at 300 or 400 K; see figure in the paper [8].

When Be was codeposited with KCl, T_c values were found to be up to 10.6 K; after codeposition with Zn etioporphyrin, T_c was up to 10.2 K whereas films prepared by alternating evaporation of Be and Zn etioporphyrin had $T_c = 7.7$ K (with T_c corresponding to the start of the transition). The films preserved their superconducting properties even after heating to room temperature [9].

With the aim of stabilizing the lattice disorder homogeneously across the Be films, Ge was used as a stabilizing impurity. Since the vapor pressures of Be and Ge are similar, a Be + 10 at% Ge alloy was evaporated from a BeO crucible and quench-condensed on a quartz substrate at 10 K. After condensation was $T_c = 8.75$ K for a 100 Å thick film. Annealing at 30 K did not change the residual resistance and the transition temperature. On annealing at 60 K the resistance irreversibly dropped and also T_c decreased (see figure in the paper). A disordered crystalline structure was assumed for the Be + Ge films as well as for very thin homogeneous films of pure Be. This was supported by tunneling studies (gap ratio $2\Delta/kT_c = 3.7$ compared with 4.5 found in amorphous structures), pointing to weak-coupling superconductivity. The residual resistance of the homogeneous Be + Ge films was proportional to their reciprocal thickness, and T_c decreased linearly with the resistance (see figure in the paper). Extrapolation to infinite film thickness led to $T_c = 9.35$ K. It was suggested that the superconducting phase corresponds to a high-temperature phase of Be which transforms to the normal room temperature phase at ~60 K [10].

Experiments on distillation-purified Be films coevaporated with additives of Ge, Cu, Ag, or Au in quantities of 10, 15, 10, and 10 at%, respectively, and condensed on a monocrystalline quartz plate at ~11 K yielded superconductive transition temperatures of 9.72, 9.4, 9.74, and 10.24 K, respectively. At 75, 76, 75, and 85 K, respectively, the films transformed irreversibly and their normal resistance was diminished to less than half of its value in the case of Au addition (see figure in the paper). It was not clear whether the high value $T_c > 10$ K is really due to the codeposition of Au because $T_c = 10.1$ K (Tessaro [11]) was also found in a pure Be film [12].

With $Be_{22}Re$ alloy films, formed by flash laser evaporation on a BeO substrate heated to 950°C, a maximum of $T_c = 9$ K was obtained; see figure in the paper [1].

Ion implantation of hydrogen, deuterium, and helium into ~10 to 20 μm thick foils and also into 0.1 to 0.2 μm thick films of Be on a substrate, carried out at temperatures below 10 K by an ion accelerator with energies of 15 to 100 kV, led to an increases of T_c. Maximum T_c values after implantation were 1.20, 1.64, and 1.64 K for H_2^+, D_2^+, and He^+ ions, respectively. Results are discussed in terms of the McMillan theory in [13, 14]. For ion implantation in superconductors, see also [15].

References:

[1] Klein, J.; Léger, A.; de Cheveigné, S.; MacBride, D.; Guinet, C.; Belin, M.; Defourneau, D. (Solid State Commun. **33** [1980] 1091/5).

[2] Klein, J.; Léger, A.; de Cheveigné, S. (Proc. 14th Intern. Conf. Low Temp. Phys., Otaniemi, Finland, 1975, Vol. 2, pp. 71/4; C.A. **84** [1976] No. 188278).

[3] Léger, A.; Klein, J.; de Cheveigné, S.; Belin, M.; Defourneau, D. (J. Phys. Letters [Paris] **36** [1975] L-301/L-304).

[4] Petersen, J. (Z. Physik B **24** [1976] 273/8).

[5] Granqvist, C. G.; Claeson, T. (Z. Physik B **20** [1975] 241/5).

[6] Lazarev, B. G.; Semenenko, E. E.; Tutov, V. I.; Kornienko, L. A.; Deineko, S. A. (Dokl. Akad. Nauk SSSR **223** [1975] 838/40; Soviet Phys.-Dokl. **20** [1975] 568/9).

[7] Lazarev, B. G.; Semenenko, E. E.; Tutov, V. I. (Dokl. Akad. Nauk SSSR **204** [1972] 837/9; Soviet Phys.-Dokl. **17** [1972] 585/7).

[8] Lazarev, B. G.; Semenenko, E. E.; Tutov, V. I.; Chupikov, A. A. (Vopr. Atom. Nauki I Tekhn. Ser. Fundament. I Prikl. Sverkhprovodimost' **1** No. 2 [1974] 12/5; C.A. **83** [1975] No. 187153).

[9] Alekseevskii, N. E.; Tsebro, V. I.; Filippovich, E. I. (Pis'ma Zh. Eksperim. Teor. Fiz. **13** [1971] 247/50; JETP Letters **13** [1971] 174/6).

[10] Comberg, A.; Ewert, S.; Wühl, H. (Z. Physik B **20** [1975] 165/8).

[11] Tessaro, G. (Diss. Univ. Karlsruhe 1976 from [12]).

[12] Müller, W. H.-G. (Z. Physik B **40** [1980/81] 203/7).

[13] Ochmann, F.; Stritzker, B. (Nucl. Instrum. Methods Phys. Res. **209/210** Pt. 2 [1983] 831/4).

[14] Ochmann, F. (JUEL-1849 [1983] 1/51; C.A. **99** [1983] No. 167827).

[15] Meyer, O. (Radiat. Eff. **48** [1980] 51/62).

13.4 Magnetoresistance

Early experiments on Be single crystals in magnetic fields at low temperatures [1] revealed a decrease of the electric conductivity with increasing field strength. The effect strongly depended on the direction of the magnetic field and of the current with respect to the crystallographic orientation. Measurements on hexagonal crystal needles (up to 1.6 cm in length and 1 mm in diameter) at about 20 and 80 K under transverse magnetic fields up to 10.9 kOe showed the anisotropy of the resistance increases with the rising field strength (see tables and figures in the paper); the anisotropy became independent of temperature for high field strengths [1]. The strong anisotropic effect of transverse magnetic fields up to 12 kOe on the electrical resistance of thin Be platelets along the hexagonal axis with the electric current being always perpendicular to the field direction was further studied at 90, 78, and 20 K. The results were discussed with regard to Kohler's theory; see figures in the paper [2]. Further measurements were performed on a Be crystal (with the prism axis parallel to the c-axis) at about 79 and 90 K with transverse magnetic fields up to 12.2 kOe and at the change from longitudinal to transverse fields up to ~6.7 kOe showing maximum field effects with transverse fields [3].

According to Kohler's theory [4] which is based on a simple two-band model, the relative resistance increase $\Delta\rho/\rho$ with a strong transverse magnetic field, H, is predicted to be quadratic. The reduced Kohler diagram, i.e., the plot of $\Delta\rho/\rho(0)$ as a function of $H\rho_\Theta/\rho(0)$, where $\rho(0)$ ist the resistivity in zero fields and ρ_Θ is the resistivity at the Debye temperature, shows Be in comparison with other metals in agreement with the theory [5]. With respect to the quadratic dependence of the relative resistance increase on strong magnetic fields at low temperatures, a series of metals including Be was discussed by Borovik [6].

The anisotropy of magnetoresistance as observed in the early studies (cf. above) was checked by further investigation [7]. The resistance dependence on the intensity of a magnetic field, H, was measured for various orientations of a Be crystal at liquid He temperature both in the dc field of an electromagnet and in pulsed fields (as described in [8]). For $H \leqq 35$ kOe the resistance increase obeyed the quadratic rule for all orientations of the field with respect to the crystal axis (thus, up to 35 kOe Be behaved like a metal with a closed Fermi surface). For $H \geqq 50$ kOe and parallel to the [0001] axis, the resistance dependence on the magnetic field showed a tendency toward saturation (this behavior was considered to be a consequence of the appearance of open trajectories parallel to the hexagonal axis). Figures and details are given in the paper [7].

The temperature dependence (in the range 0 to ~60 K) of the magnetoresistance $\rho_H(T)$ was studied on Be single crystal samples (with $\rho_{300}/\rho_0 = 76$ and 440) for field directions parallel and perpendicular ($\parallel$ and $\perp$) to the crystallographic c-axis. In both cases the current was perpendicular to the magnetic field and to the c-axis. To facilitate comparison, initially was $H = 30$ kOe in both directions; then the field was increased to $H \approx 40$ kOe. The magnetoresistance $\rho_H(T)$ was found to be proportional to T^3 for $H \parallel c$ and to T^5 for $H \perp c$. The variation of $\rho_H(T)$ was assumed to have the same temperature dependence as the electron-phonon relaxation time τ (T) as long as the phonon part of the resistance without the magnetic field is small compared to the residual resistance [9].

The temperature dependence up to $\sim$300 K of the electrical resistivity of high-purity Be ($\rho_{295}/\rho_4 = 1340$) for the current along the hexagonal c-axis under transverse magnetic fields of 1 to 12 kOe (i.e., perpendicular to the c-axis) is shown in **Fig. 13-16** together with some results of previous measurements [1, 10, 11] in zero field on less pure samples for comparison. The total resistivity is conventionally expressed by $\rho_t = \rho_i + \rho_r$, where ρ_i is the intrinsic resistivity due to scattering of electrons by phonons and ρ_r is the residual resistivity due to impurity scattering of the electrons. The measured values had an inexactness of $\sim$5% which arose mainly from geometric uncertainties. The variation of the resistivity with the angle of the magnetic field (of 10 kOe) in the basal plane for 4 and 76 K is given in figures in the paper (showing the expected 60° rotational symmetry of a hexagonal crystal). The results are discussed on the basis of the reduced Kohler diagram (see figure in the paper) with ρ_Θ estimated to be 47 $\mu\Omega \cdot$ cm at $\Theta = 1481$ K. For a given crystallographic direction (of the field) one curve in the diagram should fit data taken at any temperature and with any impurity level. The magnetoresistance depends on the direction of the magnetic field a_1 and a_2 (or a and b) in the hexagonal basal plane. However, for both field directions the magnetoresistance shows the H^2 behavior at high fields as theoretically predicted for a compensated metal with no open orbits [12].

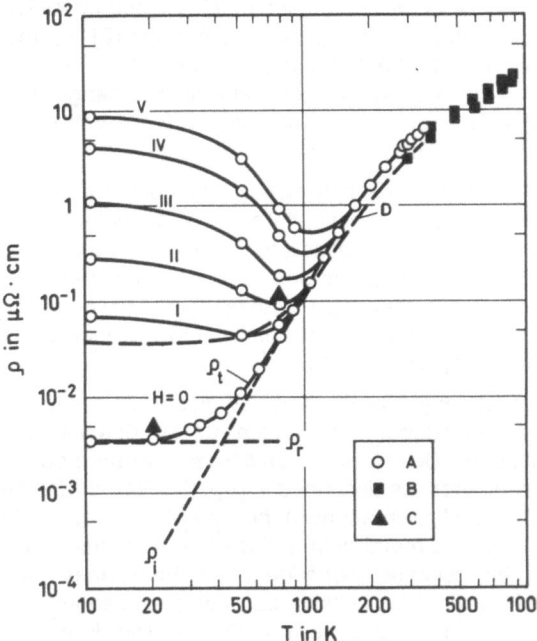

Fig. 13-16. The electrical resistivity, ρ, of high-purity Be crystals as a function of temperature, T, in various transverse magnetic fields.
I: H=1 kOe;   II: H=2 kOe;
III: H=4 kOe;   IV: H=8 kOe;
V: H=12 kOe.
Together with the measurements of the author [12]=A are given previous results from less pure samples [10]= B, [1]=C, and [11]=D for comparison.

In follow-up studies to [7] with pulsed magnetic fields (pulse duration 10^{-2} s; for the method and apparatus, see [13]) it was confirmed that in magnetic fields up to 50 kOe the resistance increased approximately as the square of the field for all field orientations with respect to the crystal axes. However, in fields stronger than 50 kOe with the directions of the electric current in the sample and of the applied magnetic field both perpendicular to the hexagonal crystal axis, a change in the law of the resistance rise with the field was observed; this was attributed to the appearance of open trajectories related to **magnetic breakdown** (Mb), see "Beryllium" Suppl. Vol. A 2, 1991, p. 273, and this volume p. 59. The angular dependence of the electrical resistance $\Delta\rho/\rho$ as measured at 4.2 and 78 K on the Be single-crystal sample ($\rho_{300}/\rho_{4.2}$=125, ρ_{300}/ρ_{78} = 50) in transverse magnetic fields of 5 to 150 kOe is shown in a figure in the paper. The dependence of the resistance change on the magnetic field intensity with the field direction perpendicular to [0001] at 4.2 and 78 K is represented in another figure in the paper [14]. The results, in agreement with previous studies [7], show clearly that at 4.2 K the deviation from the initial resistance-rise law occurred sharply in a field H = 45 kOe whereas at 78 K this transition took place more smoothly, but with the inflection point at approximately the same value of the magnetic field H = 40 to 50 kOe. This behavior of the resistance was interpreted as a consequence of magnetic breakdown. The change in the law of resistance rise occurs at 78 K in a field H = 40 to 50 kOe. When the mean free path is raised by cooling to 4.2 K, this change occurs again at the same field H = 45 kOe; i.e., the position of the inflection point in the ρ(H) dependence is determined by the intensity of the external field H (not by the effective magnetic field). At 78 K the ρ(H) change occurs in a wider range of fields. This may be due to the thermal broadening of the Fermi level. In this case Kohler's rule is not satisfied [14].

Other experiments on Be single crystals (with a residual resistivity ratio of ~2000) at 4.2 K in magnetic fields up to 100 kOe for the current along the three major crystal axes revealed a closed Fermi surface for most field directions and equal numbers of holes and electrons. Above 20 kOe open orbits directed along $\langle 10\bar{1}0\rangle$ were effective when the field direction was less than 18° from [0001] in a $\{10\bar{1}0\}$ plane. For H $\parallel$ [0001] the magnetoresistance was quadratic at low fields, showed a maximum at H $\approx$ 23 kOe, and saturated at high fields. Large-amplitude Shubnikov-de Haas oscillations observed for this field direction were concluded to result from magnetic breakdown between the "coronet" and the "cigars" of the Fermi surface (see p. 58) [15]. Further studies of the Be magnetoresistance at 1.4 to 4.2 K in fields up to 105 kOe indicated that open orbits occur for the magnetic field within a two-dimensional region (1.2° in diameter) about the c-axis and within 19.5° of the c-axis in a $\{10\bar{1}0\}$ plane as a result of magnetic breakdown. Another consequence of Mb is that Be becomes uncompensated, and large-amplitude quantum oscillations occur in the magnetoresistance for H $\parallel$ c-axis. At 4.2 K the oscillations have a single frequency (F_1) of 9.2×10^6 G which corresponds to orbits around the "waist" of the "cigars". At 1.4 K a second frequency of 0.3×10^6 G appears equal to the difference between the cigar "waist" and "hip" frequencies [16]. The large-amplitude oscillations of the high-field magnetoresistance of Be for H $\parallel$ c-axis with the frequency F_1 and the change from a quadratic to a saturating field dependence of the monotonic magnetoresistance can be understood in terms of Mb between the "cigar" and the "coronet" pieces of the Fermi surface. The difference frequency, however, which appears only below 4.2 K, can not be related to a Fermi surface area. This frequency is demonstrated to be generated through the interaction of Mb and the magnetic domains reported by Condon [17]; further details and figures are given in the paper [18].

Further studies of the Be magnetoresistance under magnetic-breakdown conditions were concerned with the oscillations observed in the dependence of the resistance on both the magnitude and direction of the magnetic field (produced by a superconducting magnet). The amplitude of the oscillations of the Shubnikov-de Haas type increased rapidly in fields H $\approx$ 50 kOe and largest amplitudes were observed for H $\parallel$ [0001] due to open trajectories in the hexagonal plane as a result of Mb. The existence of oscillations in a wide interval of field directions is discussed and figures of the angular dependence are given in the paper. Superimposed on the usual relatively smooth directional dependence of the resistance are small characteristic oscillations, the amplitude and period of which vary sharply with the angle. The positions of the oscillation maxima and minima are not connected with rational directions in the crystal. According to the Lifshitz-Onsager formula the reciprocal period, i.e., the magnetic frequency F, is proportional to the area of the extremal section of the Fermi surface; at zero angle F corresponds to the central section of the electron ellipsoid ("cigar") of the Fermi surface. The energy gap between the second and third bands in [11$\bar{2}$0] direction was estimated to be ~6×10^{-3} Ry [19]; see also [20], cf. previous studies [21].

Because in the early studies of Mb [18, 19] no attempt was made for a quantitative comparison of the results with theory, and because of the discrepancy found in the frequencies of the Shubnikov-de Haas and the de Haas-van Alphen effects in [19], magnetic breakdown in the Be basal plane was investigated in fields up to 150 kOe at 4.2 K. The current density in the single-crystal rod ($\rho_{296}/\rho_{4.2}$=100) was parallel to [11$\bar{2}$0]. Applying the theory of Falicov et al. [22] an estimate of 30 kOe was obtained for the characteristic breakdown field. The Shubnikov-de Haas frequency of the oscillations at H∥[0001] was found to be F=9.4×10^6 G which corresponds to the central cigar cross section of the Fermi surface in accordance with de Haas-van Alphen data (from the literature). For details, figures, and discussion see the paper [23].

The previously observed large-amplitude oscillations of the magnetoresistance in Be under Mb conditions, not only for H∥c-axis but also in a range of angles ϑ between H and the c-axis, were attributed to elongated Mb trajectories connecting two or three coronets through a cigar-shaped piece of the Fermi surface (see figure 5 in [24]). On this basis and also because of the discrepancy stated in [19], measurements were performed on Be under Mb conditions in fields of ~70 kOe of the oscillations of magnetoresistance and those of the susceptibility (de Haas-van Alphen effect). Magnetic breakdown was found for noncentral extremal sections of a cigar-shaped piece of the Fermi surface [25]. For the de Haas-van Alphen effect, see "Beryllium" Suppl. Vol. A 2, 1991, p. 262.

Despite the large discrepancy in the residual resistance of Be samples ($\rho_{300}/\rho_{4.2}$=130 and 1500) described in the literature, the ratio of the oscillation amplitude to the resistance was essentially of the same order (~50%). This result was attributed to the fact that the major factor in suppressing the coherence of the various effects is not impurities but dislocations. Studies of the temperature effect in the range 4.2 to 77 K revealed that the oscillations were sustained up to 20.4 K and did not disappear below 30 K [24].

The temperature dependence of the oscillation amplitudes of the magnetoresistance was studied on Be samples with $\rho_{300}/\rho_{4.2}$≈150 in the range 0 to 25 K in fields of 40 to 88 kOe (of a superconducting solenoid) and with various measuring currents. The relative amplitude of the magnetoresistance oscillations decreased with rising temperature and magnetic field (according to literature a similar temperature dependence was obtained for the de Haas-van Alphen effect). The observed dependence of the Mb probability on the field may be the consequence of the coherent effects (considered, e.g., in [24]) [26].

In connection with magnetic breakdown [24] also a "thermal breakdown" effect was theoretically predicted which results in the dependence of the resistance on temperature even when the main cause of the resistance is electron scattering by impurities [27].

Theoretical values of the transverse Be magnetoresistance, computed by path-integration in the intermediate field range on a 6-OPW (orthogonalized-plane-wave) model Fermi surface, agreed well with experimental data. Although only few experimental data were available for the Hall coefficient (see p. 40) and the longitudinal magnetoresistance, satisfactory agreement was found in comparison with the calculation [28].

For an early theoretical investigation of collective effects (carrier correlation) on the magnetoresistance of a series of metals including Be, see [29].

The influence of surface contamination on the Kapitza conductance of Be (Mo, Cu, and Ag) is discussed, based on literature data, in terms of the adsorbed impurity model for untreated surfaces and of the dislocation plus dense-layer model or the adsorbed impurity model for treated surfaces [30].

Thin Be **films** of $\sim 10^{-6}$ cm thickness and commercial (99%) purity in vacuum (10^{-6} Torr) on mica substrates at room temperature were investigated with respect to the magnetic conductivity anisotropy caused by the effect of electron localization. The films were found to have quantum corrections to the conductivity: a logarithmic increase in resistance with lowering temperature and a positive anomalous magnetoresistance or, with rising temperature, the phase shift length of the electron wave function decreases and the magnetic conductivity anisotropy (which characterizes the two-dimensionality of the localization effect) vanishes. With increasing film thickness, as well as with decreasing thickness in the case of the thinnest films, the anisotropy is reduced, corresponding to the "three-dimensionalization" effect. In summary, the anomalies of the dependence of the structural and kinetic characteristics of disordered Be films allow the detection of a transition from two-dimensional to three-dimensional localization with varying temperature and film thickness [31].

References:

[1] Grüneisen, E.; Adenstedt, H. (Ann. Physik [5] **31** [1938] 714/44, 732).
[2] Grüneisen, E.; Erfling, H.-D. (Ann. Physik [5] **38** [1940] 399/420).
[3] Erfling, H.-D.; Grüneisen, E. (Ann. Physik [5] **41** [1942] 89/99).
[4] Kohler, M. (Ann. Physik [6] **5** [1949] 89/98, 99/107).
[5] Kohler, M. (Ann. Physik [6] **6** [1949] 18/38, 31).
[6] Borovik, E. S. (Zh. Eksperim. Teor. Fiz. **23** [1952] 91/100 from C.A. **1953** 358).
[7] Alekseevskii, N. E.; Egorov, V. S. (Zh. Eksperim. Teor. Fiz. **45** [1963] 388/91; Soviet Phys.-JETP **18** [1964] 268/70).
[8] Alekseevskii, N. E.; Egorov, V. S.; Kazak, B. N. (Zh. Eksperim. Teor. Fiz. **44** [1963] 1116/9; Soviet Phys.-JETP **17** [1963] 752/4).
[9] Varyukhin, S. V.; Egorov, V. S. (Pis'ma Zh. Eksperim. Teor. Fiz. **33** [1981] 35/7; JETP Letters **33** [1981] 32/4).
[10] Powell, R. W. (Phil. Mag. [7] **44** [1953] 645/63).

[11] Reich, R.; Kinh, V. Q.; Bonmarin, J. (Compt. Rend. **256** [1963] 5558/61).
[12] Radebaugh, R. (J. Low Temp. Phys. **27** [1977] 91/105; C.A. **86** [1977] No. 164045).
[13] Alekseevskii, N. E.; Egorov, V. S. (Zh. Eksperim. Teor. Fiz. **45** [1963] 448/53; Soviet Phys.-JETP **18** [1964] 309/12).
[14] Alekseevskii, N. E.; Egorov, V. S. (Zh. Eksperim. Teor. Fiz. **46** [1964] 1205/7; Soviet Phys.-JETP **19** [1964] 815/6).
[15] Reed, W. A. (Bull. Am. Phys. Soc. [2] **9** [1964] 633).
[16] Reed, W. A. (Proc. 11th Intern. Conf. Low Temp. Phys., St. Andrews, Scot., 1968 [1969], Vol. 2, pp. 1160, 1165; C.A. **73** [1970] No. 114513).
[17] Condon, J. H. (Phys. Rev. [2] **145** [1966] 526/35).
[18] Reed, W. A.; Condon, J. H. (Phys. Rev. [3] B **1** [1970] 3504/10).
[19] Alekseevskii, N. E.; Egorov, V. S. (Zh. Eksperim. Teor. Fiz. **55** [1968] 1153/9; Soviet Phys.-JETP **28** [1969] 601/4).
[20] Alekseevskii, N. E.; Egorov, V. S. (Proc. 11th Intern. Conf. Low Temp. Phys., St. Andrews, Scot., 1968 [1969], Vol. 2, pp. 1156/9; C.A. **74** [1971] No. 17341).

[21] Alekseevskii, N. E.; Egorov, V. S.; Dubrovin, A. V. (Pis'ma Zh. Eksperim. Teor. Fiz. **5/6** [1967] 793/6; JETP Letters **6** [1967] 244/9?).
[22] Falicov, L. M.; Pippard, A. B.; Sievert, P. R. (Phys. Rev. [2] **151** [1966] 498/511).

[23] Sellmyer, D. J.; Goldstein, I. S.; Averbach, B. L. (Phys. Rev. [3] B **4** [1971] 4628/31).

[24] Alekseevskii, N. E.; Slutskin, A. A.; Egorov, V. S. (J. Low Temp. Phys. **5** [1971] 377/95, 383, 392).

[25] Egorov, V. S. (Zh. Eksperim. Teor. Fiz. **69** [1975] 2231/5; Soviet Phys.-JETP **42** [1975] 1135/7).

[26] Alekseevskii, N. E.; Dkhir, P.; Nizhankovskii, V. I. (Pis'ma Zh. Eksperim. Teor. Fiz. **14** [1971] 256/60; JETP Letters **14** [1971] 169/72).

[27] Kaganov, M. I.; Kadigrobov, A. M.; Slutskin, A. A. (Zh. Eksperim. Teor. Fiz. **53** [1967] 1135/43; Soviet Phys.-JETP **26** [1968] 670/4).

[28] Yonemitsu, K.; Sato, H. (Phys. Letters A **88** [1982] 87/9).

[29] Coldwell-Horsfall, R. A.; Ter Haar, D. (Phys. Rev. [2] **115** [1959] 891/3).

[30] Van der Sluijs, J. C. A.; Al Naimi, A. E. (Phonon Scattering Solids 2nd Intern. Conf., Nottingham, Engl., 1975 [1976], pp. 34/6 from C.A. **87** [1977] No. 12462).

[31] Butenko, A. V.; Bukhshtab, E. I.; Kashirin, V. Yu. (Fiz. Nizk. Temp. **13** [1987] 1295/8; Soviet J. Low Temp. Phys. **13** [1987] 728/9).

13.5 Hall Effect

In comparison with any other metal, **single crystalline** Be with an hcp crystal structure below ~1527 K exhibits a strong anisotropy of the Hall effect. The Hall coefficient, R_H (all values in 10^{-10} $m^3 \cdot A^{-1} \cdot s^{-1}$), shows extremely large absolute values and has two independent components: $R_\parallel$ when the magnetic induction, B, is oriented along the hexagonal c-axis, and $R_\perp$ when B is perpendicular to the c-axis. Based on earlier studies at temperatures below 300 K (with contradictory results) the anisotropy of the Hall coefficient was investigated in the temperature range 77 to 1000 K in a field B = 0.15 T. The results showed that the components $R_\parallel$ and $R_\perp$ have opposite signs ($R_\parallel$ is negative, $R_\perp$ is positive). The remarkably high absolute values of the Hall coefficient decreased considerably with increasing temperature. The strong anisotropy of the Hall effect right up to 1000 K indicated that Be retains the main anomalies of the electron spectrum of the Fermi surface to the high temperature. In **Fig.** 13-17 the curves 1 and 2 represent the results for highest purity Be (residual resistance ratio ($\rho_{300}/\rho_{4.2}$) RRR = 284), curves 3 and 4 (RRR = 5), and curves 5 and 6 (RRR = 2.13) those for less perfect samples. Only the purest sample shows a rise of $R_\perp$ with the temperature. This was attributed to the dependence of $R_\perp$ on the magnetic field strength. The rise disappeared gradually when the magnetic field was reduced from 0.15 to 0.03 T. Theoretical considerations to explain the temperature dependence of the Hall coefficients are given in the paper [1]. In Fig. 13-17 are also represented the data reported by Shiozaki [2] for 99.99% pure single crystalline platelets between 4.2 and 300 K with a field of 12.4 kOe. At 290 K is $R_\parallel = -7.64 \pm 0.4$ and $R_\perp = +14.8 \pm 0.2$, and the average $R_H = (R_\parallel + 2R_\perp)/3 = 7.3$. At 77 K is $R_\parallel = -7.56 \pm 0.4$ and $R_\perp = +13.7 \pm 0.4$ [2].

The figure shows further the results of Gladkov et al. [3], obtained for 99.6% Be from measurements between 4.2 and 253 K, revealing good agreement with respect to the sign and the values of $R_\parallel$ and $R_\perp$ [1]. The figure also gives the deviating results of Borovik [4] who reported positive values for $R_\parallel$ which decreased with the temperature in the range 2.14 to 290 K. At room temperature is $R_\parallel = 7.7$ independent of the magnetic field from about 3 to 20 kOe, whereas at lower temperatures (78, 20.4, and 2.14 K) $R_\parallel$ decreased with the field strength applied; see figure in the paper [4].

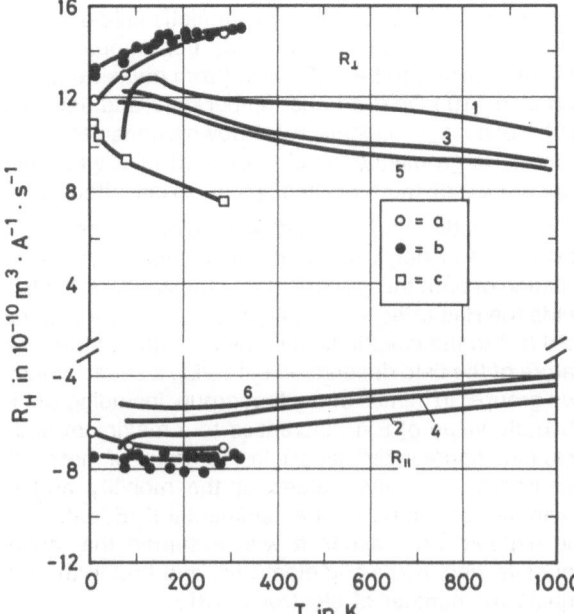

Fig. 13-17. Temperature dependence of the Hall coefficient, R_H, of Be single crystals for two orientations of the magnetic field, $R_\perp$ and $R_\parallel$: results a from [3], b from [2], c from [4]; the data for the samples of different purity from [1] are given in the curves 1 and 2 for RRR = 284, 3 and 4 for RRR = 5, and 5 and 6 for RRR = 2.13.

To explain the remarkable anisotropy and the large absolute values of the Hall coefficients of Be, the contributions from the several parts of the Fermi surface (see p. 57) were estimated. On the basis of the de Haas-van Alphen data (see "Beryllium" Suppl. Vol. A2, 1991, pp. 262/3) a simplified model of the Fermi surface was constructed (the "coronet" was replaced by a doughnut-like piece) to calculate the Hall coefficients; the results agreed with the experimental values. It was concluded that the remarkable characteristics were mainly dependent on the shape of the Fermi surface, the effective mass, and the density of states. The anisotropy of R_H occurred because only electrons (when B∥c) and only holes (when B⊥c) affect the Hall coefficient, and the large absolute values of $R_\parallel$ and $R_\perp$ are due to light electrons and light holes, respectively [5]. In further studies calculations based on a more realistic model of the Fermi surface were performed. The galvanomagnetic coefficients of Be were computed by path-integration in the intermediate field range on a multiple OPW model Fermi surface with six plane waves; the results were in satisfactory agreement with experimental data [6].

The Hall effect in (99.6%) Be single crystals was previously measured at 4.2, 77.4, and 253 K with the magnetic field parallel or perpendicular to the crystallographic c-axis (results are shown in Fig. 13-17). Independent of the magnetic field strength, the mean values were $R_\parallel = -7.3$ and $R_\perp = +16$, the latter was tentatively assumed for room temperature. The high absolute R_H values were attributed to low carrier concentration values in Be [3].

Under uniaxial compression the Hall constant of a Be crystal showed no change even when the stresses applied parallel to the c-axis were large enough (36000 psi) to cause plastic flow [7].

The Hall effect in Be was also investigated under magnetic-breakdown conditions. In a number of metals, including Be, the character of the carrier motion can undergo serious changes when the magnetic field is increased. Such changes may be connected with magnetic breakdown (cf. p. 59). The dependence of the Hall constant on the magnetic field is shown in a figure in the paper [8]. In fields up to 30 kOe, R_H does not depend on the field. In fields exceeding 30 kOe, R_H decreases with increasing field. This field dependence of R_H is attributed to the unbalance of the electron and hole volumes of the Fermi surface which results from the breakdown. On the basis of the magnetic field dependence of R_H the breakdown field strength was estimated to be ~110 kOe [8]. For comparison, in previous studies the breakdown field was determined to be ~130 kOe (from measurements of the de Haas-van Alphen effect) [9], and also 110 kOe (from the temperature and field dependence of the resistivity oscillations) [10]. Because of magnetic breakdown a considerable increase of the Hall effect in the Be basal plane in large magnetic fields of >40 kOe was stated by Egorov [11]. Conditions for the appearance of the so-called "planar" Hall effect are given in [12].

Early studies showed that the field dependence of the Hall coefficient is generally complex and differs in detail for different metals. In strong magnetic fields R_H is simultaneously a function of both the Hall effect and the variation in electrical resistance. In the region of strong fields the Hall effect can be characterized by a dimensionless parameter: the ratio of the Hall field (E_y) to the electric field in the direction of the current (E_x), i.e., by E_y/E_x. According to the nature of the field dependence of E_y/E_x, a series of investigated metals could be classified into two groups. In metals of the first group, including Be, Mg, Zn, Cd, Sn, Pb, graphite, and Bi, the absolute value of E_y/E_x increases to a maximum and then falls off with increasing magnetic field (see figure in the paper). In this group of metals the magnitude of E_y/E_x is some tenths or hundredths of unity. Values of the mobility and concentration of the mobile particles, computed on the basis of experimental E_y/E_x data, are tabulated in the paper. For explaining the experimental results it was assumed that conductivity is due to quasi-particles with opposite sign (holes and electrons) and that in the first group of metals the number of holes equals the number of electrons [13].

The scattering of R_H values of **polycrystalline** Be, published in the literature (for instance +2.4 by Ciccone [14] or +5.6 by Shklyarevskii and Yarovaya [15]), can be explained on the basis of results obtained on single crystals. Very large grains can grow in polycrystalline Be. If there is a grain of large size near the Hall probe, the measured value can take values varying from $R_{\parallel}$ to $R_{\perp}$ [2]. The same value as published in early studies [14] is reported for Be in the comparison of R_H values for a series of metals together with T_c (see pp. 21/2) values in a search for a possible correlation between superconductivity and the Hall coefficient [16]. The value published by [15] and cited in [2] was obtained (probably at room temperature) on layers from vacuum evaporation of Be from a W crucible.

The relation between the Hall constant and the film thickness was investigated on Be films condensed from the vapor on a substrate at liquid He temperature (4.2 K). The Hall voltage was measured in a field of 16 kOe at about 13 K where Be is in the normal (i.e., not superconducting) state. The Hall constant for newly condensed films changed smoothly from negative values for the thinnest (~20 Å) to positive values for the thickest (~450 Å) layers; see figure in the paper [17]. Films heated to 130 K, at which temperature the residues of amorphous phase disappeared and no superconductivity was observed, as well as films heated up to room temperature gave Hall constant values practically independent of film thickness and corresponding to those of massive polycrystalline Be metal [17]. Pure Be films quench-condensed at 10 K up to 150 Å thickness were reasonably homogeneous. Their temperature dependence of resistance agreed with the Aslamazov-Larkin theory, and they showed little current dependence of the superconductive transition (up to 300 A/cm²). Measurements of the thickness-independent Hall coefficient with a magnetic field applied up to 6 kOe led to a carrier concentration of 0.30 holes per atom assuming a one-band model for the metastable Be phase ($T_c = 9.6$ K). After annealing at room temperature, the Hall coefficient for the stable Be phase ($T_c = 0.026$ K) is thickness-dependent; it decreases with increasing film thickness as shown in **Fig. 13-18**, possibly due to a preferred crystallite orientation (texture) in the film. In this case the carrier concentration was 0.06 holes per atom. The Hall coefficient of thin, homogeneous Be films was found to differ appreciably from the coefficient predicted by the free-electron model which was expected to hold for liquid-like amorphous structures [18].

Fig. 13-18.   Dependence of the Hall coefficient, R_H, on the film thickness, d;
□ represent measurements at low temperatures on homogeneous films soon after deposition; ● represent similar measurements on films believed to be inhomogeneous; ○ represent measurements after warming the samples to room temperature (errors in these measurements are indicated).

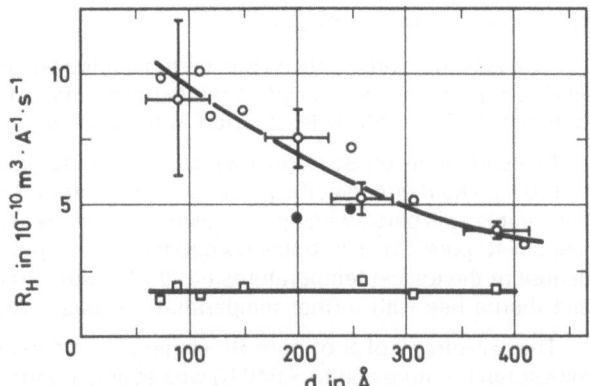

References:

[1] Khamraev, N. S.; Konstantinov, P. P.; Brukov, A. T.; Vedernikov, M. V. (Fiz. Tverd. Tela [Leningrad] **30** [1988] 1235/7; Soviet Phys.-Solid State **30** [1988] 718/9).
[2] Shiozaki, I. (Phys. Letters A **50** [1974/75] 203/4).
[3] Gladkov, V. P.; Petrov, V. I.; Protasov, E. A.; Rodionov, A. G.; Skorov, D. M.; Tulupov, I. F. (At. Energiya SSSR **37** [1974] 463/5; C. A. **83** [1975] No. 36298).
[4] Borovik, E. S. (Zh. Eksperim. Teor. Fiz. **23** [1952] 83/90; C. A. **1953** 367).
[5] Shiozaki, I. (J. Phys. F **5** [1975] 451/8).
[6] Yonemitsu, K.; Sato, H. (Phys. Letters A **88** [1982] 87/9).
[7] Gross, G. E.; Gutshall, P. L. (ASD-TDR-63-605-Pt. II [1964] 1/35; C. A. **62** [1965] 1423).
[8] Alekseevskii, N. E.; Nizhankovskii, V. I. (Zh. Eksperim. Teor. Fiz. **65** [1973] 1076/84; Soviet Phys.-JETP **38** [1974] 533/7).
[9] Alekseevskii, N. E.; Egorov, V. S. (Pis'ma Zh. Eksperim. Teor. Fiz. **8** [1968] 301/5; JETP Letters **8** [1968] 185/8).
[10] Alekseevskii, N. E.; Dkhir, P.; Nizhankovskii, V. I. (Pis'ma Zh. Eksperim. Teor. Fiz. **14** [1971] 256/60; JETP Letters **14** [1971] 169/72).

[11] Egorov, V. S. (Fiz. Metal. Metalloved. **45** [1978] 1107/9; Phys. Metals Metallog. [USSR] **45** No. 5 [1978] 185/8).
[12] Goldberg, C.; Davis, R. E. (Phys. Rev. [2] **94** [1954] 1121/5).
[13] Borovik, E. S. (Izv. Akad. Nauk SSSR Ser. Fiz. **19** [1955] 429/43; Bull. Acad. Sci. USSR Phys. Ser. **19** [1955] 383/96, 384, 390, 394).
[14] Ciccone, A. (Nature **130** [1932] 315; Nuovo Cimento [8] **10** [1933] 339/43; Atti Reale Acad. Nazl. Lincei Rend. [6] **17** [1933] 305/8).
[15] Shklyarevskii, I. N.; Yarovaya, R. G. (Opt. Spektrosk. **11** [1961] 661/6; Opt. Spectrosc. [USSR] **11** [1961] 355/7).
[16] Chapnik, I. M. (Dokl. Akad. Nauk SSSR **141** [1961] 70/3; Soviet Phys.-Dokl. **6** [1962] 988/91).
[17] Lazarev, B. G.; Kuz'menko, V. M.; Sudovtsov, A. I.; Mel'nikov, V. I. (Fiz. Metal. Metalloved. **33** [1972] 984/9; Phys. Metals Metallog. [USSR] **33** No. 5 [1972] 78/83).
[18] Yoshihiro, K.; Glover, R. E. (Low Temp. Phys.-LT 13 Proc. 13th Intern. Conf. Low Temp. Phys., Boulder, Colo., 1972 [1974], Vol. 3, pp. 547/51; C. A. **82** [1975] No. 67059).

13.6 Thermoelectric Power

Available measurements on the thermoelectric power, S (in μV/K), of Be show considerable variation, probably as a result of the experimental conditions and sample characteristics. A value of $S \approx 5$ for 298 K is reported in the review [1].

Measurements on Be **single crystals** were performed at low temperatures (in the range 4.2 to 120 K). The thermoelectric power of two different rods of reactor-grade (98.7%) purity with the sample axis parallel and perpendicular to the pressing axis, $S_\parallel$ and $_\perp$, was determined with respect to pure Cu. The obtained curves show a positive power of several micro volts per degree at the lowest temperatures (< 20 K), a smooth decrease to a minimum at about 80 K, and then a rise with further temperature increase; see figure in the paper [2].

The anisotropy of S of pure Be single crystals over a large temperature range (from the lowest temperatures up to ~540 K) was estimated from thermopower calculations in a single OPW (orthogonalized-plane-wave) approximation, using pseudopotential form factors, realistic phonon spectra, and a spherical Fermi surface as first approximation. The anisotropy of the scattering process was retained by including the Umklapp contribution and the anisotropy of the phonon spectra by using Born-von Karman lattice dynamics. At high temperatures the thermopower has a positive slope, and it is higher along the c-axis than along the perpendicular direction (due to the phonon-drag contribution). At low temperatures there is an anisotropy change (but here the calculation is not reliable). The calculated values are shown in **Fig.** 13-**19** and compared with experimental results of Powell et al. [2], which correspond to industrial single crystals (with ~2% impurities) and very low temperatures; however, these results show $S_\parallel > S_\perp$ [3].

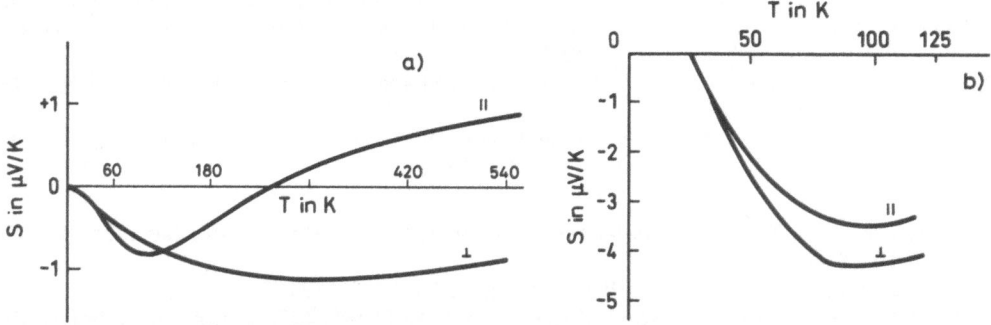

Fig. 13-19.   Thermopower, S, of beryllium; a) calculated, b) measured (from [2]).

The anisotropy of the thermoelectric power, S, of Be was experimentally studied, together with the resistivity, ρ, in vacuum of ~10^{-5} Torr in the temperature range 120 to 1480 K on three samples with the residual resistivity ratio $\rho_{300}/\rho_{4.2} = 284$ (sample 1 with the long sample axis parallel to the c-axis [0001] of the crystal) and $\rho_{300}/\rho_{4.2} = 5$ (samples 2 and 3 with the sample axis parallel and perpendicular to the crystallographic c-axis, respectively). The results are represented in **Fig.** 13-**20** and show the anomalously high anisotropy of S (curves 2 and 3): the difference is $S_\parallel - S_\perp \approx 20$ μV/K at high temperatures. Moreover, the temperature dependences $S_\parallel$ (T) and $S_\perp$ (T) exhibited a number of singularities: different signs of $S_\parallel$ and $S_\perp$ below 700 K, a reversal of the sign of $S_\perp$ near 700 K, a strong rise of $S_\parallel$ (T) with temperature, "humps" of $S_\parallel$ (T) and $S_\perp$ (T) below 500 and 800 K, respectively, and a considerable change in the slope of the $S_\parallel$(T) dependence at 1200 K. The high thermoelectric power of Be, its strong anisotropy, and the complex nature of the temperature dependence of S without any change in the crystal structure are unusual for normal metals. Some of the pronounced singularities of S(T) could be attributed to singularities of the strongly anisotropic Fermi surface of Be (cf. p. 57). Finally, the coexistence of a strong anisotropy of S and the disappearance of the anisotropy of ρ at high temperatures as well as the practically complete similarity in the temperature dependences $S_\parallel$ (T) and $\rho_\parallel$ (T) of the samples 1 and 2, i.e., for crystals of different purity degree, should be mentioned [4]. The strong anisotropy of the Be thermopower was already observed in early studies [5]. Measurements against Cu were performed at 0°C on a pure Be crystal needle with the needle axis along the hexagonal crystal axis and a residual resistivity about 0.0001 of the resistivity at 0°C, giving $S_\parallel = +10.3_3$; $S_\perp = -3.3_7$ was measured on thin platelets of other Be specimens of somewhat less purity (taking $\rho_{20K} \approx 0.00396$ and 0.00243 as residual resistivities). The following S values (in µV/K) of Be were measured vs. manganin (Mng) at various temperatures [5]:

T in °C	−250	−194	−182	−78	0	+20
T in K	23	79	91	195	273	293
Be$_\parallel$–Mng ...	−1.69	+6.28	+7.40	+11.5	+11.2	+11.7
Be$_\perp$–Mng ...	−1.03	−1.84	−1.80	(−1.9)*)	−2.50	−2.64
Be$_\parallel$–Be$_\perp$ ...	−0.66	+8.12	+9.20	(+13.4)*)	+13.7	+14.34

*) Values in parantheses are interpolated.

In the same study the influence of transverse **magnetic fields** up to 12 kOe on the Be thermopower was investigated at different temperatures between 22 and 91 K; results are shown in figures in the paper [5]. Follow-up studies on a Be crystal (rod axis∥c-axis) with transverse and longitudinal magnetic fields at about 80 and 91 K revealed maximum changes of the thermopower at H = 4.85 kOe with transverse fields. Little effect was found with the field direction longitudinal to the sample axis; see figures in the paper [6].

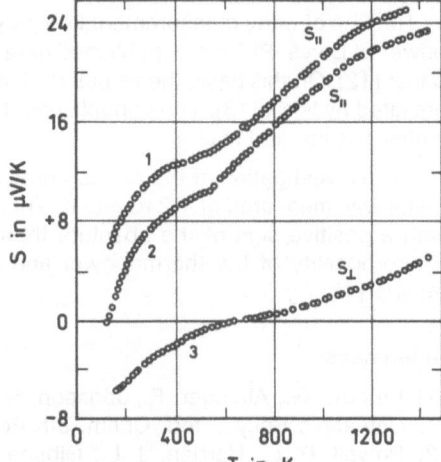

Fig. 13-20. Temperature dependence of the absolute thermoelectric power, S, of beryllium for three different samples 1, 2, and 3 described in the text.

Based on theoretical considerations of the behavior of the thermoelectric power in metals at low temperatures in strong magnetic fields, it was found that the thermopower S(H) tends in any case toward saturation. On the other hand, quantum oscillations were expected to occur in the dependence of the thermopower on a magnetic field similar to those of the electrical resistance as a result of the Shubnikov-de Haas effect (such a correlation in the quantum oscillations of the magnetoresistance and thermopower was observed). Under magnetic-breakdown (Mb, see p. 59) conditions the oscillations of the magnetoresistance become gigantic. In this case the conductivity is almost entirely determined by the electrons in the narrow magnetic-breakdown band, and gigantic oscillations occur also in the thermoelectric power in the magnetic field (cf. "Beryllium" Suppl. Vol. A 2, 1991, p. 273). Giant quantum oscillations of the thermopower in Be were briefly reported by Egorov [7] with an amplitude many times larger than the monotonic components. These oscillations were observed in a Be crystal at 4.2 K in a magnetic field H∥c-axis along with the usual magnetic-breakdown oscillations with the same periodicity as the resistance, but with a phase shift of a quarter period relative to the field [7].

More detailed follow-up studies of the thermopower in magnetic fields up to H ≈ 80 kOe on single crystals mainly of Be (and some other metals such as Mg, Zn, and Al) revealed generally nearly the same behavior: a tendency toward saturation in strong effective fields and oscillations. Giant quantum oscillations of the thermopower were observed in liquid He under magnetic breakdown conditions when the crystal samples were heated with an electric current of ~0.1 A. The Be samples used in this investigation had resistance ratios $\rho_{300}/\rho_{4.2}$ of 11.5 (I), 500 (II), and 1000 (III) (the Be sample used in [7] had a resistance ratio of 100). For the purer sample (II) the effective magnetic fields were much stronger and magnetic breakdown occurred already in relatively weak magnetic fields. The magnetic frequency of the oscillations corresponded to the extremal sections of the Fermi surface, i.e., it is determined by the central section of the "cigar" (see p. 58), and the phase of the oscillations is shifted by one quarter of a period relative to the Shubnikov-de Haas oscillations. Compared with the theory of Blatt et al. [8], the results obtained for Be are discussed in the paper [9]. In further studies the relative shifts of the zeros of the Mb oscillations of the resistance and of the thermopower were measured for [10$\bar{1}$0] and [1$\bar{2}$10] Be crystals rotated about the longitudinal axis through an angle $\vartheta \leqq 2.2°$ from the initial H∥[0001] position [10]; the results of this study are outlined in detail in the section of magnetic breakdown on p. 60.

Results of early measurements on **polycrystalline** metal are compiled in [11]. The thermo-power of Be vs. Pt from unpublished data in the range 300 to 900°C is shown in a figure in the paper [12]. On this basis the values S = 9.4, 13.8, and 18 at 400, 600, and 800°C, respectively, are cited by Weik [13]. The unpublished data shown in [12] are also cited, together with other earlier results, in [14].

In an investigation of the phonon-drag effect in Be, the thermopower of the polycrystalline metal was measured at 4.2 to 300 K. The results are rather scattered (see figure in the paper) with a positive sign of the absolute thermopower throughout (values below about 1 μV/K). Proportionality of the thermopower and the temperature could not be concluded from the results [15].

References:

[1] Petzow, G.; Aldinger, F.; Jönsson, S.; Preuss, O. (Beryllium and Beryllium Compounds, Ullmann's Encycl. Ind. Chem. 5th Ed. A **4** [1985] 11/33, 12).
[2] Powell, R. L.; Harden, J. L.; Gibson, E. F. (J. Appl. Phys. **31** [1960] 1221/4).
[3] Pecheur, P.; Toussaint, G. (J. Phys. Chem. Solids **40** [1979] 1123/8).
[4] Burkov, A. T.; Vedernikov, M. V. (Fiz. Tverd. Tela [Leningrad] **28** [1986] 3737/9; Soviet Phys.-Solid State **28** [1986] 2105/6).
[5] Grüneisen, E.; Erfling, H.-D. (Ann. Physik [5] **36** [1939] 357/67).
[6] Erfling, H.-D.; Grüneisen, E. (Ann. Physik [5] **41** [1942] 89/99, 96).
[7] Egorov, V. S. (Pis'ma Zh. Eksperim. Teor. Fiz. **22** [1975] 86/9; JETP Letters **22** [1975] 38/9).

[8] Blatt, F. J.; Chiang, C. K.; Smrcka, L. (Phys. Status Solidi A **24** [1974] 621/9).

[9] Egorov, V. S. (Zh. Eksperim. Teor. Fiz. **72** [1977] 2210/23; Soviet Phys.-JETP **45** [1977] 1161/8).

[10] Alekseevskii, N. E.; Nizhankovskii, V. I. (Zh. Eksperim. Teor. Fiz. **83** [1982] 1163/9; Soviet Phys.-JETP **56** [1982] 661/5).

[11] Nyström, J. (Landolt-Börnstein 6th Ed. Vol. 2 Pt. 6 [1959] 929/83, 933).

[12] Lillie, D. W. (in: White, D. W., Jr.; Burke, J. E.; The Metal Beryllium, Vol. 1, ASM, Cleveland, Ohio, 1955, pp. 304/27, 309, 311).

[13] Weik, G. (Metall **13** [1959] 202/13, 204).

[14] Darwin, G. E.; Buddery, J. H. (Beryllium, Metallurgy of the Rarer Metals, Vol. 7, Butterworth, London 1960, p. 173).

[15] Yamaguchi, M.; Takahashi, Y.; Takasaki, Y.; Ohta, T. (Bull. Fac. Eng. Yokohama Natl. Univ. **23** Pt. 2 [1974] 175/8; C.A. **82** [1975] No. 79302).

14   Electronic Properties

Some general information on the peculiar type of bonding in beryllium metal and the anisotropy that is a consequence of its electronic structure is given in "Beryllium" Suppl. Vol. A 2, 1991, pp. 117 ff.

14.1   Core Level

Experimental values of the K-shell (1s) binding energy of bulk Be metal with respect to the Fermi level, E_b^F, are (in eV): 111.3 [1], 111.7 [2], and 111.8 [3]. These values were obtained from photoelectric spectra. For additional values, also from X-ray spectra, see "Beryllium" Suppl. Vol. A 1, 1986, pp. 140, 145. Frequently, E_b values are related to the vacuum level, $E_b^V = E_b^F + \Phi$; see, e.g. [4, 5], but the value taken for the work function Φ was suspected to be low (3.9 to 4.0 eV instead of ~5 eV); see p. 72.

The 1s binding energy in Be metal is lower than in Be atoms; the solid state shift amounts to about 7 eV. The free-atom K-shell binding energy, $E_b^V = E(1s2s^2\,^2S) - E(1s^22s^2\,^1S) = 123.6 \pm 0.1$ eV (and $E_b^F = 123.6 - 5.1 = 118.5$ eV) corresponding to the lowest possible ionization energy, was obtained by applying Auger spectroscopy to fast Be beams being core-excited in single collisions with CH_4 and He [6]. This value is higher than the free-atom values previously derived from measurements on solid samples, e.g., $E_b^V = 119.3$ (solid state shift 3.7 eV) [4], whereas good agreement was found with the more recent theoretical free-atom values listed in [6].

State-specific, many-electron calculations were performed based on the cluster (self-consistent-field difference) ΔSCF theory, combined with calculations of atomic electron correlation, to predict the 1s binding energy of Be metal [5]. A Be_{13} hcp cluster was used for this calculation since Be metal is well represented by Hartree-Fock calculations on such a cluster (see [7]); the Be cluster 1s binding energy was found to be 115.0 eV [5], in close agreement with experimental values, e.g., $E_b^V = 115.2$ eV [1], 115.6 eV [4]. The contribution of atomic electron correlation and relativistic effects, 0.4 eV ($115.0 + 0.4 = 115.4$ eV) [8], improves the coincidence [5]. Calculations with an "excitonic" model gave $E_b = 114.0$ eV [9].

The solid state shift for the X-ray photoelectron or binding energy, $\Delta E_{XPS} = 8.00$ eV, was calculated for Be using configurational averages and the spin-density functional formalism [10].

Upon creation of the 1s hole, the p character of the valence band increases significantly. According to a Mulliken population analysis, the bulk atom's valence shell shows 59% p character and only 41% s character (with an overall electron density of 4.46) for the whole excited state, whereas for the ground state the ratio is 62% s to 38% p character (with an overall density of 3.53). Thus, a charge redistribution among the nearly degenerate s–p orbitals occurs, caused by the effective nuclear charge increase created by the hole. This explains the discrepancy between the experimental solid state fluorescence yield of Be and the theoretical prediction for the atomic state [5]; cf. "Beryllium" Suppl. Vol. A 1, 1986, p. 161. The magnitude of the yield depends on atomic and on solid state (hybridization) effects; see [8] and pp. 131/2.

References:

[1] Höchst, H.; Steiner, P.; Hüfner, S. (Phys. Letters A **60** [1977] 69/71).

[2] Nyholm, R.; Flodström, A. S.; Johansson, L. I.; Hörnström, S. E.; Schmidt-May, J. N. (Surf. Sci. **149** [1985] 449/59).

[3] Fuggle, J. C.; Mårtensson, N. (J. Electron Spectrosc. Relat. Phenom. 21 [1980] 275/81).

[4] Shirley, D. A.; Martin, R. L.; Kowalczyk, S. P.; McFeely, F. R.; Ley, L. (Phys. Rev. [3] B 15 [1977] 544/52).

[5] Nicolaides, C. A.; Zdetsis, A. D.; Andriotis, A. N. (Solid State Commun. 50 [1984] 857/60); Zdetsis, A. D.; Nicolaides, C. A. (J. Phys. Colloq. [Paris] 48 [1987] C 9-1071/ C 9-1074).

[6] Bisgaard, P.; Bruch, R.; Dahl, P.; Fastrup, B.; Rødbro, M. (Phys. Scr. 17 [1978] 49/52).

[7] Zdetsis, A. D.; Miliotis, D. (Solid State Commun. 42 [1982] 227/30).

[8] Nicolaides, C. A.; Komninos, Y.; Beck, D. R. (Phys. Rev. [3] A 27 [1983] 3044/52).

[9] Beck, D. R.; Nicolaides, C. A. (in: Nicolaides, C. A.; Beck, D. R.; Excited States in Quantum Chemistry, Reidel, Dordrecht 1979, p. 329 from [5, 8]).

[10] Rantala, T. T. (Phys. Rev. [3] B 28 [1983] 3182/92).

14.2 Electron Energy Bands

14.2.1 Hexagonal Beryllium

Several **methods** were used to calculate the electronic band structure of hcp Be, for instance the orthogonalized-plane-wave (OPW) method [1 to 3], the pseudopotential (PS) method [4, 5], the augmented-plane-wave (APW) method [6 to 9], and a cellular method [10]. Taut [11] investigated the influence of the nonlocality of Shaw's optimized model potential (OMP) on the band structure of Be. Radwan and Ali [12] also performed model potential calculations based on Shaw's OPM; three different sets [13 to 15] were used for the model potential.

Chatterjee and Sinha [16], cf. [17], used the composite wave variational method in conjunction with the quantum defect (QD) method. Bhokare and Yussouff [18] applied Green's function method. A modification of the exact Korringa-Kohn-Rostoker (KKR)-Green function method was presented by Yussouff and Zeller [19] using hcp Be as an example. Dovesi et al. [20] employed a Hartree-Fock linear combination of atomic orbitals (HF-LCAO) self-consistent-field method in an ab initio study. Chou et al. [21] presented an ab initio calculation of the structural and electronic properties of Be using a self-consistent pseudopotential approach within the local-density functional scheme (PS-LD); see also Jensen et al. [22]. Blaha and Schwarz [23] performed self-consistent band structure calculations by means of the full-potential linearized APW method on the basis of local-density approximation (LDA). A "singular factor" method for the electronic structure of Be was given by Holzwarth et al. [24].

Experimental results for the band structure along Δ (= ΓA) were obtained by angle-resolved photoemission spectroscopy (ARUPS) [22]. For the determination of the occupied band width Γ_1^+ or $E_F - \Gamma_1^+$ the usual XPS (see p. 95) measurements can also be used; see, e.g. [25, 26].

Some **results** are discussed in the following section. The band structure of hcp Be from an ab initio self-consistent pseudopotential LDA calculation is represented in **Fig. 14-1**, p. 50 (for the first Brillouin zone, see Fig. 9-35 in "Beryllium" Suppl. Vol. A 2, 1991, p. 128, for the basic segment of the hcp reciprocal lattice cell). This band structure is quite similar to the results of other pseudopotential calculations [23, 27], but differs from the HF band structure found by Dovesi et al. [20], in which the Γ_4^- state is occupied. According to the present calculation, the first energy band does not deviate too much from that obtained assuming a free-electron gas, but the higher bands split and generate crossings and other complexities [21]. An essential property of the Be energy bands is the lowering of p states relative to s states, as seen at $\Gamma_3^+(p)$ and $\Gamma_4^-(s)$ or at $H_2(s)$ relative to $H_3(p)$ and $H_1(p)$ [12]. Although the atomic valence configuration of Be is (2s)², a large amount of electrons appear to have p character in the solid [20, 21]. For the dominant character of the symmetry points (Γ_1^+ is the lowest s state, H_2 is the highest s state, Γ_3^+ is the lowest p state, and Γ_2^- is the highest p state), see, e.g. [7]. The 2s and 2p bands hybridize in the solid state. The atomic 2s and 2p levels begin to overlap at about $a/a_0 = 1.6$ and the p state drops below the s state in the vicinity of the Fermi energy (E_F) leading to metallic behavior and a wide sp band. Due to the strong hybridization of s and p wave functions an energy separation between bonding and antibonding orbitals seems to occur [7]; cf. pp. 52 ff.

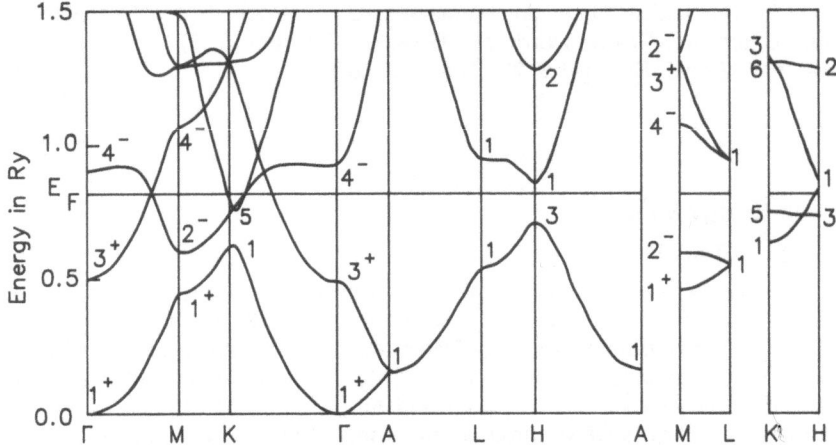

Fig. 14-1.   Calculated band structure for hcp Be. The Fermi energy is 0.82 Ry.

The unoccupied bands remain simple (little d or f character) for some distance above E_F [22].

Figures of energy bands from different calculations were compiled by Cracknell [28].

Radwan and Ali [12] listed and compared energy values at the points of high symmetry, calculated on the basis of Shaw's OMP method and obtained by other authors using different methods; they also listed the deviation of the energy bands from free-electron behavior for $\Gamma_4^- - \Gamma_3^+$, $M_2^- - M_1^+$, and $K_5 - K_1$ (which are all zero for free electrons), and the occupied band width $E_F - \Gamma_1^+$ (which is 0.529 a.u. = 14.39 eV for free electrons; 1 atomic unit = 1 Hartree = 2 Ry = 27.21 eV).

Experimental and theoretical energy values at Γ (in eV) relative to E_F are:

method	PS-LD	APW	APW	ARUPS	XPS
$\Gamma_1^{+\,*)}$	−11.16	−11.16	−11.96	−11.1	−11.9
Γ_3^+	−4.32	−4.35	−4.42	−4.8	—
Γ_4^-	—	1.19	0.57	—	—
Ref.	[21, 22]	[9]	[8]	[22]	[25]

*) The occupied band width is characterized by the Γ_1^+ symmetry point energy.

Other theoretical values for the occupied band width are (in eV): 11.24 from APW [9], 11.42 from OMP [11], 16.19 from QD [16], 19.05 from HF-LCAO [20]; the last two values seem to be too large.

The lowest-energy unoccupied band, determined in ARUPS experiments, is in good agreement with the PS-LD calculation with the exception of some extra structure near the plasmon threshold. This extra structure is associated with interaction between the excited electron and plasmons and is thus a many-body distortion [22, 29]. For deficiencies in the single-particle picture of UPS valence band emission (fairly good agreement between theory and ARUPS results), see [30].

The splitting between bonding and antibonding states is a characteristic feature of the electronic structure of Be, and is in part reflected in the X-ray Raman spectrum (see p. 100). The splitting of bonding–antibonding states was suggested by Inoue and Yamashita [7]. The existence of two (b_1 and b_2) bonding–antibonding bands in the long wavelength Raman band region was detected, e.g., by Papademetriou et al. [31], see p. 99. The bands were assigned to a $3sp_xp_y$–$3sp_z$ splitting of the 3ps hybridized band.

Ab initio cluster calculations of metallic Be were performed to study the bonding–antibonding character of the states [32]; see p. 54.

The band structure and the wave functions of positrons, annihilating with electrons in hcp Be, from the thermalized state $k^+ = 0$ were calculated according to the APW method. The $k^+ = 0$ state, being the absolute minimum of the lowest energy band, is 0.3566 Ry relative to the positron muffin-tin zero [33]; cf. p. 58. Another band structure calculation by the APW method without self-consistency was performed with respect to the relation with the structure stability of Be by Dobrovol'skii et al. [34].

References:

 [1] Herring, C.; Hill, A. G. (Phys. Rev. [2] **58** [1940] 132/62, 154, 157).
 [2] Jacques, R. (Cahiers Phys. **10** No. 70 [1956] 1/30, No. 71/72 [1956] 23/46).
 [3] Loucks, T. L.; Cutler, P. H. (Phys. Rev. [2] **133** [1964] A 819/A 829).
 [4] Cornwell, J. F. (Proc. Roy. Soc. [London] A **261** [1961] 551/64, 559).
 [5] Tripp, J. H.; Everett, P. M.; Gordon, W. L.; Stark, R. W. (Phys. Rev. [2] **180** [1969] 669/78).
 [6] Terrell, J. H. (Phys. Rev. [2] **149** [1966] 526/34).
 [7] Inoue, S. T.; Yamashita, J. (J. Phys. Soc. Japan **35** [1973] 677/83).
 [8] Nilsson, P.-O.; Arbman, G.; Gustafsson, T. (J. Phys. F **4** [1974] 1937/50, 1938, 1943, 1949).
 [9] Wilk, L.; Fehlner, W. R.; Vosko, S. H. (Can. J. Phys. **56** [1978] 266/79, 271).
[10] Altmann, S. L.; Bradley, C. J. (Proc. Phys. Soc. [London] **86** [1965] 915/31).

[11] Taut, M. (Phys. Status Solidi B **54** [1972] 149/57).
[12] Radwan, A. M.; Ali, M. G. S. (Acta Phys. Polon. A **72** [1987] 645/57).
[13] Shaw, R. W., Jr. (Phys. Rev. [2] **174** [1968] 769/81).
[14] Ese, O.; Reissland, J. A. (J. Phys. F **3** [1973] 2066/74).
[15] Appapillai, M.; Heine, V. (Tech. Rept. No. 5 Solid State Theory Group, Cavendish Laboratory, Cambridge 1972 from [12]).
[16] Chatterjee, S.; Sinha, P. (J. Phys. F **5** [1975] 2089/97).
[17] Chatterjee, S.; Sinha, P.; Chatterjee, J. (Proc. Nucl. Phys. Solid State Phys. Symp. C **17** [1974] 48/51 from C.A. **84** [1976] No. 35616).
[18] Bhokare, V. V.; Yussouff, M. (Nuovo Cimento B **19** [1974] 149/60; Lettere Nuovo Cimento Soc. Ital. Fis. [2] **5** [1972] 470/2 from C.A. **77** [1972] No. 168989).
[19] Yussouff, M.; Zeller, R. (J. Phys. F **11** [1981] 1771/4).
[20] Dovesi, R.; Pisani, C.; Ricca, F.; Roetti, C. (Phys. Rev. [3] B **25** [1982] 3731/8).

[21] Chou, M. Y.; Lam, P. K.; Cohen, M. L. (Phys. Rev. [3] B **28** [1983] 4179/85).

[22] Jensen, E.; Bartynski, R. A.; Gustafsson, T.; Plummer, E. W.; Chou, M. Y.; Cohen, M. L.; Hoflund, G. B. (Phys. Rev. [3] B **30** [1984] 5500/7).

[23] Blaha, P.; Schwarz, K. (J. Phys. F **17** [1987] 899/911).

[24] Holzwarth, N. A.; Azhar, S.; Kerr, T. J. (Phys. Rev. [3] B **38** [1988] 9409/16).

[25] Höchst, H.; Steiner, P.; Hüfner, S. (J. Phys. F **7** [1977] L309/L314; Z. Physik B **30** [1978] 145/54).

[26] Skinner, H. W. B. (Phil. Trans. Roy. Soc. [London] A **239** [1946] 95/134, 127, 129).

[27] Barth, von, U.; Pedroza, A. C. (Phys. Scr. **32** [1985] 353/8).

[28] Cracknell, A. P. (Landolt-Börnstein New Ser. Group III **13c** [1984] 1/462, 56/9, 453/4).

[29] Jensen, E.; Bartynski, R. A.; Gustafsson, T.; Plummer, E. W. (Phys. Rev. Letters **52** [1984] 2172/5).

[30] Plummer, E. W. (Surf. Sci. **152/153** [1985] 162/79).

[31] Papademetriou, D. K.; Katsanos, D.; Doukas, A. G. (Phys. Status Solidi B **133** [1986] 223/8).

[32] Zdetsis, A. D.; Miliotis, D. (Solid State Commun. **42** [1982] 227/30).

[33] Oriade, J. O. (Intern. J. Quantum Chem. **20** [1981] 891/6).

[34] Dobrovol'skii, V. D.; Lisenko, A. A.; Maiboroda, V. P.; Morozov, M. M. (Metallofizika Akad. Nauk Ukr. SSR Otd. Fiz. **6** No. 4 [1984] 16/8 from C. A. **101** [1984] No. 79046).

14.2.2 Body-Centered Cubic Beryllium

The band structure of bcc Be (or β-Be, see "Beryllium" Suppl. Vol. A 2, 1991, pp. 2, 23), existing above 1250°C, was calculated both at the Hartree and exact-exchange Hartree-Fock level. While the α-Be band structure exhibits strongly overlapping s- and p-type valence bands, explaining its semimetallic character, β-Be has no gap immediately above the valence band (anywhere near E_F). This is consistent with the resistivity which rises sharply, but not discontinuously, above the hcp→bcc transition at ~1254°C.

Monkhorst, H. J.; Pack, J. D. (Solid State Commun. **29** [1979] 675/6).

14.3 Density of States

Theoretical Studies

Density of states (DOS) curves were evaluated from energy bands calculated by different methods, e.g., OPW [1, 2], PS [3], APW [4, 5], also for a (hypothetical) fcc Be modification and a variation of the lattice parameter a with fixed c/a ratio [6], Green's function method [7], a QD method [8]; they are pictured in [9].

Total and partial (contributions from s, $p_x + p_y$, p_z states) DOS obtained by an ab initio HF self-consistent-field (SCF) LCAO method are shown in **Fig.** 14-2 from Dovesi et al. [10]. Total DOS together with the individual contributions to the DOS from different angular moments of the wave functions were obtained by a self-consistent PS-LD approach in another ab initio study and are shown in a figure in the paper by Chou et al. [11]. Nearly the same results for total and partial DOS were obtained from the full-potential LAPW method by Blaha and

Schwarz [12]. For a DOS calculation using biquadratic quasi-analytical methods (QUALIN, LIN, BIQ) for Brillouin zone integration with Be as an example, see Oriade [13]. In connection with an ab initio HF calculation a new expansion technique (orthogonal expansion set) was applied to Be [14]. The DOS according to an ab initio unrestricted HF cluster calculation by Zdetsis, Miliotis [15] is represented in **Fig.** 14-**3**, p. 54. It shows remarkable agreement with other calculated DOS [2, 6] in their common energy regions.

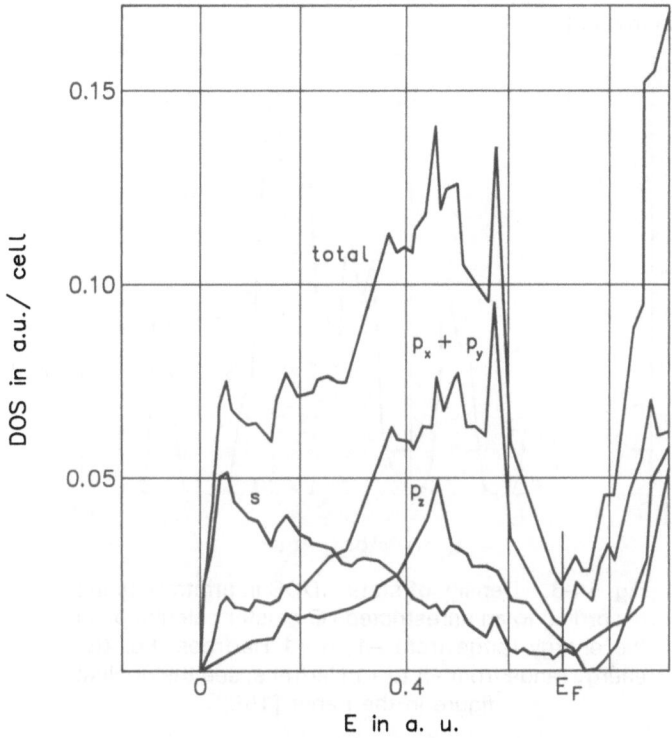

Fig. 14-2.   Total and projected density of states of Be.

A characteristic feature of the DOS is the dip near the Fermi level found by Herring and Hill [1] and later, e.g., by [4, 5, 7, 8, 10] for the hcp Be modification (but not so expressed in fcc Be [6]), and also shown clearly in the photoemission spectra of polycrystalline films [16, 17]. The dip appears in both s and p bands [10, 11].

According to the decomposition (made in PS-LD), the number of electrons with s, p_x (or p_y), and p_z character are 0.63, 0.41, and 0.44 per atom, respectively, where z is chosen along the c-axis. Although the atomic valence configuration of Be is $2s^2$, a large amount of electrons appear to have p character in the solid. Both the band structure and the density of states suggest that there is a large deviation from the free-electron model near the Fermi level [11]. The dip in the DOS appears somewhat below the Fermi level, as found by [5]. The DOS at E_F is larger in hypothetical fcc Be than in hcp Be [6]. The density of states at the Fermi level is still very low: 0.85 [18] to 1.07 states·atom^{-1}·Ry^{-1} [5].

On the basis of the HF cluster calculations [15], which can take into account hybridization and the bonding–antibonding character of the states, and regarding the partial DOS's the following conclusions were drawn: the bottom of the valence band consists mainly of 2s bonding orbitals, with a very small admixture of p character. As the energy is increased, larger mixing with p orbitals occurs, intercepted by the 2s antibonding orbitals which have a comparatively small p character. The p_z-type orbitals lie lower in energy relative to the p_x and p_y orbitals, due to the bonding charge along the z-axis. This is also reflected in the relative Mulliken populations per atom over the p_x, p_y, p_z, and s occupied orbitals. The Mulliken population over the p_z orbitals was found to be 0.478 electrons/atom, whereas the average Mulliken population over the p_x or the p_y orbitals was 0.399 electrons/atom with an s population of 2.724 electrons/atom [15].

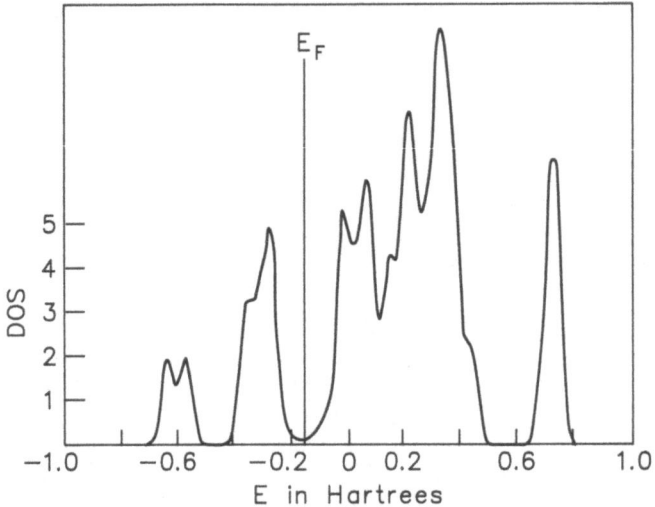

Fig. 14-3.   Density of states (DOS in arbitrary units) according to an unrestricted HF cluster calculation in the energy range from −1 to +1 Hartrees. For the energy range from −5 to +1 Hartrees, see the original figure in the paper [15].

Just below the Fermi level the $2sp_yp_z$ orbitals dominate, whereas at the Fermi level the p_y admixture dominates. The Fermi level coincides with the $3sp_y$ bonding orbital. The first excited state has $2sp_yp_z$ character with a higher p_y admixture than the occupied state (see figure in the paper). The dip in the DOS at the Fermi level appears to be due to the large $p_y - p_z$ bonding–antibonding splitting, which results from a bonding charge concentration along the z-direction. The two main maxima in the DOS below the Fermi level correspond to the p_x, p_z orbitals of the 2sp hybrids, with a secondary maximum at the antibonding 2s orbitals. The region above the Fermi level is dominated by the 2sp antibonding orbitals and then by the large 3s peak at about 0.25 Hartrees (Hy). The hybridized 3s3p orbitals appear at energies larger than 0.30 Hy, with energies pushed higher as the $3p_x$, p_y character increases. One emerging feature is that the antibonding $3s3p_zp_x$ orbitals lie lower in energy than the bonding $3s3p_xp_y$ orbitals, with an energy gap of 0.20 Hy. This gap appears as a wide dip in the DOS curve between the 0.4 Hy peak, which corresponds to the $3s3p_zp_x$ orbitals, and the peak at about 0.75 Hy, which is due to the $3s3p_xp_y$ orbitals. Figures showing a typical orbital in the high-energy tail of the $3s3p_zp_x$ peak and the bonding and antibonding orbitals of the $3s3p_xp_y$ band, respectively, corresponding to the lowest and highest energy orbitals of the 0.75 Hy peak, are given in the paper [15].

Experimental Results

Theoretical results for DOS have to be compared with experimental data from photoemission spectra [16, 19], soft X-ray absorption [20 to 23] and emission spectra [24 to 27], appearance potential spectra [28], and so on. The agreement between the XPS valence band spectrum [19] and the calculated total DOS of the occupied bands [5, 6, 8], however, is not good. While the XPS curve shows a peak at about 8 eV below E_F, the calculated total DOS has a peak around 2 eV below E_F. On the other hand, the peak in the s partial DOS roughly coincides with the XPS-DOS peak, which indicates that the XPS spectrum is dominated by the s DOS curve. With an s to p photoexcitation cross-section ratio of $\sigma_s/\sigma_p = 5$, and some other corrections concerning the role played by plasmons (see p. 86), a coincidence between the "theoretical" XPS spectrum evaluated in this manner and the theoretical DOS from the band structure can be achieved [19].

The X-ray K-emission spectrum of Be originates from an s level and measures only the p DOS. Nilsson et al. [5] compared their APW-DOS results with the results of Wiech [26]. For a similar comparison using the experimental states of [27, 29], see [21]. Such comparisons show that in the occupied region of the band not far from E_F the proportion of s states is small [21]. The peak in the total DOS near 3 eV (2.8 eV) below E_F, which has been found experimentally and theoretically by most authors, occurs close to the energy of the lowest K state (K_1) which should contribute to the structure [5]. A hump at about 1 eV occurring at the energy of the lowest H state (H_3) and the second lowest K state (K_5) was also found by [26]; the calculated structure at about 4.5 eV, caused by the critical points at the lowest M and Γ states (M_1^+ and Γ_3^+, respectively) [5], was not observed in the K-emission spectrum [26].

Polarized K-emission valence band spectra [24, 25] allow a distinction between p_x-, p_y-, and p_z-like states. A more unambiguous correlation of spectral structures with van Hove singularities of the band structure provides a better experimental check of the band structure calculations [4 to 6]. This is clearly demonstrated for discontinuities at the points Γ_3^+ (p_z-like) between -4 and -4.5 eV and M_1^+ ($p_{x,y}$-like) at -4.9 eV [24]. The increased intensity of the π band, $(I(E\|c) - I(E\perp c)) = I(\sigma + \pi$ component$) - I(\sigma$ component$)$, points to a higher number of occupied states of p_z-like symmetry compared with the number of occupied p_x- and p_y-like states (existence of bonding charges) [24]. Starting with a pseudopotential band structure calculation [30], the intensities of the polarized (K) X-ray emission spectrum and partial DOS were calculated to a special angular momentum, l, and for a special representation, Γ, of the point group of the atomic site where the hole is located, D_l^Γ ($\Gamma = \Gamma_2, \Gamma_5$) [25, 31].

The unoccupied DOS above E_F was compared with soft X-ray K-absorption spectra and appearance potential spectra by Nilsson et al. [5, 28]. The theoretical peak at 2 eV above E_F was found in both spectra, also the theoretical peak at 7 eV, further peaks at 10 and 14 eV [20], or 13 eV [28]. A weak shoulder just above E_F, implying that E_F is located at a lower energy than the DOS minimum, was only observed in APS. The pronounced peak at about 2 eV was associated with the nonfree electron bands, extending out from Γ into the ΓKM plane and from L into the LHKM plane, the second peak with bands close to the HKM line [5]. The experimental APS spectrum [5] also correlated well – concerning the position of the two peak maxima and the minimum – with a calculation [31] based on Taut's pseudopotential method [30].

The older SXA spectrum of Sagawa [23] showed much fine structure, attributed to the existence of quasi-stationary states within the energy gaps in particular directions of the k space due to perturbation produced by the creation of a vacancy in the K level of the ion core [23].

The single particle (APW) density of states at the Fermi level is 0.079 states $\cdot$ eV$^{-1} \cdot$ atom^{-1} ($= 1.074$ states $\cdot$ atom$^{-1} \cdot$ Ry^{-1}) [5], see above. From specific heat data Gmelin [32] obtained

0.078 states·eV⁻¹·atom⁻¹ (=1.061 states·atom⁻¹·Ry⁻¹), and Ahlers [33] found 0.073 states·
eV⁻¹·atom⁻¹(=0.993 states·atom⁻¹·Ry⁻¹) in excellent agreement with the calculation. This
signifies that the electron-phonon enhancement of the effective mass is small. $N(E_F) =$
5.8×10^{-21} states·eV⁻¹·cm⁻³·spin⁻¹ was deduced for an ion-beam-sputtered amorphous Be
film via the superconductor upper critical field H_{c2} (which led to $\gamma = 46$ J/m³·K²) [34].

Matrix elements were calculated with OPW for initial and final electronic states. Their
effects of the plane DOS of simple metals were studied. A comparison of the plane DOS with
the calculated spectra of the ΓM and ΓK symmetry lines of Be revealed agreement between
the DOS peak positions and the calculated spectra [35].

In a study to detect oscillations of the chemical potential (μ) of Be associated with
variations in the DOS in a quantized magnetic field, no μ-oscillations were found. The result
was attributed to cancellation of the magnetostriction-associated changes in the chemical
potential by oscillations caused by variations in the DOS (this conclusion is valid if the
compressibility of Be is determined by conduction electrons) [36].

References:

 [1] Herring, C.; Hill, A. G. (Phys. Rev. [2] **58** [1940] 132/62, 154).
 [2] Loucks, T. L.; Cutler, P. H. (Phys. Rev. [2] **133** [1964] A819/A829).
 [3] Cornwell, J. F. (Proc. Roy. Soc. [London] A **261** [1961] 551/64).
 [4] Terrell, J. H. (Phys. Rev. [2] **149** [1966] 526/34).
 [5] Nilsson, P.-O.; Arbman, G.; Gustafsson, T. (J. Phys. F **4** [1974] 1937/50).
 [6] Inoue, S. T.; Yamashita, J. (J. Phys. Soc. Japan **35** [1973] 677/83).
 [7] Bhokare, V. V.; Yussouff, M. (Nuovo Cimento Soc. Ital. Fis. B [11] **19** [1974] 149/60).
 [8] Chatterjee, S.; Sinha, P. (J. Phys. F **5** [1975] 2089/97).
 [9] Cracknell, A. P. (Landolt-Börnstein New Ser. Group III **13c** [1984] 1/462, 59/60).
[10] Dovesi, R.; Pisani, C.; Ricca, F.; Roetti, C. (Phys. Rev. [3] B **25** [1982] 3731/8).

[11] Chou, M. Y.; Lam, P. K.; Cohen, M. L. (Phys. Rev. [3] B **28** [1983] 4179/85).
[12] Blaha, P.; Schwarz, K. (J. Phys. F **17** [1987] 899/911).
[13] Oriade, J. (Intern. J. Quantum Chem. Quantum Chem. Symp. No. 14 [1980] 597/605;
     C.A. **94** [1981] No. 127584).
[14] Angonoa, G.; Dovesi, R.; Pisani, C.; Roetti, C. (Phys. Status Solidi B **122** [1984] 211/20).
[15] Zdetsis, A. D.; Miliotis, D. (Solid State Commun. **42** [1982] 227/30).
[16] Gustafsson, T.; Brodén, G.; Nilsson, P.-O. (J. Phys. F **4** [1974] 2351/8).
[17] Höchst, H.; Steiner, P.; Hüfner, S. (Phys. Letters A **60** [1977] 69/71).
[18] Tripp, J. H.; Everett, P. M.; Gordon, W. L.; Stark. R. W. (Phys. Rev. [2] **180** [1969] 669/78).
[19] Höchst, H.; Steiner, P.; Hüfner, S. (J. Phys. F **7** [1977] L309/L314).
[20] Haensel, R.; Keitel, G.; Sonntag, K.; Kunz, C.; Schreiber, P. (Phys. Status Solidi A **2**
     [1970] 85/90).

[21] Fabian, D. J.; Watson, L. M.; Marshall, C. A. W. (Rept. Progr. Phys. **34** Pt. 2 [1972]
     601/96, 639/41).
[22] Johnston, R. W.; Tomboulian, D. H. (Phys. Rev. [2] **94** [1954] 1585/9).
[23] Sagawa, T.; Iguchi, Y.; Sasanuma, M.; Ejiri, A.; Fujiwara, S.; Yokota, M.; Yamaguchi, S.;
     Nakamura, M.; Sasaki, T.; Oshio, T. (J. Phys. Soc. Japan **21** [1966] 2602/10).
[24] Dräger, G.; Brümmer, O. (Phys. Status Solidi B **78** [1976] 729/35).
[25] Rennert, P.; Schelle, H.; Gläser, U. H. (Phys. Status Solidi B **121** [1984] 673/84).
[26] Wiech, G. (in: Fabian, D. J.; Soft X-Ray Band Spectra and the Electronic Structure of
     Metals and Materials, Academic, London 1968, pp. 59/70).

[27] Lindsay, G. M.; Watson, L. M.; Fabian, D. J. (unpublished results from [21]).
[28] Nilsson, P.-O.; Kanski, J. (Surf. Sci. **37** [1973] 700/7).
[29] Henke, B. L.; Smith, E. N. (J. Appl. Phys. **37** [1966] 922/3).
[30] Taut, M. (Phys. Status Solidi B **54** [1972] 149/57).

[31] Rennert, P.; Dörre, T.; Gläser, U. H. (Phys. Status Solidi B **87** [1978] 221/6).
[32] Gmelin, E. (Compt. Rend. **259** [1964] 345/61).
[33] Ahlers, G. (Phys. Rev. [2] **145** [1966] 419/23).
[34] Okamoto, M.; Takei, K.; Kubo, S. (J. Appl. Phys. **62** [1987] 212/5).
[35] Nguyen Van Hung (Tap Chi Vat Ly **6** No. 3 [1981] 6/10 from C. A. **97** [1982] No. 44 535).
[36] Alekseevskii, N. E.; Nizhankovskii, V. I. (Zh. Eksperim. Teor. Fiz. **88** [1985] 1771/9;
     Soviet Phys.-JETP **61** [1985] 1051/5).

14.4 Fermi Surface. Magnetic Breakdown

The shape and dimensions of the Fermi surface (FS) are the most important parameters to be investigated. A preliminary description of the shape of the FS of Be was given in "Beryllium" Suppl. Vol. A 2, 1991, pp. 117/8.

The FS of Be has been studied experimentally using the de Haas-van Alphen (dH-vA) effect, see for instance [1 to 3], cf. "Beryllium" Suppl. Vol. A 2, 1991, pp. 262/3, the oscillatory magnetostriction [4], quantum oscillations in the magnetoresistance [5, 6], sound velocity (Landau) quantum oscillations [7], magnetoacoustic geometric oscillations (MAGO) [8], magnetoacoustic quantum oscillations (MAQO) [9], and Doppler-shifted acoustic cyclotron resonances (DSACR) [10]; see also "Beryllium" Suppl. Vol. A 2, 1991, pp. 271/2.

Further, the FS was the subject of theoretical calculations by means of the OPW [11, 12], APW [13 to 15], and pseudopotential (PS) methods [3, 16]. The effect of temperature was theoretically investigated by the PS method [17], and the effect of hydrostatic pressure by OPW [3] and an approximate analytical OPW method [18, 19].

Experimental results of the pressure dependence were obtained from the dH-vA effect [20], the Landau quantum oscillations of the sound velocity [7], and the oscillatory magnetostriction [4]. The effects of uniaxial stress [4, 7, 18, 19], uniaxial strain [4, 7, 20], and angular shear [7] were investigated by the same methods. The influence of dilute alloying with Cu was studied experimentally by the dH-vA effect and theoretically by means of a nonlocal pseudopotential [21].

Results for unstrained Be were compiled by Cracknell [22] and those for stressed and strained Be by Joss et al. [23].

The **shape** of the Fermi surface of Be, along with the first Brillouin zone, is shown schematically in **Fig.** 14-4, p. 58, from Vozenilek and Reed [8]. This figure is also reproduced in [9, 10, 22]. An earlier representation from [4], also reproduced in [3, 7], is given in [23]. The FS consists of a large hole surface, called the "coronet" and located in the second zone (h_2), and six equivalent third-zone electron surfaces (e_3) which can form two "cigars" per coronet. Of several theoretical models of the FS the semiempirical nonlocal pseudopotential model of Tripp et al. [3] gave the best agreement with available experimental data [8, 10]. The cross section of the cigar in the basal plane is approximately triangular in shape and has a vertex of the triangle adjacent to a lobe of the coronet. The waist or midsection α_1 of the cigar in the ΓKM plane is characterized by a cross-sectional area that is 3% smaller than the hip cross section

α_2, which is displaced up or down by ~1/3 of the distance KH. The hip cross sections are believed to be more nearly circular in shape. The energy gap between the cigar and the coronet is small in the basal plane, and magnetic breakdown can occur when the applied magnetic field, H, is along the [0001] direction, i.e., the electron orbits are parallel to the basal plane. The six large pieces of the coronet, sometimes called bellies, are disposed about the ΓK lines and joined by thin, almost cylindrical necks (the frequency branches β and γ refer to these bellies and necks, respectively). The volumes of the electrons and holes are equal, about 0.8% of the volume of the first Brillouin zone or 0.016 states per atom [3]; cf. [9].

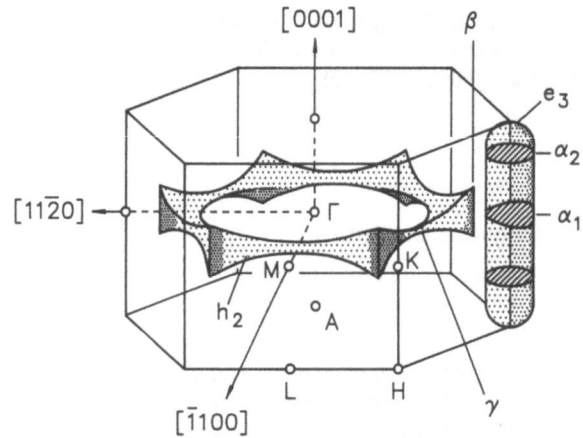

Fig. 14-4.   Fermi surface of Be: h_2 coronet, centered at Γ;
e_3 cigar, centered at K.

According to a full-potential LAPW study the FS topology is similar to previous pseudo-potential calculations [24, 25], but contrasts with the HF band structure [26] in which the Γ_4 state is occupied. The first band exhibits free-electron character at the bottom of the band; close to E_F, however, deviations from the free-electron behavior occur. The wave functions are dominated by s character at the bottom of the bands whereas at higher energies the main contributions come from the p states [27]; cf. p. 49.

The linear **dimensions** of the FS and areas of cross section, given in reciprocal atomic units (1 a.u. = 5.29×10^{-9} cm), from OPW [11, 12, 14] and APW [13] calculations but also from dH-vA [1, 3] and MAGO [8] measurements are compiled in [22]. Among these the shortest distance between the coronet and the cigar is 0.01 [11], 0.022 [14], 0.08 [1], 0.0098 [3] and the neck diameter is 0.04 [11], 0.019 [14], 0.02 [1], 0.0196 [3]. The cross-sectional areas of the FS, deduced from MAQO [9], are in good agreement with data from the dH-vA effect [3]. Studies of DSACR yielded extremal derivatives of the FS cross-sectional areas [10], which were compared with data calculated using the pseudopotential model of Tripp et al. [3]. The experimental data were typically in agreement with the theoretical values to better than 15%; yet these results show that the pseudopotential model does not yield as good agreement with DSACR data as it does with data derived from the dH-vA effect [3] or MAQO [9, 28] which predict extremal cross-sectional areas. In these latter cases the agreement is essentially within the experimental error.

The APW calculation of Oriade [15] was claimed to be in excellent agreement with Watts' dH-vA measurements [1], better than Terrell's results [13], or those found by Loucks and Cutler [11].

The caliper dimensions of the cigar and coronet sections of the FS were both derived from magnetoacoustic geometric oscillations observed in the absorption of longitudinal ultrasound by Be at 1 K and sonic frequencies up to 1.85 GHz. The absence of magneto-acoustic geometric oscillations for certain calipers could not be explained, since strong oscillations from closely related calipers were observed. The experimental data from this study [8] agree well with the semiempirical pseudopotential model of the FS of Be [3]. Caliper dimensions from several authors are compared in [14]. With a multiple OPW Fermi surface having six plane waves as the model FS, c/a=1.56886, c=3.5814 Å, and E_F=1.28417 Ry for Be, the galvanomagnetic coefficients were computed by path integration in the intermediate field range. The results are in satisfactory agreement with experimental data [29] (see figure in the paper [30]). The calculated caliper dimensions of the model FS were compared with data of Tripp et al. [3] in a table in [30].

In addition to the cross-sectional areas of the FS (see above) MAQO due to magnetic breakdown orbits were observed between the cigar and coronet pieces of the FS. Such orbits were not reported by authors using other techniques (such as the dH-vA effect). The pseudo-potential model [3] erroneously predicts that magnetic breakdown (Mb, see p. 37 and below) in the basal plane between the cigar and the coronet should take place at $\sim H_0$=120 kOe, but the MAQO results show that H_0 must be at least 50 kOe [9].

The **effect of temperature** on the Be Fermi surface was treated by a straightforward extension of the pseudopotential model originally developed for pure Be at 0 K. Because of the high Debye temperature ("Beryllium" Suppl. Vol. A 2, 1991, p. 252), the changes are not large even at room temperature. There are no changes in the topology of the FS with rising temperature and the probability of breakdown between orbits in a magnetic field is not significantly enhanced. From 0 K up to 300 K the cigar-coronet gap decreases from 0.0098 to 0.0092 a.u. and the neck diameter rises from 0.0196 to 0.0208 a.u. [17].

Hydrostatic pressure and volume derivatives of extremal cross-sectional areas of the FS obtained experimentally [4, 7, 20] and theoretically [3, 18, 19] are compiled in [23]. The most striking result from dH-vA effect measurements is that for H∥[0001] the waist of the cigar decreases whereas the hips increase [20]. Uniaxial stress, uniaxial strain, and angular shear derivatives of the extremal cross-sectional areas are also compiled in [23]. The agreement between theoretical and experimental data is often poor; the calculation by Watts and Mayers [18] seems to be satisfactory.

The effect of alloying with Cu on the FS of Be was studied in some low-Cu (<1 at%) alloys experimentally by means of the dH-vA effect and theoretically by the nonlocal pseudopotential model derived for random substitutional alloys. The changes in the FS of these alloys arise principally from the shift of the Fermi level due to variation of the electron density and the changes in lattice spacings, whereas the effect of explicitly including the solute pseudopoten-tial appears to be unimportant. With increasing Cu concentration the cigar-coronet separation decreases practically linearly; at 1.2 at% Cu the two sheets should touch. The magnetic breakdown field decreases and the absence of one of the α-frequency branches should be explainable by magnetic breakdown. The FS topology is more easily changed by altering the electron concentration (by alloying) than by applying pressure alone [21].

The effect of **magnetic breakdown** (Mb) consists in that the FS in magnetic experiments depends on whether the magnetic field is low or high (for preliminary information see "Beryllium" Suppl. Vol. A 2, 1991, p. 273). At low fields the "single zone" picture must be used (see figure 5.20 on p. 217 in the paper [31]) because of the spin-orbit splitting across the plane ALH. When the field is high enough, magnetic breakdown can occur; the electrons are able to cross the small energy gaps in the plane and the "double zone" picture becomes appropriate. For intermediate fields the situation is more complicated [31].

In early studies it was shown that in magnetic fields up to 30 kOe Be behaved like a metal with a closed Fermi surface (with $n_1 = n_2$), but in stronger fields a sharp increase in anisotropy of the magnetoresistance indicated the appearance of open trajectories which were suggested to be the result of Mb [32]. Magnetic breakdown in Be leads to the onset of resistance oscillations and thermopower oscillations with giant amplitudes, see for instance [31, 33, 34].

Magnetic breakdown was observed for noncentral extremal sections of the cigar-shaped FS when a Be single-crystal sample was subjected to fields of ~70 kOe with H∥[0001]. Measurements were made of oscillations in susceptibility (dH-vA effect) and magnetoresistance [35]. Earlier studies had revealed that large-amplitude oscillations of the transverse magnetoresistance occurred in Be under Mb conditions not only for field directions close to H∥[0001], but also in a fairly wide range of angles ϑ between H and the c-axis. This was attributed to the appearance of elongated Mb trajectories connecting two or three coronets through a cigar-shaped FS whose contribution to the conductivity for relatively short lifetimes was comparable with the conductivity due to open trajectories [36].

The principal orbit around the outside of the coronet has never been observed. Watts [1] suggested that Mb between the coronet and the cigar would account for the absence of a dH-vA signal from this orbit [18].

In a study of the phases of Mb oscillations of the resistance and of the thermoelectric power the relative shifts of the zeros were measured for [10$\bar{1}$0] and [1$\bar{2}$10] Be single crystals rotated about the longitudinal axis through an angle $\vartheta \leqq 2.2°$ from the initial position H∥[0001]. The Mb oscillations are determined by the extremal sections of the electronic cigars, and the phase shift between them is $-\pi/2$. For $\vartheta \rightarrow 0$, an additional shift of the oscillation zeros was observed which could not be attributed to geometric singularities of the shape of the Be Fermi surface nor to the influence of the Shoenberg effect. The results were regarded as a manifestation of the two-dimensional character of the phase-coherent magnetic breakdown in Be [37]. Previously, a convincing proof of coherent Mb in Be was found experimentally in the angular dependence (between hexagonal specimen axis and magnetic field direction) of the phase difference between oscillations of the diagonal tensor components of the magnetoresistance [38].

For studies of the anomalous amplitude of the dH-vA effect at a field of 60 kOe where the Mb influence is not so important, see Egorov [39].

Amplitudes expected for magnetization and chemical potential oscillations can be estimated from the known FS of Be. From an unsuccessful attempt to detect oscillations of the chemical potential associated with variations in the density of states in a quantized magnetic field, it was deduced that the amplitude of the oscillations must be at least an order of magnitude less than expected. The result was attributed to cancellation of changes in the chemical potential associated with state density effects and magnetostriction [40]. A more fundamental reason for the negative results [40] (and similar results for Bi [41], based on Kelvin's method) was suggested in another study. The quantum oscillations of the contact potential difference in a magnetic field are responsible for the chemical potential oscillations in the bulk of the metals (Be, Bi). The potential oscillations decreased because of screening; the screening factor depended on the diameter of the electron orbit in the magnetic field and the screening radius [42].

In addition to the FS, Fermi wavevectors on the coronet and on the cigars and Fermi velocities are calculated by the APW method ($E_F = 0.872$ Ry); the results are tabulated by Sano [43].

References:

[1] Watts, B. R. (Phys. Letters **3** [1963] 284/5; Proc. Roy. Soc. [London] A **282** [1964] 521/46, 532).

[2] Condon, J. H. (Phys. Rev. [2] **145** [1966] 526/35).

[3] Tripp, J. H.; Everett, P. M.; Gordon, W. L.; Stark, R. W. (Phys. Rev. [2] **180** [1969] 669/78).

[4] Chandrasekhar, B. S.; Fawcett, E.; Sparlin, D. M.; White, G. K. (Tr. Mezhdunar. 10th Konf. Fiz. Nizkikh Temp., Moscow 1966/67, Vol. 3, pp. 328/32; Proc. 10th Intern. Conf. Low Temp. Phys., Moscow 1966 [1967], Vol. 3, pp. 328/32; C. A. **70** [1969] No. 24 246; Bull. Am. Phys. Soc. [2] **10** [1965] 350), Chandrasekhar, B. S.; Fawcett, E. (Advan. Phys. **20** [1971] 775/94, 790).

[5] Reed, W. A.; Condon, J. H. (Phys. Rev. [3] B **1** [1970] 3504/10).

[6] Sellmyer, D. J.; Goldstein, I. S.; Averbach, B. I. (Phys. Rev. [3] B **4** [1971] 4628/31).

[7] Testardi, L. R.; Condon, J. H. (Phys. Rev. [3] B **1** [1970] 3928/42; Phys. Acoust. **8** [1971] 59/94).

[8] Vozenilek, E. F.; Reed, R. W. (Phys. Rev. [3] B **12** [1975] 1140/5).

[9] Reed, R. W.; Vozenilek, E. F. (Phys. Rev. [3] B **13** [1976] 3320/8), Vozenilek, E. F.; Reed, R. W.; Brickwedde, F. G. (Ultrason. Symp. Proc. **1974** 453/6; C. A. **84** [1976] No. 37 973).

[10] Campbell, P. M.; Reed, R. W. (Phys. Rev. [3] B **22** [1980] 4550/7).

[11] Loucks, T. L.; Cutler, P. H. (Phys. Rev. [2] **133** [1964] A 819/A 829).

[12] Loucks, T. L. (Phys. Rev. [2] **134** [1964] A1618/A1620).

[13] Terrell, J. H. (Phys. Letters **8** [1964] 149/51).

[14] Inoue, S. T.; Yamashita, J. (J. Phys. Soc. Japan **35** [1973] 677/83).

[15] Oriade, J. (Intern. J. Quantum Chem. **19** [1981] 721/7).

[16] Tripp, J. H.; Gordon, W. L.; Everett, P. M.; Stark, R. W. (Phys. Letters A **26** [1967] 98/9).

[17] Tripp, J. H. (Phys. Rev. [3] B **1** [1970] 550/1).

[18] Watts, B. R.; Mayers, J. (J. Phys. F **10** [1980] 1693/718).

[19] Watts, B. R.; Sundström, L. J. (J. Phys. F **3** [1973] 1941/55).

[20] Schirber, J. E.; O'Sullivan, W. J. (Phys. Rev. [2] **184** [1969] 628/34), O'Sullivan, W. J.; Schirber, J. E. (Phys. Letters A **25** [1967] 124/5).

[21] Tripp, J. H.; Everett, P. M.; Fiske, J. M.; Gordon, W. L. (Phys. Rev. [3] B **2** [1970] 1556/63).

[22] Cracknell, A. P. (Landolt-Börnstein New Ser. Group III **13 c** [1984] 1/462, 60/3).

[23] Joss, W.; Griessen, R.; Fawcett, E. (Landolt-Börnstein New Ser. Group III **13 b** [1983] 1/405, 41/6).

[24] Chou, M. Y.; Lam, P. K.; Cohen, M. L. (Phys. Rev. [3] B **28** [1983] 4179/85).

[25] Barth, von, U.; Pedroza, A. C. (Phys. Scr. **32** [1985] 353/8).

[26] Dovesi, R.; Pisani, C.; Ricca, F.; Roetti, C. (Phys. Rev. [3] B **25** [1982] 3731/9).

[27] Blaha, P.; Schwarz, K. (J. Phys. F **17** [1987] 899/911).

[28] Reed, W. A. (Phys. Rev. [3] B **15** [1977] 2416/9).

[29] Alekseevskii, N. E.; Egorov, V. S. (Zh. Eksperim. Teor. Fiz. **46** [1964] 1205/7; Soviet Phys.-JETP **19** [1964] 815/6).

[30] Yonemitsu, K.; Sato, H. (Phys. Letters A **88** [1982] 87/9).

[31] Shoenberg, D. (Magnetic Oscillations in Metals, Cambridge Univ. Press, Cambridge 1984, pp. 1/570, 216/21, 331/68; J. Low Temp. Phys. **56** [1984] 417/40).

[32] Alekseevskii, N. E.; Egorov, V. S. (Zh. Eksperim. Teor. Fiz. **45** [1963] 388/91; Soviet Phys.-JETP **18** [1964] 268/70).

[33] Alekseevskii, N. E.; Egorov, V. S. (Pis'ma Zh. Eksperim. Teor. Fiz. **8** [1968] 301/5; JETP Letters **8** [1968] 185/8).
[34] Alekseevskii, N. E.; Egorov, V. S.; Dubrovin, A. V. (Pis'ma Zh. Eksperim. Teor. Fiz. **5/6** [1967] 793/6; JETP Letters **6** [1967] 244/9?).
[35] Egorov, V. S. (Zh. Eksperim. Teor. Fiz. **69** [1975] 2231/5; Soviet Phys.-JETP **42** [1975] 1135/7).
[36] Alekseevskii, N. E.; Egorov, V. S. (Zh. Eksperim. Teor. Fiz. **55** [1968] 1153/9; Soviet Phys.-JETP **28** [1969] 601/4).
[37] Alekseevskii, N. E.; Nizhankovskii, V. I. (Zh. Eksperim. Teor. Fiz. **83** [1982] 1163/9; Soviet Phys.-JETP **56** [1982] 661/5).
[38] Alekseevskii, N. E.; Nizhankovskii, V. I. (Fiz. Met. Metalloved. **38** [1974] 1105/8; Phys. Metals Metallog. [USSR] **38** No. 5 [1974] 196/9).
[39] Egorov, V. S. (Fiz. Tverd. Tela [Leningrad] **30** [1987] 1253/6, **28** [1986] 318/20; Soviet Phys.-Solid State **30** No. 4 [1987] 730/2, **28** No. 1 [1986] 177/8).
[40] Alekseevskii, N. E.; Nizhankovskii, V. I. (Zh. Eksperim. Teor. Fiz. **88** [1985] 1771/9; Soviet Phys.-JETP **61** [1985] 1051/5).

[41] Nizhankovskii, V. I.; Mokerov, V. G.; Medvedev, B. K.; Shaldin, Yu. V. (Zh. Eksperim. Teor. Fiz. **90** [1986] 1326/35; Soviet Phys.-JETP **63** [1986] 776/81).
[42] Semenchinskii, S. G.; Édel'man, V. S. (Fiz. Nizk. Temp. [Kiev] **13** [1987] 979/81; Soviet J. Low Temp. Phys. **13** [1987] 558/9; C.A. **108** [1988] No. 85884).
[43] Sano, H. (J. Phys. Soc. Japan **39** [1975] 1268/76, 1272).

14.5 Effective Masses

Cyclotron effective masses in the unstrained Be lattice were obtained from associated dH-vA frequencies which were measured using both the field modulation and the torque methods and calculated by means of a nonlocal pseudopotential model. The experimental results, with an estimated uncertainty of the mass of ±3%, are compared with the calculated data in a table given by Tripp et al. [1]; cf. preliminary results [2] where the difference between experimental and calculated masses of 10 to 20% was attributed to mass enhancement due to electron-phonon interaction. The results [1] are in general agreement with those found by Watts [3]. Similar results were obtained for dilute (<1 at% Cu) alloys [4].

The electronic thermal effective mass (including electron-electron interactions) was calculated based on the spin-density functional theory. The necessary exchange-correlation functionals were treated in the local-spin-density approximation. The calculated thermal effective masses (tabulated in the paper) agree well with semiempirical results from fits to the Fermi surface [5].

References:

[1] Tripp, J. H.; Everett, P. M.; Gordon, W. L.; Stark, R. W. (Phys. Rev. [2] **180** [1969] 669/78, 675).
[2] Tripp, J. H.; Gordon, W. L.; Everett, P. M.; Stark, R. W. (Phys. Letters A **26** [1967] 98/9).
[3] Watts, B. R. (Proc. Roy. Soc. [London] A **282** [1964] 521/46, 532, 543).
[4] Tripp, J. H.; Everett, P. M.; Fiske, J. M.; Gordon, W. L. (Phys. Rev. [3] B **2** [1970] 1556/63).
[5] Wilk, L.; Fehlner, W. R.; Vosko, S. H. (Can. J. Phys. **56** [1978] 266/79, 271).

14.6 Electronic Charge-Density Distribution

Survey

In order to explain the properties of metallic beryllium, a detailed description of its spatial electronic structure leading to bonding is required. This information may also be useful for choosing a model as basis for the calculations of the band structure. In order to plot an electron charge-density map, accurate values of the crystal structure factor, F_{cr}, are needed. In addition, good values of the free-atom structure factor, F_{fa}, are often helpful. Unfortunately, F_{cr} values obtained by X-ray diffraction are frequently not accurate enough for plotting a finely scaled electron charge-density map. However, low-angle values of F_{cr} obtained by electron diffraction, combined with the best higher-angle values from X-ray measurements, and/or band structure calculations yield sufficiently accurate data. Commonly used presentations of the electron density distribution $\rho(\vec{r})$ are deformation and valence-electron density maps which are calculated as differences between the crystal structure factors, F_{cr}, and the structure factors corresponding to the free-atom model, F_{fa}, or the core-electron part of F, F_{core}. From F atomic scattering (form) factors, f, can be calculated. Since half of the charge density in Be is provided by bonding electrons, X-ray and electron scattering should be relatively sensitive to the form of the valence-electron wave functions. Low-angle structure factors give important information on the vibrational anisotropy. Some preliminary information about the charge density of Be is given in the chapter on bonding in "Beryllium" Suppl. Vol. A 2, 1991, pp. 117 ff.

Structure and Form Factors

Early experimental determinations of the electron density distribution in Be are based on the precision X-ray diffraction studies at room temperature (Ag Kα) of Brown [1] who produced a set of 27 low-angle structure factors and found no evidence for vibrational anisotropy and that F_{cr} was considerably lower than F_{fa}. These data were reanalyzed by Stewart [2] and Yang, Coppens [3]. They obtained for the Debye-Waller factors the values $B_{33}(\|c) = 0.53$ Å² and $B_{11}(\perp c) = 0.58$ Å² which were, together with the isotropic Debye-Waller factor (B = 0.54 Å²) derived by [1], much too high. Better values were obtained by neutron diffraction: $B_{33} = 0.395$ Å² and $B_{11} = 0.435$ Å² [4]. The possibility that the difference in the values could be due to an expansion of the core electrons was examined by Manninen and Suortti [5], but a careful analysis of the integrated X-ray intensities from high-angle reflections, using the best F_{fa} values, gave $B_{33} = 0.415$ Å² and $B_{11} = 0.460$ Å², in good agreement with the neutron diffraction results. Collins and Whitehurst [6] renormalized the results of [1], using the Debye-Waller factors of Larsen et al. [4], and found a set of structure factors which were, on the average, some 6% higher than those of [1]. More recently, Hansen et al. [7] and Larsen, Hansen [8] measured a series of structure factors both by X-ray and γ-ray (0.03 Å) diffractometry, and the results were on the average 8% higher than those of [1], yet in good agreement with each other. From the X-ray data $B_{33} = 0.416$ Å² and $B_{11} = 0.489$ Å² were obtained [8]. In addition, Dovesi et al. [9] and Chakraborty et al. [10] performed theoretical calculations of the atomic form factors of Be; at low angles they were significantly higher than the experimental values of [7, 8]. A newer large set of structure factors, measured by Hansen et al. [11] with a 0.12 Å γ-ray diffractometer, agreed well with the X-ray and γ-ray diffraction results of [7, 8].

The data of Brown [1] were again subjected to a precision treatment by Tsirel'son et al. [12]. The new set of X-ray structural amplitudes thus obtained was used to investigate the peculiarities of the electronic spatial distribution in Be.

From the early studies there were several efforts to produce an accurate set of atomic (scattering) form factors for Be. Fox and Fischer [13] placed the X-ray structure factors for the 27 lowest-angle Bragg reflections measured by [1] on an absolute scale, using the very accurate $10\bar{1}1$ and 0002 low-angle structure factors determined by high-energy electron diffraction. The atomic form factors calculated from these results appear to be much more accurate than those deduced from the original X-ray measurements. At low angles the new structure factors show excellent agreement with the values obtained by X-ray and γ-ray diffractometry (except for 0002). At higher angles the experimental atomic scattering factors agree closely with the best theoretical free-atom values, whereas the low-angle form factors deviate considerably from the free-atom results. The deviations can be explained as corresponding to $(sp^2)^a(sp)^b$ hybridization of the Be atoms with $b > a$ [13] although other authors (see below) suggested that sp^3-like hybrids can also explain the structure-factor data.

Form factors, obtained theoretically by several authors using different methods, are compiled, e.g., by Blaha, Schwarz [14]. They made calculations by means of the full-potential linearized augmented-plane-wave (LAPW) method on the basis of LDA (1 s core form factors derived in the LDA and HF approximations, respectively). The results compared in [14] are the APW (LDA 1 s) data of Redinger et al. [15], the LCAO (HF 1 s) data of Dovesi et al. [9], and the PS (HF) data of Chou et al. [16] which differ in some reflections from the experimental values of [7].

The X-ray scattering factors of hcp Be were calculated using Wannier functions, which were constructed from 2 s and 2 p atomic wave functions. The basic idea is to apportion the charge density of a crystalline solid to the atomic sites of the crystal lattice and to the superlattices formed by the midpoints of nearest, next-nearest, and so on neighbor bonds [17].

Static structure factors for conduction electrons can be directly measured by X-ray scattering using synchrotron radiation. Measurements on Be as an example are described by Eisenberger et al. [18] and Contini, Sacchetti [19], but result in significant differences. New measurements of the static structure factor of Be, employing a different experimental setup and a more sophisticated data-reduction procedure, are performed by Mazzone et al. [20]. The conditions under which the electronic contribution (static structure factor) can be extracted are discussed together with the approximations which have to be introduced in order to distinguish between core and band contributions. The static structure factor and the exchange and correlation energy of conduction electrons in Be, determined in this study, agree within experimental accuracy ($\sim 5\%$) with the current electron-gas theory [20].

Deformation and Valence-Electron Density Distribution

Most of the authors who evaluated structure and form factors also constructed deformation ($\Delta\rho(r)$) and valence-electron ($\rho(r)$) density distribution maps, see, e.g. [8, 11] and [8, 21], respectively. The valence-charge density distribution in the $(11\bar{2}0)$ and (0001) plane according to Chou et al. [21] is shown in Fig. 9-32 (a and b) in "Beryllium" Suppl. Vol A 2, 1991, p. 119.

Considering deformation density distribution, the most conspicuous features are a surplus of electrons (positive $\Delta\rho(r)$) in the bipyramidal space around the tetrahedral hole and a deficiency of electrons (negative $\Delta\rho(r)$) in den channel formed by adjoining octahedral holes along the c-axis of the hcp structure [8] in agreement with previous observations [3, 6]. Charge integration over the bipyramidal space of two adjacent tetrahedral holes amounts to 0.013 electrons [8]. An increase in electronic density towards atomic nuclei was found by Tsirel'son et al. [12]; at the same time, there is a flow of electrons away from the region of octahedral voids and continuous channels with lower density are formed, joining them. However, the three different deformation density maps evaluated by [8], e.g., from their X-ray values of the structure factors with both their own Debye-Waller factors and with those of [4], showed some differences near the atomic positions. Particularly the $\Delta\rho(r)$ map [8] obtained from their X-ray structure factors normalized to the γ-ray results of [7], also with Debye-Waller factors of [4], agreed closely with the theoretical map obtained from an LCAO calculation by Dovesi et al. [22]; this latter showed a negative region around the nucleus which was interpreted as charge transfer from s- to p-type functions.

The accurate low-angle structure factors of Fox, Fisher [13] also agree well with the γ-ray values of [7]; therefore the γ-ray normalized X-ray $\Delta\rho(r)$ maps of Larsen, Hansen [8] are considered to be a good representation of the deformation density in Be.

The form of all valence-electron density maps constructed by Inoue, Yamashita [23] (theoretical APW), Stewart [2], Yang, Coppens [3], Dovesi et al. [22] (theoretical LCAO), Holzwarth et al. [24] (singular factor method), and Larsen, Hansen [8] (from γ-ray normalized X-ray structure factors) is very similar with a buildup of valence electrons around the tetra-hedral holes as described above. However, the agreement between the valence $\rho(r)$ maps of [8] obtained from the high-accuracy experimental structure factors with theoretical (LCAO) valence $\rho(r)$ maps of [22] is very close to quantitative except near the center of the channel of the octahedral holes. From these valence maps and the accurate deformation maps described in the previous section, Larsen and Hansen [8] suggested sp^3-type bonding as an alternative to the $(sp^2)^a(sp)^b$ hybridization (with $b > a$) proposed by Yang, Coppens [3]. Arguments for the $(sp^2)^a(sp)^b$ bonding scheme are discussed by Fox, Fisher [13].

A single-determinant experimental Be hybrid-atom wave function was obtained by Massa et al. [25] from the single-crystal X-ray diffraction data [8] using the quantum formalism of Clinton, Massa [26]. A superposition of these Be hybrid atoms (each Be at its position in the lattice) yielded a model electron-density distribution [25] which by construction is close to the experimental one and also to that obtained in the theoretical (full-potential LAPW) calculation by Blaha, Schwarz [14].

Bonding and Hybridization Schemes

Two hybridization schemes can describe the electronic density in the Be crystal: a) sp^2 orbitals partially filled with electrons in the atomic layers and sp orbitals between the layers parallel to the c-axis, the latter having the greater density; i.e., $(sp^2)^a(sp)^b$ with $b > a$ [3, 13]; b) sp^3 orbitals [8] accompanied by the appearance of two kinds of hybrids with lobes along the z- (or c-) axis, having the form $(s + p_z)$ and $(s - p_z)$. The latter are also partially filled with electrons and their linear combination describes the details of the electronic distribution, see for instance [12]. The diffraction results can not discriminate between these models. The hybrid's charge buildup along c is consistent with the Be lattice parameter ratio $c/a = 1.568$ which is 3% below the ideal value for hexagonal close packing (1.633). If bonding along c is stronger than along a or b (which is indicated by the deformation density distribution) then shortening along c (and hence a reduced c/a value) is the consequence. Since charge flows into the tetrahedral hole regions and out of the nuclear regions and octahedral channels [8], bonding is directed through the tetrahedral holes. However, the results of Massa et al. [25] are not consistent with an sp^3 hybrid scheme (sp^3 orbitals directed toward tetrahedral holes); they are consistent with an orbital having a much larger p_z character than the sp^3 hybrid.

A calculation of the surface core level shift of the Be 1s level with a metallic $(sp)^2$ initial state and an Al $(sp)^3$ final state led to a shift of the right direction and magnitude [27].

Core Expansion

For more than a decade there was a discussion about a possible expansion of the core electrons with respect to the free-atom state as first suggested by Yang, Coppens [3] and other authors [2, 17, 28] from structure factors [2, 3, 8, 17, 28], but also indicated in calculations [22, 23]. The possibility of such an effect was investigated by Blaha and Schwarz [14]. They compared free-atomic and crystalline form factors, both derived in the local density approximation, and found a core expansion of about 0.1 to 0.3% which was too small to be detected previously, but is in agreement with more recent experiments [7, 8, 11]. However, the valence-electron density contains a core-like cusp which depends on the state of hybridization. This effect overshadows the modifications of the core itself [11].

References:

[1] Brown, P. J. (Phil. Mag. [8] **26** [1972] 1377/94, 1385, 1391).
[2] Stewart, R. F. (Acta Cryst. A **33** [1977] 33/8).
[3] Yang, Y. W.; Coppens, P. (Acta Cryst. A **34** [1978] 61/5).
[4] Larsen, F. K.; Lehmann, M. S.; Merisalo, M. (Acta Cryst. A **36** [1980] 159/63).
[5] Manninen, S.; Suortti, P. (Phil. Mag. [8] B **40** [1979] 199/207).
[6] Collins, D. M.; Whitehurst, F. W. (Acta Cryst. A **37** [1981] 848/50).
[7] Hansen, N. K.; Schneider, J. R.; Larsen, F. K. (Phys. Rev. [3] B **29** [1984] 917/26).
[8] Larsen, F. K.; Hansen, N. K. (Acta Cryst. B **40** [1984] 169/79).
[9] Dovesi, R.; Pisani, C.; Ricca, F.; Roetti, C. (Phys. Rev. [3] B **25** [1982] 3731/9).
[10] Chakraborty, S.; Manna, A.; Ghosh, A. K. (Phys. Status Solidi B **129** [1985] 211/20, 213/4).

[11] Hansen, N. K.; Schneider, J. R.; Yelon, W. B.; Pearson, W. H. (Acta Cryst. A **43** [1987] 763/9).
[12] Tsirel'son, V. G.; Lobanov, N. N.; Ozerov, R. P. (Fiz. Metal. Metalloved. **63** [1987] 24/30; Phys. Metals Metallog. [USSR] **63** No. 1 [1987] 18/24; Acta Cryst. A **40** [1984] Suppl. C-168).
[13] Fox, A. G.; Fisher, R. M. (Phil. Mag. [8] B **57** [1988] 197/208).
[14] Blaha, P.; Schwarz, K. (J. Phys. F **17** [1987] 899/911, 906/7).
[15] Redinger, J.; Schwarz, K.; Hansen, N. K.; Bauer, G. E. W.; Schneider, J. R. (Hahn-Meitner Inst. Kernforsch. Berlin Ber. HMI-B No. 412 [1984] 79/99 from [14]).
[16] Chou, M. Y.; Lam, P. K.; Cohen, M. L. (Phys. Rev. [3] B **28** [1983] 4179/85).
[17] Matthai, C. C.; Grout, P. J.; March, N. H. (J. Phys. F **10** [1980] 1621/6; Phys. Letters A **68** [1978] 351/4).
[18] Eisenberger, P.; Marra, W. C.; Brown, G. S. (Phys. Rev. Letters **45** [1980] 1439/42).
[19] Contini, V.; Sacchetti, F. (J. Phys. F **11** [1981] L1/L6).
[20] Mazzone, G.; Sacchetti, F.; Contini, V. (Phys. Rev. [3] B **28** [1983] 1772/80).

[21] Chou, M. Y.; Lam, P. K.; Cohen, M. L. (Phys. Rev. [3] B **28** [1983] 4179/85).
[22] Dovesi, R.; Angonoa, G.; Causa, M. (Phil. Mag. [8] B **45** [1982] 601/6).
[23] Inoue, S. T.; Yamashita, J. (J. Phys. Soc. Japan **35** [1973] 677/83).
[24] Holzwarth, N. A. W.; Azhar, S.; Kerr, T. J. (Phys. Rev. [3] B **38** [1988] 9409/16).
[25] Massa, L.; Goldberg, M.; Frishberg, C.; Boehme, R. F.; La Placa, S. J. (Phys. Rev. Letters **55** [1985] 622/5).
[26] Clinton, W. L.; Massa, L. (Phys. Rev. Letters **29** [1972] 1363/6), Frishberg, C.; Massa, L. J. (Phys. Rev. [3] B **24** [1981] 7018/24).
[27] Nyholm, R.; Flodström, A. S.; Johansson, L. I.; Hörnström, S. E.; Schmidt-May, J. N. (Surf. Sci. **149** [1985] 449/59).
[28] Weiss, R. J. (Phil. Mag. [8] B **37** [1978] 659/62).

14.7   Electron Momentum Distribution. Compton Profiles

The momentum distribution of electrons can be examined by means of Compton profiles (CPs) which measure the integrated electron momentum distribution over a plane perpendicular to a specific direction. Experimentally these profiles can be measured directly from the photon scattering pattern, and theoretically they can be calculated using wave functions obtained from band-structure calculations. A comparison between experimental and theoretical profiles then serves as a test of the accuracy of the calculated wave functions.

Information on the electron momentum distribution in metallic Be was obtained from a number of Compton scattering experiments (cf. p. 103) using X-rays [1 to 7], γ-rays [8, 9], and, in addition, using positron annihilation techniques [10]; see also [11, 12].

The modulation of the CP for Be (and other light elements) was studied in the presence of a laser field. Low-frequency lasers were found to be well-suited for the experimental investigation of the electron momentum distribution [13].

Because of improvements in the experimental facilities, the accurate CPs in the three crystallographic directions [10$\bar{1}$0], [11$\bar{2}$0], and [0001], obtained by Hansen et al. [8] using 412 keV γ-rays from a [198]Au source, appeared to be far the most reliable. Loupias et al. [1] used 10 keV synchrotron radiation to obtain seven directional profiles characterized by high resolution. Their tabulations and discussions, however, concerned only the three directional profiles considered in [8].

In early experiments [4, 6] no corrections were made for multiple Compton scattering, and the long tails found in the experimental profiles could be attributed to this effect. The influence of multiple scattering on the X-ray Compton data was confirmed by Phillips and Chin [3]. Since that time it has become routine to correct experimental profiles for this systematic error [8]. Improvements in the Monte Carlo multiple scattering correction for CPs are described with the application for synchrotron-source photons by Chomilier et al. [14]. The multiple scattering correction program for almost totally polarized incident photons, when considering the attenuation of double-scattered photons due to the analyzing crystal, decreases the amount of double scattering. For Be this contribution was in good agreement with experimental results from samples of two thicknesses [14].

Compton profiles of Be have also been computed using several different models for the wave functions of valence electrons, for instance, plane waves filling the first plus second Brillouin zone [4, 8], orthogonalized Bloch functions from s-type atomic orbitals (LCAO) [15], an OPW method [16], orthogonalized plane waves as resulting from a pseudopotential calculation [17], a pseudopotential method improved by considerations of orthogonalization corrections [18], Hartree-Fock ab initio techniques with an extended basis set [19].

Most of the available models were not able to describe quantitatively the details of the momentum anisotropy observed in the experimental studies [1, 8]. A comparison of CPs from an accurate HF computation (using an extended AO basis) with experimental directional CPs revealed close agreement, in particular with the profiles from γ-ray experiments. The differential CPs could also be semiquantitatively reproduced, better on average than with all previous calculations. Concerning the Fourier transforms of the CPs, all features of the experimental curves were accurately reproduced, especially the position of the zero passages which is a test of a theoretical model to account for the metallic nature of the solid [19].

One serious error in most theoretical calculations of CPs in solids was illustrated in the LCAO calculation of Aikala [15] for Be metal. The kinetic energy difference between the HF free atom and the calculated CP yielded a decrease of 23 eV in the metal whereas the cohesive energy indicated an increase of $\sim$3 eV. Most theoretical solid state calculations of CPs did not give a reasonable cohesive energy [5].

Calculations of the anisotropic behavior of CPs agree qualitatively with experimental results, except for a discrepancy in the amplitude [1, 8]. Since directional CPs measure the momentum distribution integrated over a plane perpendicular to a specific direction, they are very sensitive to the accuracy of the wave functions [20]. The anisotropic CPs from an ab initio calculation using a self-consistent pseudopotential approach within the local-density functional scheme show excellent agreement with experimental profiles. The small discrepancy of the directional profiles can be improved by inclusion of correlation effects in an approximative way. For example, the anisotropy of CPs in $\langle 0001 \rangle$ and $\langle 10\bar{1}0 \rangle$ as well as in $\langle 0001 \rangle$ and $\langle 11\bar{2}0 \rangle$ directions, compared with experiments and other calculations, is shown in **Fig.** 14-5 a) and b). This agreement suggests that the theoretical approach used can also give an accurate description of the valence wave functions [21]; cf. [20, 22]. The main characteristics of the anisotropy arise from the special shape of the second Brillouin zone [4, 22]. The difference between theoretical pseudopotential and experimental CPs allows one to estimate the electron correlation in metals. For Be it is almost directionally independent (whereas for graphite it is anisotropic). The correlation effects evaluated for a homogeneous electron gas (in Be) reduced the difference by almost a factor of 2 [23]. A comparison of experimental [1] and calculated [18, 22] directional CPs furnished a check of the quality of the wave functions used in computation [23].

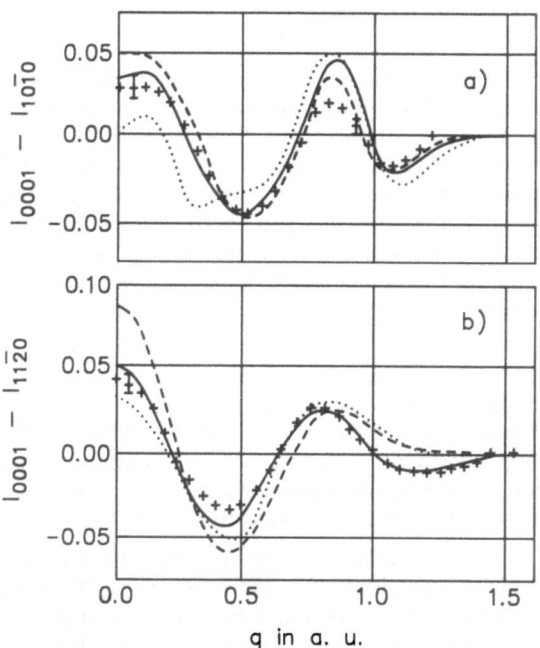

Fig. 14-5. Comparison a) of the measured X-ray anisotropy $I_{0001} - I_{10\bar{1}0}$ (crosses) with theoretical values (folded with the experimental resolution function): calculated [22] (solid line); linear combination of atomic orbitals [15] (dotted line), and pseudopotential calculation [17] (dashed line). Comparison b) of the γ-ray anisotropy $I_{0001} - I_{11\bar{2}0}$ (crosses) [8] with folded theoretical values as in a). Momentum q in atomic units.

Good agreement was obtained between experimental [1, 8] and calculated [20] total directional profiles (along $\langle 0001 \rangle$; see figure in the paper). However, detailed comparisons with the anisotropic profiles are more appropriate, since the systematic deviations inherent in both theory and experiment can be removed when one directional profile is subtracted from another [20].

References:

[1] Loupias, G.; Petiau, J.; Issolah, A.; Schneider, M. (Phys. Status Solidi B **102** [1980] 79/95).

[2] Berko, S. (in: Williams, B. G.; Compton Scattering, McGraw-Hill, New York 1977, p. 273).

[3] Phillips, W. C.; Chin, A. K. (Phil. Mag. [8] **27** [1973] 87/93).

[4] Currat, R.; DeCicco, P. D.; Kaplow, R. (Phys. Rev. [3] B **3** [1971] 243/51).

[5] Weiss, R. J. (Phil. Mag. [8] B **37** [1978] 659/62).

[6] Phillips, W. C.; Weiss, R. J. (Phys. Rev. [2] **171** [1968] 790/800).

[7] Schülke, W.; Berg, U. (Phys. Status Solidi **23** [1967] K87/K91).

[8] Hansen, N. K.; Pattison, P.; Schneider, J. R. (Z. Physik B **35** [1979] 215/25; Hahn-Meitner-Inst. Kernforsch. Berlin Ber. HMI-B No. 310 [1979] 69/76 from C. A. **93** [1980] No. 155938).

[9] Manninen, S.; Suortti, P. (Phil. Mag. [8] B **40** [1979] 199/207).

[10] Stewart, A. T.; Shand, J. B.; Donaghy, J. J.; Kusmiss, J. H. (Phys. Rev. [2] **128** [1962] 118/9).

[11] Berko, S. (Phys. Rev. [2] **128** [1962] 2166/8).

[12] Lynn, K. G.; Goland, A. N. (Solid State Commun. **18** [1976] 1549/52).

[13] Sharma, B. S.; Singh, G. S.; Tripathi, A. N. (J. Phys. B **14** [1981] 979/84).

[14] Chomilier, J.; Loupias, G.; Felsteiner, J. (Nucl. Instrum. Methods Phys. Res. A **235** [1985] 603/6; C. A. **102** [1985] No. 228883).

[15] Aikala, O. (Phil. Mag. [8] **33** [1976] 603/11).

[16] Chaddah, P.; Sahni, V. C. (Phys. Letters A **56** [1976] 323/4).

[17] Rennert, P.; Dörre, T.; Gläser, U. (Phys. Status Solidi B **87** [1978] 221/6).

[18] Rennert, P. (Phys. Status Solidi B **105** [1981] 567/75).

[19] Dovesi, R.; Pisani, C.; Ricca, F.; Roetti, C. (Phys. Rev. [3] B **25** [1982] 3731/9; Z. Physik B **47** [1982] 19/26).

[20] Chou, M. Y.; Lam, P. K.; Cohen, M. L. (Phys. Rev. [3] B **28** [1983] 1696/700)

[21] Chou, M. Y.; Lam, P. K.; Cohen, M. L. (Phys. Rev. [3] B **28** [1983] 4179/85).

[22] Chou, M. Y.; Lam, P. K.; Cohen, M. L.; Loupias, G.; Chomilier, J.; Petiau, J. (Phys. Rev. Letters **49** [1982] 1452/5).

[23] Issolah, A.; Chomilier, J.; Garreau, Y.; Loupias, G. (J. Phys. Colloq. [Paris] **48** [1987] C9-851/C9-854).

14.8 Surface Electronic Structure of (0001)Beryllium

Surface Core Level Shift

The (0001)Be surface core level shift (SCS) of the Be 1s level is -0.50 eV (the bulk binding energy relative to E_F is $E_b = 111.7$ eV, see p. 48). Surface-sensitive photoelectron spectroscopy using monochromatic synchrotron radiation in the photon energy range of 120 to 150 eV (see figures in the paper) reveals for the (0001) single-crystal surface two 1s core peaks, one from the bulk atoms and the other from the first layer surface atoms, separated by -0.50 eV in binding energy (the surface peak is located at the lower binding energy). This experimental SCS is compared to the calculated SCS using different assumptions about the character of the initial and final state of the photoionized Be metal atoms in the bulk and at the surface. A final state having the Al(sp)^3 valence configuration yields an SCS of -0.53 eV, i.e., of the right direction and magnitude [1]. An SCS of −0.55 eV, i.e., $\Delta E_b^{BS} = E_b^{B,V} - E_b^{S,V} - (\Phi^B - \Phi^S)$ with the binding energies referred to vacuum $E_b^{B,V} = 115.4$ eV, $E_b^{S,V} = 114.7$ eV and the work functions $\Phi^B = 3.50$ eV, $\Phi^S = 2.25$ eV, was calculated for a monolayer Be film (after adjustment of the Fermi level) using the unrestricted Hartree-Fock (UHF) method, in excellent agreement with experimental data [2].

From LDA calculations the surface core level 1s shift was estimated to be −0.21 eV from a three-layer film (the thinnest system said to be able to exhibit such a shift) [3].

Valence-Electron Surface States

The surface electronic structure of polyvalent metals is closely connected with their crystal structure. Experimentally the surface electronic structure of (0001)Be was defined using angle-resolved photoemission spectroscopy [4, 5]. A one-layer (n=1) (0001)Be film was studied within the framework of the local-density functional formalism in [6 to 8]. These calculations were extended to the case of a two-layer (n=2) film in [9] and up to a three-layer (n=3) film in [3]. The one- to four-layer films of Be were calculated in [10] using the Hartree-Fock LCAO method. Six- and ten-layer films were calculated in [11] self-consistently using the pseudopotential method in the local-density functional approximation; cf. [12]. **Fig. 14-6** shows the projection of the calculated one-electron spectrum E (k) of bulk Be onto the surface Brillouin zone (SBZ) along symmetric directions. The dashed lines show the calculated surface and resonance states. The surface states are understood as states that fall into the band gaps. The resonance states are the states in which the charge density concentrates mainly in the surface layer, but their energy lies in the range of allowed bulk states. The main feature of the

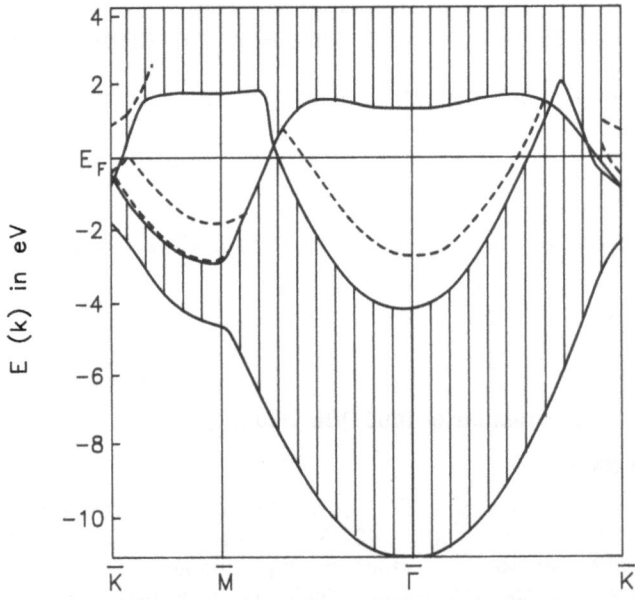

Fig. 14-6. Surface electronic structure of (0001)Be along the symmetry directions of the SBZ: projection of the volume band structure and the surface and resonance states (dashed lines).

Be bulk state projection, in which it differs from other polyvalent metals, is the presence of wide band gaps around the Fermi level [11].

According to the calculation [11] there is a surface state with $E = -2.6$ eV at the $\overline{\Gamma}$ point within the gap. This is in good agreement with the experimental result of a surface state in the Γ_3^+-Γ_4^- band gap with a binding energy of -2.8 ± 0.1 eV [4, 5]. The LDA calculation for a three-layer film gave $E = -2.0$ eV [3]. According to the calculation (ten-layer film) of [11] the pseudocharge distribution of this state is mainly localized in the surface layer which holds 65% of the state charge. The two upper atomic layers contain 90% of the charge of this state. Experimentally, for photon energies $\hbar\omega > 40$ eV, the energy dependence of the photoexcitation cross section of this state shows only a weak structure (caused by changes in the local electromagnetic field at the surface). This high-energy behavior is quite different from the large resonances observed for surface states on other metals and is associated with the short penetration depth of this Be surface state [5]. Away from $\overline{\Gamma}$ the surface state disperses parabolically towards E_F with an effective mass $m^* \approx 1.5 \cdot m_e$ [5], (theoretically $m^* = 1.03 \cdot m_e$ [3]). Towards $\overline{M}$, E_F is reached at about 55% of the distance to the SBZ boundary [4]. The surface character becomes less explicit when moving away from $\overline{\Gamma}$ [11].

The main distinction of the surface electronic structure of Be compared to other polyvalent metals (in particular that of Mg) is the presence of two surface states at the $\overline{M}$ point (in the M_2^--M_4^- gap). Experimentally these two surface states, found in a small range of $k_{||}$ (momentum parallel to the surface) around $\overline{M}$, have the binding energies (-1.8 ± 0.1) and (-3.0 ± 0.1) eV [5], which are in good agreement with the theoretical values -1.9 and -2.9 eV, respectively [11]. The upper state has a back-bond character; 95% of this state is located in the two surface layers. When the wave vector varies along $\overline{M}\overline{\Gamma}$, the state turns quickly in a bulk one. The lower surface state at $\overline{M}$ is localized at the bottom of the energy gap and the electron density distribution of this state is nontypical (distribution peak in the third layer) for surface states [11]. The resonance states theoretically found at $\overline{K}$ are, as well, above and below the Fermi level: 0.9, -0.6, -0.4 eV. The resonance state with the energy $E = 0.9$ eV is localized in the surface layer up to 82% [11].

The density of states (DOS) of the central layer (n = 4) is typical for bulk Be where E_F is in the minimum of the DOS, see pp. 52 ff. The sharp peak in the DOS above E_F splits with decreasing layer number [11]. For calculated occupied band widths for one- and two-layer Be films (values ranging from 5.93 to 9.52 eV), see [7, 9].

An emission signal, observed in Be and a series of other metals after an intense microwave pulse (having a resonance in the magnetic field with $\sim 10^{-4}$ s decay time), apparently resulted from the appearance of electron surface states on the metal samples in high vacuum [13].

The experimental evidence obtained by LEED measurements [14, 15] that the (0001) surface of Be does not reconstruct (see, for instance, "Molybdenum" Suppl. Vol. A 2a, 1985, pp. 115 ff.) can be related to the intrinsic stability of planar sections of the hcp crystal lattice [16].

References:

[1] Nyholm, R.; Flodström, A. S.; Johansson, L. I.; Hörnström, S. E.; Schmidt-May, J. N. (Surf. Sci. **149** [1985] 449/59).

[2] Zdetsis, A. D. (J. Phys. Colloq. [Paris] **48** [1987] C 9-839/C 9-842).

[3] Boettger, J. C.; Trickey, S. B. (Phys. Rev. [3] B **34** [1986] 3604/9).

[4] Karlsson, U. O.; Flodström, S. A.; Engelhardt, R.; Gädeke, W.; Koch, E. E. (Solid State Commun. **49** [1984] 711/4).

[5] Bartynski, R. A.; Jensen, E.; Gustafsson, T.; Plummer, E. W. (Phys. Rev. [3] B **32** [1985] 1921/6).

[6] Mintmire, J. W.; Sabin, J. R.; Trickey, S. B. (Phys. Rev. [3] B **26** [1982] 1743/53).

[7] Wimmer, E. (J. Phys. F **14** [1984] 681/90).

[8] Boettger, J. C.; Trickey, S. B. (J. Phys. F **14** [1984] L151/L153).

[9] Boettger, J. C.; Trickey, S. B. (Phys. Rev. [3] B **32** [1985] 1356/8).

[10] Angonoa, G.; Koutecky, J.; Pisani, C. (Surf. Sci. **122** [1982] 355/70).

[11] Chulkov, E. V.; Silkin, V. M.; Shirykalov, E. N. (Surf. Sci. **188** [1987] 287/300; Fiz. Metal. Metalloved. **64** [1987] 213/36; Phys. Metals Metallog. [USSR] **64** No. 2 [1987] 1/23, 11).

[12] Silkin, V. M.; Chulkov, E. V. (Poverkhnost No. 4 [1987] 126/32 from C. A. **107** [1987] No. 84102).

[13] Smolyakov, B. P.; Khaimovich, E. P. (Pis'ma Zh. Eksperim. Teor. Fiz. **37** No. 2 [1983] 95/7; JETP Letters **37** No. 2 [1983] 116/8).

[14] Zimmer, R. S.; Robertson, W. D. (Surf. Sci. **43** [1974] 61/76).

[15] Baker, J. M.; Blakely, J. M. (J. Vac. Sci. Technol. **8** [1971] 4).

[16] Pacchioni, G.; Pewestorf, W.; Koutecky, J. (Chem. Phys. **83** [1984] 261/74, 273).

14.9 Work Function

Clean (oxygen-free) Be surfaces have a work function Φ (all values in eV) of about 5.0. On (0001)Be $\Phi = 5.10 \pm 0.02$ was measured [1]; 5.08 ± 0.08 [2] and 4.98 ± 0.10 [3] were measured on UHV-deposited polycrystalline films according to the Fowler method (photoelectron current) [1, 3] or by the Kelvin method (contact potential difference against Au, $\Phi = 5.22 \pm 0.05$) [2].

Earlier reported lower values, e.g., 3.67 ± 0.03 as the effective thermionic work function [4] and 3.22 ± 0.08 as the Richardson work function [5], may have resulted from oxygen-contaminated surfaces. When the (pure) polycrystalline Be surfaces mentioned above were exposed to O_2, the work function decreased to lower values reaching saturation after about 600 L exposure at 3.6 [2] and after about 80 L exposure at 3.2 [3]. This is consistent with the limiting value of 3.3 eV on the (0001) surface, not reached, however, before 3000 L [1]. On the other hand, the high value $\Phi = 5$, measured for thick Be layers deposited on a W field emitter [6], was questioned with respect to its assignment to clean bulk Be (determined 4.1, recommended value from previous literature 3.9 eV) on the basis of average and single-plane field-emission work function measurements for Be adsorbed on W [7]. Additional discussions of this result are reported in [1, 8].

Theoretical Φ values for thin Be layers [9 to 12] are consistent with a high work function of Be near 5 eV. For instance, the LDF pseudopotential method gave for a ten-layer (n = 10) film of (0001)Be a value $\Phi = 5.4$ [12], and the LDF (Hedin-Lundquist) value $\Phi = 5.19$ [10] was obtained for the monolayer with c/a = 1.69 and a = 4.10 a.u. For Be a sufficiently strong dependence of the monolayer work function upon the lattice parameter ($\partial \Phi_{n=1}/\partial a \approx -0.7$ eV/a.u.) was found [9] to raise the possibility that quantum size effects (QSE) in Be n-layers might be obscured or even obliterated by lattice relaxation effects. However, whether QSE appears in the work function of ultrathin Be films is difficult to determine from existing calculations. In principle QSE Φ-oscillations can be as much as 0.5 eV, but the expected values for relaxed n-layer films are within 2% of the bulk value [13]. Theoretical work (SCF-Xα-SW method) on Be clusters gave for $Be_{13}(D_{3h})$ $\Phi = 8.4$, for $Be_{19}(D_{3h})$ $\Phi = 10.4$, and for bulk Be $\Phi \approx 4.5$ eV [14].

The surface dipole barrier of metals, $D = \Phi + E_F$, was calculated by a quantum mechanical approach. For hcp Be with c/a = 1.57 it is anisotropic with $D_{1000} = 4.54$, $D_{2210} = 4.23$, $D_{1010A} = 4.29$, and $D_{1010B} = 4.18$ eV. In the case of the (1010) surface of the hcp structure there is the stacking sequence ABCDABCD .... and the distance d_{AB} (between A and B) is not equal to d_{BC} ($d_{BC} = 2 d_{AB}$). All planes have two different and alternating distances. Therefore D depends on whether the top layer is of either A or B type (E_F does not depend on the surface plane; $D_x - D_y = \Phi_x - \Phi_y$) [15]. For further calculations of the surface dipole barrier, see [16, 17].

The temperature coefficient of the jellium work function, calculated for all metals including Be, is negative with the sign of the total work function coefficient depending on the core radius of the (Ashcroft) model potential and on the crystallographic plane [18].

References:

[1] Green, A. K.; Bauer, E. (Surf. Sci. **74** [1978] 676/81).
[2] Dixon, R. D.; Lott, L. A. (J. Appl. Phys. **40** [1969] 4938/9)
[3] Gustafsson, T.; Brodén, G.; Nilsson, P.-O. (J. Phys. F **4** [1974] 2351/8).
[4] Wilson, R. G. (J. Appl. Phys. **37** [1966] 2261/7).
[5] Jerner, R. C.; Magee, C. B. (Oxid. Metals **2** [1970] 1/9 from C.A. **73** [1970] No. 8401), Jerner, R. C. (Diss. Univ. Denver 1965, pp. 1/130 from Diss. Abstr. **26** [1966] 5963).
[6] Komar, A. P.; Savchenko, V. P.; Shrednik, U. N. (Radiotekh. Elektron. [Moscow] **5** [1960] 1211/7 from C.A. **1961** 18231).
[7] Polanski, J.; Sidorski, Z.; Zuber, S. (Acta Phys. Polon. A **49** [1976] 299/305).
[8] Bauer, E. (J. Phys. Colloq. [Paris] **38** [1977] C4-146/C4-154).
[9] Boettger, J. C.; Trickey, S. B. (J. Phys. F **14** [1984] L151/L153, F **16** [1986] 693/706; Phys. Rev. [3] B **32** [1985] 1356/8).
[10] Wimmer, E. (J. Phys. F **14** [1984] 681/90, 2613/24).

[11] Boettger, J. C.; Trickey, S. B. (J. Phys. Condensed Matter **1** [1989] 4323/38).
[12] Chulkov, E. V.; Silkin, V. M.; Shirykalov, E. N. (Surf. Sci. **188** [1987] 287/300, 297).
[13] Vicente, J. L.; Paola, A.; Razzitte, A.; Mola, E. E.; Trickey, S. B. (Phys. Status Solidi B **155** [1989] K93/K98].
[14] Kolesnikov, V. V.; Polozhentsev, E. V.; Sachenko, V. P.; Kovtun, A. P. (Fiz. Tverd. Tela [Leningrad] **19** [1977] 1510/1; Soviet Phys.-Solid State **19** [1977] 883/4).
[15] Taut, M.; Schubert, M. (Phys. Status Solidi B **107** [1981] K139/K144), Taut, M.; Eschrig, H.; Schubert, M. (Phys. Status Solidi B **100** [1980] 243/50).
[16] Heine, V.; Hodges, C. H. (J. Phys. C **5** [1972] 225/30).
[17] Alonso, J. A.; Gonzales, D. J.; Inignez, M. P. (J. Phys. F **10** [1980] 1995/2008).
[18] Kiejna, A.; Wojciechowski, K. F.; Zebrowski, J. (J. Phys. F **9** [1979] 1361/6, **11** [1981] 2495).

14.10   Electron-Phonon Interaction

In his review of the electron-phonon interaction (EPI) in normal metals, Grimvall [1] gives as the recommended value of the electron-phonon mass enhancement parameter (McMillan parameter) or electron-phonon coupling constant $\lambda = 0.24 \pm 0.05$ for Be. This value is based on estimates from the superconducting transition temperature (cf. p. 21) according to the McMillan equation ($\lambda = 0.24 \pm 0.01$ [1], $\lambda = 0.23$ [2, 3]) or from a generalized form of this equation ($\lambda = 0.24$ [3]), using for the reduced Coulomb coupling constant the value $\mu^* = 0.10$, and from calculations using a pseudopotential model ($\lambda = 0.26$ [3, 4]). An estimation from the ratio of the specific heat density of states to a calculated band density of states gave $\lambda = 0.25$ [3, 4].

The EPI function obtained by point-contact (pc) measurements [5] in the form of the point-contact spectrum $g_{pc}(\omega)$ in comparison with the phonon density of states $F(\omega)$, determined experimentally by the method of coherent neutron scattering from a polycrystal [6] and theoretically using the dispersion curves for phonons in the principal crystallographic directions [7], is shown in **Fig.** 14-7 from the review [8]. Hence the pc EPI spectrum depends primarily on the phonon frequencies rather than on the electronic properties and exhibits singularities at $\hbar\omega = 52$ (maximum peak), 64, 71.5, and 77 meV. The value of $\lambda_{pc} = 0.23 \pm 0.04$ [5, 8], obtained within the free-electron model by integrating $g_{pc}(\omega)$, practically coincides with the cited λ values from transport properties, see above.

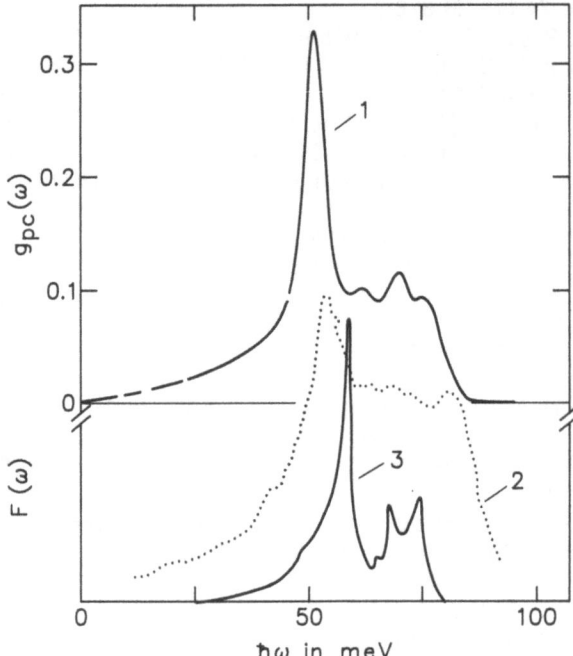

Fig. 14-7.   Point-contact EPI function of Be (curve 1) and the phonon density of states, determined experimentally by inelastic coherent neutron scattering from a polycrystal (curve 2), and theoretically using the dispersion curves for phonons in the principal crystallographic directions (curve 3).

In a study of electron-phonon interaction in disordered Be films, the inelastic scattering of electrons was investigated experimentally by measuring the localization corrections to conductivity. The existence of the temperature dependence of the phase breakdown time of the electron wave function, $t_\phi \sim T^{-2}$, above a characteristic temperature T_0 (i.e., $T > T_0$), which varies with the film thickness, and its absence for $T < T_0$ was confirmed; this dependence was observed earlier for semimetal films. The temperature T_0 corresponds to the equality of the thermal phonon wavelength to the film thickness. A comparison of the experimental results with the theory and with calculated data of the sound velocity indicated that the inelastic electron scattering by transverse phonons plays a decisive role [9].

References:

[1] Grimvall, G. (Phys. Scr. **14** [1976] 63/78, 70).
[2] McMillan, W. L. (Phys. Rev. [2] **167** [1968] 331/44, 337).
[3] Garland, J. W.; Allen, P. B. (Physica **55** [1971] 669/77).
[4] Allen, P. B.; Cohen, M. L. (Solid State Commun. **7** [1969] 677/80).

[5] Naidyuk, Yu. G.; Shklyarevskii, O. I. (Fiz. Tverd. Tela [Leningrad] **24** [1982] 2631/5; Soviet Phys.-Solid State **24** [1982] 1491/3).

[6] Bulat, I. A. (Fiz. Tverd. Tela [Leningrad] **21** [1979] 1001/8; Soviet Phys.-Solid State **21** [1979] 583/7).

[7] Raubenheimer, L. J.; Gilat, G. (Phys. Rev. [2] **157** [1967] 586/99).

[8] Yanson, I. K. (Fiz. Nizk. Temp. **9** [1983] 676/709; Soviet J. Low Temp. Phys. **9** [1983] 343/60, 349, 353).

[9] Butenko, A. V.; Bukhshtab, E. I.; Kashirin, V. Yu.; Komnik, Yu. F. (Fiz. Nizk. Temp. **14** [1988] 421/4; Soviet J. Low Temp. Phys. **14** [1988] 233/4; C.A. **109** [1988] No. 120429).

14.11 Beryllium Clusters Be_n (n = 2 to 73)

Survey

Metal clusters, such as those of Be, have been the subject of numerous, mainly theoretical investigations. Some were to identify features that are related to the convergence of cluster properties from those of the free atom to those of the bulk metal. Other studies were performed because of the importance of metal clusters as models of highly dispersed catalysts and of chemisorption sites as well as of point defects on surfaces. Some preliminary information on Be_n clusters is given in the section of "Cohesive Energy" in "Beryllium" Suppl. Vol. A 2, 1991, p. 123.

In order to understand the nature of bonds present in small clusters of metal atoms, rules about the cluster stability as a function of cluster size and geometry are required. General rules which may govern the electronic and geometric structures of clusters are formulated based on quantum mechanical investigations of Be_n (n = 3 to 14) and other metal clusters. The topology of the framework of the nuclei in a cluster influences strongly the electronic structure. Three factors determine the cluster geometries: the compactness of a cluster contributes favorably to its stability; the nodal properties, degeneracies, and occupancies of the relevant one-electron functions have a decisive influence on the stability of small cluster geometries, and the polarization functions in the AO basis set play an important role in determining and understanding the cluster stability [1]. Current experimental methods are not able to fully answer questions about the geometry of small clusters. Because of its closed-shell ^1S ground state, Be is expected to be especially suitable for larger cluster calculations.

n = 2

As was already pointed out in "Beryllium" Suppl. Vol. A 2, 1991, p. 123, the dimer Be_2 is a special case in that it is not a typical molecule, see also Rao et al. [2, figure 1], and the nature of bonding in Be_2 is not of the simple van der Waals type. The ground state of the dimer has been the subject of a great deal of theoretical interest and calculation for many years, but the various theories and methods applied predicted potential curves of sometimes widely differing nature. The discrepancies were attributed to partial inclusion or summation of terms involving triple and quadruple excitations, the incomplete description of the sp quasi-degeneracy in the Be atom, and to basis set deficiencies. More definitive calculations with larger basis sets and a more thorough treatment of electron correlation agree upon a relatively deep well of the order of 8.37 kJ/mol (2 kcal/mol) at 4.7 to 5.0 a.u. (2.49 to 2.65 Å). A compilation of previous investigations on Be_2 is given by Harrison, Handy [3]. They investigated the ground-state potential curve within the frozen-core approximation at the full configuration interaction (CI) level of accuracy with a series of Gaussian basis sets [4], the largest of which gave an equilibrium distance of 4.75 a.u. (2.51 Å) and a dissociation energy of 7.78 kJ/mol (1.86 kcal/mol), and extrapolated the results, by estimating the effects of g basis functions, to predict 9.33 to 9.79 kJ/mol (2.23 to 2.34 kcal/mol). The problem of bonding is connected with the near $2s,2p$ degeneracy of Be. The qualitative nature of the ground-state curve may be accounted for by avoiding crossing between two states of $^1\Sigma_g^+$ symmetry, one expected to be repulsive, the other expected to be strongly binding in simple MO theory. This model may also account for the importance of d functions which are required to obtain a qualitatively correct potential curve [3].

From CI calculations, according to the interacting correlated fragments (ICF) method developed for weakly interacting systems, in order to obtain an ab initio potential curve for the $Be_2(^1\Sigma_g^+)$ ground state, a binding (or dissociation) energy of 0.10 eV (9.62 kJ/mol or 2.3 kcal/mol) was obtained at an equilibrium separation of 4.7 a.u. (2.49 Å) [5]. The binding energy of [5] was cited as a value per atom (1.16 kcal/mol) by Bagus et al. [6]. Previously computed Hartree-Fock (HF) potential energy curves agreed with the prediction of the simple MO theory that the $^1\Sigma_g$ ground state of Be_2 has a repulsive character. The absence of a minimum in the SCF curve agreed with other SCF studies on the electronic structure of Be_2 [7 to 9]. Full CI treatment with the larger ([4s2p1d]) of two basis sets gave an equilibrium distance of 8.1 a.u. (4.3 Å) and a binding energy of 2.383 kJ/mol (0.57 kcal/mol). The data indicated that a much larger basis set is required to obtain more realistic results [10]. Purely repulsive HF potential curves for Be_2 were also expectably found by Stoll et al. [11] and their density functional approximation for the correlation energy led to a bound ground state with a 9.1 a.u. (4.8 Å) bond length and 0.335 kJ/mol (0.08 kcal/mol) dissociation energy.

The problematic nature of the Be_2 potential energy curve, showing how different kinds of theories (particularly the nonvariational approaches) can produce very different results, is discussed by Chiles and Dykstra [12].

Ab initio calculations indicated that Be_2 can form a stable anion Be_2^- with an equilibrium bond length of 4.62 a.u. (2.45 Å) [8]; cf. previous studies [13]. While Be^- is not stable, it appears that all Be_n clusters with $n > 2$ will possess stable anions [8].

n = 3

For the next smallest cluster, the trimer Be_3, two forms are possible: linear and triangular (equilateral). Restricted and unrestricted Hartree-Fock (RHF and UHF) calculations indicated that the $X^1\Sigma_g$ state of linear Be_3 is repulsive over the range of internuclear distances studied, whereas the ground state of triangular Be_3 is weakly bound. The predicted van der Waals minimum (of roughly 4 to 8 kJ or 1 to 2 kcal) for symmetric linear Be_3 was estimated to occur for a bond length of ~5 a.u. (2.65 Å). The equilateral triangular Be_3 was found to possess a minimum-energy configuration with bond lengths of 4.32 a.u. (2.29 Å) which is stable with respect to dissociation by 0.06 eV. A total of five negative ion states was found for linear Be_3. The anions of triangular Be_3 are more stable than the corresponding anions of linear Be_3. Further details and figures such as charge-density maps and potential energy curves of the two Be_3 forms are given in the paper [8].

Optimized equilibrium geometries and electronic structures of neutral Be_3 and other metal clusters are shown and discussed in [2]. The triangular (D_{3h} symmetry) structure of Be_3 is the most compact one with a binding energy of 0.586 kJ/mol (0.14 kcal/mol) at an equilibrium distance of 9.5 a.u. (5.03 Å), only slightly higher than that found for Be_2. Bonding is said to be a pure correlation effect [11]. Linear and triangular Be_3 do not appear to be appreciably bound [8, 11, 14], if d functions are not included in the AO basis set. The small stability attributed purely to correlation effects [11] was confirmed by the data of Pacchioni, Koutecky [10] using a [4s2p] basis set for calculation. On the contrary, a stability of 25.1 kJ/mol (6 kcal/mol) was predicted for triangular Be$_3$ using a basis set including d polarization functions [15]; also a dissociation energy of 29.1 kJ/mol (7.1 kcal/mol) at a distance of 4.2 a.u. (2.2 Å) was found for triangular Be$_3$ using a [4s2p1d] basis set [10]. Potential energy curves obtained with the [4s2p] basis set for the two $^1\Sigma_g^+$ and $^3\Sigma_u^+$ states of linear Be$_3$ closely resemble those for diatomic Be and exhibit a slightly bonding $^1\Sigma_g^+$ ground state [10].

According to the majority of previous calculations at the SCF level (without polarization functions in the basis) Be$_3$(D$_{3h}$) seemed to be only weakly bound (if at all). The results [10] indicated that Be$_3$ is akin to Be$_2$ and that considerable care has to be taken in constructing the wave functions for Be$_3$. The problems of ab initio calculations are discussed and the results of CI calculations, yielding 92 to 105 kJ/mol (22 to 25 kcal/mol) for the Be$_3$ binding energy, are compared with other results by Harrison, Handy [3].

Similarities and discrepancies obtained at the SCF and CI levels in calculations of Be$_n$ clusters (with $3 \leqq n \leqq 7$) and the importance of the electron correlation effect are discussed by Rubio et al. [16].

n = 4

Tetramers or Be$_4$ clusters, in the form of a planar rhombus with D$_{2h}$ symmetry and a pyramid with T$_d$ symmetry (another tetrahedron would have D$_{2d}$ symmetry), were considered in early chemisorption studies by Bauschlicher et al. [17]. Near-tetrahedral Be$_4$ was suggested to be the most stable species in the vapor of metallic Be. In addition, other possible Be$_4$ cluster configurations are a square (D$_{4h}$), a linear configuration (D$_{\infty h}$), and a triangular configuration with another Be atom added to one apex (C$_{2v}$); see figure 1 in the paper [10]; cf. [1].

Stability for nearly tetrahedral Be$_4$ is predicted to be between 125.5 and 167.4 kJ/mol (30 and 40 kcal/mol) [8, 11, 15, 18] and the inclusion of d functions in the basis set increases this value by about 50% [14]. The effect of correlation was considered by Dykstra et al. [7]. The large difference observed when the two basis sets ([4s2p] and [4s4p/4s2p]) are used in calculation, demonstrates the importance of accounting for the interaction of s and p orbitals in the Be$_4$ tetrahedron. Several authors [8, 14, 18] pointed out that the large stability of the tetrahedral arrangement is due to effective s-p hybridization. This hybridization is emphasized by the results [10]. In a study of two interacting Be$_2$ fragments from the tetrahedral to the square-planar arrangement, the binding energy of the system clearly shows that the tetrahedral structure represents a minimum with respect to the other "buckled" clusters. The transition from the T$_d$ to the D$_{4h}$ structure is characterized by the appearance of an energy barrier generated by the avoided crossing of the two electronic configurations. The height of the energy barrier depends on the detailed geometry of the intermediate D$_{2d}$ structure, and the square structure represents a nonbonding local minimum in the reaction path. In the [4s2p] pseudopotential treatment, the two geometries of rhombus and square Be$_4$ exhibit a bonding situation more similar to that of Be$_2$ than to that of the Be$_4$ tetrahedron. Details of the electronic structure are discussed in [10]. Another comparison of the planar D$_{4h}$ configuration with tetrahedral Be$_4$ of T$_d$ symmetry also revealed a great difference with the bond length becoming smaller by a factor of 2 and the binding energy increasing drastically. The difference in the binding energy from SCF calculation, with 178.24 kJ/mol (42.6 kcal/mol) being somewhat larger than that reported by Brewington et al. [14], was attributed to core-valence correlation [11]. The bond length of [11] is in fair agreement with that (3.99 a.u. or 2.11 Å) of Dykstra et al. [7].

Clusters of Be$_4$ with T$_d$ symmetry were also treated in other theoretical studies. The dissociation energy was computed from configuration-interaction wave functions to be 247.7 kJ/mol (59.2 kcal/mol). A summary of dissociation energies, calculated with different basis sets and for different wave functions, is given in [19]. The dissociation energy for tetrahedral Be_4 [19] is cited as a value per atom of 61.9 kJ/mol (14.8 kcal/mol) for an equilibrium distance of 3.92 a.u. (2.07 Å) by Bagus et al. [6].

A comparison of tetramers of different metals shows that Be_4 is five times more strongly bound than Mg_4 whereas solid Be is only twice as stable as Mg. The difference can be (crudely) related to different coordination numbers of the nearest neighbors (leading to different s-p hybridization) [6]. For the relatively large stability of Be_4 due to hybridization, see also [16].

Studies by the SCF–LCAO–MO technique of the equilibrium geometries and electronic structures of small metal clusters such as tetrahedral Be_4 showed that the equilibrium geometries depend on the number of valence electrons and their bonding character. The possibilities of Be_4 and Mg_4 of being "magic numbers" are discussed in [2].

In a study of the effects of basis set expansion and electron correlation on the structure and dissociation energy of Be_4 (and Be_{13}), the resulting dissociation energy per bond of the same cluster indicates, with respect to both factors, convergence to the bulk cohesive energy. Dissociation energies of tetrahedral Be_4 in the $^1A_1$ electronic state, calculated with various basis sets, are tabulated for comparison in the paper [20]. The sequence Be_2, Be_3, Be_4 is interesting in its own right because of the remarkable increase in stability as the binding mechanism changes due to increased s-p hybridization. Contrary to an earlier belief, electron correlation contributes substantially to the binding energy, and since it is still impossible to perform correlated calculations for larger clusters, accurate results for small clusters are valuable in estimating corrections for larger systems. The best estimated values of the equilibrium Be–Be bond length, r, and binding energy, E_b, for small Be_n clusters (n = 2, 3, 4) according to Harrison and Handy [3] are:

cluster	symmetry	r in a.u.	E_b in kJ/mol	(E_b in kcal/mol)
Be_2	$D_{\infty h}$	4.65	9.4	(2.25±0.08)
Be_3	D_{3h}	4.2 (±0.1)	100.4	(24±2)
Be_4	T_d	3.9 (±0.1)	313.8	(75±2)

The smallest cluster to show appreciable bonding is Be_4 and thus provides a test for any method to be applied to larger clusters. Equally important as the effect of cluster size on the preferred structure and cohesive energy (with respect to convergence to bulk properties) are basis set size and correlation effects. Be_2 and Be_4 clusters were considered at both SCF and CI levels in order to calibrate larger clusters. At the SCF level, Be_2 is unbound whereas Be_4 is bound by ~167 kJ/mol (40 kcal/mol) with respect to the dissociation to four Be atoms. For Be_2 two Be atoms must be hybridized to form only one bond, while for Be_4 six bonds are formed from the hybridization of four Be atoms. This leads to the observed change in bonding, and Be_4 becomes bound at the SCF level. The inclusion of correlation and basis set improvement increases the binding energy by a factor of 1.8. It was concluded that the bonds in Be_4, Be_{13} (see p. 82), and bulk Be are similar. Therefore, the difference in cohesive energy between the bulk and small clusters is the number of bonds and not the nature of the bonding [21].

n = 5

For Be_5 clusters two planar and three tridimensional configurations were considered: C_{2v}, D_{5h}, C_{4v}, C_s, and D_{3h} (see figure 1 in the paper [10]; cf. [1]). On the basis of the discussion of the $Be_4(T_d)$ configuration it is not unexpected that the bipyramidal trigonal form (D_{3h} symmetry), obtained by joining two regular tetrahedra, shows a higher stability with respect to the other Be_5 clusters. Moreover, on going from $Be_4(T_d)$ to Be_5 (D_{3h}) the vertical ionization potential, computed as the difference between the neutral and the charged species, decreases from 7.68 to 6.41 eV. The other two nonplanar configurations are also part of the hcp Be crystal structure. The electronic structure of square-pyramidal Be$_5$(C$_{4v}$) looks very similar to that found for Be$_4$. The energy minimum of the best SCF electronic configuration of 13 kJ/mol (3.1 kcal/mol) is localized around a distance of 4 a.u. (2.12 Å). The Be$_5$(C$_s$) configuration is obtained by capping one triangular part of the Be$_4$ rhombus. The CI stability of 54.8 kJ/mol (13.1 kcal/mol) for a distance of 4.15 a.u. (2.2 Å) indicates that the addition of one atom to rhombic Be$_4$ in a tetrahedral coordinate position stabilizes the system much more than the planar extension of the structure to give the Be$_5$(C$_{2v}$) cluster. On the contrary, the pentagonal Be$_5$(D$_{5h}$) configuration was found to be the most stable Be$_5$ cluster at the Hartree-Fock level with a binding energy of 112.55 kJ/mol (26.9 kcal/mol) in agreement with [15]. The stability and the shorter equilibrium distance of 4.2 a.u. (2.22 Å) of this cluster with respect to other planar structures can be rationalized on the basis of smaller geometrical strains of the pentagonal arrangement, where the Be–Be–Be angles of 128° are not too far from the classical angle of the sp^2 hybrid [10].

Generally it seemed that the binding energy per atom is an increasing function of the number of Be atoms in the clusters which are sections of the hcp crystal lattice. The average binding energy for Be$_5$(D$_{3h}$) was found to be higher than for Be$_4$(T$_d$) which demonstrates that the number 8 (=2×1+2×3) is not exceptionally favorable for the stability of the Be clusters. Strong hybridization among s and p atomic orbitals is known to represent the main reason for the stability of Be$_n$ clusters with n ≧ 4. Consequently simple counting of valence electrons is not a sufficient measure of cluster stability [1]. Previous pseudopotential calculations indicated that in trigonal-bipyramidal Be$_5$(D$_{3h}$) the nearest-neighbor distance (of 3.81 a.u. or 2.02 Å) as well as the binding energy per atom (exceeding that of Be$_2$) are virtually the same as for the Be$_4$ tetrahedron (62.55 kJ/mol or 14.95 kcal/mol). Be$_5$ is bound with respect to Be$_4$ + Be by 59.83 kJ/mol (14.3 kcal/mol) [11]. The trigonal-bipyramidal (D$_{3h}$) configuration was found to be the most stable form of a Be$_5$ cluster, but the reported binding energy of 41.0 kJ per bond was smaller than for tetrahedral Be$_4$ with 47.2 kJ per bond and still much smaller than the experimental heat of sublimation of 54 kJ per bond [15].

For Be$_5$(D$_{3h}$) the possibility of a stable dianion was suggested [8].

Using nonempirical pseudopotentials for the Be atoms, Be$_5$(C$_s$) was studied at the SCF and CI levels. In Be$_5$(C$_s$; 4,1), i.e., 4 atoms in the first layer and 1 atom in the second layer, it appears that while at the SCF level the double open shell of triplet multiplicity is the ground state, the CI calculation shows that the stablest one corresponds to a closed-shell configuration (for n ≦ 4 the calculated ground state always coincides with that found at the CI level) [16].

n = 6

For n = 6 seven configurations with the symmetries D$_{3h}$ planar, C$_{5v}$, O$_h$, D$_{3h}$ prismatic, C$_{2v}$, D$_{2h}$, and C$_{2h}$ have been depicted in the cluster topologies of Koutecky, Fantucci [1]. For Be$_6$ cluster structures (D$_{2h}$, O$_h$, C$'_{2v}$, C$_{2v}$) were considered in an ab initio study using the [4s4p/4s2p] basis set so ensure a good description of the s-p interaction. Three of these configurations, Be$_6$(D$_{2h}$), Be$_6$(O$_h$), and Be$_6$(C$'_{2v}$), are regular fragments of the hcp lattice, whereas Be$_6$(C$_{2v}$) is obtained by twisting the Be$_6$(D$_{2h}$) cluster to obtain three condensed identical tetrahedra. All these clusters are characterized by a short equilibrium distance and also by a considerable stability in the SCF treatment. The Be$_6$(D$_{2h}$) cluster, obtained by growing Be$_5$(C$_s$) in a symmetrical way, is the most stable Be$_6$ cluster with a CI binding energy of 228.03 kJ/mol (54.5 kcal/mol). Despite the fact that the basis sets used for Be$_5$(C$_s$) and Be$_6$(D$_{2h}$) are not comparable, the large stability of the Be$_6$(D$_{2h}$) structure is due to the possibility of the edge atom of the pentamer to form strong bonds when the added atom is in a tetrahedrally coordinated position. The same observation holds for the octahedron $Be_6(O_h)$, where the sixth atom added to the square-pyramidal Be_5 plays a fundamental role in the cluster stability. It makes possible a perfectly uniform participation of all p atomic orbitals in the bond. In the more compact C_{2v} cluster, the higher electronic energy is compensated by the stronger nuclear repulsion, and this cluster is 12.55 kJ/mol (3 kcal/mol) less stable than the corresponding $Be_6(D_{2h})$. The second Be_6 cluster of C_{2v} symmetry can be considered as an intermediate step of growing planar clusters from the rhombic to the symmetrical $Be_7(D_{6h})$. The relative binding energy of 151.9 kJ/mol (36.3 kcal/mol) in CI indicates the increasing stability of planar structures with the increasing number of nearest neighbors. The final order of stabilities of the configurations considered in this study, $D_{2h} > C_{2v} > O_h > C'_{2v}$, supports once more the conclusion that tetrahedral arrangements are preferred. The two clusters $Be_6(D_{2h})$ and $Be_6(C'_{2v})$ have their triplet state 41.84 and 20.92 kJ/mol (10 and 5 kcal/mol) above the corresponding singlets for the interatomic distances 3.8 a.u. (2.01 Å) and 4.0 a.u. (2.12 Å), respectively [10].

n = 7

For n = 7 three configurations are represented in the cluster topologies with the symmetries D_{5h}, C_{3v}, and D_{6h}, see figure in the paper [1]. The planar structure of $Be_7(D_{6h})$ was investigated using a [4s4p/3s2p] basis set in which the two inner s Gaussian functions have been contracted. The geometry of this cluster, representing a part of the hcp lattice, is characterized by the six peripheral atoms in hexagonal configuration, interacting with a single central atom, which reflects the situation in a Be monolayer. For this arrangement the triplet SCF energy is much lower than the singlet one (the binding energies are 332 and 214 kJ/mol or 79.4 and 51.2 kcal/mol, respectively), but when correlation is considered the two electronic states are nearly degenerated. The analysis of the electronic structure shows an important participation of the p orbitals to the occupied MOs. The presence of the central atom is crucial for the stabilization of the structure and seems to produce an effect similar to that of the apical atoms in $Be_6(D_{2h})$ or $Be_6(O_h)$. A calculation of the same Be_7 cluster with a nearly minimal basis set, [4s4p/2s1p], showed that the ordering and composition of the occupied MOs were very similar for both basis sets [10].

Planar Be_7 is the first cluster with considerable stability containing no tetrahedral unit. This cluster is a part of the (0001) crystal plane of the hcp Be crystal. Bonding in Be_7 can be understood as being due to a large electronic redistribution between the s and p orbitals [22]. In the case of $Be_7(D_{6h})$ the ground state appeared to be a double open shell triplet at both SCF and CI levels. However, at the SCF level the ground state appeared to be a $^3A_{2u}$ state while $^3B_{1g}$ was found at the CI level [16]. The average bond strength (here called "cohesive energy" and defined as the negative value of the binding energy per atom), obtained at the CI level, of 36.3 kJ/mol or 8.68 kcal/mol [16] is in agreement with that (35.9 kJ/mol or 8.57 kcal/mol) found by Pacchioni et al. [22]. The interatomic distances for the equilibrium geometry of $Be_7(D_{6h}; 7,0)$ are very close to the Be–Be distance in the hcp crystal lattice. The average distance of 4.2 a.u. (or 2.2 Å) between neighboring atoms in Be_n clusters seems to be independent of the cluster geometry and can be rationalized as the same kind of chemical bonding present in clusters and in the bulk metal. Pseudopotential SCF and CI potential energy curves for low-lying states of $Be_7(7,0)$ are shown in figures and SCF and CI energies and equilibrium bond distances for the six molecular states, obtained with all-electron and pseudopotential methods, are tabulated in [23].

UHF calculations (using a so-called STO-3G minimum basis set [32]) on Be_7 in the simple cubic symmetry gave an equilibrium "lattice constant" of 2.03 Å, a total energy of −100.3816, a binding energy of −0.0116, and an ionization potential of 0.0695 (all values in a.u. or Hartrees). The Be_7 cluster in sc symmetry is not bound at the STO-3G level [24].

n = 8, 9, 10

For n = 8 three configurations with T$_d$, C$_{2v}$, and O$_h$ symmetry are described in [1], but no literature was found treating a Be$_8$ cluster.

For n = 9 one optimized configuration is represented in [1]. In the simple cubic symmetry the equilibrium "lattice constant" for Be$_9$ is 2.34 Å (from UHF calculations using the STO-3G minimum basis set). The total energy is −129.3163, the binding energy 0.0166, and the ionization potential is 0.1398 (all values in a.u. or Hartrees) [24].

For n = 10 two configurations with D$_{2d}$ and C$_{3v}$ symmetry and one optimized structure are represented in [1]. The two configurations of Be$_{10}$, here designated as D$_{2h}$ and C$_{3v}$, are shown in **Fig.** 14-8. The Be$_{10}$(10,0) cluster is a planar extension of Be$_7$(7,0), and Be$_{10}$(7,3) serves to represent tridimensional clusters; it contains three tetrahedral subunits with one common vertex. For investigation the interatomic distances have not been optimized and were taken equal to the value of the crystal (4.2 a.u. or 2.2 Å). Because the cluster geometry reproduces sections of the hcp Be lattice, the stability of the structures is (expectedly) high: average bond strength or binding energy per atom E$_b$/n = 51.5 kJ/mol for D$_{2h}$ and 49.8 kJ/mol for C$_{3v}$ [22]. Data for low-lying electronic states are listed in [23].

Some further information on Be$_7$ and Be$_{10}$ in comparison with Be$_{13}$ is given below.

Fig. 14-8. Geometrical structures of computed Be$_{10}$ clusters: ○ atoms in the first layer, ● atoms in the second layer.

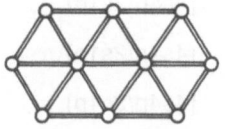

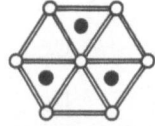

Be$_{10}$ (10,0) D$_{2h}$ Be$_{10}$ (7,3) C$_{3v}$

n = 13

For n = 13 four configurations with the symmetries T$_h$, O$_h$, D$_{3h}$, and D$_{2h}$ are described in the cluster topologies and the clusters Be$_7$(D$_{6h}$), Be$_{10}$(C$_{3v}$), and Be$_{13}$(D$_{2d}$) are compared because of their very stable planar structures by Koutecky, Fantucci [1].

A major role of coordination is to lead to a different s → p hybridization which has significant consequences for the directed bonding between the metal atoms and hence to the cluster stability. In order to model the bulk situation, a Be$_{13}$(D$_{3h}$; 3,7,3) cluster with 3 atoms in the first and third layers and 7 atoms in the center layer with geometry and bond distance of the bulk crystal was used. The central atom of Be$_{13}$(D$_{3h}$) has the 12-fold coordination of an atom in bulk Be. To characterize the extent of s-p hybridization, the Mulliken gross atomic p populations for SCF wave functions for Be$_4$ and Be$_{13}$ clusters are tabulated in the paper. The Be bond is sufficiently strong that the 2p character in Be$_4$ is already large and does not increase much for the more highly coordinated central atom in Be$_{13}$. A comparison showed that the properties, in particular the binding, of Be clusters will converge more rapidly to the bulk values than those of Mg and Ca clusters [6].

Dramatic changes in bond strength are observed when proceeding from the more or less weakly bonded species Be$_2$, Be$_3$, Be$_4$, Be$_5$ to larger clusters Be$_7$, Be$_{10}$, Be$_{13}$ with a rapid increase of cluster stability, especially on considering planar configurations. Electronic properties of some Be$_n$ clusters (n = 2 to 13), as obtained from ab initio calculations according to the pseudopotential multireference double excitation CI method and compiled in [22], are given in the table on p. 82.

cluster	symmetry	basis set[*]	ground state	d_{min} in a.u.	E_b/n in kJ/mol
Be_2	$D_{\infty h}$	[4s2p]	$^1\Sigma_g^+$	9.3	0.4
		[4s2p1d]	$^1\Sigma_g^+$	8.3	0.8
Be_3	$D_{\infty h}$	[4s2p]	$^1\Sigma_g^+$	9.1	0.4
		[4s2p1d]	$^3\Sigma_u^+$	4.1	11.7
Be_4	T_d	[4s4p/2s1p]	$^1A_1$	4.3	13.8
		[4s4p/4s2p]	$^1A_1$	4.0	35.9
Be_5	D_{3h}	[4s4p/2s1p]	$^1A_1'$	4.2	11.7
		[4s4p/4s2p]	$^1A_1'$	4.2	36.8
Be_6	D_{2h}	[4s4p/2s1p]	^1A_g	4.1	17.6
		[4s4p/4s2p]	^1A_g	3.8	38.0
Be_7	D_{6h}	[4s4p/2s1p]	$^3B_{1g}$	4.15	35.9
		[4s4p/4s2p]	$^3B_{1g}$	4.0	48.9
Be_{10}	D_{2h}	[4s4p/2s1p]	^1A_g	4.2[**]	51.5
Be_{10}	C_{3v}	[4s4p/2s1p]	$^1A_1$	4.2[**]	49.8
Be_{13}	D_{2h}	[4s4p/2s1p]	^1A_g	4.2[**]	64.9
Be_{13}	D_{3h}	[4s4p/2s1p]	$^3E''$	4.2[**]	71.6

[*] The description of the basis set is given in [10]. – [**] Distance not optimized.

Metal clusters are of interest not only because of their own properties and reactivity, but also as models of more extended metallic systems (e.g., representing part of a metal surface or bulk). The Be_{13} cluster is studied for both D_{3d} (fcc-like) and D_{3h} (hcp-like) geometries. The neutral cluster as well as the positive and negative ions are considered for idealized (all bonds equal) and distorted geometries. The D_{3d} geometry was found to be the lowest for Be_{13}; this is the reverse of what is expected based upon the hcp metal structure. Large distortions are found for both the D_{3d} and D_{3h} geometries. The ions have geometries similar to those of the neutral system. The dissociation energy per atom for Be_{13} (0.77 eV) is twice that of Be_4 (0.39 eV). Since in Be_{13} there are only about three bonds per Be as compared to six bonds per Be in the bulk metal, a difference of about a factor of 2 was expected, but the dissociation energy per atom for Be_{13} is 0.8 to 0.9 eV while the bulk cohesive energy is 3.38 eV per atom. Considering the different number of bonds in the clusters and in the bulk, it was concluded that the bonds are similar in Be_4, Be_{13}, and the bulk metal. The difference in cohesive energy between the bulk and small clusters is the number of bonds and not the nature of the bonding. A summary of equilibrium bond distances, binding energies, and distortions, mainly for Be_{13} clusters and ions, is tabulated in the paper [21].

The preferred structures and dissociation energies of Be_4 and Be_{13} are determined by ab initio methods, including Møller-Plesset perturbation theory, coupled-cluster theory, and multireference CI. Both species serve as calibration points for basis set extension and electron correlation effects. For the hcp configuration of Be_{13} two electronic states were considered. The first, $^1A_1'$, is the lowest-lying state (D_{3h} symmetry) examined by Bauschlicher and Pettersson [21]. The second, $^5A_1''$, is the lowest state (D_{3d} symmetry) found by Ermler et al. [25]. For the fcc structure (D_{3d} symmetry), a $^3A_{1g}$ state was considered. Trends in basis set extension and correlation were similar for Be_4 and Be_{13}. Dissociation energies of the several Be_{13} clusters obtained with various basis sets are tabulated in [20].

By ab initio Hartree-Fock calculations on the electronic states of Be_{13} with D_{3h} symmetry, corresponding to a central Be with 12 atoms situated at the hcp nearest-neighbor positions, a total of 14 states were identified below 2.0 eV. The lowest state at the HF level of theory is $^5A_1''$ and the first excited state is $^5E''$ lying 1.12 eV higher. The $^5A_1''$ HF ground state has a binding energy of 50.208 kJ/mol (12 kcal/mol) relative to the separated atoms and an ionization potential of 0.54 eV (52.1 kJ/mol). The cluster calculations gave also a reasonable estimate for the Sternheimer correction to the electric field gradient of the bulk metal [25].

n = 2 to 73 (comparisons)

Low-lying electronic states of Be_n clusters (n = 7, 10, 13) were studied by means of a pseudopotential technique. Correlation effects have been taken into account through multireference double excitation CI procedure. The molecular ground state is different in SCF and CI procedures for many Be_n clusters. Already in Be_7 the energies of some states are separated only by a few kJ/mol. Particular attention was paid to Be_n clusters with many nearly degenerated states of singlet and triplet spin multiplicity. The SCF ordering of these states is frequently reversed in CI. The separation between the lowest singlet and lowest triplet states decreases with increasing cluster size. The existence of energetically low-lying triplet states is probably connected with the increasing metallic character and can not be attributed to edge or corner effects in a simple way. Planar and nonplanar clusters show comparable high stabilities caused by a strong s-p hybridization [23].

Electron states of small periodic aggregates of Be (and Li) atoms in the form of chains were calculated according to the SCF–MO–LCGO method using two basis sets (optimized by Preuss [28] and by Chu, Frost [29]). The internuclear distance of the Be chains (n = 2 to 12) was 4.32 a.u. (2.29 Å). The binding energy $E_b = -E_{tot}/n$, calculated with basis set 2 (given in the paper), was about 14.55 a.u. (or Hartrees). It decreased with increasing n; therefore Be chains are not stable [30].

In model chemisorption studies (atomic H on (0001)Be) Be_n clusters with n = 3 to 14 were investigated and results are compiled in tables in the paper. A larger cluster studied, $Be_{22}(14,8)$, is shown in a schematic representation in the paper [31].

An optimized configuration for n = 14 is represented in [1].

Equilibrium "lattice constants", a_e, total energies, E_{tot}, binding energies, E_b, and ionization potentials, IP, for Be_n clusters with n = 13, 15, 19 in the simple cubic symmetry from UHF calculations, using the so-called STO-3G basis set [32], are given in the following table:

cluster	a_e in Å	E_{tot} in a.u.	E_b in a.u.	IP in a.u.
Be_{13}	2.94	− 187.3250	0.0577	0.1570
Be_{15}	2.35	− 216.1776	0.0600	0.1660
Be_{19}	2.04	− 273.4660	0.0408	0.1303

For Be_{19} with fcc symmetry the corresponding values are 2.96 Å, −273.8658, 0.0621, and 0.1371 a.u. (or Hartrees), respectively [24].

A nonself-consistent calculation using a one-electron multiple scattering model (with a muffin-tin potential) for Be_n clusters indicated the formation of the conduction band bottom with increasing cluster size from 12 to 50 atoms [26]. A follow-up study of the effect of short-range order and screening of a deep vacancy on the shape of the X-ray spectra of metallic Be by numerical methods within the framework of multiple-screening theory gave qualitative agreement for clusters with 13 and 57 atoms (D_{6h} symmetry) with experimental X-ray spectra when only the first 3 to 5 coordination spheres were taken into account. Substantial improvement in the agreement between calculated and experimental emission curves with increasing cluster size was observed [27].

Ab initio SCF calculations using effective core potentials are reported for electronic states of Be_{51} and Be_{57} clusters with full D_{3h} point symmetry. The calculated binding energy per atom for the ground states is 25.8 and 25.4 eV, respectively (the ground states in both Be_{51} and Be_{57} are closed-shell and of $^1A_1'$ symmetry). The electric field gradient and the quadrupole moment were found to be sensitive to the choice of electronic state. These properties give a measure of distortion from spherical symmetry; Be_{57} is more spherically symmetric with respect to charge than Be_{51}. The results (compiled in tables in the paper) show that ionization potentials and binding energies compare well to those calculated for Be_{55} [33] and approach the values for the bulk metal, in contrast to those calculated for states of Be_{13} [25]. Overall net charges are also found to be quite constant. Finally, it was anticipated that calculations on larger clusters will show even closer behavior to the bulk metal [34].

Calculations at the SCF level revealed that in comparison of fcc and hcp structures, the clusters Be_{13} and Be_{55} both prefer the fcc over the bulk hcp structure, but the energy difference per atom decreases for Be_{55} relative to Be_{13}. The binding energy per atom, 0.8 to 0.9 eV for Be_{13} and 1.3 eV for Be_{55}, reflects the greater total number of bonds in the larger cluster rather than a difference in bonding. The energies per bond in the range of 0.30 to 0.34 eV are much more similar for both clusters. An extended p-basis set was found to be very important in order to describe the bonding in these clusters [35]. Considering previous works [21, 35], bonding in Be_4, Be_{13}, and Be_{55} was found to be essentially the same and differences between the clusters and the bulk were attributed to the number of bonds and not to the nature of bonding. The dissociation energy per bond (at SCF level) converged rapidly within the series Be_4 to Be_{55}, but to a value much smaller than that of the bulk. Calculations were performed on Be_4 and Be_{13} clusters, including Møller-Plesset perturbation theory, coupled cluster theory, and multireference CI. The results were used to interpret the binding energies for the larger clusters and they showed that, by treating also electron correlation and larger basis sets, the dissociation energy per bond of the same clusters does indicate convergence to the bulk cohesive energy of Be [20].

The electronic structure of thin Be films of one to four atomic layers was investigated by an ab initio LCAO-HF method; band structure, electronic density, and density of states were determined. A comparison of the electronic structure in the films of various thickness with the situation found for Be clusters and bulk metal showed strong modifications in the surface region and a progressive reconstruction of the bulk bonds. The s-p hybridization is the determining factor for the stability of Be slabs analogous to the binding in Be clusters. The specific electronic structure for Be monolayer and planar clusters helped to explain the properties of the "surfaces" of the slab models: the missing bonds of the Be atoms in the "surface" layers straighten the bonds parallel to the surface in these layers providing (at first sight unexpected) stability of the surface. This phenomenon may be the reason for the lack of surface reconstruction of Be crystals [36]; cf. p. 71.

Some experimental studies concerned with surface states are reported, e.g., LEED from (0001)Be [37, 38], LEED-Auger analysis of (0001)Be [39], ionization spectroscopy (Si on Be) [40].

Further studies of surface states and properties are mostly performed by theoretical calculations on large Be clusters, for instance, surface relaxation and induced stresses accompanying the adsorption of atomic H on (0001)Be by a series of SCF-MO calculations on Be_7 to Be_{15} [41]. Chemisorption of atomic H on (0001)Be was studied by MO calculations on the clusters Be_5, Be_{10}, and Be_{73} using the extended Hückel theory, Anderson's modification of the extended Hückel theory, and the complete neglect of the differential overlap method [42]. Mixed basis set calculations for atomic H on (0001)Be were reported on Be_6 to Be_{22} by [43]; cf. previous studies [31, 44].

References:

[1] Koutecky, J.; Fantucci, P. (Z. Physik D **3** [1986] 147/53).
[2] Rao, B. K.; Khanna, S. N.; Jena, P. (Phys. Rev. [3] B **36** [1987] 953/60).
[3] Harrison, R. J.; Handy, N. C. (Chem. Phys. Letters **123** [1986] 321/6).
[4] Harrison, R. J.; Handy, N. C. (Chem. Phys. Letters **98** [1983] 97/101).
[5] Liu, B.; McLean, A. D. (J. Chem. Phys. **72** [1980] 3418/9).
[6] Bagus, P. S.; Nelin, C. J.; Bauschlicher, C. W., Jr. (Surf. Sci. **156** [1985] 615/22).
[7] Dykstra, C. E.; Schaefer, H. F., III; Meyer, W. (J. Chem. Phys. **65** [1976] 5141/6).
[8] Jordan, K. D.; Simons, J. (J. Chem. Phys. **67** [1977] 4027/37).
[9] Jones, R. O. (J. Chem. Phys. **71** [1979] 1300/8).
[10] Pacchioni, G.; Koutecky, J. (Chem. Phys. **71** [1982] 181/98, 184).

[11] Stoll, H.; Flad, J.; Golka, E.; Krüger, T. (Surf. Sci. **106** [1981] 251/7).
[12] Chiles, R. A.; Dykstra, C. E. (J. Chem. Phys. **74** [1981] 4544/56; Chem. Phys. Letters **85** [1982] 447/50).
[13] Jordan, K. D.; Simons, J. (J. Chem. Phys. **65** [1976] 1601/2).
[14] Brewington, R. B.; Bender, C. F.; Schaefer, H. F., III (J. Chem. Phys. **64** [1976] 905/6).
[15] Whiteside, R. A.; Krishnan, R.; Pople, J. A.; Krogh-Jespersen, M.-B.; Ragué Schleyer, von, P.; Wenke, G. (J. Comput. Chem. **1** [1980] 307/22, 314, 318, 321).
[16] Rubio, J.; Illas, F.; Ricart, J. M. (J. Chem. Phys. **84** [1986] 3311/6).
[17] Bauschlicher, C. W., Jr.; Liskow, D. H.; Bender, C. F.; Schaefer, H. F., III (J. Chem. Phys. **62** [1975] 4815/25).
[18] Jordan, K. D.; Simons, J. (J. Chem. Phys. **72** [1980] 2889/90).
[19] Bauschlicher, C. W., Jr.; Bagus, P. S.; Cox, B. N. (J. Chem. Phys. **77** [1982] 4032/8).
[20] Rohlfing, C. M.; Binkley, J. S. (Chem. Phys. Letters **134** [1987] 110/4).

[21] Bauschlicher, C. W.; Jr.; Pettersson, L. G. M. (J. Chem. Phys. **84** [1986] 2226/32).
[22] Pacchioni, G.; Plavšić, D.; Koutecky, J. (Ber. Bunsenges. Physik. Chem. **87** [1983] 503/12).
[23] Pacchioni, G.; Pewestorf, W.; Koutecky, J. (Chem. Phys. **83** [1984] 261/74).
[24] Ray, A. K. (J. Phys. B **19** [1986] 1253/9).
[25] Ermler, W. C.; Kern, C. W.; Pitzer, R. M.; Winter, N. W. (J. Chem. Phys. **84** [1986] 3937/43).
[26] Sachenko, V. P.; Polozhentsev, E. V.; Kovtun, A. P.; Migal, Yu. F.; Kolesnikov, V. V.; Vedrinskii, R. V. (Phys. Fenn. **9** Suppl. S1 [1974] 129/31 from C. A. **82** [1975] No. 147728).
[27] Polozhentsev, E. V.; Sachenko, V. P.; Kovtun, A. P. (Izv. Akad. Nauk SSSR Ser. Fiz. **40** [1976] 240/2; Bull. Acad. Sci. [USSR] **40** No. 2 [1976] 14/6).
[28] Preuss, H. (Z. Naturforsch. **20a** [1965] 1290/8).
[29] Chu, S. Y.; Frost, A. A. (J. Chem. Phys. **54** [1971] 764/8).
[30] Stoll, H.; Preuss, H. (Phys. Status Solidi B **53** [1972] 519/25).

[31] Bauschlicher, C. W., Jr.; Bagus, P. S.; Schaefer, H. F., III (IBM J. Res. Develop. **22** [1978] No. 3, pp. 213/34, 215, 220).

[32] Hehre, W. J.; Stewart, R. F.; Pople, J. A. (J. Chem. Phys. **51** [1969] 2657/64).

[33] Pettersson, L. G. M.; Bauschlicher, C. W., Jr. (private communication) cited in [34].

[34] Ross, R. B.; Ermler, W. C.; Pitzer, R. M.; Kern, C. W. (Chem. Phys. Letters **134** [1987] 115/20).

[35] Pettersson, L. G. M.; Bauschlicher, C. W., Jr. (Chem. Phys. Letters **130** [1986] 111/4).

[36] Angonoa, G.; Koutecky, J.; Pisani, C. (Surf. Sci. **121** [1982] 355/70, 359, 368/9).

[37] Zimmer, R. S.; Robertson, W. D. (Surf. Sci. **43** [1974] 61/76).

[38] Baker, J. M.; Blakely, J. M. (J. Vac. Sci. Technol. **8** [1971] 4).

[39] LeJeune, E. J., Jr. (J. Vac. Sci. Technol. **8** [1971] 9).

[40] Musket, R. G. (Surf. Sci. **44** [1974] 629/34).

[41] Cox, B. N.; Bauschlicher, C. W., Jr. (Surf. Sci. **102** [1981] 295/311).

[42] Hoflund, G. B.; Merrill, R. P. (J. Phys. Chem. **85** [1981] 2037/41).

[43] Bauschlicher, C. W., Jr.; Bagus, P. S. (Chem. Phys. Letters **90** [1982] 355/8).

[44] Bauschlicher, C. W., Jr.; Bender, C. F.; Schaefer, H. F., III; Bagus, P. S. (Chem. Phys. **15** [1976] 227/35).

14.12 Plasmons and Plasmon Scattering

A plasmon is the term used to describe the collective quantized oscillations of electrons in a solid. The energies of plasmons are measured directly from the characteristic energy loss suffered by fast charged particles passing through a thin foil of material. Numerous investigators studied the correlation between the fine structure in X-ray absorption spectra and the results obtained from characteristic energy loss spectra. On this basis Ferrell [1] and Nozières, Pines [2] suggested that the X-ray transition in metals could be accompanied by the excitation of one or more plasmons. Such plasmons can be excited by electrons, X-rays, and light. A review on plasmon excitation is given by Raether [3].

Early investigations of (volume) plasmons in usually polycrystalline Be samples by electron energy loss spectroscopy (EELS) with fast and slow electrons in transmission and reflection are reviewed by Powell [4]. For further studies by electron scattering, see [5 to 8], and by X-ray scattering [9 to 13]. Evidence for volume and surface plasmons is confirmed by the peaks in the dielectric loss funcions found in optical studies [14].

Plasmons are usually interpreted on the basis of the Bohm-Pines electron gas theory [15]. The theoretical value for the volume plasmon energy (volume plasma loss) assuming free electrons (Be: two valence electrons), gained by the random phase approximation (RPA) and neglecting electron-electron correlation, is $E_p = \hbar\omega_p = 18.4$ eV and the corresponding surface plasmon energy is $\hbar\omega_p/\sqrt{2} = 13.8$ eV (the band structure can not be neglected since it determines important properties of the plasmons, e.g. their lifetime), see [4, 16]. From the optical dispersion $\hbar\omega_p = (19.2 \pm 0.4)$ eV for the volume plasmon and $\hbar\omega_p/\sqrt{2} = 13.6$ eV for the surface plasmon were derived [14]. Another optical study gave $E_p = 18.4$ eV [17, 18].

Using the reflection technique and 750 and 1500 eV primary electrons, a prominent volume plasma loss at (19.9 ± 0.2) eV as well as multiples of this loss, but also a lowered plasma loss at (11.9 ± 0.2) eV were found. These results agree fairly well with data of earlier studies cited in [4]. Other values for E_p (in eV) are 18.7 [6, 22], 18.5 [19], 18.9 [20]. From X-ray scattering experiments the plasmon energy for a scattering angle $\vartheta = 0$ was found to be 19.4 [10, 11], 20.7 [9].

A table comparing E_p data (in the range 18.4 to 19.4 eV) of several authors is given in [14]. By means of X-ray-induced photoemission spectroscopy (XPS, see p. 95) a bulk plasmon energy of 19.5 was obtained, whereas a surface plasmon energy of 11.5 was found as compared to a calculated value of 13.7, indicating that Be is not a free-electron metal [21]. The theoretical surface plasmon energy as calculated from the electron density of the Be metal was 12.87 compared to 12.31 estimated through a chemical approach and 14.07 obtained from the experimental bulk value [46].

The half width of the plasmon peak (in eV) for $\vartheta = 0$ is relatively large for Be, $\Delta E_{1/2} = 4.4 \pm 0.7$ [6], 4.7 ± 0.1 [17], 5.0 ± 0.2 [19], 5.4 [8].

In a study of the angular dependence of the half width of the plasmon peak on several metals (classified into two groups), electron-electron interaction was found to be an important factor of plasmon damping for the group of metals comprising Be, Si, and Ge [19].

In the case of X-ray emission, a plasmon can be excited at the same time as the filling of the core vacancy, and in absorption the plasmon is excited when the core vacancy is created. The probability of simultaneous creation of a plasmon with either emission or absorption of a photon was expected to be small because neither process involves a large modification of the valence-electron distribution. This was supported by experiments. The plasmon satellite is weak relative to the intensity of the parent emission band. When the processes do occur simultaneously, the X-ray photon is deprived of an amount of energy equal to $\hbar\omega_p$, where ω_p is the angular frequency of the plasmon. Thus a satellite band occurs in emission at an energy $\hbar\omega_p$ below the parent band. Plasmon satellites in the emission bands of Be (and Mg) observed by Watson et al. [23] are shown in figures in the review [24].

In absorption, the fine structure is due to the superposition on the true absorption spectrum of the same curve translated by an amount $\hbar\omega_p$ to higher energies. The probability of observing a high-energy plasmon satellite in emission is expected to be extremely small, particularly in the soft X-ray region, because the lifetime of the inner core vacancy is long ($\sim 10^{-14}$ s) compared with the lifetime of the plasmon ($\sim 10^{-16}$ s). Within the discussion of theoretical interpretation of plasmon satellite bands and of related many-body effects in soft X-ray emission (SXE) by several investigators, the existence of a "plasmaron" contributing to the plasmon satellite band was suggested [24].

Generally, X-ray transitions in metals can be accompanied by the excitation of one or more plasmons. A plasmon can transfer its energy on decay to a conduction electron which subsequently fills the core valency with the emission of an X-ray photon. Plasmons can also be excited during Auger electron transitions. Evidence for plasmon excitation during X-ray or Auger transitions was reported by several investigators. Instead of being excited, a plasmon can also be observed during X-ray or Auger transitions if it preexists, and this gives rise to the emission of a high-energy X-ray satellite. Several authors suggested that the plasmon may be involved in the production of high-energy X-ray satellites. The most likely electrons of the atom to interact with both the plasmon and the inner hole are the valence electrons. If a valence electron is involved in the simultaneous annihilation of an inner hole and a plasmon, it may give rise to the emission of a satellite of the normal valence band line shifted to a higher energy by the amount of plasmon energy. On this basis it was of great interest to study theoretically plasma oscillations during the process of high-energy X-ray emission satellites. For Be it was found that the transition probability for plasma oscillations is greater than predicted by the Kronig-Hayasi theory. The fine structure observed at an energy distance of 20.5 eV from the parent line can be assigned to plasmon oscillations [25].

High-energy satellites observed in the X-ray emission spectra (K spectra for Li and Be, L spectrum for Na) were attributed to double-ionization of the relevant core levels and were said not to involve volume plasmons, whereas the low-energy satellites were associated with the creation of volume plasmons. For Be the main K-emission peak was observed at 108.5 eV, the low-energy satellite at 90 eV, and the high-energy satellite at 143 eV [26]. Similar events were observed in Auger spectra, see for instance [27]; cf. p. 128. In later studies the high-energy satellites observed at an energy distance of $2\hbar\omega_p$ (with $\hbar\omega_p$ = plasmon energy) on the high-energy side from the main X-ray emission line were ascribed to double plasmon satellites. The calculated relative intensity of the high-energy plasmon satellites of Li, Na, Be, and Cu were in fair agreement with observed values; see table in the paper [22].

The electron mean free path for volume plasmon excitation in Be was evaluated from the increase of the peak intensities with the primary energy, in the range 100 to 1000 eV, in reflection electron energy loss (REEL) spectra (minimum at ~70 eV); see figure in the paper [28].

An expression derived analytically for the total number of electrons participating in plasmon oscillations in metals gave for Be Z_{eff} = 1.95. Calculated plasmon energies for several metals using Z_{eff} were in fair agreement with observed characteristic electron energy losses (from literature) [29].

In early studies of X-ray plasmon scattering (using $CuK\alpha$ radiation), the spectral profile of plasmon scattering by thin polycrystalline Be slabs was found to be quite similar to that of the $K\alpha$ spectra. Also the half width of each was the same, leading to the conclusion that the spectrum of plasmon scattering is approximately of the δ-function type, as predicted by the random-phase-approximation (RPA) theory [30]. The intensity of this plasmon scattering was comparatively strong and amounted to about one-half of the nonloss scattering intensity. The observed plasmon peaks were separated from the nonloss scattering peaks by (22.5 ± 1.5) eV. The $\alpha_1\alpha_2$ separation was 19.8 eV. The results satisfied the dispersion relation very well. Using $CrK\alpha$ radiation, the plasmon peaks P_1 and P_2 (corresponding to $K\alpha_1$ and $K\alpha_2$, respectively) were separately observed at scattering angles of 10°, 15°, and 20°. The spectral width of the plasmon scattering increased with the scattering angle, and above 30° the two plasmon peaks could not be distinguished. At higher scattering angles the plasmon scattering disappeared and Compton scattering took place [10]. Follow-up studies using $CuK\beta_1$ and $CrK\beta_1$ radiation confirmed the excitation of plasma oscillations by X-rays as well as by electron impact. The X-rays suffered an energy loss equal to the plasmon energy of Be (a table comparing the plasmon energy values in Be of several investigators is given in the paper). The inelastic scattering was still observed beyond the critical angle estimated by theory. At lower scattering angles the spectral breadth of the scattering was ~4 eV corresponding to a plasmon lifetime of 1.5×10^{-16} s. With an increasing angle, the breadth rapidly broadened and the scattering disappeared gradually [31]. For plasmon excitation observed in Compton scattering experiments, see [9].

Usually the plasmon satellite is smeared out because the plasmon energy obeys a dispersion relation. The discrepancy in the form of the dispersion curve of X-ray bulk plasmons in Be has been a matter for discussion. Especially in the early studies by Tanokura et al. [10, 31] the dispersion curve of the plasmon energy (measured as the plasma loss $\Delta E(k)$) was found to be a linear function of k^2 (where k corresponds to the momentum transferred to the sample) for k values up to 2 Å^{-1} which was considered to be the critical value of the wave vector. Theoretically it was computed as $|k_c| = 1.3$ Å^{-1}. Miliotis [11] found that for $|k| \leq |k_c|$ the plasmon energy is a linear function of k^2 and the slope of this line agreed with the theoretically expected one. From there up to $|k| = 1.95$ Å^{-1} the plasmon energy increased very little, so it appeared that a break in the dispersion curve occurs around k_c. Kliewer and Raether [32] then showed theoretically, within the context of RPA, that the results [11] could be understood qualitatively. On this basis further studies were performed (using $CuK\beta$ radiation) at scattering angles 6.5°, 12°, 14°, 17.5°, and 22° (at higher angles the appearance of the Compton band introduced an uncertainty) to show the break in the plasmon dispersion curve of Be around k_c [33].

According to RPA theory, plasmons can be excited when the transferred momentum k is smaller than the critical plasmon cut-off wave vector k_c ($k < k_c$), where $k = (4\,\pi/\lambda)\sin\vartheta/2$ and $k_c = 0.47\sqrt{r_s}k_F$, where λ = wavelength of the incident radiation, ϑ = scattering angle, r_s = mean interelectron distance (in units of the Bohr radius), and k_F = Fermi wave vector. The plasmon energy $\hbar\omega_k$ depends on k and is given for $k < k_c$ by the Bohm-Pines dispersion relation $\hbar\omega_k = \hbar\omega_p\,[1+(6/5)(E_F/\hbar\omega_p)^2(k/k_F)^2]$, where $\hbar\omega_p$ (or E_p) is the plasmon energy corresponding to zero transferred momentum and E_F is the Fermi energy. Experiments on metals including Be ([11, 12]) have shown, however, that a) plasmons continued to persist as an excitation for $k > k_c$ and b) that for $k < k_c$ the plasmons follow the above dispersion relation while for $k > k_c$ they have a dispersion much smaller than that given by the relation (for Be, see figure in the paper [34]). With $k_F = 1.94$ Å^{-1}, $E_p = 18.45$ eV, and $r_s = 1.87$ a.u. (without exchange correction), the value $k_c = 1.25$ Å^{-1} was the same as that given by the equation $k_c = 0.47\sqrt{r_s}k_F$ (cf. above). After correction for the effects of exchange and correlation a new value of k_c (=1.15 Å^{-1}) was obtained from another dispersion relation (according to Vashishta, Singwi [35]) which is smaller than the value calculated without corrections. Thus the existing disagreement between theory and experiments regarding the appearance of plasmons for $k > k_c$ can not be explained, in contrast to the proposal of Srivastava [36], as due to the effect of exchange [34]. Follow-up studies of the volume plasmon dispersion in Be excited by X-rays for k up to 1.5 k_F revealed a flattening of the dispersion curve for k near k_F. For $k \approx 1.2\,k_F$ the slope of the curve became negative. These results, consistent with previous X-ray and some electron scattering experiments, could not be explained in the context of RPA theories. The dispersion curve, represented in **Fig. 14-9**, shows a linear dependence of the volume plasmon energy $E_p(k)$ with k^2 up to approximately k_c. In the region $k_c < k < k_F$ there is still a linear dependence of $E_p(k)$ with k^2, but the slope is 2.3 times smaller than in the region before and the two lines intersect at $k \approx k_c$ [13].

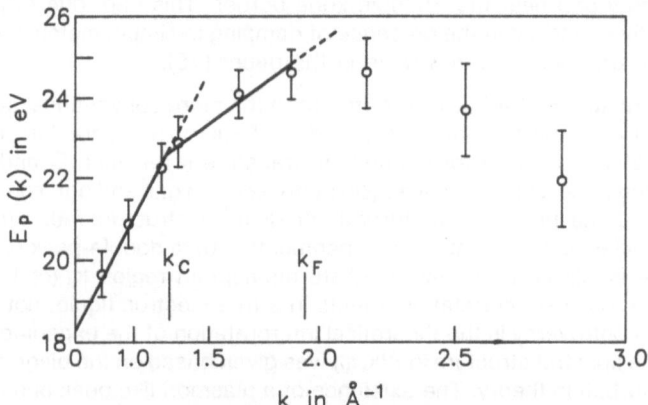

Fig. 14-9.   Dispersion curve $E_p(k)$ measured up to
1.5 k_F (quadratic scale of the wave vector k).

Plasmon dispersion was measured in polycrystalline Be at wave vectors up to k = 1.29 Å^{-1} by electron energy loss spectroscopy (EELS, 30 keV primary energy). The plasmon energy after correction was (18.8 ± 0.1) eV in good agreement with the average value of (18.8 ± 0.2) eV of previous EELS measurements. The fitted quadratic dispersion coefficient (defined by $\omega_p(k) = \omega_p(0) + \hbar/m\cdot\alpha\cdot k^2$ with m = electron mass) is $\alpha = 0.36 \pm 0.03$ for $0 < k < 1.29$ Å^{-1}. Assuming two intervals of different quadratic dispersion, the dispersion coefficient at lower k values is $\alpha_L = 0.52 \pm 0.04$ for $0 < k < 0.55$ Å^{-1}, and the dispersion coefficient at higher k values is $\alpha_H = 0.38 \pm 0.03$ for $0.55 < k < 1.29$ Å^{-1} (see figures in the paper) [37]. Previous studies gave $\alpha = 0.34 \pm 0.03$ for a strongly oriented monocrystalline sample and $\alpha = 0.42 \pm 0.02$ for polycrystalline Be [38] in agreement with earlier results [5, 6]; see also [19].

Plasmon scattering in single-crystal Be was studied using CuKα radiation and an X-ray flux from a 60 kV, 1000 mA source. The momentum-transfer region was 0.64 to 1.55 Å^{-1} with the momentum transfer along the c- or a-axis in hcp Be. Plasmon dispersion was different in the two directions. Even though the $k = 0$ plasmon frequencies were nearly identical, at momentum transfers of ~1 Å^{-1} they differed by 1 eV or ~5% with that along the c-axis being higher. In addition, the widths and shapes of the plasmon peaks differed with the general feature in that the plasmon peak along the c-axis was narrower. By comparing the data with an RPA-type theory which includes band structure, the contributions from the lattice and from the true many-body interactions to the temporarily and spatially dependent response of an interacting electron gas at metallic densities could be separated. At the largest scattering angles studied, the spectrum had a rather peculiar shape that was not understood. For details and figures, see the paper [12].

The surface plasmon dispersion relation for a metal was computed within RPA by use of a variational principle. For all metals considered (Na, Be, Mg, Al), the dispersion relation exhibits a minimum for nonzero momenta [39].

There were suggestions that plasmons in metal crystals should undergo Bragg scattering from the periodic lattice and from plasmon bands. These expectations were not confirmed by either inelastic X-ray scattering (for Be, see [11, 12]) or by EELS experiments which so far gave no evidence of Bragg splitting around G/2 (where G is a reciprocal lattice vector) though often a damped plasmon peak persisting beyond G/2 was observed. Within RPA for a weak periodic potential a theoretical evidence was found for the existence of two plasmon bands separated by a complex energy gap near the Brillouin zone border. This gap, due to plasmon Bragg diffraction, is rapidly quenched in the presence of damping in simple metals such as Be. Data for several metals are compared in a table in the paper [16].

The **dynamic structure factor** $S(k, \omega)$ is different to the more conventional structure factors measured in typical X-ray experiments. X-ray studies of $S(k, \omega)$ in Be, graphite, or Al revealed a new type of excitation spectrum for momentum transfers k between k_F and $2k_F$. The new excitation was found to be a property of a solid state electron gas and not a one-electron band structure effect [40]. The interesting features in the dynamic structure factor found for metals such as Be in the inelastic X-ray scattering experiments [40], a double-peak or one-peak-one-shoulder structure in $S(k, \omega)$ in the intermediate momentum region k_F (or k_c) $< k < 2k_F$, are commonly believed to reflect correlation effects in a free-electron liquid, not band structure effects. There is a controversy in the theoretical interpretation of the experiments, but a clear interpretation of the spectral structure in $S(k, \omega)$ was given based on the diagrammatic method of many-body perturbation theory. The existence of a plasmon-like peak and a broad peak in the intermediate momentum region was ascribed to the striking damping effect of one-electron states originating from virtual plasmon emission under the influence of strong short-range correlations at metallic densities. An anomalous plasmon dispersion around the cut-off wave number observed in electron scattering experiments was also interpreted by comparing $k_c = 0.73 \, k_F$ from RPA and $k_c = 0.68 \, k_F$ calculated from the dielectric function with the local-field factor $G(k)$ included, i.e., the magnitude of k_c is reduced by the local-field correction [41]. Previously, the experimental information on the dynamic form factor $S(k, \omega)$ [11, 12] was explained at least semiquantitatively starting with a simple Gaussian ansatz for the imaginary part of the density-density response function and using the zeroth, the first, and the third frequency moments of the density spectral function to determine the unknown k-dependent parameters. The mean field theories of an electron liquid at metallic densities (going beyond RPA) were inadequate for phenomena involving high frequencies [42].

Later inelastic X-ray scattering experiments for a wide k region to determine $S(k, \omega)$ revealed that the dispersion curve deviated from RPA predictions, especially for $k > k_c$, while for $k > k_F$ it split into two branches. The lower branch corresponding to the plasmon-like component showed a negative slope for $0.9\ k_F < k < 1.5\ k_F$ whereas for larger k values it remained nondispersive. The upper branch showed considerable dispersion and tended to become parallel to the recoil energy curve. The observed discrepancies between experimental results and RPA were attributed, according to the theory of Awa et al. [41], to short-range correlation effects and the existence of higher-order excitations such as two-pair excitation and one-pair plasmon excitation [43].

Results of further measurements of $S(k, \omega)$ of electrons in Be metal with 1 eV resolution by inelastic X-ray scattering spectroscopy (IXSS), both on single crystals for a large number of different k directions and on polycrystalline samples, were compared with model calculations beyond RPA. The overall shape of experimental $S(k, \omega)$ was understood when taking into account both short-range fluctuations (local-field correlations) and multiple excitations via a state-dependent quasi-particle lifetime. The strongly k-orientation-dependent fine structure of $S(k, \omega)$ was interpreted as due to excitation gaps which are generated by transitions to final states on Bragg planes. Also the less k-orientation-dependent fine structure of $S(k, \omega)$ appears to be due to ion-electron interaction although an electron-correlation-induced origin could not be completely ruled out [44].

Direct evidence for dynamic electron-electron Coulomb correlations in metals such as Be, generating the twin peaks of the $S(k, \omega)$ fine structure, was found by calculations [45].

It should be born in mind that the dynamic and static structure factors for the conduction-electron liquid in metals have to be carefully distinguished from the more conventional structure factors measured in typical X-ray experiments [40].

References:

[1] Ferrell, R. (Rev. Mod. Phys. **28** [1956] 308/37, 316).

[2] Nozières, P.; Pines, D. (Phys. Rev. [2] **113** [1959] 1254/67).

[3] Raether, H. (Springer Tracts Mod. Phys. **88** [1980] 1/196, 45).

[4] Powell, C. J. (Proc. Phys. Soc. [London] **76** [1960] 593/610, 596/8).

[5] Watanabe, H. (J. Phys. Soc. Japan **11** [1956] 112/9).

[6] Sueoka, O. (J. Phys. Soc. Japan **20** [1965] 2203/11).

[7] Bronshtein, I. M.; Krainskii, I. L.; Libenson, B. N. (Fiz. Tverd. Tela [Leningrad] **19** [1977] 958/63; Soviet Phys.-Solid State **19** [1977] 558/61).

[8] Swanson, N. (J. Opt. Soc. Am. **54** [1964] 1130/3).

[9] Priftis, G. (Phys. Rev. [3] B **2** [1970] 54/9).

[10] Tanokura, A.; Hirota, N.; Suzuki, T. (J. Phys. Soc. Japan **27** [1969] 515, **28** [1970] 1382).

[11] Miliotis, D. M. (Phys. Rev. [3] B **3** [1971] 701/5).

[12] Eisenberger, P.; Platzman, P. M.; Pandy, K. C. (Phys. Rev. Letters **31** [1973] 311/4).

[13] Priftis, G. D.; Boviatsis, J. (Phys. Status Solidi B **104** [1981] 673/7).

[14] Seignac, A.; Robin, S. (Solid State Commun. **19** [1976] 343/5).

[15] Bohm, D.; Pines, D. (Phys. Rev. [2] **92** [1953] 609/25); Pines, D. (Phys. Rev. [2] **92** [1953] 626/36; in: Seitz, F.; Turnbull, D.; Solid State Phys. **1** [1955] 367/450, 382, 400; Physica **26** [1960] S103/S123).

[16] Girlanda, R.; Parrinello, M.; Tosatti, E. (Rev. Phys. Letters **36** [1976] 1386/9).

[17] Toots, J.; Fowler, H. A.; Marton, L. (Phys. Rev. [2] **172** [1968] 670/6).

[18] Powell, C. J. (J. Opt. Soc. Am. **59** [1969] 738/43, **60** [1970] 214/20).

[19] Aiyama, T.; Yada, K. (J. Phys. Soc. Japan **36** [1974] 1554/62).

[20] Bakulin, E. A.; Balabanova, L. A.; Bredow, M. M.; Ostroumova, E. G.; Stepin, E. V.; Shcherbinina, V. V. (Fiz. Tverd. Tela [Leningrad] **11** [1969] 685/9, **12** [1970] 72/8; Soviet Phys.-Solid State **11** [1969] 549/52, **12** [1970] 57/61).

[21] Höchst, H.; Steiner, P.; Hüfner, S. (Phys. Letters A **60** [1977] 69/71).

[22] Srivastava, K. S.; Harsh, O. K. (Phys. Status Solidi B **104** [1981] K35/K38).

[23] Watson, L. M.; Dimond, R. K.; Fabian, D. J. (in: Fabian, D. J.; Soft X-Ray Band Spectra and the Electronic Structure of Metals and Materials, Academic, London 1968, pp. 45/58).

[24] Fabian, D. J.; Watson, L. M.; Marshall, C. A. W. (in: Ziman, J. M.; Rept. Progr. Phys. **34** Pt. 2 [1971] 601/96, 686/90).

[25] Harsh, O. K.; Kaushik, Y. D. (Indian J. Phys. B **57** [1983] 89/91).

[26] Arakawa, E. T.; Williams, M. W. (Phys. Rev. [3] B **8** [1973] 4075/8).

[27] Jenkins, L. H.; Zehner, D. M. (Solid State Commun. **12** [1973] 1149/51).

[28] Grzeszczak, A. (Z. Naturforsch. **42a** [1987] 1372/3).

[29] Goel, J. P.; Harsh, O. K.; Srivastava, K. S. (Physica B+C **144** [1987] 190/2).

[30] Ohmura, Y.; Matsudaira, N. (J. Phys. Soc. Japan **19** [1964] 1355/60).

[31] Suzuki, T.; Tanokura, A. (J. Phys. Soc. Japan **29** [1970] 972/8).

[32] Kliewer, K. L.; Raether, H. (Phys. Rev. Letters **30** [1973] 971/4).

[33] Marinos, D.; Miliotis, D. (Phys. Status Solidi B **68** [1975] K133/K135).

[34] Boviatsis, J.; Priftis, G. D. (Solid State Commun. **33** [1980] 577/9).

[35] Vashishta, P.; Singwi, K. S. (Phys. Rev. [3] B **6** [1972] 875/87).

[36] Srivastava, K. S. (Solid State Commun. **30** [1979] 19/20); Srivastava, K. S.; Harsh, O. K.; Kumar, V. (Proc. Nucl. Phys. Solid State Phys. Symp. C **21** [1978] 64/6 from C.A. **92** [1980] No. 188427).

[37] Diekmann, W.; Eickmans, J.; Otto, A. (Z. Physik B **65** [1986] 39/41).

[38] Cazaux, J.; Vilanove, R. (Compt. Rend. B **263** [1966] 460/2).

[39] Beck, D. E.; Celli, V. (Surf. Sci. **37** [1973] 48/58).

[40] Platzman, P. M.; Eisenberger, P. (Phys. Rev. Letters **33** [1974] 152/4; Solid State Commun. **14** [1974] 1/3).

[41] Awa, K.; Yasuhara, H.; Asahi, T. (Solid State Commun. **38** [1981] 1285/8; Phys. Rev. [3] B **25** [1982] 3670/86, 3687/702); Yasuhara, H. (NATO Advan. Study Inst. Ser. B **81** [1983] 411/21 from C.A. **98** [1983] No. 132553).

[42] Kalia, R. K.; Mukhopadhyay, G. (Solid State Commun. **15** [1974] 1243/7).

[43] Vradis, A.; Priftis, G. D. (Phys. Rev. [3] B **32** [1985] 3556/61).

[44] Schülke, W.; Nagasawa, H.; Mourikis, S.; Kaprolat, A. (Phys. Rev. [3] B **40** [1989] 12215/28); Schülke, W.; Bonse, U.; Nagasawa, H.; Mourikis, S.; Kaprolat, A. (Phys. Rev. Letters **59** [1987] 1361/4).

[45] Green, F.; Lowy, D. N.; Szymanski, J. (Phys. Rev. Letters **48** [1982] 638/41).

[46] Vijh, A. K. (Surf. Technol. **10** [1980] 277/82).

14.13 X-Ray Spectra and Scattering

14.13.1 X-Ray Spectra

Data on the X-ray spectra of the free atoms and of elemental Be in its standard state as well as of the ions Be^+ and Be^{2+} are given in "Beryllium" Suppl. Vol. A 1, 1986, pp. 141, 159, and 173 ff., respectively.

Soft X-ray emission and absorption (SXE, SXA) spectra have been studied on metallic Be by numerous investigators.

The Be **emission** band spectrum is easy to obtain. It shows a sharp rise in intensity at the high-energy (short wavelength) limit of the band (K edge) with a slight change in gradient about one third of the way up to the peak. This agrees with the calculated density of states (see p. 52). The hump observed on the low-energy side of the spectrum has been questioned as being real, but it does exist and also agrees with the calculated density of states. The K spectrum reflects the density of states of p symmetry, and the favorable comparison with the calculated curve suggests that the proportion of states with s and d symmetry in the occupied band region is small. The measured second order K-emission spectrum for metallic Be (from Lindsay et al. [1]) with a higher resolution than that measured in first order is shown in a figure in the paper [2, p. 640]. An interesting measurement of the Be K-emission band using fluorescence excitation was performed by Henke and Smith [3]; the spectrum did not appear to differ greatly in band width or shape from those reported by other investigators.

The intensity of the emission band, being much less on the short wavelength side than on the long wavelength side, was explained by the effect of self-absorption on the shape of the spectrum. The short-wave limit (Fermi energy) was determined from the inflection point of the sharp short-wave edge of the emission band. The energy difference between the Fermi edge and the beginning of the emission band determines the width of the filled part of the valence band [4]. The shape of the emission band was nearly the same as that described in early studies [5].

As was mentioned in "Beryllium" Suppl. Vol. A 2, 1991, p. 119, an early band calculation suggested the existence of localized bonding charges concentrated along the c-axis connecting next-nearest Be atoms. The existence of such charges was confirmed by X-ray diffraction measurements of structure amplitudes [6]. The localized charges were explained in a tight-binding LCAO approach that more occupied p_z-like states exist than p_x and p_y states and that the former overlap which results in covalent bonds. This concept was proved by measuring polarized K-emission bands emitted by single crystalline Be (for K-emission bands especially image partially p-like DOS). Measurements of polarized K spectra of single crystals increase the information content (via selection rules) and allow the distinction between p_x-, p_y-, and p_z-like states. From the polarized Be K-emission valence bands, conclusions were drawn on the existence of bonding charges. The anisotropy of the Be crystal is reflected in different π and σ bands. A figure comparing theoretical and X-ray spectroscopic data on the metallic Be valence band is given in the paper [7]. Intensities for polarized X-ray emission spectra are given by the elements of the density matrix, expanded with respect to lattice harmonics, for Be, Mg, and Zn starting from a pseudopotential calculation. The anisotropic spectra can be excellently described by located density of states. The spectra are broadened and compared with experimental K spectra [7, 8]. The comparison between measured and calculated intensities is shown in figures in the paper [9]. All structures lie at the same energy. This holds especially for Be where several structures can be seen. There are only small differences in the relative height of the peaks. There is a difference in the magnitudes of the anisotropy of the K spectra (greater than the experimental error). The ratio of the intensities at the maximum is $I_\pi/I_\sigma = 1.06$ for Be in the experimental spectrum; this ratio is 1.02 in the calculated spectrum [9].

In a study of the effect of short-range order and screening on emission bands of metals, qualitative agreement was found between the X-ray spectra of metallic Be and calculations within the framework of multiple screening theory, when only the first 3 to 5 coordination spheres were taken into account in the Be cluster calculation [10].

The Be K-emission spectrum was supposed to be highly sensitive to the effect of alloying of the constituent atoms and, therefore, to be very suitable for obtaining information on the electronic structure and binding mechanisms in Be-containing alloys. Experimental results of SXE K spectra of pure Be and some Be-Cu alloys were compared with those of electronic structure calculations performed in the cluster approximation. The density of Be 2p states changed from a single peak for pure Be to the two-peak structure when a small amount of Cu was added. Since the two peaks are not related to the two components of the alloy and originate, instead, from the same Be 2p band, the energy separation between them is practically insensitive to the concentration. The effects of disorder and of direct Be-Be interaction also have a rather small effect on the Be K-emission spectra [11].

Satellites observed in the X-ray K-emission spectra of Be (and Li) in the low-energy region were attributed to the creation of volume plasmons (see p. 86), whereas those in the high-energy region were said to result from double ionization of the relevant core levels (they do not involve volume plasmons). The K-emission spectrum for Be obtained with an electron energy of 2 keV and an electron beam current of 4.6 mA (see figure in the paper) shows the main K-emission peak at 108.5 eV, a low-energy satellite at 90 eV, and a high-energy satellite at 143 eV [12]. A plasmon band was observed in the K-emission spectrum with its edge about 18 eV below the Fermi edge of the parent band. This satellite can be clearly seen, after subtraction of the extrapolated low-energy tail of the parent band, in the Be K-emission spectrum shown in a figure in the paper [2, p. 689]. In later studies the high-energy X-ray satellites observed at an energy distance of twice the characteristic electron energy loss value from the main X-ray emission line were ascribed to double plasmon satellites. An expression was derived for the relative intensity I/I_0 of the high-energy plasmon satellites and fair agreement was obtained between the calculated and experimental data from literature for Be (Li, Na, and Cu) [13]. For plasmon bands, see also p. 99.

The widths of the conduction bands in metals such as Be were calculated by using experimental plasma-loss values. The calculated band widths are in fair agreement with those observed experimentally for K, L, and M X-ray emission [14].

For an early interpretation of SXE spectra of Be (and Li), see [15]. The K-emission bands of Be (and C) were illustrated in experimental studies using a flow proportional counter [16].

The K-**absorption** spectrum measured on a 2000 Å Be film using bremsstrahlung [4] clearly showed the Kronig fine structure. Only three broad maxima were observed between 110 and 200 eV. These maxima agreed well with theoretical calculations [17], but did not agree with the fine structure found in early studies [18] using a capillary discharge. In further studies of SXA spectra using synchrotron radiation, the K-absorption spectrum of Be films (620 and 2000 Å thick) agreed in large with the results reported by [4] except for some details; larger differences are found in the results [18]. A sharp edge at 111 Å corresponded to the Fermi surface and coincided with the short wavelength limit of the K-emission band. An extended fine structure just above the edge was characteristic of the X-ray absorption spectra. The theoretical density of states calculated by Loucks, Cutler [19], cf. p. 52, also agreed surprisingly well with the general behavior of the Be K-absorption spectrum [20]; see also [2].

The SXA spectra of Be (and BeO) near the K edge were measured using (NBS 180 MeV) synchrotron radiation, and the energy-loss spectra of 20 keV electrons transmitted through thin Be films at zero scattering angle were measured in the corresponding 100 to 170 eV energy-loss range. The K edge of Be, as observed by SXA in first and second order, is at (112.1±0.1) eV in good agreement with the energy-loss edge at (111.7±0.4) eV and with the edge position (111.5 eV) found by Lukirskii and Brytov [4] and that (111.7 eV) found by Sagawa et al. [20]. The shoulder observed in the present study and the energy-loss maximum at 118.7 eV may possibly be due to an oxide layer corresponding to the K edge of Be in BeO at 118.4 eV [21].

An improved method is described for extracting absorption information from SXE spectra using a Gaussian expression proposed for the ionization as a function of depth for an electron beam in an X-ray target. Quantitative agreement was found with attenuation coefficients measured directly. However, for evaluating the absorption coefficient from self-absorption data in this study, Mg proved to be more suitable than Be or Li where self-absorption spectra run out of information at the high-energy side of the K edge and the edge is not precisely determined. Calculated parameters for the intensity of X-ray production as a function of mass depth (at 1 and 3 kV) for several metals including Be are compared in a table in the paper [22].

Nonthermal ultrasoft X-ray spectra were studied by thin-foil differential transmission measurements of ultrasoft X-rays (on ZT-40M at Los Alamos Laboratory). Time histories of the ratio of intensities seen through Be foils were in reasonable agreement with Thomson scattering measurements 3 ms into the discharge, but the intensity vs. foil thickness showed evidence of pronounced nonthermal characteristics. Predictions based on the theory of emission for a runaway electron distribution function were capable of reproducing the observed behavior [23].

Soft X-ray or extreme UV (XUV) radiation is produced, besides vacuum UV (or VUV) radiation, both as line emission as well as continua by laser-produced plasmas. Generally, the emission extends over a considerable range in the wavelength, but its peak is related to the energy of the laser installation used to produce the plasma. Such XUV radiation is used for several applications, e.g., X-ray diffraction or VUV absorption studies, with the advantage of the laser-produced plasma as XUV light source. The emission spectrum of a Be plasma in the range $\lambda = 20$ to 80 Å is shown in a figure and discussed in the paper [24]. As an example of experimental performance, the absorption spectra, both discrete and continuous, of the 1s electron of the Be IV, III, II, I ions obtained with two laser-produced plasmas are shown together with the corresponding computed photoionization cross-section spectra in the XUV range (between 45 and 110 Å). The 1s electron is an inner electron in the Be I and II ions, but it is an optical electron in the Be III and IV ions. The spectra show clearly a discrete series of lines followed by the photoionization jump and the continuum spectrum. For further details and discussions, see the paper [25].

Photoabsorption was studied in the XUV range from the onset of K absorption up to photon energies of 170 eV, and the K absorption of Be (at room temperature) was found very similar to that of Li. At the onset, the K absorption of Be had a prominent peak and a broad structured hump extending from 115 to 130 eV. The gross structure of the absorption spectrum agreed with the structure of the transmission curves given by Sagawa et al. [20] and by Swanson and Codling [21]. The positions of the absorption peaks tabulated in [21] were in good agreement with the present results, as well as the comparison of [20] with the theoretical [19] density of states curve [26].

The **X-ray photoelectron** spectroscopy (XPS) is used to measure the 1s spectrum of Be metal (evaporated in situ on Au) and the plasmons accompanying it. The analysis, primarily of the 1s line shape, yielded a value for the singularity index $\alpha = 0.06$ which is smaller than $\alpha = 0.13$ predicted theoretically [27] using a free-electron model. Comparing measured and calculated plasmon energies, Be was found not to be a free-electron metal. The strength of the hole-conduction-electron coupling, as given by the singularity index α, is strongly influenced by band structure effects. An analysis of the creation rate for intrinsic and extrinsic plasmons in Be revealed that they were of about equal magnitude; i.e., the plasmons were created with about equal probability by direct energy loss and by conduction-electron-hole coupling [28]. The energy of photoemitted electrons measures the difference between the initial (ground state) energy and a final state energy which may contain contributions from highly excited states. For using this technique, an exact analysis of the excited state is mandatory. The plasmon structure in XPS experiments on Be (Na, Mg, and Al) can be analyzed from a superposition of intrinsic and extrinsic plasmons. In each case, intrinsic and extrinsic plasmons were present and the creation rate for intrinsic plasmons could be described by a simple power law. Assuming that the extrinsic rate is constant for adjacent plasmon peaks, the probability of intrinsic plasmon creation is proportional to b^n ($b_n = b^n$, where b is the creation rate for a single plasmon and n is the plasmon number), deviating from a Poisson distribution [29].

The **appearance potential** spectroscopy (APS) is used to study the unoccupied part of the valence band of Be, Mg, and Al. The samples were prepared in situ ($\sim 10^{-10}$ Torr) by evaporation from a W wire and bombarded with monochromatic electrons. Usually the first derivative of the total X-ray yield is measured. All of the spectra were characterized by a step which occurs at an electron energy equal to the binding energy of a core electron. A comparison revealed a strong correlation between APS and SXA and very good agreement in the results for Be. The onset at 112 eV was ascribed to excitation from the K core level. APS peaks were observed at 2, 7, 10, and 13 eV, whereas the SXA K spectrum [26] gave 2, 7, 10, and 14 eV. There is also a good correlation to the density of states [19]; the slow onset in the APS spectrum is connected with a density of states minimum close to the Fermi level. The edge singularity considered theoretically shows up in SXA and SXE as sharp peaks at the Fermi edge (see [26]). The peak observed in APS at the onset for all three metals studied can be explained for Be, at least partially, by a density-of-states effect [19]. This seems, however, not to be the case for Mg and Al [30]. Based on a pseudopotential calculation for metallic Be the APS is calculated (in addition to the K-emission spectra of different polarizations and Compton profiles). The APS is proportional to the derivative of the self-convolution of the density of the unoccupied states and compared with the experimental APS [30]. The experimental APS is shifted by 1.5 eV to higher energies. Besides this shift there is a good correlation between the expression evaluated with the density of states of Be [31] and the experiment concerning the position of the two maxima and the minimum; the linear shape in the last part of the experimental curve is also well described. The slow onset of the spectrum is due to a minimum of the density of states near the Fermi level which is further diminished by the convolution [32].

Comparing the plasmon satellite structure in APS and XPS spectra of Be and Li, the plasmon coupling constant in APS seemed to be smaller than in corresponding XPS experiments. The observed differences were concluded to be primarily intrinsic in origin, but no simple explanation was found for this result [33].

References:

[1] Lindsay, G. M.; Watson, L. M.; Fabian, D. J. (unpublished results) cited in [2].
[2] Fabian, D. J.; Watson, L. M.; Marshall, C. A. W. (in: Ziman, J. M.; Rept. Progr. Phys. **34** Pt. 2 [1971] 601/96, 639/41).
[3] Henke, B. L.; Smith, E. N. (J. Appl. Phys. **37** [1966] 922/3).
[4] Lukirskii, A. P.; Brytov, I. A. (Fiz. Tverd. Tela [Leningrad] **6** [1964] 43/53; Soviet Phys.-Solid State **6** [1964] 33/41).
[5] Skinner, H. W. B. (Phil. Trans. Roy. Soc. [London] A **239** [1940] 95/134, 127/9).
[6] Brown, P. J. (Phil. Mag. [8] **26** [1972] 1377/94).

[7] Dräger, G.; Brümmer, O. (Phys. Status Solidi B **78** [1976] 729/35).
[8] Dräger, G.; Brümmer, O.; Bonitz, J. (Phys. Status Solidi B **94** [1979] K111/K114).
[9] Rennert, P.; Schelle, H.; Gläser, U. H. (Phys. Status Solidi B **121** [1984] 673/84).
[10] Polozhentsev, E. V.; Sachenko, V. P.; Kovtun, A. P. (Izv. Akad. Nauk SSSR Ser. Fiz. **40** [1976] 240/2; Bull. Acad. Sci. [USSR] Phys. Ser. **40** No. 2 [1976] 14/6).

[11] Kozlenkov, A. I.; Shulgin, A. I.; Postnikov, A. V.; Kurmaev, E. Z.; Ivanovskii, A. I.; Gubanov, V. A. (J. Phys. C **18** [1985] 3581/9).
[12] Arakawa, E. T.; Williams, M. W. (Phys. Rev. [3] B **8** [1973] 4075/8).
[13] Srivastava, K. S.; Harsh, O. K. (Phys. Status Solidi B **104** [1981] K35/K38).
[14] Srivastava, K. S. (J. Electron Spectrosc. Relat. Phenom. **20** [1980] 319/22 from C.A. **93** [1980] No. 194550).
[15] Catterall, J. A.; Trotter, J. (Phil. Mag. [8] **3** [1958] 1424/31).
[16] Holliday, J. E. (Rev. Sci. Instrum. **31** [1960] 891/5).
[17] Kozlenkov, A. I. (Izv. Akad. Nauk SSSR Ser. Fiz. **27** [1963] 364/77; Bull. Acad. Sci. [USSR] Phys. Ser. **27** [1963] 373/86).
[18] Johnston, R. W.; Tomboulian, D. H. (Phys. Rev. [2] **94** [1954] 1585/9).
[19] Loucks, T. L.; Cutler, P. H. (Phys. Rev. [2] **133** [1964] A819/A829).
[20] Sagawa, T.; Iguchi, Y.; Sasanuma, M.; Ejiri, A.; Fujiwara, S.; Yokota, M.; Yamaguchi, S.; Nakamura, M.; Sasaki, T.; Oshio, T. (J. Phys. Soc. Japan **21** [1966] 2602/10).

[21] Swanson, N.; Codling, K. (J. Opt. Soc. Am. **58** [1968] 1192/4).
[22] Crisp, R. S. (J. Phys. F **13** [1983] 1325/32).
[23] Watt, R. G.; Thomas, K. S.; Little, E. M. (LA-9588 MS [1982] 1/13 from C.A. **99** [1983] No. 30302).
[24] Nicolosi, P.; Jannitti, E.; Tondello, G. (Appl. Phys. B **26** [1981] 117/24).
[25] Jannitti, E.; Nicolosi, P.; Tondello, G. (Phys. Scr. **36** [1987] 93/8).
[26] Haensel, R.; Keitel, G.; Sonntag, B.; Kunz, C.; Schreiber, P. (Phys. Status Solidi A **2** [1970] 85/90).
[27] Minnhagen, P. (Phys. Letters A **56** [1976] 327/9).
[28] Höchst, H.; Steiner, P.; Hüfner, S. (Phys. Letters A **60** [1977] 69/71).
[29] Steiner, P.; Höchst, H.; Hüfner, S. (Phys. Letters A **60** [1977] 410/2).
[30] Nilsson, P.-O.; Kanski, J. (Surf. Sci. **37** [1973] 700/7).

[31] Taut, M. (Phys. Status Solidi B **54** [1972] 149/57).
[32] Rennert, P.; Dörre, T.; Gläser, U. (Phys. Status Solidi B **87** [1978] 221/6).
[33] Bradshaw, A. M.; Wyrobisch, W. (J. Electron Spectrosc. Relat. Phenom. **7** [1975] 45/53).

14.13.2 Photon Cross Section. Mass Attenuation Coefficients

Some fundamental information and comprehensive tabulations of total photon cross sections and mass attenuation coefficients in extended energy regions (with allocated bibliography) are given in "Beryllium" Suppl. Vol. A 1, 1986, pp. 154/7.

Beryllium absorbs very little X-rays, γ-rays, electron or other electromagnetic radiation. Therefore it is used in X-ray tubes as a window through which the X-rays pass. The transmitted X-ray intensity is described by $I = I_0 exp(-\mu/\rho)x\rho$, where I is the intensity at thickness x, I_0 is the initial intensity, ρ is the density, and μ/ρ is the mass absorption (or attenuation) coefficient, which not only depends upon the absorbing material, but also upon the X-ray wavelength. For a 0.010 in (0.0254 cm) thick Be window and CuKα radiation is $\mu/\rho=1.007$ cm²/g and the transmitted intensity 95% (the energy which is not transmitted is converted to heat). The advantage of using Be is obvious by comparing it with other materials such as Al or Ti [1].

In early studies the total X-ray attenuation coefficients for a series of metals including Be were measured in the range E = 40 to 412 keV. The uncertainties in the values due to impurities, statistics, etc., were expected to be less than 2%. Selected values for Be are given in the table from [2]:

E in keV	40.12	45.40	50.00	57.52	67.87
μ/ρ in cm²/g	0.169	0.162	0.154	0.153	0.144

E in keV	86.54	105	158	208	412
μ/ρ in cm²/g	0.137	0.127	0.119	0.109	0.0828

Experimental mass absorption coefficients for low-energy photons in the range E = 13.43 to 59.6 keV in high-purity Be (>99.99%) are given together with theoretical values from Hubbell [3] in the following table (with errors in the range 0.002 to 0.005) according to Nathuram et al. [4]:

E in keV	13.432	17.218	20.163	46.520	59.20
$(\mu/\rho)_{exp}$ in cm²/g	0.341	0.258	0.225	0.154	0.127
$(\mu/\rho)_{th}$ in cm²/g	0.35	0.24	0.22	0.15	0.14

The value at 46.52 keV is comparable to that found by Hsu and Dowdy [5] in their comparison of theoretical and calculated γ-ray attenuation coefficients (in the range 0.04 to 15 MeV) for several metals including Be.

The effect of scattered photons on attenuation coefficient measurements was studied using the Monte Carlo method. The relative deviation $\Delta\mu/\mu$ in the attenuation coefficient due to scattered photons was calculated for small collimation angles from 0.4° to 6°, photon energies in the 8 to 60 keV region, and elements up to atomic number 13. The $\Delta\mu/\mu$ can not be ignored in some experimental situations. The effect of scattered photons can not generally be accounted for by a calibration procedure. It can be minimized by placing the material as far as possible from the detector or by using a multihole collimator on the detector. Figures of $\Delta\mu/\mu$ for Be as functions of the collimation angle and 8 keV X-rays, of the photon beam energy at several collimation angles, of the photon beam collimation angle for several photon beam energies and distances between detector and material, and of the material thickness are given in the paper [6].

For measurements of the continuous photoelectric absorption cross section for Be for photon energies extending from the K edge (at 110 eV) to 180 eV, see Peterson et al. [7].

References:

[1] Marder, J. M. (J. Mater. Energy Syst. **8** No. 1 [1986] 17/26, 21; C.A. **105** [1986] No. 138276).
[2] Wiedenbeck, M. (Phys. Rev. [2] **126** [1962] 1009/10).
[3] Hubbell, J. H. (NSRDS-NBS-29 [1969] 1/80, 14, 42, 78/80).
[4] Nathuram, I. S.; Rao, S.; Mehta, M. K. (Indian J. Phys. A **58** [1984] 300/4).
[5] Hsu, H.-H.; Dowdy, E. J. (Nucl. Instrum. Methods **204** [1983] 505/9).
[6] Gayer, A.; Bukshpan, S. (Nucl. Instrum. Methods Phys. Res. A **220** [1984] 525/30).
[7] Peterson, T. J., Jr.; McGuire, E. J.; Tomboulian, D. H. (Phys. Rev. [2] **129** [1963] 674/7).

14.13.3 X-Ray Scattering

Several interesting phenomena arise in X-ray inelastic scattering by free electrons in a solid with the decrease of the scattering angle. These phenomena involve excitation of longitudinal plasma oscillations arising from Coulomb interactions among the electrons. Plasmon scattering is a combination process in which the frequency of the scattered quanta is smaller than that of the incident quanta by the frequency of the plasma oscillations. In addition to Rayleigh (R), Compton (C), Raman (RS) bands, and two bonding and antibonding bands (b_1, b_2), two more bands were observed in X-ray inelastic scattering spectra of polycrystalline Be (using CuKβ radiation), designated as P_1 and P_2 and attributed to first and second order plasmons. These bands are due to the energy loss of the primary radiation creating collective oscillations of the electron gas in the metal. The position of the plasmon peaks and the corresponding energy loss, E_p, of the primary radiation are tabulated assuming that the CuKβ coherently scattered primary radiation is located in the spectrum at an energy loss equal to zero [1]:

scattering angle	50°	55°	57°	77°
E_{p1} in eV	21	22	21	22
E_{p2} in eV	43	43	42.5	44

It was suggested that for momentum transfer (k) much higher than the critical value (k_c) the dispersion of the plasmon energy has ceased completely. At the energy positions of the spectrum where the plasmon peaks were observed, none of the existing elements give any fluorescence band. The two bonding–antibonding bands (b_1, b_2) can be clearly distinguished at low scattering angles (50°) with an energy separation of $\sim$6 eV. At higher scattering angles (77°) both bands merge into one; see figures in the paper [1]. In follow-up studies (using CuKα radiation) plasmon bands were observed for $k > k_c$ at scattering angles of 47° and 50° at about the same position where the theory provides for small scattering angles $\vartheta > \vartheta_c = 16°$. The plasmon band produced by Kα_2 is due to loss energy of the primary radiation and the shift of the plasmon band is (21 ± 1) eV from the position of Kα_2. The plasmon band produced by Kα_1 is not observed because of overlapping with the intense Kα_2 band. According to the random phase approximation (RPA), the plasmons do not decay into excitations when the momentum transfer (k) is smaller than the critical value $k_c = \omega_p / v_F$ (where ω_p is the frequency of the plasmon oscillations and v_F is the Fermi velocity). For the two scattering angles 47° and 50° the values of k are 32 and 34 nm^{-1}, respectively. For $k > k_c$ the plasmons decay rapidly by Landau damping into electrons and holes. This means a considerable broadening and disappearance of the plasmon band. At the long-wavelength region of the Raman bands R_1, R_2, four peaks of bonding–antibonding bands b_1, b_2, b_1', b_2' were found and assigned to the $3sp_xp_y$-$3sp_z$ splitting of the 3ps hybridized band. The values of energy shift, ΔE (in eV), from the position of the primary Kα_1 line are given in the table [2]:

scattering angle	R_1	R_2	b_1	b_2	b_1'	b_2'
47.0°	121	142	156	161	176	181
50.0°	122	142	156	162	175	181

The existence of the b_1', b_2' bands about 20 eV from b_1 and b_2 were predicted in a previous fundamental study [3].

X-ray scattering factors for Be were calculated from approximate Wannier functions constructed from 2s and 2p atomic wave functions plus Schmidt orthogonalization. They gave a useful representation of the Bragg reflection intensities (except at the first reciprocal lattice vector). The good agreement between the calculations and experimental scattering factors justified the model used and the approximations made [4].

As a follow-up paper to the compilation of low-energy X-ray interaction coefficients [5], f_1 and f_2 parameters of the atomic scattering factor are given for Be (and 93 other elements) at 125 values of photon energy in the range 100 to 2000 eV and wavelength in the range 124 to 6.2 Å. The approximate K-absorption edge position was identified within the table. The f_1 and f_2 parameters, the values of which have been interpolated at regular intervals, can be used to calculate the low-energy X-ray interactions (absorption, scattering, specular and Bragg reflection) [6].

In the study of the influence of electronic correlations on the scattering of X-rays and fast electrons by Be and some other materials it is necessary to distinguish between coherent and elastic differential cross sections (which are related to the mean values of one-electron operators) and incoherent and inelastic cross sections (which are given by the expectation values of two-electron operators). For the first two quantities the influence of electronic correlations is small but not negligible. The second quantities are much more sensitive to correlation effects. It was found necessary to take into account the electron correlations to compare theory and experiment [7].

Mean square amplitudes of thermal motion were obtained from an X-ray diffraction measurement on a Be single crystal (AgKα radiation) by comparison with theoretical atomic scattering factors. Close agreement with the theoretical scattering factors of the free atom was obtained with the thermal parameters $B_{11} = 0.46$ Å^2 and $B_{33} = 0.415$ Å^2 (each ± 0.01 Å^2) which are almost identical with neutron results. This perfect agreement between X-ray and neutron data indicated that the free-atom calculation is also valid for the 1s core in Be metal, and that there is no need to introduce core expansion on these grounds [8].

References:

[1] Papademetriou, D. K.; Miliotis, D. M. (Solid State Commun. **48** [1983] 799/801).
[2] Papademetriou, D. K.; Katsanos, D.; Doukas, A. G. (Phys. Status Solidi B **133** [1986] 223/8).
[3] Papademetriou, D. K.; Miliotis, D. M. (J. Phys. Soc. Japan **51** [1982] 2966/72).
[4] Matthai, C. C.; Grout, P. J.; March, N. H. (J. Phys. F **10** [1980] 1621/6; C.A. **93** [1980] No. 195823).
[5] Henke, B. L.; Lee, P.; Tanaka, T. J.; Shimabukuro, R. L.; Fujikawa, B. K. (At. Data Nucl. Data Tables **27** [1982] 1/144, 7, 25, 141/4).
[6] Henke, B. L.; Lee, P.; Tanaka, T. J.; Shimabukuro, R. L.; Fujikawa, B. K. (AIP Conf. Proc. **75** [1981] 340/88, 341/2; C.A. **95** [1981] No. 228450).
[7] Naon, M.; Cornille, M. (J. Phys. B **6** [1973] 1347/56).
[8] Manninen, S.; Suortti, P. (Phil. Mag. [8] B **40** [1979] 199/207).

14.13.4 Raman Scattering

Early spectroscopic analyses of X-ray beams scattered from some elements with a low atomic number verified the existence of Rayleigh (unchanged frequency) and Compton scattering but not the phenomenon of the partial absorption of photons of the Smekal-Raman type. Based on calculations of the modified lines of Smekal-Raman type for carbon as the scatterer (Cu and Mo radiation), the modified lines of Smekal-Raman type could be observed in the calculated positions in scattering experiments on Be (and graphite) using Kα and Kβ radiation of Cu or Mo. For Be as the scatterer the modified line appeared (in addition to the

Rayleigh lines and a strong Compton band) at a 116 eV-gap on the low-energy side of the primary Kα of Cu or Mo [1]. Further studies on Be and other light elements showed that at low-angle scattering the Raman scattering appeared as a band, not as a line. The shape of the band is quite similar to the spectrum of the K absorption of the same material, and the wavelength of the critical edge of the band shifts from the incident X-ray wavelength corresponding to the lowest excitation energy of the K electron of the scatterers. The Raman band almost always overlapped the Compton band, except for low-angle scattering. The shape of the whole spectrum consisting of Rayleigh, Compton, and Raman bands depends upon the scattering angle and the element. Spectra of Be at scattering angles of 30°, 60°, 80°, and 120° show that the Rayleigh and Raman peak positions do not change, but the Compton peak position depends upon the scattering angle. The intensity of Raman scattering was confirmed as dependent upon the scattering angle as predicted by theory. Without knowledge of this angular dependence of the scattering intensity, the Raman scattering overlapped by Compton scattering was not recognized previously. The intensity of Raman scattering is too weak to be observed at lower scattering angles where Raman scattering can be separated from Compton scattering, while at higher angles ($>60°$) the Compton scattering completely overlaps the Raman scattering (or vice versa) while the Rayleigh scattering becomes weaker [2]. Follow-up studies on Be (Li, B, and C) using CrKα radiation confirmed the results [2]: the Raman band could be distinctly observed except when the Compton band overlapped it. The shape of the Raman band was similar to that of the K-absorption spectrum of the solids. The short-wavelength edge of the Raman band is displaced from the Rayleigh scattering peak, showing an energy loss corresponding to the K-electron binding energy. For Be the Raman band is observed at 30° and 60°; at 90° overlapping of the Compton band begins. The energy loss denoted by the edge of the Raman band is about 112 eV [3].

Based on the assumption that X-ray Raman scattering can be understood in terms of excitations from discrete ground state levels to an energy band with the Fermi level as bottom, the quantum-mechanical theory of X-ray Raman scattering was developed (employing the dipole approximation) and checked with experimental observations on Be by Mizuno, Ohmura [4]. The theory was generalized by Babushkin [5] who introduced a hydrogenic type of wave functions in order to describe the K-shell electrons. It was concluded that the angular dependence of the scattering intensity upon the scattering angle, ϑ, has a Compton-like profile with a finite amplitude at $\vartheta = 0$, in contrast to [4]. Further studies were concerned with the discrepancy in the angular dependence of X-ray Raman scattering intensity from polycrystals between calculations using the dipole approximation and those calculations taking electromagnetic interaction completely into account. The discrepancy was found to be due to the way the polycrystalline limit is considered. Application to polycrystalline Be showed the results [6] to be closer to the findings [2], whereas Babushkin's results [5] may be applicable to a perfect crystal as scatterer.

The energy resolution (~ 20 eV) in early X-ray scattering experiments was so poor that it was almost impossible to compare the characteristics of the X-ray Raman spectrum, either with near-edge fine structure measurements, or with the overall shape of the highly accurate absorption experiments (~ 0.1 eV resolution), made using synchrotron radiation. In an X-ray inelastic scattering experiment with 0.8 eV resolution, the near-edge fine structure of the X-ray Raman spectra of Be (Li and graphite) was measured, and the Raman spectrum was confirmed to be equivalent to the corresponding soft X-ray absorption spectrum. The Be Raman spectra exhibited an anisotropy (see figures in the paper), in contrast to Li spectra which were independent of the q-direction (the X-ray Raman spectrum is identical to the dynamical structure factor $S(\vec{q}, \omega)$ of the inner-shell electrons for $qa < 1$, where $\vec{q}$ is the transferred momentum, $\hbar\omega$ is the energy transfer, and a is the orbital radius of the inner-shell electrons); for details see the paper [7].

The angular dependence of X-ray Raman scattering bands of polycrystalline Be (99.99% pure) was studied at scattering angles in the range 51.5° to 81° using $CuK\alpha_1$, $K\alpha_2$ radiation and a flat single-crystal spectrometer. Definitely detected were the Raman scattering bands resolving the RS_1 and RS_2 bands originally found by Das Gupta [8] and examined by many investigators. The two separated scattering bands, with an energy shift of (120 ± 1) eV from the position of the corresponding coherently scattered radiation, exist at all scattering angles used. The position of the RS_1 and RS_2 peaks is independent of the scattering angle, ϑ, in contrast to the behavior of the Compton peaks. The full widths at half maximum of the RS_1 band are the same for all scattering angles. The integrated intensities of the Raman bands are proportional to $(1 \pm \cos^2\vartheta)\sin^2\vartheta/2$, as far as the lower scattering angles are concerned. At these angles the intensities of the Raman bands agree well to the theoretical curve [4] deduced from the dipole approximation. At the long wavelength side of the Raman bands, particularly at lower scattering angles, there appeared two other small bands beyond RS_1 and RS_2 (see figure in the paper), named b_1 and b_2 with an energy separation of about 6 eV [9]. In further X-ray spectra taken at scattering angles of 50°, 55°, 77°, and 83° (using $CuK\beta$ radiation), the following bands could be distinguished (see figures in the paper): Rayleigh (R), Compton (C), Raman (RS), plasmons (P_1 and P_2), bonding (b_1), and antibonding (b_2) [10].

An ab initio UHF cluster calculation was performed to investigate the bonding character of the Be electronic structure and to identify the "anomalous" peaks observed in the X-ray Raman spectrum. The long wavelength limit of this spectrum of polycrystalline Be (experimental energy range 1 to 180 eV) shows the two anomalous absorption bands similar to those observed for graphite which were suggested to be due to the energy separation between bonding and antibonding orbitals. The splitting between bonding and antibonding states is also a characteristic feature in the Be electronic structure which is, in part, reflected in the X-ray Raman spectrum, and the "anomalous" peaks were attributed to the large $3s3p_zp_x$-$3s3p_yp_x$ splitting [11]. A similar $2s2p_zp_x$-$2s2p_yp_x$ splitting was found to be responsible for the dip near the Fermi level, cf. p. 53. A comparison of the bonding characteristics and the inner core 1s binding energy of a monolayer Be film und of bulk metal revealed that the "anomalous" peaks observed in the bulk Be X-ray Raman spectrum are not present in the theoretical X-ray Raman spectrum of the thin film in accordance with the predicted origin of these peaks as due to the large $3sp_xp_y$-$3sp_z$ splitting of the 3sp band. This type of splitting is not present in the thin films, and as a consequence, the X-ray Raman spectrum of a Be film is predicted not to show any anomalous peaks similar to those observed in bulk metal [12].

References:

 [1] Das Gupta, K. (Phys. Rev. Letters **3** [1959] 38/40; Nature **166** [1950] 563/4).
 [2] Suzuki, T. (J. Phys. Soc. Japan **22** [1967] 1139/50, **21** [1966] 2087).
 [3] Suzuki, T.; Kishimoto, T.; Kaji, T.; Suzuki, T. (J. Phys. Soc. Japan **29** [1970] 730/6).
 [4] Mizuno, Y.; Ohmura, Y. (J. Phys. Soc. Japan **22** [1967] 445/9).
 [5] Babushkin, F. A. (Acta Phys. Polon. A **40** [1971] 183/7; Izv. Vyssh. Uchebn. Zaved. Fiz. No. 9 [1971] 26/8; Soviet Phys. J. **14** [1971] 1182/4).
 [6] Andriotis, A. N.; Londos, C. A. (Solid State Commun. **49** [1984] 213/6).
 [7] Nagasawa, H. (J. Phys. Colloq. [Paris] **48** [1987] C9-863/C9-866).
 [8] Das Gupta, K. (Phys. Rev. [2] **128** [1962] 2181/8).
 [9] Papademitriou, D. K.; Miliotis, D. M. (J. Phys. Soc. Japan **51** [1982] 2966/72).
[10] Papademitriou, D.; Miliotis, D. (Solid State Commun. **48** [1983] 799/801).

[11] Zdetsis, A. D.; Miliotis, D. (Solid State Commun. **42** [1982] 227/30).
[12] Zdetsis, A. D. (J. Phys. Colloq. [Paris] **48** [1987] C9-839/C9-842).

14.13.5 Compton Scattering

Scattering from a single (localized) electron in a potential well created self-consistently by other electrons and ion cores consists of two parts, the Raman and the Compton scattering. Raman scattering is discrete. It consists of the excitation of the electron to the next excited state in the well and is believed to be the analog of the plasmon band. This feature of the spectrum has no dispersion, i.e., it is strictly a local excitation. Compton scattering consists of the excitation of the electron to the continuum. The spectrum in this case is a broad smear characterized by the Doppler shift associated with the motion of the electron in its ground state [1].

Numerous Compton scattering experiments have been performed with Be using X-rays and γ-rays in order to obtain information on the electron momentum distribution, see p. 67.

The Compton intensity profile of valence electrons is partially superimposed by the profile of the core electrons. Since the scattered intensity from bound electrons is zero for $h\omega < E_b$ (where ω is the frequency difference between the incident and scattered wave and E_b is the binding energy of the core electrons), the scattering angle can be chosen in such a manner that one branch of the parabola is not influenced by the intensity scattered from the core electrons. Compton profiles (CPs) obtained from early measurements of the intensity profile of the Compton band on Be single crystals for the (wave vector) direction $k\|[10\cdot0]$ and $k\|[00\cdot1]$, shown in a figure in the paper, revealed two important features. They were deviations of the valence-electron part of the Compton curve from a free-electron parabola beyond the statistical error limit and some intrinsic differences between the two orientations with respect to the valence-electron parabola. The deviations and the orientation dependence could be qualitatively understood from the shape of the Be Fermi surface (cf. p. 58) in the extended zone scheme (as calculated by Loucks and Cutler [2]). The maximum of the Compton band was found at a value of $h\omega$ smaller by 2.9 eV than that predicted by the theory for free electrons. The separated CP of bound electrons began at $h\omega = 113$ eV which is in good agreement with the ionization energy of K electrons of Be (111.8 eV from literature). A sharp line as found in [3] was not observed at this position. A clear maximum of scattered intensity from K electrons for larger $h\omega$ was assumed to arise from a maximum of the density of states above the Fermi surface. In this case the K-electron curve of the Compton scattered intensity should be similar to the structure of the K-absorption spectrum [4].

In a previous study of the double Compton effect on Be and other scatterers of thickness varying from 40 to 400 mg/cm², the double Compton cross section, integrated over the energy range 80 to 530 keV, was estimated to be 3×10^{-3} of the single Compton cross section. For the coincidence rate per recorded quantum a value of 0.4×10^{-4} was found [5].

The long tails observed in early experimental profiles [6, 7] were attributed to the effect of multiple Compton scattering which had not been taken into account in the studies before. The influence of multiple scattering on X-ray Compton data was confirmed by Phillips and Chin [8]. From that time on experimental profiles were corrected for this systematic error. However, the results [7] concerning the anisotropy of CPs of Be were in good agreement with CPs calculated in momentum approximation for the three directions $\Delta(\Gamma A)$, $\Sigma(\Gamma M)$, and $\Pi(\Gamma K)$ [9].

The Compton scattering experiments of Loupias et al. [10] using 10 keV synchrotron radiation and measuring seven directional CPs and the experiments of Hansen et al. [11] using 412 keV γ-rays from a ^{198}Au source and measuring CPs along $\langle0001\rangle$, $\langle10\bar{1}0\rangle$, and $\langle11\bar{2}0\rangle$ of Be are looked upon as the most reliable, and good agreement was found with theoretical CPs from an accurate HF computation [12].

For Compton scattering in connection with plasmon excitation and scattering, see Priftis [13], Tanokura et al. [14]; cf. p. 86.

Using the Coulomb field of a nucleus as the photon source (with the target materials Be, Al, C, Pb, Fe, and Cu), the elastic scattering of π^--meson on the photons was studied at 40 GeV/c in the region of small four-momentum transfer. The measured differential cross sections were in agreement with theoretical calculations. The dependence of the Compton effect cross section on the π^--meson vs. the incident photon energy was obtained in the range 120 to 600 MeV in the pion rest frames; figures are given in the paper [15].

References:

[1] Platzman, P. M.; Eisenberger, P. (Phys. Rev. Letters **33** [1974] 152/4).
[2] Loucks, T. L.; Cutler, P. H. (Phys. Rev. [2] **133** [1964] A 819/A 829).
[3] Das Gupta, K. (Phys. Rev. [2] **128** [1962] 2181/8).
[4] Schülke, W.; Berg, U. (Phys. Status Solidi **23** [1967] K 87/K 91).
[5] Cavanagh, P. E. (Phys. Rev. [2] **87** [1952] 1131).
[6] Phillips, W. C.; Weiss, R. J. (Phys. Rev. [2] **171** [1968] 790/800).
[7] Currat, R.; DeCicco, P. D.; Kaplow, R. (Phys. Rev. [3] B **3** [1971] 243/51).
[8] Phillips, W. C.; Chin, A. K. (Phil. Mag. [8] **27** [1973] 87/93).
[9] Rennert, P.; Dörre, T.; Gläser, U. (Phys. Status Solidi B **87** [1978] 221/6).
[10] Loupias, G.; Petiau, J.; Issolah, A.; Schneider, M. (Phys. Status Solidi B **102** [1980] 79/95).

[11] Hansen, N. K.; Pattison, P.; Schneider, J. R. (Z. Physik B **35** [1979] 215/25; Hahn-Meitner-Inst. Kernforsch. Berlin Ber. HMI-B No. 310 [1979] 69/76 from C. A. **93** [1980] No. 155938).
[12] Dovesi, R.; Pisani, C.; Ricca, F.; Roetti, C. (Z. Physik B **47** [1982] 19/26).
[13] Priftis, G. (Phys. Rev. [3] B **2** [1970] 54/9).
[14] Tanokura, A.; Hirota, N.; Suzuki, T. (J. Phys. Soc. Japan **27** [1969] 515, **28** [1970] 1382).
[15] Antipov, Yu. M.; Batarin, V. A.; Bessubov, V. A.; Budanov, N. P.; Gorin, Yu. P.; Denisov, S. P.; Klimenko, S. V.; Kotov, I. V.; Lebedev, A. A.; Petrukhin, A. I.; Polovnikov, S. A.; Roinishvili, V. N.; Stoyanova, D. A. (Z. Physik C **24** [1984] 39/44).

15 Optical Properties. Emission and Impact Phenomena

15.1 Raman Effect

In hcp beryllium with two atoms per unit cell and the space group D_{6h}^4–$P6_3$/mmc, No. 194 (see "Beryllium" Suppl. Vol. A 2, 1991, p. 17), E_{2g} is the only Raman-active optical mode. At 300 K on polycrystalline Be the Raman line was observed at 455 cm^{-1} (line width $\Delta\omega \approx 16$ cm^{-1}) with 4880 Å (Ar ion laser) excitation as the Stokes line [1]. At room temperature on single crystal Be with the polished surface in the crystallographic x-z plane, the line was observed at 463 cm^{-1} (also with 4880 Å excitation). The three components of the Raman tensor, α_{xx}, α_{xz}, and α_{zz}, could be measured. The Raman scattered light was found to follow the symmetry of E_{2g} of the space group D_{6h}^4, assuming a symmetric tensor $\alpha_{xy} = \alpha_{yx}$ [2]. The values agreed quite well with the frequency 458 cm^{-1} for the Raman mode (k = 0) obtained by inelastic neutron scattering [3]; cf. p. 99.

Using the method of Raman light scattering (4880 Å laser) the temperature dependence of the frequency, ω, and the line width, $\Delta\omega$, of the E_{2g} phonons for k∥[0001] of a cleaved Be surface were investigated between 100 and 650 K: ω decreased from ~460 cm^{-1} at 100 K to ~442 cm^{-1} at 600 K while $\Delta\omega$ increased from 5 to >11 cm^{-1} in the same region; at room temperature ω was 455.7 cm^{-1}. The temperature dependence of the damping of E_{2g} phonons in Be was of anharmonic origin with the main mechanism governing the anharmonic damping assumed to be the decay of an optical phonon into two accoustic phonons, each with half the frequency [4].

The E_{2g} phonon frequency at 80 and 300 K was determined by the method of inelastic neutron scattering. At room temperature a value of 455 cm^{-1} was found [5] in agreement with the data of [4]. The shift of 2.7 cm^{-1} [5] was half the value found in [4] and the discrepancy was explained by the large errors of the measurements [5].

The optical frequency at q = 0, Γ_5^+, which is 13.64 ± 0.03 THz at 300 K [5], may be compared with results for the same temperature from Raman spectrometry 13.63 THz [1] and 13.88 THz [2].

References:

[1] Feldman, D. W.; Parker, J. H., Jr.; Ashkin, M. (Phys. Rev. Letters **21** [1968] 607/8).
[2] Fraas, L. M.; Porto, S. P. S.; Loh, E. (Solid State Commun. **8** [1970] 803/5).
[3] Schmunk, R. E.; Brugger, R. M.; Randolph, P. D.; Strong, K. A. (Phys. Rev. [2] **128** [1962] 562/73).
[4] Ponosov, Yu. S.; Bolotin, G. A. (Fiz. Tverd. Tela [Leningrad] **27** [1985] 2636/9; Soviet Phys.-Solid State **27** [1985] 1581/3).
[5] Stedman, R.; Amilius, Z.; Pauli, R.; Sundin, O. (J. Phys. F **6** [1976] 157/66).

15.2 Optical Constants

The optical properties of beryllium have been the subject of numerous experiments. However, early studies of Be films were performed either after evaporation in poor vacuum or exposure of the samples to contaminating atmospheres before or during measurements, see, for instance [1 to 7]. Later, the optical constants were measured on Be layers at 82 and 290 K in the IR from 0.11 to 1.5 eV by Shklyarevskii and Yarovaya [8] and at room temperature in the VUV from 10.5 to 26 eV by Toots et al. [9].

Based on Swanson's [10] determination of the line shape of the characteristic electron-energy loss (CEL) peak of Be, LaVilla and Mendlowitz [11] subsequently analyzed the data by means of the Kramers-Kronig (KK) dispersion relation to obtain the optical constants and the reflectivity in the photon energy range from 12 to 28 eV.

Measurements of the optical absorptivity of Be single crystals were carried out at 4.2 K in the energy range 0.12 to 4.5 eV with polarized light in order to show the anisotropy of the optical data (gained by KK analysis) for the electric field vector, $\vec{E}$, of the incident radiation parallel and perpendicular to the c-axis [12].

The optical properties of Be films were further investigated by measuring the reflectance and transmittance in the visible and near IR [13], by absorption (using a UHV ellipsometric technique) from about 0.5 to 4 eV at room temperature and at 20 K (the spectra are largely independent of the substrate temperature; see figures in the paper) [14], and by reflectance in the UV between 2 and 25 eV at room temperature [15]. For the transmittance of Be films in the VUV below 1000 Å, see also [16].

The nonrelativistic APW band calculations of Chatterjee and Sinha [17] were used to calculate the imaginary part $\varepsilon_2(\omega)$ of the dielectric function via the joint density of states (JDOS) and the double density of states (DEDOS) for the optical transitions [18]. The theoretical optical conductivity, $\sigma_{opt}(0,\omega)$, for zero wave vector was also obtained from an APW band structure calculation neglecting the k-variation of the optical transition elements [19]. A calculation of the frequency-dependent dielectric constant, $\varepsilon(\omega) = \varepsilon_1(\omega) + i\,\varepsilon_2(\omega)$, using Shaw's model potential was reported by Taut, Skokan [20]. The imaginary part was calculated directly from the inter- and intraband contributions of one-electron wave functions and energies. The real part was obtained by using the KK relation for conductivity [20]. In continuation of this calculation of the longitudinal (polarization q parallel to the hexagonal axis) dielectric function $\varepsilon(q,\omega)$, the optical properties including anisotropy, plasmon dispersion relation, and half width of the plasmon peak were compared with experiments and the relaxation times for the inclusion of damping were calculated [21].

A procedure for simultaneously fitting reflectance data, obtained for various photon energies and angles of incidence by using a simple physical model for $\varepsilon(\omega)$, was given by Powell [23] and applied to the data of Toots et al. [9].

In the **infrared** spectral region from 11 to 0.8 μm (0.113 to 1.550 eV) all optical constants decrease with increasing photon energy: the reflectivity R (at normal incidence and 290 K) from 0.922 to 0.498, $|\varepsilon_1|$ from 304 to 0.55 (ε_1 being negative), ε_2 from 502 to 15, the refractive index n from 11.9 to 2.7 (it has a minimum of 2.4 around 1 eV), and the absorption coefficient k from 21.1 to 2.8; these data are given from a table of Foiler [24] based on the measurements of [8]. The decrease of R with increasing hν (measured from 40 to 0.4 μm on polished Be) is smooth above 2 μm; the strongest decrease appears in the 2 to 1 μm range [25].

According to Shklyarevskii, Yarovaya [8] the optical properties of Be in the 0.8 to 11 μm region are entirely determined by free current carriers. Using the formulas of the normal skin effect taking account of the quantum character of the interaction of free carriers with the IR radiation, the concentration (N), effective mass (m*), and frequency of collisions (ν_{coll}) among carrier particles were determined to be $N/m^* = 1.9 \times 10^{49}$ g^{-1} cm^{-3} (with $N = 1.1 \times 10^{22}$ cm^{-3} from the Hall effect) which gives $m^*/m = 0.64$, $\nu_{coll} = 3.6 \times 10^{14}$ s^{-1}; (from the data [24] is $\hbar\omega_p = 4.87$). The investigations of Weaver et al. [12] on Be single crystals from 0.12 to 4.5 eV revealed that already in the spectral range studied by [8] interband effects and anomalous skin-effect conditions to the free-electron term are important. While the increase in conductivity near 0.4 eV did herald interband effects, such effects could not be excluded at lower energies. The slope of the absorptivity curves indicated that such effects do play a significant part in the absorption process at energies as low as 0.12 eV [12]. **Fig. 15-1** pictures the absorptivity (A) results obtained on Be single crystals at 4.2 K and a 15° angle of incidence for both $\vec{E}\|c$ and $\vec{E}\perp c$. The absorptivities rise sharply from ~0.2 eV, the $\vec{E}\|c$ rise being more precipitous. Even in the range 0.12 to 0.20 eV, both curves show a steady rise. At 0.9 eV the difference $A_\| - A_\perp$ is nearly 21%. For $\vec{E}\|c$, A levels off at 1.3 eV, but rises again at 1.5 eV. This shoulder has no counterpart in $\vec{E}\perp c$. For a comparison with the results of [9], see the original figure in the paper [12]. The optical conductivity is shown for both polarizations in **Fig. 15-2**. Both curves drop from large, free-electron values before turning up at about 0.5 and 0.7 eV for $\vec{E}\|c$ and $\vec{E}\perp c$, respectively [12]. On evaporated films with highly disordered structure the σ minimum lies at ~1.0 eV (300 K) and at ~1.3 eV (20 K) [14].

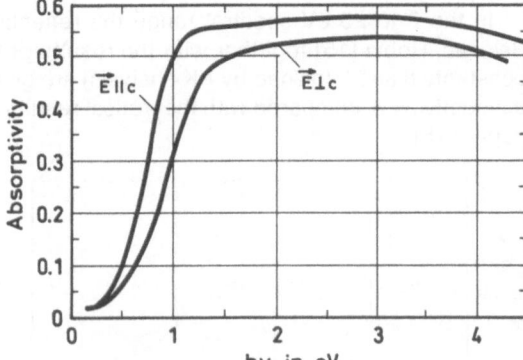

Fig. 15-1.   Absorptivity, A, of Be at 4.2 K and a 15° angle of incidence.

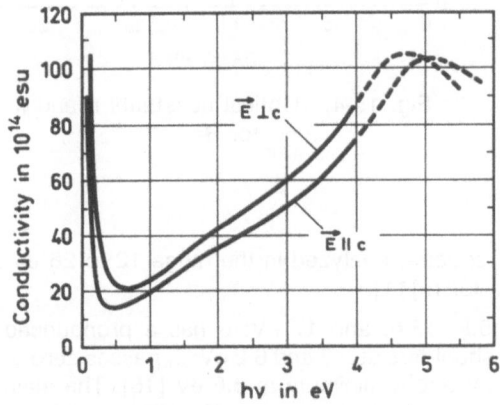

Fig. 15-2.   Optical conductivity, σ, for Be. Above 4 eV, the magnitudes are not reliable due to interpolation in the range 4.5 to 10.5 eV (the peaks persist for all interpolations used).

The real and imaginary parts of the dielectric function are shown for both polarizations in a figure in the paper [12].

The reflectivity, R, of Be is high, particularly in the IR region. The reflection of IR radiation (10.6 μm) was measured at 20 and 300 K [26], giving absorption coefficients of about 0.3 and 1.2%, respectively, which are shown in a study of IR reflection from metal surfaces by Gordeev et al. [27]. Analytical expressions were derived for the IR absorption coefficients (using a diagram technique), and the temperature dependence was calculated up to 400 K for Be crystals with the hexagonal c-axis parallel and perpendicular to the surface. The linear temperature dependence existing at high temperature was found to be retained well below the Debye temperature ($\Theta_D = 1450$ K) [27].

The frequency dependence of the optical function was described in a wide range from IR to X-rays on the basis of the Drude-Sellmeier model by Bakulin [28]. The preparation of optical surfaces of Be mirrors was studied over the spectral region from 0.3 to 40 µm, see Bloxsom, Schroeder [29]. The construction and testing of an IR refractometer by measuring R (in the range 1 to 4 µm) of Be films of various thickness is reported in [7].

In the **visible** range (1.7 to 3.2 eV) and beyond it the absorptivity, A, rises to a maximum at 3.0 eV; above 3.0 eV, A drops off gradually. The optical conductivity, $\sigma_\perp$ and $\sigma_\parallel$, shows maxima between 4.5 and 5.1 eV. For $\vec{E}\parallel c$ and $\vec{E}\perp c$, the ε_2 curves reach minima at about 2.9 and 3.1 eV, respectively, and ε_1 rises from large negative values, passes through zero at 1.0 and 1.36 eV, respectively, showing finally broad maxima at about 2.2 and 2.4 eV, respectively [12].

In the 2 to 25 eV spectral range the reflectivity of Be films is shown in **Fig.** 15-3 from Seignac, Robin [15] together with the results of the measurements [9] and [14]. The optical constants n and k (gained by KK analysis) are given in **Fig.** 15-4, and the experimental optical conductivity, σ, compared with theoretical results of [19, 20], is shown in a figure in the original paper [15].

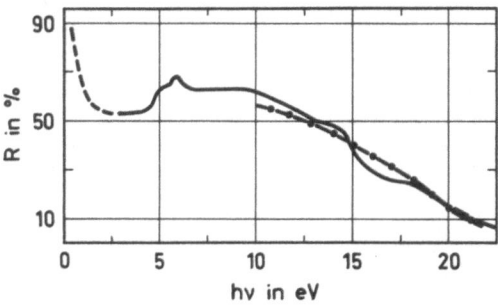

Fig. 15-3.   Reflectivity, R, of Be films for a 20° angle of incidence: —— measurements [15], –·–·– [9], --- from optical constants [14].

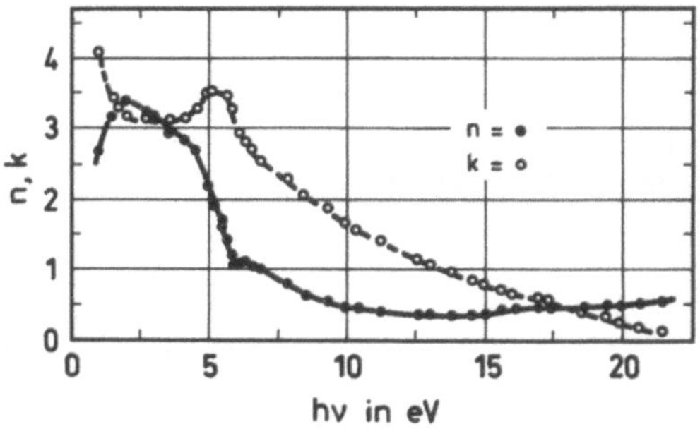

Fig. 15-4.   Optical constants n and k for Be.

The optical constants as derived from CEL spectra, analyzed in the range 12 to 28 eV, show a strong decrease for k and an increase for n [11].

The reflectivity, R(hν), has a structure at 5.8, 13.6, and 17 eV; σ has a pronounced maximum at 4.8 eV (already found by [12]) with shoulders at ~3 and 6.2 eV; ε_1 passes zero at 1.6, 3.2, and 17.4 eV with a maximum at 2.2 eV and a minimum at 5.6 eV [15]. The main structure at hν = 4.8 eV (σ maximum) was attributed to L_1–L_1 transitions ($E(L_1)$–$E(L_1) \approx 4.8$ eV) which are allowed for both polarizations, but transitions from Γ could also contribute to absorption for $\vec{E}\parallel c$ near 5 eV ($\Gamma_3^+ \rightarrow \Gamma_4^-$) [12]. The shoulder at 6.2 eV may have its origin in transitions $M_2^- \rightarrow M_4^-$ [20]; cf. [15]. The attributions suggested in [18] are more complex. The random phase approximation (RPA, cf. p. 88), neglecting the off-diagonal elements, however including realistic wave functions and energy values, was found to be capable of describing the gross features of the optical properties and loss functions of Be. Reasonable agreement between calculation and experiments was obtained by fitting the relaxation times [21].

The dielectric loss function $-\mathrm{Im}(\varepsilon^{-1})$ has a maximum at hν = 19.2 ± 0.4 eV which is attributed to plasma oscillations; from the peak width the relaxation time for the oscillations was obtained as $\tau = 1.23 \times 10^{-16}$ s [15]. This is in fair agreement with the theoretical oscillation frequency for the two free electrons, $\hbar\omega_p = 18.5$ eV [15], and also with other experimental values for $\hbar\omega_p$ (in eV) and τ (in 10^{-16} s), respectively: 18.4 and 1.3 [9]; 18.9 and 1.2 [10], see p. 86. The maximum of the function $-\text{Im}(\varepsilon+1)^{-1}$ corresponds fairly well to the theoretical value of surface plasmons, $\hbar\omega_p/\sqrt{2}=13.6$ eV [15]. Near 19 eV, transmittance of Be films begins [16].

In the extreme **ultraviolet** range (1000 to ~100 Å) the transmittance of Be increases with decreasing wavelength, cutting off sharply at 115 Å (111.7 eV), which is the K-absorption edge (see p. 95). Linear absorption coefficients, μ_l, as a function of wavelength (300 to 20 Å range) according to different investigators [16, 30 to 33] are given in **Fig.** 15-5 from [33]. Values for the mass attenuation coefficients are given in "Beryllium" Suppl. Vol. A 1, 1986, pp. 154/7; see also this volume pp. 97/8.

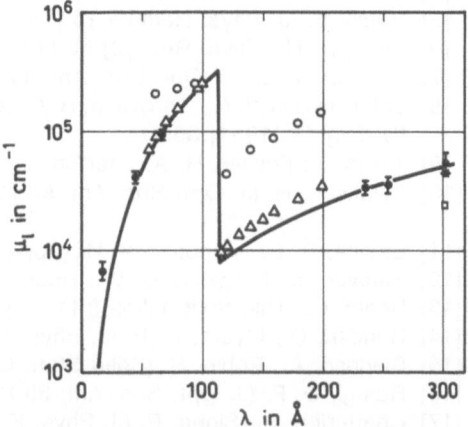

Fig. 15-5.   Linear absorption coefficient, μ_l, of Be as a function of wavelength, λ. Measurements [33] ● are compared with those of [30] ○, [31] △, [32] ★, and [16] □. The solid curve represents the calculated μ_l.

For X-ray absorption coefficient measurements in the λ region 8 to 14 Å and comparison of the values with various semiempirical estimates, see Senemaud [34]. An improved method for extracting absorption information from soft X-ray emission spectra is applied to Be (and other metals); results for 1 and 3 kV incident electron energy are tabulated in the paper [35]. Reflection curves for several X-ray mirrors, e.g. Be, for the energy range 1 to 15 keV and angles 0.5° to 20° are given in [36]. Theoretical X-ray reflectance from optical surfaces is compared with experimental data, e.g. on Be, at $\lambda=8.34$ Å in [37].

Reflection coefficients of radiation in the range of wavelength 23.6 to 113 Å for a number of substances including Be are given and the refractive index $n=1-\delta-i\cdot\beta$ and the absorption coefficient are determined by Lukarskii et al. [38]; some values are given for Be as example:

λ in Å	$\delta\times10^{-3}$	$\beta\times10^{-3}$	$\mu\times10^4$ in cm^{-1}
23.6	1.22	0.29	1.56
44	3.52	1.48	4.23
67	5.80	2.90	5.43

The real decrement, δ, of the X-ray refractive index, n (see above), of Be was measured noninterferometrically for the K$\bar{\alpha}$ lines of Mo, Cu, and Ag to an accuracy $\leq0.1\%$, using a thin-wafer monolithic Laue-Laue diffractometer, to be $\delta(\times10^6)=1.1244$ for MoK$\bar{\alpha}$ ($\lambda=0.7107$ Å), 5.296 for CuK$\bar{\alpha}$ ($\lambda=1.54177$ Å), and 0.7007 for AgK$\bar{\alpha}$ ($\lambda=0.5608$ Å). The results are in excellent agreement with the theory assuming for the dispersion correction $f'=0$ [39].

Anomalous dispersion calculations predicted f' values of 0, 3, and -1 millielectrons for Be at the MoKα_1, CuKα_1, and AgKα_1 energies, respectively [40]. The value of $\delta(\times10^6)=5.296\pm$ 0.16% was measured for CuKα radiation by using an X-ray interferometer and film detection

[41] and compared with a theoretical value of 5.2947 (without dispersion correction); the measurements [41] are also in good agreement with the data [39].

References:

[1] Givens, M. P. (Phys. Rev. [2] **61** [1942] 626/30; Solid State Phys. **6** [1958] 313/52, 348/50).
[2] Sabine, G. B. (Phys. Rev. [2] **55** [1939] 1064/9).
[3] Banning, M. (J. Opt. Soc. Am. **32** [1942] 98/102).
[4] Conn, G. K. T.; Eaton, G. K. (J. Opt. Soc. Am. **44** [1954] 477/83).
[5] Robin, S. (J. Phys. Radium **14** [1953] 427/8).
[6] Bock, R. O. (Phys. Rev. [2] **68** [1945] 210/3).
[7] Oldham, M. S. (J. Opt. Soc. Am. **41** [1951] 673/5).
[8] Shklyarevskii, I. N.; Yarovaya, R. G. (Opt. Spektrosk. **11** [1961] 661/6; Opt. Spectrosc. [USSR] **11** [1961] 355/7).
[9] Toots, J.; Fowler, H. A.; Marton, L. (Phys. Rev. [2] **172** [1968] 670/6).
[10] Swanson, N. (J. Opt. Soc. Am. **54** [1964] 1130/3).

[11] LaVilla, R. E.; Mendlowitz, H. (Appl. Opt. **4** [1965] 955/60).
[12] Weaver, J. H.; Lynch, D. W.; Rosei, R. (Phys. Rev. [3] B **7** [1973] 3537/41).
[13] Reale, C. (Thin Solid Films **9** [1972] 395/407).
[14] Hunderi, O.; Myers, H. P. (J. Phys. F **4** [1974] 1088/93).
[15] Seignac, A.; Robin, S. (Solid State Commun. **19** [1976] 343/5).
[16] Rustgi, O. P. (J. Opt. Soc. Am. **55** [1965] 630/4).
[17] Chatterjee, S.; Sinha, P. (J. Phys. F **5** [1975] 2089/97).
[18] Sinha, P.; Chatterjee, S. (Indian J. Phys. A **52** [1978] 503/10).
[19] Nilsson, P.-O.; Arbman, G.; Gustafsson, T. (J. Phys. F **4** [1974] 1937/50).
[20] Taut, M.; Skokan, R. (Phys. Status Solidi B **70** [1975] K157/K160).

[21] Taut, M.; Hanke, W. (Phys. Status Solidi B **77** [1976] 543/50).
[22] Tripp, J. H.; Everett, P. M.; Gordon, W. L.; Stark, R. W. (Phys. Rev. [2] **180** [1969] 669/78).
[23] Powell, C. J. (J. Opt. Soc. Am. **60** [1970] 214/20).
[24] Foiler, C. L. (Landolt-Börnstein New Ser. Group III **15** Pt. b [1985] 210/490, 236).
[25] Blickensderfer, R.; Deardorff, D. K.; Lincoln, R. L. (J. Less-Common Metals **51** [1977] 13/23).
[26] Guadagnoli, M. D.; Saito, T. T. (Appl. Opt. **14** [1975] 2806/8).
[27] Gordeev, V. F.; Kucherov, R. Ya.; Malkhozov, M. F. (Poverkhnost **1984** No. 6, pp. 15/21; C.A. **101** [1984] No. 80860; Phys. Chem. Mech. Surf. **3** [1985] 1607/20, 1618).
[28] Bakulin, E. A. (Fiz. Tverd. Tela [Leningrad] **24** [1982] 3635/41; Soviet Phys.-Solid State **24** [1982] 2072/5).
[29] Bloxsom, J. T.; Schroeder, J. B. (Appl. Opt. **9** [1970] 539/43).
[30] Johnston, R. W.; Tomboulian, D. H. (Phys. Rev. [2] **94** [1954] 1585/9).

[31] Henke, B. L.; Elgin, R. L. (Advan. X-Ray Anal. **13** [1970] 639/65).
[32] Mulder, B. J.; Vrakking, J. J. (J. Phys. E **11** [1978] 743/4).
[33] Barstow, M. A.; Lewis, M.; Petre, R. (J. Opt. Soc. Am. **73** [1983] 1220/2).
[34] Senemaud, G. (J. Phys. [Paris] **30** [1969] 811/8).
[35] Crisp, R. S. (J. Phys. F **13** [1983] 1325/32).
[36] Seward, F. D. (UCID-17505 [1977] 25 pp. from C.A. **88** [1978] No. 128903).
[37] Neergaard, J. R.; Reynolds, J. M.; Fields, S. A. (NASA-TN-D-8366 [1976] 55 pp. from C.A. **86** [1977] No. 130465).

[38] Lukirskii, A. P.; Savinov, E. P.; Ershov, O. A.; Shepelev, Yu. F. (Opt. Spektrosk. **16** [1964] 310/9; Opt. Spectrosc. [USSR] **16** [1964] 168/72).

[39] Deutsch, M.; Hart, M. (Phys. Rev. [3] B **30** [1984] 643/6).

[40] Cromer, D. T.; Liberman, D. A. (J. Chem. Phys. **53** [1970] 1891/8; Acta Cryst. A **37** [1981] 267/8).

[41] Bonse, U.; Hellkötter, H. (Z. Physik **223** [1969] 345/52).

15.3 Thermal Radiation. Emissivity

A value frequently cited for the spectral emissivity at $\lambda = 0.65\ \mu m$ is $\varepsilon_\lambda = 0.61$ for solid and liquid Be between 900 and 2000°C [1 to 3]. Obviously, this value stems from Burgess and Waltenburg [4], cf. "Beryllium" 1930, p. 57.

The total normal emittance, ε_{tn}, was measured on polished Be from 400 to 850 K. The emittance was very low, ranging from 0.023 (at 400 K) to 0.076 (at 850 K). Emittance values calculated for 300 K from reflectivity data agreed with the emission data [5]; the data were compared with those of [6].

The emissivity of anodized Be was studied in the range 400 to 1000 K in relation to the thickness (studied up to 92 μm) and structure of the oxide layer. A stable emissivity was observed for anodized Be bearing an oxide film annealed in air and consisting of the equilibrium oxide BeO. The integral normal degree of blackness increased continuously with the oxide film thickness, indicating that oxide films consisting of equilibrium BeO are to a certain extent transparent at all thicknesses [7]. For a smooth surface and thin oxide film $\varepsilon_{tn} = 0.177$ was reported previously [8].

References:

[1] Udy, M. C.; Shaw, L.; Boulger, F. W. (Nucleonics **11** No. 5 [1953] 52/9).

[2] Weast, R. C.; Lide, D. R. (CRC Handbook of Chemistry and Physics, 70th Ed., Boca Raton, Florida, 1989/90, p. E-407).

[3] Weik, H. (Metall **13** [1959] 202/13, 204).

[4] Burgess, G. K.; Waltenburg, R. G. (J. Washington Acad. Sci. **4** [1914] 566/7).

[5] Blickensderfer, R.; Deardorff, D. K.; Lincoln, R. L. (J. Less-Common Metals **51** [1977] 13/23, 17).

[6] Adams, J. G. (AD-274558 [1962] 259 pp. from C.A. **60** [1964] 15302).

[7] Zhorov, G. A.; Al'movskii, R. M.; Urazbaev, M. I.; Gornyi, D. S. (Teplofiz. Vysok. Temp. **14** [1976] 42/6; High Temp. [USSR] **14** [1976] 37/40).

[8] Zhorov, G. A. (Teplofiz. Vysok. Temp. **5** [1967] 450/7; High Temp. [USSR] **5** [1967] 403/9).

15.4 Particle Channeling and Channeling Radiation

For theory and general references see "Tungsten" Suppl. Vol. A 3, 1989, pp. 223/6 and 270/2.

In a computer simulation of the channeling of protons through a thin Be crystal in $\langle 10\bar{1}0 \rangle$ direction the angular dependence of the momentum density was computed using the particle trajectory approximation. The results, given as transmission spectra, were characteristic of the

Be hcp crystal structure. In obtaining the spectra, the energy loss suffered by the protons due to electron multiple scattering was considered and the effect of thermal vibrations treated separately (there was no remarkable effect on the spectra). The positions of the major dips in the spectra correlated well with the directions of neighboring strings. Variations in the angle of incidence of the beam and in its initial azimuthal angle modified the spectra depending on the transverse kinetic energy of the incident particles and the Be crystal structure [1].

The channeling radiation from relativistic electrons (using a 45 MeV betatron as electron source) in single crystals was studied and compared with calculations using a many-beam formalism. Radiation spectra (counts per channel vs. photon energy) are given for the 45.0 MeV electrons channeled along the $(01\bar{1}0)$ and $(\bar{1}2\bar{1}0)$ plane and the $\langle 0001 \rangle$ axis in Be; calculated interplanar potentials and the $\langle 0001 \rangle$ axial potential as well as eigenvalues for the 45.0 MeV electrons are also given (in figures in the paper). Good agreement was found between theory and experiment for the planar channeling in Be; for axial channeling there is a large number of bound states in the potential, and it was not possible to associate a line energy in the spectrum with a single transition. Therefore, no detailed comparison was made of the data and theory, but the calculations seem to reproduce well the shape of the spectrum [2].

References:

[1] Rajasekharan, K.; Neelakandan, K. (Pramana **31** [1988] 399/412; C.A. **110** [1989] No. 121734).

[2] Buschhorn, G.; Diedrich, E.; Kufner, W.; Pollmann, D. (Nucl. Instrum. Methods Phys. Res. B **30** [1988] 29/33; C.A. **108** [1988] No. 139266).

15.5 Phonon Fluorescence Spectra

Phonon fluorescence of Be films, deposited on an $SrF_2:Sm^{2+}$ crystal, was investigated; the phonon fluorescence was excited by electric current pulses passing through the films. The emission spectra of the phonons penetrating into the supporting crystal from the film were determined by the method of phonon spectroscopy based on an analysis of the luminescence corresponding to vibronic transitions between the states of the Sm^{2+} ions in SrF_2 (for the method based on the study [1] and applied to other cases, see [2 to 4]). The emission of phonons from a defect-free film is confined to a surface layer of a certain thickness. Including the emission of both longitudinal and transverse phonons by the films in the study, it was found that the Rayleigh scattering of phonons by defects resulted in a clear separation of the spectrum into two parts: first, low-frequency, characterized by ballistic propagation of phonons in the film and low emissivity of phonons, and second, high-frequency, characterized by diffuse propagation of phonons and a high emissivity of both longitudinal and transverse phonons. The latter part was attributed to the "mixing" of the phonon polarizations in the course of elastic scattering. Good agreement between the experimental results and calculated spectra made it possible to attribute the gradual changes in the phonon emission spectra to the process of film aging and a consequent increase in the mean free path of phonons in the case of elastic scattering by defects; for further details and figures, see the paper [5].

For theoretical considerations, see [6, 7].

References:

[1] Dynes, R. C.; Narayanamurti, V. (Phys. Rev. [3] B **6** [1972] 143/71).

[2] Bron, W. E.; Grill, W. (Phys. Rev. [3] B **16** [1977] 5303/14, 5315/20).

[3] Akimov, A. V.; Basun, S. A.; Kaplyanskii, A. A.; Titov, R. A. (Fiz. Tverd. Tela [Leningrad] **21** [1979] 231/3; Soviet Phys.-Solid State **21** [1979] 136/7).

[4] Abramov, A. P.; Abramova, I. N.; Gerlovin, I. Ya.; Razumova, I. K. (Fiz. Tverd. Tela [Leningrad] **22** [1980] 946/7; Soviet Phys.-Solid State **22** [1980] 556/7).

[5] Gerbshtein, Yu. M.; Nikulin, E. I. (Fiz. Tverd. Tela [Leningrad] **23** [1981] 1022/8; Soviet Phys.-Solid State **23** [1981] 591/5).

[6] Kaganov, M. I.; Lifshits, I. M.; Tanatarov, L. V. (Zh. Eksperim. Teor. Fiz. **31** [1956] 232/7; Soviet Phys.-JETP **4** [1957] 173/8).

[7] Shklovskii, V. A. (Zh. Eksperim. Teor. Fiz. **78** [1980] 1281/93; Soviet Phys.-JETP **51** [1980] 646/53).

15.6   Laser-Plasma Interactions

Disk targets of Be (CH, Ti, and Au) were irradiated with 600-ps near-Gaussian pulses of 0.53-μm laser light (from a frequency-doubled Argus Nd-glass laser) and laser energies of 3 to 35 J at intensities from 3×10^{13} to $\sim 4 \times 10^{15}$ W/cm². The measured absorption A (increasing with decreasing intensity and increasing target Z as expected for inverse bremsstrahlung absorption), hard X-ray fluxes, and sub-keV emission properties were compared with detailed computer simulations of hydrodynamics. The results showed strong collisional absorption, some Brillouin scattering, little suprathermal electron production, and efficient conversion of absorbed energy into sub-keV X-rays, in general accord with wavelength-scaling predictions; for figures representing the results, see the paper [1]. Time-resolved observations of stimulated Raman scattering from 0.532-μm laser-produced plasma ($0.72 < \lambda < 0.9$ μm, < 50-ps pulses) were reported in [2].

Plasmas produced by irradiating plane targets of low Z elements, e.g. Be, with moderate energy (3 to 10 J) lasers emit considerable amounts of free-free (bremsstrahlung) and free-bound (recombination) continua in the soft X-ray region (10 to 100 Å) of the spectrum. With intensity-calibrated grazing incidence spectrographs an intensity of the order 10^8 W·sr^{-1}·Å^{-1}·cm^{-2} was determined, emitted by the densest portion of the plasma inside the crater formed at the interaction. The result was in good agreement with theoretical predictions for a plasma of predetermined parameters; for details of the experiments, see the paper [3]. With the aim to determine plasma parameters as electron and ion densities, bound state populations, and distribution of velocities in a laser-produced plasma, observed line profiles were compared with a suitable model of the plasma. The region of interaction between a focused 1 GW ruby-laser beam and a plane Be target was observed end-on with a high-resolution stigmatic grazing incidence spectrograph. The profile of the Lyα line of Be IV was made up of two components: a broad asymmetric one, spatially corresponding to the radiation emitted by the high-density ablation region, and a narrow one superimposed, emitted by the expanding plasma. The adopted model that solves the radiative transfer equation for a moving plasma with adjustable parameters reproduced the observed features very well, and the plasma parameters can be determined unambiguously [4]; see also p. 95.

The second harmonic emission from laser-plasma interaction was studied by irradiating various targets including Be with a high-intensity pulsed laser (1.05 μm, 100 ps, <1 Å band width). The second harmonic emission spectra were measured as a function of the laser intensity (10^{12} to 10^{15} W/cm²) and incidence angle (0°, 20°, 45°, etc.) [5].

Based on the separate excitation zones that exist for virtually all elements, studies of background-corrected ion and atom emission profiles of the group IIa elements in the two-electrode direct current plasma show a great variation in both vertical and horizontal emission zones. An anodic shift of the ion zone with respect to the atom zone was observed additionally for Be, Mg, and Ca (this shift points to an electrostatic interaction between the ion and the dc potential between the electrodes). Figures of the spatial distributions of the ion and atom emission of the elements studied are given in the paper [6]. The emission spectra originating from an inductively-coupled plasma excitation source operating under a standard set of conditions were classified and tabulated for the alkaline earth elements. The nature of the spectra, obtained by means of a scanning, echelle monochromator (measuring the wavelength position to 0.01 Å) in the spectral range 2075 to 6005 Å, is discussed in [7].

Experimental and theoretical investigations of the effect of various solid obstacles on the radiation of a recombining laser plasma showed that the obstacles affect the radiation of the plasma by changing its hydrodynamic parameters such as the density and temperature of the electrons. An amplification of recombination emission of the plasma was observed behind a shock-wave front. The measurements were performed in the X-ray ($\lambda = 1.1$ to 1.7 nm) and UV-visible ($\lambda = 250$ to 600 nm) spectral regions, using Be as target as well as obstacle (in addition to other materials) and an Nd-glass laser with about 20 to 50 J energy, $\sim 2 \times 10^{10}$ W/cm² flux density, and ~ 10 ns pulse duration [8].

An experimental procedure was developed for the investigation of the VUV emission ($\lambda = 20$ to 80 nm) from a recombining laser-induced Be plasma in the far expansion zone. The plasma-emission spectra (produced by an Nd laser with 50 J energy, 10^{12} to 10^{13} W/cm² flux density, and 50 ns pulses) were recorded at distances $d \approx 5$ and 10 cm from the target; the structure of the spectrum and the relative line intensities were found to be practically independent of d. In the far expansion zone ($d \gg 1$ cm) only one intense process takes place in the plasma: a three-particle recombination of ions with electrons. The electron temperature of the plasma was estimated to be ~ 0.1 eV. The spectrum consists of the transitions $n \rightarrow 2$ ($n = 3$ to 7) of Be III and $n \rightarrow 2$ ($n = 3$ to 7) and $n \rightarrow 3$ ($n = 5$ to 8) of Be IV ions; tables and figures are given in the papers [9].

Two methods for absolute measurement of soft X-ray spectra emitted by laser-produced plasmas (of Be and other metals) are described in [10].

References:

[1] Mead, W. C.; Campbell, E. M.; Estabrook, K. G.; Turner, R. E.; Kruer, W. L.; Lee, P. H. Y.; Pruett, B.; Rupert, V. C.; Tirsell, K. G.; Stradling, G. L.; Ze, F.; Max, C. E.; Rosen, M. D. (Phys. Rev. Letters **47** [1981] 1289/92).

[2] Turner, R. E.; Phillion, D. W.; Campbell, E. M.; Estabrook, K. G. (Phys. Fluids **26** [1983] 579/81 from C.A. **98** [1983] No. 117909).

[3] Nicolosi, P.; Jannitti, E.; Tondello, G. (Appl. Phys. B **26** [1981] 117/24).

[4] Jannitti, E.; Nicolosi, P.; Tondello, G. (J. Phys. [Paris] **43** [1982] 1043/7).

[5] Xu, Zh-zh.; Xu, Y-g.; Yin, G-g.; Zhang, Y-zh.; Yu, J-j.; Li, H-y. (Kexue Tongbao **27** No. 9 [1982] 575 from C.A. **97** [1982] No. 153616).

[6] Williams, R. R.; Coleman, G. N. (Appl. Spectrosc. **35** [1981] 312/7).

[7] Anderson, T. A.; Forster, A. R.; Parsons, M. L. (Appl. Spectrosc. **36** [1982] 504/9).

[8] Boiko, V. A.; Bryunetkin, B. A.; Bunkin, F. V.; Derzhiev, V. I.; Dyakin, V. M.; Maiorov, S. A.; Skobelev, I. Yu.; Faenov, A. Ya.; Fedosimov, A. I.; et al. (Fiz. Plazmy [Moscow] **10** [1984] 999/1009, 1187/94 from C.A. **102** [1985] No. 36237, No. 86878).

[9] Afrosimov, V. V.; Bobashev, S. V.; Golubev, A. V.; Simanovskii, D. M.; Shmaenok, L. A. (Zh. Eksperim. Teor. Fiz. **91** [1986] 485/92; Soviet Phys.-JETP **64** [1986] 284/8; Pis'ma Zh. Tekh. Fiz. **10** [1984] 1017/20 from C.A. **102** [1985] No. 35822).

[10] Kishimoto, T. (Max-Planck-Inst. Quantenopt. Ber. MPQ **108** [1985] 131 pp. from C.A. **104** [1986] No. 233973).

15.7 Electron Emission

15.7.1 Thermionic Emission

For measuring the thermionic work function, Φ, between 900 and 1200 K a small sample of 99.7% pure Be was used; the solid sample was polycrystalline with a moderate preferred (100) orientation, relative to the more predominant (101) orientation in a random sample, and was pretreated for 6 h at 1180 K (outgassing). The effective thermionic vacuum work function was found to be 3.67 ± 0.03 eV and a slight temperature dependence $\Phi(T) = 3.75 - 8 \times 10^{-5}$ T. The upper temperature limit of 1200 K was dictated by vaporization of the Be surface [1]. Beryllium surfaces covered with Cs were also studied, and a work function of 1.94 eV was determined for the "heavily caesiated" Be surface (constant work function at coverages greater than unity). Comparing figures for a series of metals including Be are given in the paper [2]. Further studies were concerned with the effect of surface oxidation on the thermionic work function of Be (five different samples). From visual and X-ray analyses it was concluded that the lowest work function determined for any of the Be samples of 3.22 ± 0.08 eV is the closest to the (Richardson) work function of pure polycrystalline Be metal [3]. However, all these values are too low, and the most reliable value of the work function for clean Be surfaces is $\Phi \approx 5$ eV, see p. 72. Surface cleaning of BeO layers, generated on polycrystalline Be foils in UHV by Ar bombardment, caused a reduction of the thermally stimulated exoelectron emission (TSEE) with an increase of the Richardson work function [4]; see pp. 118/9.

References:

[1] Wilson, R. G. (J. Appl. Phys. **37** [1966] 2261/7).
[2] Wilson, R. G. (J. Appl. Phys. **37** [1966] 3161/9).
[3] Jerner, R. C.; Magee, C. B. (Oxid. Metals **2** [1970] 1/9 from C.A. **73** [1970] No. 8401); Jerner, R. C. (Diss. Univ. Denver 1965, pp. 1/130 from Diss. Abstr. **26** [1966] 5963).
[4] Euler, M.; Scharmann, A. (Z. Physik B **22** [1975] 373/5).

15.7.2 Photoemission and Photoelectron Spectra

Early investigations of the threshold for photoelectron emission from "clean" Be surfaces provided a photoelectric work function Φ (in eV) of 3.3 [1], 3.17 [2], or 3.92 [3]. According to later investigations applying the Fowler method (photoelectric current), the work function of clean (oxygen-free) Be surfaces was found to be ~5.0 eV, see [4, 5], cf. p. 72.

Mann and DuBridge [3] who measured the absolute photoelectric yields for vapor-deposited Be surfaces, determined the value of alpha (the proportional constant involving the probability of light absorption by electrons at the surface) in the Fowler equation to be 25×10^{-34} cm$^2 \cdot$ s/quantum for the clean surface. Exposure of the surface to oxygen glow discharges changed the threshold from 3150 Å ($\Phi = 3.92$ eV) to less than 2200 Å and reduced the alpha value to 4.6×10^{-34} cm$^2 \cdot$ s/quantum. In another experiment Schulze [2] used an evaporated surface of Be to determine the absolute photoelectric yield as a function of the wavelength of the incident light; the yield decreased smoothly with increasing wavelength. Likewise, Suhrmann and Schallamach [1] investigated the effect of temperature on the photoelectric yield, using a vapor-deposited Be film at 293, 83, and 20 K. They found a yield about 20% less at lower temperatures.

The photoelectric yield in the vacuum UV (wavelength range 1200 to 200 Å) was measured by Cairns and Samson [6], and for oblique incidence of extreme UV radiation (1.3 to 13 Å) by Heroux et al. [7], see also [8]. The photoelectric yield of ultrasoft X-ray radiation (1.39 to

13.33 Å) was measured by Lukirskii et al. [9]. For the photoemission properties in the soft X-ray region, see also Ivanov and Chigak [10]; cf. pp. 97/8. The investigation of the thickness dependence of the photoeffect excited by soft X-rays allowed the estimation of the emission region of slow secondary electrons (formed by fast photoelectrons) and to find the effective depth of photoemission. Results for some metals including Be are tabulated in the paper [11].

The continuous photoelectric absorption cross section of Be was measured for photon energies extending from the K edge (at 110 eV) to 180 eV [12]. For the photoionization absorption cross section of Be in this range with a discussion of previous results (including [12]), see [13].

Angle-resolved photoemission (ARPE) spectra from (0001)Be in the photon energy range 10 to 120 eV were used to evaluate the bulk band dispersion [14] and the surface states of (0001)Be [15] below the Fermi energy (E_F), see pp. 49/51 and 69/71. The distortion of the lowest-energy unoccupied band (above E_F) in Be by excited electron-plasmon interaction was also studied by ARPE [16].

For nearly free-electron metals such as Be, Mg, and Al the agreement between band theory and measured dispersion for both the occupied and unoccupied bands is fairly good. Yet quantitative discrepancies exist, arising from finite hole and electron lifetimes, electron-plasmon coupling, and inadequacies of local-density theory. Many of these effects can qualitatively be explained with an interaction electron-gas theory. The two-dimensional dispersion of intrinsic and extrinsic surface states (or resonances) is fairly well reproduced by single-particle calculations [17].

Photoemission spectra from evaporated polycrystalline Be films (grain size ~100 Å) provided results for the density of states (DOS) from the electron distribution curve [5, 18, 19]. The XPS valence band spectra were found to be dominated by the DOS curve of the s states [18], see p. 55. The 1s core level spectrum of Be metal and the plasmons accompanying it were measured by the XPS method, see pp. 48 and 95. The line shape of the spectrum is asymmetrical, tailing out to high binding energies; this is a consequence of the response of the conduction-electron system to the deep core hole which creates electron-hole pairs, and their energy is reflected in the tail of the zero loss line [18, 19].

Spin-polarized density-functional conduction-electron screening calculations for core ionization and Auger transition for the elements Be (Li, Na, Mg, Ca, and Al) were performed and screening was described by conduction-electron scattering phase shifts obtained from self-consistent-field ab initio calculations. Both spins were treated separately to investigate the spin-polarization effects. The core-electron states were resolved by a Hartree-Fock-Slater method in order to apply the same local-density exchange and correlation approximation to all electrons, including the conduction electrons. X-ray edge exponents and X-ray photoelectron (XPS) line singularity indices were extracted from the phase shifts, and Auger line singularity indices were also evaluated. The screening of Be and Li was found to be strongly spin-polarized, which increases the singularity index and decreases the edge exponents (for Na, Mg, and Al the results agreed with previous spin-symmetric calculations and experimental data) [20]. The measured X-ray photoemission singularity index of Be is $\alpha = 0.06$ [18], whereas theoretical values are 0.132 (SDF) [20], 0.09 (using a nonseparable scattering potential in calculation) [21], 0.13 and 0.14 (with variations in the calculation method) [22].

The (0001)Be surface core level shift was also determined by XPS [23], see p. 69.

Inverse photoemission spectroscopy (IPES) is used to study unoccupied electronic structure at surfaces. A series of submonolayer coverages of Li and Cs on (0001)Be was used to vary the position of the vacuum level (E_V) systematically, and the steplike contribution to the inverse photoemission spectrum at E_V was found to shift uniformly with E_V. The steplike intensity onset at E_V is common to almost all IPE spectra from metals in normal incidence, and was observed with a cross section similar in magnitude to many bulk-derived features. The step was assigned to emission into surface continuum final states; (the mechanism of this emission is still unexplored) [24].

References:

[1] Suhrmann, R.; Schallamach, A. (Z. Physik **91** [1934] 775/91, 780).
[2] Schulze, R. Z. (Z. Physik **92** [1934] 212/27, 221).
[3] Mann, M. M., Jr.; DuBridge, L. A. (Phys. Rev. [2] **51** [1937] 120/4).
[4] Green, A. K.; Bauer, E. (Surf. Sci. **74** [1978] 676/81).
[5] Gustafsson, T.; Brodén, G.; Nilsson, P.-O. (J. Phys. F **4** [1974] 2351/8).
[6] Cairns, R. B.; Samson, J. A. (J. Opt. Soc. Am. **56** [1966] 1568/73).
[7] Heroux, L.; Manson, J. E.; Hinteregger, H. E.; McMahon, W. J. (J. Opt. Soc. Am. **55** [1965] 103/4).
[8] Hinteregger, H. E. (Phys. Rev. [2] **96** [1954] 538/9).
[9] Lukirskii, A. P.; Rumsh, M. A.; Smirnov, L. A. (Opt. Spektrosk. **9** [1960] 511/5; Opt. Spectrosc. [USSR] **9** [1960] 265/7; Rumsh, M. A.; Lukirskii, A. P.; Shchemelev, V. N. (Dokl. Akad. Nauk SSSR **135** [1960] 55/7; Soviet Phys.-Dokl. **5** [1960] 1231/3).
[10] Ivanov, S. A.; Chigak, F. F. (App. Metody Rentgenovskogo Anal. No. 4 [1969] 197/200 from C. A. **73** [1970] No. 114314).

[11] Nakhodkin, N. G.; Mel'nik, P. V. (Fiz. Tverd. Tela [Leningrad] **5** [1963] 1732/4; Soviet Phys.-Solid State **5** [1963] 1259/61).
[12] Peterson, T. J., Jr.; McGuire, E. J.; Tomboulian, D. H. (Phys. Rev. [2] **129** [1963] 674/7).
[13] Brytov, I. A.; Lukirskii, A. P. (Opt. Spektrosk. **16** [1964] 363; Opt. Spectrosc. [USSR] **16** [1964] 199).
[14] Jensen, E.; Bartynski, R. A.; Gustafsson, T.; Plummer, E. W.; Chou, M. Y.; Cohen, M. L.; Hoflund, G. B. (Phys. Rev. [3] B **30** [1984] 5500/7).
[15] Bartynski, R. A.; Jensen, E.; Plummer, E. W. (Phys. Rev. [3] B **32** [1985] 1921/6).
[16] Jensen, E.; Bartynski, R. A.; Gustafsson, T.; Plummer, E. W. (Phys. Rev. Letters **52** [1984] 2172/5).
[17] Plummer, E. W. (Surf. Sci. **152/153** [1985] 162/79).
[18] Höchst, H.; Steiner, P.; Hüfner, S. (Phys. Letters A **60** [1977] 69/71; J. Phys. F **7** [1977] L309/L314).
[19] Steiner, P.; Höchst, H.; Hüfner, S. (Phys. Letters A **61** [1977] 410/2; Z. Physik B **30** [1978] 129/43, 135).
[20] Rantala, T. T. (Phys. Rev. [3] B **28** [1983] 3182/92).

[21] Leiro, J. A. (Solid State Commun. **41** [1982] 97/8).
[22] Minnhagen, P. (Phys. Letters A **56** [1976] 327/9).
[23] Nyholm, R.; Flodström, A. S.; Johansson, L. I.; Hörnström, S. E.; Schmidt-May, J. N. (Surf. Sci. **149** [1985] 449/59).
[24] Bruhwiler, P. A.; Watson, G. M.; Plummer, E. W.; Sagner, H.-J.; Frank, K.-H. (Europhys. Letters **11** [1990] 573/9).

15.7.3   Field Emission

Adsorption of Be on (110), (111), and (211) planes of W single crystals was studied using probe-hole field electron microscopy (FEM). The dependence of the work function on the time of deposition was measured for these planes; the average work function was also determined. On (110) the work function varied regularly, i.e., it decreased at low coverage, passed through a minimum and saturated for a thick layer at 4.1 eV, which is the value for bulk Be. On (111) and (211) the work function increased for low coverage and saturated for a thicker layer at high values of 5.3 and 6 eV, respectively. The irregular behavior of the work function was attributed to the loose structure of these planes and the small atomic radius of Be [1]. The behavior of Be on a W single crystal, primarily surface diffusion, was previously studied by means of an electron projector. The activation energy for the Be migration varied in relation to a succession of crystallographic W faces, and an analogous succession of faces was also obtained for the evaporation rate of Be from W. The study of the emission of the monoatomic Be film revealed a simultaneous decrease of the W work function by Be in one crystallographic region and an increase in other regions. Extreme values of the W work function were obtained with different signs of the faces (110) and (111) with the largest Be concentration in a monolayer. For thick Be layers, the work function had a value of 5 eV [2].

Preliminary results were reported of the experimental observation of the oscillation of the field emission current in a magnetic field. A new method was described for achieving the field emission from individual whiskers of any metal. Measurements were made at 4.2 K for magnetic fields of 0 to 40 kOe with the field vector coinciding with the axis of the 6th order for Be whiskers. Under the given conditions, the field emission current did not depend on the magnetic field [3].

References:

[1] Polanski, J.; Sidovski, Z.; Zuber, S. (Acta Phys. Polon. A **49** [1976] 299/305).

[2] Komar, A. P.; Savchenko, V. P.; Shrednik, V. N. (Radiotekh. Elektron. **5** [1960] 1211/7 from C.A. **1961** 18231).

[3] Grohman, A.; Wojda, L. (Tr. 10th Mezhdunar. Konf. Fiz. Nizkikh Temp., Moscow 1966 [1967], Vol. 3, pp. 346/50; Proc. 10th Intern. Conf. Low Temp. Phys., Moscow 1966 [1967], Vol. 3, pp. 346/50; C.A. **70** [1969] No. 52041).

15.7.4   Exoelectron Emission

Exoemission is a new type of emission of negative charges (slow electrons and ions) together with thermal emission, autoemission, photoemission, and secondary emission. In contrast to the known types of emission, it appears after a solid has been subjected to various influences inducing the formation of excited states, for instance, by physicochemical reactions in the surface layers accompanying adsorption (desorption), oxidation, corrosion, and catalytic processes, phase transformations, etc., or as a result of external influences such as mechanical effects, radiation, etc. The emission exhibits an appreciable inertia. On subsequent heating in the pre-Richardson region, the emission shows a number of peaks characterizing the energy spectrum of the charge localization levels on the surface. Since emission occurs at low temperatures, the emitting surface is assumed to be in a thermodynamic nonequilibrium state and to retain the properties resulting from its previous history. Photostimulated exoelectron emission takes place under the influence of light with a wavelength exceeding the red limit of the photoeffect. Thermostimulated exoelectron emission is a result of thermal liberation of negative charges in the pre-Richardson region when the energy of the electrons in the state of thermodynamic equilibrium is insufficient to overcome the surface potential. For further details of the complex physicochemical phenomenon, as well as the confusion in terminology, and discussion of lacking theory, see the review [1].

Photostimulated exoelectron emission during and after tensile deformation of thin (~1000 Å) films of Be and other metals, vacuum-deposited on Cu substrates at room temperature, was detected at 2 to 3% strain. Decay curves of the emission from thin films of Be, Mg, and Al (shown in a figure in the paper) after a 10% deformation of the substrate may be associated with the adsorption of oxygen on the active sites which are created by the diffusion of vacancies introduced by deformation [2].

Previously, electron emission from several metal surfaces, including Be, after mechanical working was observed by Lohff and Raether [3 to 5]. The intensity of the electron beam depended on the oxygen but not on the nitrogen pressure in the surrounding gas atmosphere, and the rate of emission increased with rising temperature from abraded surfaces of Be, Mg, Ca, and Al, as was found by studies of Ku and Pimbley [6]; this temperature dependence is in contrast to the findings of [3].

In further investigations of the exoelectron emission from evaporated Be and Al films, the Al showed no emission in vacuum (4×10^{-7} Torr). Addition of oxygen caused an emission which decreased with time, whereas nitrogen had no observable effect on the electron emission rates. Similar results were obtained with Be [7].

Oxidized (0001)Be surfaces were submitted to fractional heating up to 965°C in UHV (5×10^{-10} Torr) and the surface composition was monitored by Auger spectroscopy and contact potential measurements. Intensity variations of the thermally stimulated exoelectron emission (TSEE) could be observed, although the work function remained constant. The additionally observed decrease of exoelectron emission was caused by the reduction of the surface oxygen and density of TSEE centers. This was accompanied by an increase of work function [8], which had been observed previously on a BeO layer on polycrystalline Be for the Richardson work function. Thus, an applicable emission model must involve recombination processes [9].

References:

[1] Krylova, I. V. (Usp. Khim. **45** [1976] 2138/67; Russ. Chem. Rev. **45** [1976] 1101/18).
[2] Yamamoto, S. (Japan. J. Appl. Phys. **15** [1976] 1573/4).
[3] Lohff, J. (Z. Physik **146** [1956] 436/46).
[4] Raether, H. (J. Chim. Phys. **54** [1957] 48/52).
[5] Lohff, J.; Raether, H. (Naturwissenschaften **42** [1955] 66/7).
[6] Ku, T. C.; Pimbley, W. T. (J. Appl. Phys. **32** [1961] 124/5).
[7] Wüstenhagen, J. (Z. Naturforsch. **14a** [1959] 634/41; Naturwissenschaften **44** [1957] 228/9).
[8] Scharmann, A.; Wiessler, U. (Vakuum-Technik **27** [1978] 235/7).
[9] Euler, M.; Scharmann, A. (Z. Physik B **22** [1975] 373/5).

15.8 Electron Impact Phenomena

15.8.1 Secondary Electron Emission (SEE)

When a surface is bombarded with primary electrons (PE), electrons are released. **Fig. 15-6**, p. 120, from [1] shows schematically the energy distribution of such electrons released by PE with energies in the range 100 eV $< E_{PE} <$ 1 keV. With increasing energy of the PE, the elastic peak decreases and a broader maximum of inelastically reflected (backscattered) electrons (RE) can be observed. According to their energy, the electrons can be classified into two groups: first, electrons with energies $E \leqq 50$ eV: secondary electrons (SE), and second, electrons with energies 50 eV $< E \leqq E_{PE}$: inelastically backscattered and elastically reflected electrons (RE). On the basis of the two groups, the following parameters were defined: 1) SE yield δ: number of SE/number of PE; 2) backscattering coefficient η: number of RE/number of PE; 3) total electron yield $\sigma = \delta + \eta$; 4) coefficient of elastically reflected electrons η_E: number of elastically RE/number of PE. Superimposed on the energy distribution there are often several peaks, some of which are Auger electrons (having an energy depending on the surface material) and others are electrons which are backscattered with an energy loss by volume or surface plasmon excitation or ionization of surface atoms. A comprehensive survey on secondary electron emission from surfaces bombarded by primary electrons, including a compilation of previous reviews, is given by Seiler [1].

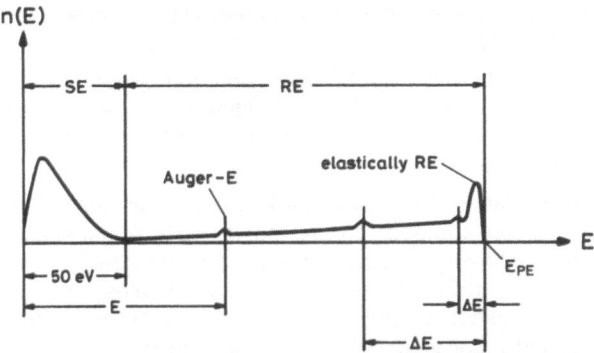

Fig. 15-6. Schematic energy distribution of electrons emitted from a surface bombarded with primary electrons. The secondary electron yield is up to 10^4 times higher than the Auger electron yield.

The general shape of the SE yield (δ) curve as a function of primary energy (E_{PE}) is the same for all materials: δ increases with E_{PE}, reaches a maximum value δ^m at E_{PE}^m, and then decreases, see **Fig.** 15-7 from [1], who quotes $\delta^m = 0.5$ to 0.9 for Be at $E_{PE}^m = 200$ to 300 eV. Some care has to be taken in comparing δ^m values of different authors because often the maximum total yield σ^m is given instead of δ^m (for metals δ^m values lie between 0.35 and 1.6 at 100 eV $\leqq E_{PE}^m \leqq 800$ eV) [1].

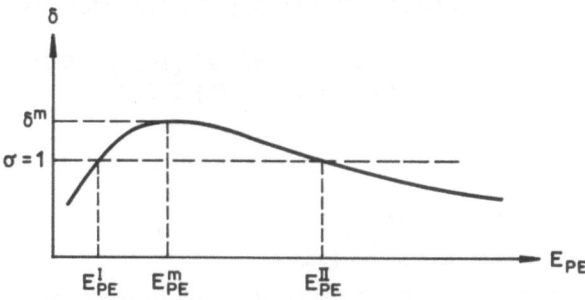

Fig. 15-7.   Secondary electron yield, δ, as a function of primary energy, E_{PE}.

Kollath [2] reported $\delta^m = 0.5$ at $E_{PE}^m = 200$ eV for polycrystalline Be from early studies [3 to 5], and Makhov, Gutkin [6] measured $\delta^m = 0.49$ at $E_{PE} = 700$ eV.

For $E_{PE} > 100$ keV the backscattering coefficient is constant $\eta = 0.08$ for Be; for 20 keV $\delta = 0.18$ was calculated on the basis of a semiempirical theory [1]. For η values of Be at E_{PE}^m, see [7].

The energy distribution of the secondary electrons released by PE with $E_{PE} > 100$ eV is essentially independent of E_{PE}, and is characterized by the most probable energy, E_{SE}^m, and the full width at half maximum, HW [1]. For metals is 1.3 eV $\leq E_{SE}^m \leq$ 2.5 eV and 4 eV $\leq$ HW $\leq$ 7 eV [2], and for Be evaporated on Mo was $E_{SE}^m = 1.9$ to 2.2 eV ($E_{PE} > 85$ eV) measured by [8]. An increase of E_{SE}^m with E_{PE} up to a maximum $E_{SE}^m = 4.8$ eV at $E_{PE} \approx 50$ eV and a decrease to a limiting value $E_{SE}^m = 2.6$ eV, reached at $E_{PE} \approx 600$ eV, was observed by Bronshtein et al. [9]; see also previous studies [10]. The dependence of HW of SE energy spectra on the layer thickness was studied with a Be film. Evaporating Be at a constant rate on Cu, the HW initially became smaller with increasing Be film thickness, and then became broader as the thickness increased beyond a certain level. Satisfactory agreement was obtained between the experimental results and a Monte Carlo simulation [11].

Certain laws governing the SE emission from thin layers of metals, e.g. Be on Ca, and semiconductors are discussed in [12]. Earlier fundamental studies of SE emission in thin Be layers on Ni or Ag targets such as the dependence of σ on E_{PE} for various layer thicknesses and the effect exerted by the backing on the SE emission properties are reported in [13].

High-energy SE emission caused by high-energy PEs for atomic numbers up to 80 is described and shown in a comparing figure in the first paper of [14]. Monte Carlo calculations of differential and integral characteristics of SE emitted from various monoatomic targets such as Be, Al, W, and Fe, which were irradiated with electrons of 2 to 12 MeV energy, are presented in [15].

In the true secondary electron spectrum from a clean Be surface (see figure 3 in the paper [16]) minima were observed at 1.0, 2.7, 7.6, and 12.0 eV. A comparison of this spectrum with the characteristic loss spectra (given in figure 2 in the paper [16]) showed that an approximate correlation between the two types of spectra is possible if the inelastic peak is at 1.0 eV, for then events are detected at 1.7, 6.6, and 11 eV intervals above the inelastic peak. The peaks in these spectra were discussed in terms of characteristic energies related to excited electron states in the solids [16].

For path lengths of slow SEs in Be and other metals, see the fundamental studies of [17].

Characteristics of the inner SE generation in Be, Al, Cu, and Au targets are described by [18]; cf. pp. 124/5.

The angular distribution of SE from polycrystalline surfaces is generally a cosine distribution independent of the angle of PE incidence [1]. Some results were presented of the angular distribution of medium-energy (0.1 to 2 keV) electrons which were elastically scattered by Be, Al, Cu, Ag, and Au films; the data refer to a 75° angle of incidence on the target. In the case of light materials, the discontinuous form of the angular distribution curves decreased with increasing energy of the PE and the curves became quite smooth in the case of Be, Al, and Cu at $E_{PE} \gtrsim 0.3$, 0.8, and 2.5 keV, respectively. When the atomic number decreased, the distribution curve became nonmonotonic at smaller electron energies [19]; see also [20].

Using an oscillographic method, SE emission was studied in Be and Pb under large angles of incidence (up to 89°) of the primary beam in the energy range 0.5 to 5 keV. The resulting $\sigma(E_{PE})$ curves revealed a strong increase of the angular dependence of σ with increasing E_{PE} for Be. The $\delta(E_{PE})$ curves basically behaved similarly to the $\sigma(E_{PE})$ curves with an increasing angle of incidence. The average efficiency (S) of the inelastically RE scattered below the exit region decreased with an increasing angle of incidence above 60°. Further details and figures are given in [21].

The results of the follow-up studies [21] are based on previous fundamental studies of SE emission from solids such as Be, Si, Al, Pb, and Ni with the primary beam at oblique incidence [22]. The electron backscattering at various angles of incidence on Be and Au targets was also measured by [23].

For nonelastic scattering (δ-η diagrams), see [10, 17, 24].

The energy distribution of β-particles passing through foils of Be and other materials of varying thickness was measured and is graphically shown in [25].

The coefficients of total (σ) and true (δ) SE emission were analyzed as functions of primary electron energy (E_{PE}) for Be and some other solids. Linear sections of these dependences can be described by the general expression $P = P_o - mX$, where $X = \ln E_{PE}$ and the slope m is 0.25 to 0.92 depending on the material; P_o is also a constant for a given material, and is generally different for the $\sigma(E_{PE})$ and $\delta(E_{PE})$ curves [26].

The cross sections of elastic and inelastic scattering of electrons are related to the atomic scattering factor, f, and the incoherent (Compton) X-ray scattering function, S. Because the calculation of f and S entails considerable labor, simple approximation formulas were derived. The results of the study demonstrated a high accuracy of approximation and interpolation of the electron scattering amplitudes [27].

A theory for the reflection of high-energy (10 to 1000 keV) electrons at normal incidence on a film free surface was developed, and the angular and energy distributions of the reflected electrons were studied; for details of the considerations including Be, see [28].

Calculations of the elastic reflection of electrons (E_{PE} = 100 to 1500 eV) from disordered targets such as Be, Al, Cu, Ag, and Au, including spatial distributions and the dependences on E_{PE} and the angle of incidence [29], gave qualitative agreement with the experimental data [19].

The effect of film structure, e.g., Be films sputtered on glass at 0° to 80° incidence, on the SE emission characteristics (σ, η), was studied by the electron microscope method. For obliquely sputtered Be films σ and η were smaller than for films sputtered at normal incidence [30].

The dependence of the coefficient of inelastically reflected primaries η (A/B) from thin layers of a given metal (A) deposited on a substrate (B) on the deposited thickness d(A) was investigated. An empirical equation was derived fitting the experimental data. The combinations Ti/Be and Be/Pt were studied as examples [31]. For Be on W and Si, Pt on Be, see [32].

In a study of SE emission from thin metal foils bombarded with 70 MeV electrons, the SE emission coefficient from several elements was measured. It was found that the SE emission coefficient/target electrons in the metal was larger for light elements such as Be and Al. This indicated that metal oxide on the foil surface (Malter effect, i.e., electron emission caused by field emission at the bulk material through the insulating layer; the escape possibility of the SE is increased) plays a dominant role in SE emission of these metals [33].

For elastic scattering of high-energy (116 MeV) electrons at various angles of incidence on Be and other materials and the method of nuclear recoil, see [34], for elastic scattering of (125 to 190 MeV) electrons by Be, see also [35].

The SE emission properties of conducting surfaces for use in multistage depressed collectors were measured (e.g. on Be) in an Auger spectrometer which also allowed the determination of the surface chemical constituents [36].

The SE emission yield (δ) and other electron spectroscopic techniques were applied to study slow oxidation of Be (and Mg) surfaces from residual oxygen. The most significant results were related to changes in δ during oxidation. With continuing oxidation the maximum yield δ^m of Be increased from 0.65 to 4.73 (similarly for Mg). This increase of δ^m was discussed and interpreted as a reduction of potential barriers caused by band bending [37]. The SE emission characteristics such as δ, first crossover, and voltage at δ^m, were studied on oxidized Be cathodes after several hours of operation as well as their dependence on various manufacturing procedures [38].

After exposure to oxygen at room temperature for various intervals, characteristic changes due to a thin BeO film overlaying the clean Be surfaces occurred in all three energy regions of the SE spectra [16].

For the influence of film ordering (deposition temperatures 293 and 83 K) and adsorbed oxygen on the SE emission, see [39].

References:

[1] Seiler, H. (J. Appl. Phys. **54** [1983] R1/R18).
[2] Kollath, R. (in: Flügge, S.; Handbuch der Physik, Vol. 21, Springer, Berlin 1956, pp. 232/303, 267).
[3] Kollath, R. (Ann. Physik [5] **33** [1938] 285/99).
[4] Bruining, H.; de Boer, J. H. (Physica **5** [1938] 17/30, 24).
[5] Schneider, E. G. (Phys. Rev. [2] **54** [1938] 185/8).
[6] Makhov, A. F.; Gutkin, A. A. (Izv. Vyssh. Uchebn. Zaved. Fiz. **1958** No. 1, pp. 113/9 from C. A. **1960** 17057).
[7] Seiler, H. (Z. Angew. Physik **22** [1967] 249/63, 255).
[8] Kollath, R. (Ann. Physik [6] **1** [1947] 357/80, 363, 377).
[9] Bronshtein, I. M.; Karasik, B. S.; Krainskii, I. L.; Khinich, I. I. (Fiz. Tverd. Tela [Leningrad] **16** [1974] 3472/4; Soviet Phys.-Solid State **16** [1974] 2248/9).
[10] Bronshtein, I. M.; Segal', R. B. (Dokl. Akad. Nauk SSSR **123** [1958] 639/42; Soviet Phys.-Dokl. **3** [1958] 1184/7).

[11] Koshikawa, T.; Goto, K.; Shimizu, R.; Ishikawa, K. (J. Phys. D **7** [1974] L174/L177).
[12] Bronshtein, I. M.; Fraiman, B. S. (Dokl. Akad. Nauk SSSR **135** [1960] 1097/100; Soviet Phys.-Dokl. **5** [1960] 1273/6).
[13] Bronshtein, I. M.; Smorodina, T. A. (Zh. Eksperim. Teor. Fiz. **27** [1954] 215/23, **29** [1955] 495/9; Soviet Phys.-JETP **2** [1956] 410/3).
[14] Miller, B. L. (J. Chem. Phys. **23** [1955] 599), Miller, B. L.; Porter, W. C. (Phys. Rev. [2] **85** [1952] 391).
[15] Chernov, G. Ya.; Akkerman, A. F.; Botvin, V. A. (At. Energiya SSSR **43** No. 2 [1977] 124/6; C. A. **87** [1977] No. 126036).
[16] Jenkins, L. H.; Zehner, D. M.; Chung, M. F. (Surf. Sci. **38** [1973] 327/40, 329, 334).
[17] Bronshtein, I. M.; Fraiman, B. S. (Fiz. Tverd. Tela [Leningrad] **3** [1961] 1371/2, 1638/49; Soviet Phys.-Solid State **3** [1961] 995/6, 1188/96).
[18] Fitting, H.-J.; Glaefeke, H.; Wild, W.; Neumann, G. (J. Phys. D **9** [1976] 2499/510).
[19] Bronshtein, I. M.; Pronin, V. P. (Fiz. Tverd. Tela [Leningrad] **17** [1975] 2431/3; Soviet Phys.-Solid State **17** [1975] 1610).
[20] Segal', R. B. (Izv. Vyssh. Uchebn. Zaved. Fiz. **1961** No. 6, pp. 27/9 from C. A. **57** [1962] 1667).

[21] Bronshtein, I. M.; Dolinin, V. A. (Fiz. Tverd. Tela [Leningrad] **9** [1967] 683/4, 2718/27; Soviet Phys.-Solid State **9** [1967/68] 535/6, 2133/40).

[22] Bronshtein, I. M.; Denisov, S. S. (Fiz. Tverd. Tela [Leningrad] **7** [1965] 1846/55, **9** [1967] 938/9; Soviet Phys.-Solid State **7** [1965] 1484/91, **9** [1967] 731/2).

[23] Fitting, H.-J.; Technow, R. (Phys. Status Solidi A **76** [1983] K151/K154).

[24] Bronshtein, I. M.; Segal', R. B. (Fiz. Tverd. Tela [Leningrad] **1** [1960] 1489/99; Soviet Phys.-Solid State **1** [1960] 1365/74).

[25] Pertsev, A. N.; Khodasevich, V. V.; Kas'ko, I. V.; Ermolkevich, E. S. (Vestsi Akad. Navuk BSSR Ser. Fiz.-Energ. Navuk **1984** No. 2, pp. 29/32 from C. A. **101** [1984] No. 45221).

[26] Makhov, A. F. (Fiz. Tverd. Tela [Leningrad] **17** [1975] 2408/10; Soviet Phys.-Solid State **17** [1975] 1589/90).

[27] Pilyankevich, A. N.; Vereshchaka, V. M. (Zavodsk. Lab. **49** No. 10 [1983] 26/9; Ind. Lab. [USSR] **49** [1983] 1018/22).

[28] Tilinin, I. S. (Poverkhnost **1986** No. 3, pp. 13/22; C. A. **104** [1986] No. 156219).

[29] Kanchenko, V. A.; Kryn'ko, Yu. N.; Mel'nik, P. V.; Nakhodkin, N. G. (Fiz. Tverd. Tela [Leningrad] **25** [1983] 1448/53; Soviet Phys.-Solid State **25** [1983] 832/5).

[30] Kryn'ko, Yu. N.; Koval, I. F.; Mel'nik, P. V.; Nakhodkin, N. G.; Shaldervan, A. I. (Ukr. Fiz. Zh. [Russ. Ed.] **17** No. 2 [1972] 302/6; C. A. **76** [1972] No. 118732).

[31] Grais, K. I. (Japan. J. Appl. Phys. **15** [1976] 1973/5).

[32] Bronshtein, I. M.; Shchuchinskii, Ya. M. (Radiotekh. Elektron. **6** [1961] 670 from C. A. **56** [1962] 2985; Izv. Vyssh. Uchebn. Zaved. Fiz. **1962** No. 4, p. 182 from C. A. **63** [1965] 10704).

[33] Blankenburg, S. A.; Cobb, J. K.; Muray, J. J. (IEEE Trans. Nucl. Sci. **12** [1965] 935/42 from C. A. **63** [1965] 10925).

[34] Hofstädter, R.; Fechter, H. R.; McIntyre, J. A. (Phys. Rev. [2] **91** [1953] 422/3).

[35] Advani, M. K.; Shah, G. Z.; Gatha, K. M. (Current Sci. [India] **24** [1955] 367/8).

[36] Forman, R. (IEEE Trans. Electron. Devices **25** No. 1 [1978] 69/70 from C. A. **88** [1978] No. 179851).

[37] Fadavi, M. (Diss. Univ. Keele, Staffs., Engl., 1981 from Phys. Abstr. **85** [1982] No. 1777, p. 737, entry 9230).

[38] Ritz, V. H.; Thomas, R. E.; Gibson, J. W.; Klebanoff, J. (SIA Surf. Interface Anal. **11** No. 6/7 [1988] 389/97 from C. A. **109** [1988] No. 120933).

[39] Suhrmann, R.; Kundt, W. (Z. Physik **120** [1943] 363/82, **121** [1943] 118/32).

15.8.2 Stopping Power (Energy Loss of Electrons)

The interaction of fast electrons with matter depends on the target material and on the energy of the primary electrons, particularly for initial energies E_0 below 5 keV. The energy distributions of backscattered and transmitted electrons (providing information about the elastic or inelastic character of collisions and interaction processes) were investigated by many authors, mostly with regard to the effect of varying the atomic number Z of the target. But the ratio of the mean energy $\bar{E}/E_0$ (x/R) and the normalized penetration characteristic η_T (x/R) depend on the initial energy E_0 as well as the backscattering fractions η_{RO} of several bulk materials which are also functions of E_0. It was shown in the early literature that the behavior of the η_{RO} (Z) curve is quite different for an initial energy of $E_0 = 0.5$ keV than for $E_0 = 2$ keV. Because of the increased use of scanning electron microscopy (SEM), electron probe microanalysis (EPMA), and electron energy-loss spectroscopy (EELS), considerable interest has arisen in multiple scattering and the energy dissipation of fast electrons within solids [1].

The energy distributions of electrons transmitted through thin free films of Be (Al and Ge) with initial energies of $E_0 = 0.5$ to 4.0 keV were studied with a spherical retarding-field analyzer. The most probable (E^m/E_0) and mean ($\overline{E}/E_0$) energies as a function of the penetration depth x depend on the Z of the target material and the initial energy E_0. The energy loss of transmitted electrons at the average penetration depth $\overline{x}$ is equal to the backscattered electrons. A relationship between the penetration parameter (p), the backscattering fraction (η_{RO}), and the energy dissipation ($\overline{E}/E_0$) of fast electrons in solid materials, including Be, is shown in one of the figures given in the paper [1].

Using a new film-bulk method, the absorption depth distribution of fast electrons within Be, Al, Cu, and Au targets as well as their inner SE generation was directly measured. The primary electrons (PE) had initial energies of $E_0 = 1$ to 10 keV. The most probable and average absorption depths (x^m and $\overline{x}$, respectively) were determined as functions of the maximum range $R(Z, E_0)$ in the various target materials of atomic number Z (the maximum range R is defined by $\eta_T(R) = 1\%$, where η_T is the transmission fraction). The characteristic of the inner SE generation had a maximum near the surface and agreed qualitatively with the depth dependence of the energy transfer of scattered PE to the solid obtained by other authors from Monte Carlo calculations. By integration of the measured SE generation functions, the total number of SE within the bulk targets, as well as the mean excitation energy for a single SE, was determined. For a bulk Be target and $2 \leqq E_0 \leqq 10$ keV, the penetration parameter p = 2.7, the re-emission fraction $\eta_{RO} = 0.06$, and both the most probable absorption depth x^m/R and the average absorption depth $\overline{x}/R = 0.53$ [2].

For the elements Be, Mg, K, Al, C, Si, Sb, Bi, and Ge calculations of the mean free paths in inelastic interactions and stopping powers of low-energy electrons up to 20 keV were performed (in the region of low primary energy the processes of electron interaction with matter are complicated by bound electrons). In the computational model, Lindhard's formalism of the dielectric response function for pair and plasmon excitation and the classical cross section for ionization processes were used. Total mean free paths and stopping powers are tabulated in the paper. The results were compared with theoretical values based on a statistical model and experimental values (from literature cited in the paper). For energies above 10 keV the calculated stopping power dE/dx agreed well with values calculated from the Bethe-Bloch formula. A Monte Carlo method was used for the calculation of the energy loss distribution of electrons passing through thin targets. Blunck-Leisegang's theory failed to describe the energy straggling at electron energies below 10 keV [3].

A general method for calculation of the electron mean free path for material of any atomic number using experimental results for Mo, W, and Cr (from literature) was developed and applied, for example, to Be and other metals. A comparison of several calculation methods revealed close agreement between experimental data and those calculated with the newly developed method [4].

Monte Carlo calculations of keV electron-energy loss distributions were performed using a simple model of the generalized oscillator strength density to simulate inelastic collisions. Comparison was made with experimental results reported for 2- and 3-keV electron beam transmission through Be [1] and for other materials (Al, Fe) [5].

Electronic energy-loss moments were calculated as functions of impact parameter for $E_0 = 1$ to 7 MeV incident upon Be, Al, Ta, Cu, and Ag targets [6].

Parameters of the energy distribution (e.g., probable energy) of β-particles during passage through matter (e.g., Be foils) were experimentally determined as a function of foil thickness, target material, etc. [7].

For early investigations on electron scattering and energy dissipation (straggling, distribution) on the passage of impacted electrons or β-particles through Be foils, see for instance [8 to 23]; for theory, see [18, 24, 25]. The influence of the polarization effect on the energy loss of low-energy electrons passing through Be (Al and C) foils is described in [26].

References:

[1] Fitting, H.-J. (J. Phys. D **8** [1975] 1480/6).
[2] Fitting, H.-J.; Glaefeke, H.; Wild, W.; Neumann, G. (J. Phys. D **9** [1976] 2499/510).
[3] Akkerman, A. F.; Chernov, G. Ya. (Phys. Status Solidi B **89** [1978] 329/33).
[4] Tokutaka, H.; Nishimori, K.; Hayashi, H. (Surf. Sci. **149** [1985] 349/65).
[5] Liljequist, D. (J. Appl. Phys. **57** [1985] 657/65).
[6] Winterbon, K. B. (Radiat. Eff. **79** [1983] 251/63; C.A. **100** [1984] No. 41446).
[7] Pertsev, A. N.; Khodasevich, V. V.; Kas'ko, I. V.; Ermolkevich, E. S. (Vestsi Akad. Navuk BSSR Ser. Fiz.-Energ. Navuk **1984** No. 2, pp. 29/32 from C.A. **101** [1984] No. 45221).
[8] Ruthemann, G. (Ann. Physik [6] **2** [1948] 113/34, 124).
[9] Paul, W.; Reich, H. (Z. Physik **127** [1950] 429/42).
[10] Chen, J. J. L.; Warshaw, S. D. (Phys. Rev. [2] **84** [1951] 355/61).

[11] Kalil, F.; Birkhoff, R. D. (Phys. Rev. [2] **91** [1953] 505/9).
[12] Ford, G. W.; Mullin, C. J. (Phys. Rev. [2] **110** [1958] 520/5).
[13] Husain, S. A.; Putman, J. L. (Proc. Phys. Soc. [London] A **70** [1957] 304/5).
[14] Agu, B. N. C.; Burdett, T. A.; Matsukawa, E. (Proc. Phys. Soc. [London] **72** [1958] 727/32).
[15] Westermark, T. (Arkiv Kemi **17** [1961] 101/38).
[16] Wright, K. A.; Trump, J. G. (J. Appl. Phys. **33** [1962] 687/90).
[17] Nakai, Y. (Japan. J. Appl. Phys. **2** [1963] 743/56).
[18] Ziegler, B. (Z. Physik **151** [1958] 556/62).
[19] Hall, C. E. (NBS-C-527 [1954] 95/100 from C.A. **1954** 8036).
[20] Mozley, R. F.; Smith, R. C.; Taylor, R. E. (Phys. Rev. [2] **111** [1958] 647/9).

[21] Hudson, A. M. (Phys. Rev. [2] **105** [1957] 1/6).
[22] Hanson, A. O.; Lanzl, L. H.; Lyman, E. M.; Scott, M. B. (Phys. Rev. [2] **84** [1951] 634/7).
[23] Schweizer, F.; McKinley, W. A. (Bull. Am. Phys. Soc. [2] **6** [1961] 520).
[24] Pilyankevich, A. N. (Zh. Tekh. Fiz. **30** [1960] 232/8 from C.A. **1960** 19150).
[25] Spencer, L. V. (Phys. Rev. [2] **98** [1955] 1597/615).
[26] Hsiao, Chen-Hsi; Chou, Teh-Lin (Wu Li Hsueh Pao **16** No. 2 [1960] 98/106 from C.A. **60** [1964] 7546).

15.8.3 Characteristic Electron Energy Losses

Electron energy-loss spectroscopy (EELS) investigations concerning plasmons (volume plasmons ~18 eV below the elastic peak) and plasmon scattering are compiled in the Section 14.12 on pp. 86/92. In addition, Jenkins et al. [1] investigated the characteristic loss spectra from clean and oxidized Be surfaces for $E_{PE} = 600$, 250, and 100 eV (normal incidence). Multiple losses attributed to bulk plasmons were observed at 36 and 54 eV below the elastic peak, in addition to the surface loss at 12 eV, and minima at 3 and 5 eV which can be attributed to excitation of interband transitions [1]; cf. p. 55. A comparison of the characteristic energy losses of electrons with the fine structure found on the short wavelength side of the X-ray absorption edge for several metals including Be indicates a correlation between the two phenomena [2].

Theoretical and experimental studies were performed on the inner-shell edge profiles in electron energy-loss spectroscopy, i.e., in the case of Be the edge profile for the K edge (1s). Using a Hartree-Slater central field model, theoretical spectra were calculated and compared with experimental spectra (shown in figures in the paper) fitting in all cases from 50 eV from the threshold onwards. According to simple theory, K shells should show a sharp rise at the threshold followed by a slow decay. In experimental spectra this is often complicated by fine structure effects near the threshold from transitions to σ^* and π^* antibonding orbitals. In Be transitions to σ^* and π^* states account for peaks in the experimental data ($E_{PE} = 200$ keV) occurring up to 20 eV beyond the threshold energy of 111 eV [3].

The structural sensitivity of electron energy-loss near-edge structure (ELNES) was studied theoretically by Spence [4]. Calculations are reported using the multiple scattering Green function method for the transmission ELNES of inner-shell excitations. A high degree of sensitivity to the local crystallographic environment was revealed. An experimental K-edge spectrum ($E_{PE} = 120$ keV) from Be was used to demonstrate the removal of plasmon multiple inelastic scattering artifacts from ELNES by the method of logarithmic deconvolution without adjustable parameters. Figures showing the experimental transmission energy-loss spectrum from a thin Be foil in the region of the K edge, the low-loss region of the spectrum showing plasmons, and the single scattering energy-loss spectrum obtained by logarithmic deconvolution are given in the paper [4].

The conditions were studied for the validity of the "dipole approximation" (electric dipole selection rules relate the angular-momentum quantum numbers of the core and ejected electrons) in EELS of atomic K-shell excitations on an analytic model that takes account of the wave function of the ejected core electrons, as well as that of the core state. A closed-form expression for the limiting magnitude (q_d) of the momentum-transfer vector as a function of both the atomic number and the energy (ε) of the ejected core electrons was derived. A few selected elements including Be were treated for example [5].

For characteristic energy losses of electrons in Be and Mg, see [6].

Ionization-loss spectra were produced by the interaction of a 260 eV primary electron beam (normal incidence) with both clean and oxidized Be surfaces. On the clean Be surface only one peak 113 eV below the 260 eV elastic peak was observed, whereas oxidation produced an additional structure with peaks at 116 and 122 eV below the elastic peak [1].

References:

[1] Jenkins, L. H.; Zehner, D. M.; Chung, M. F. (Surf. Sci. **38** [1973] 327/40, 329, 333/4).
[2] Leder, L. B.; Mendlowitz, H.; Marton, L. (Phys. Rev. [2] **101** [1956] 1460/7).
[3] Ahn, C. C.; Rez, P. (Ultramicroscopy **17** [1985] 105/15; C.A. **104** [1986] No. 80936).
[4] Spence, J. C. H. (Ultramicroscopy **18** [1985] 165/72; C.A. **104** [1986] No. 78297).
[5] Saldin, D. K.; Yao, J. M. (Phys. Rev. [3] B **41** [1990] 52/61).
[6] Pronin, V. P. (Vzaimodeistvie Elektronov i Fotonov s Tverd. Telom L. **1984** 47/50 from C.A. **102** [1985] No. 213689).

15.8.4 Auger Spectrum

Information on the Auger spectrum (excitation) of solid Be is given together with the description of the Auger spectra from neutral free Be atoms and ions (Be$^+$, Be^{2+}) in "Beryllium" Suppl. Vol. A 1, 1986, pp. 143, 160, and 174 ff., respectively.

In the early studies, agreement existed about the main Auger peak at about 104 to 105 eV attributed to a KVV transition from pure Be. Discussions were concerned with a second strong peak near 92 eV observed by some authors and reported to be weak or absent by other authors. Similarly differing were the assignments of this peak as either an intrinsic property of pure Be or due to impurities such as oxygen or silicon. The satellites found in the energy regions below and above the main peak were also under discussion. In many cases the knowledge of the energy separation of the major Auger peak and high-energy satellites was insufficient to identify the mechanism involved. The possible mechanisms responsible for emission of low- and high-energy satellites were extensively discussed by Watts [1] and other authors cited in [1] in terms of a one- or two-step model (not unambiguously resolved experimentally).

In the Be KVV Auger spectrum of Jenkins et al. [2] with the main peak at 105 eV (see figures in the papers), first and second order volume plasmon losses by Auger electrons ($\hbar\omega_p \approx 19$ eV) can be seen at the lower energy side of the spectrum at about 86 and 67 eV, respectively. A shoulder at 91 eV was attributed to a surface plasmon loss. Two relatively weak, higher energy satellites were observed at 18 and 38 eV above the main peak; the lower energy satellite at 123 eV was assigned to coupling of energy from bulk plasmon de-excitation ($\hbar\omega_p \approx 18$ eV) with Auger electrons (KVV + $\hbar\omega_p$ events) and the higher energy event at 143 eV to Auger electrons ejected from Be atoms with doubly ionized K levels (that is (K)$_2$VV emission). These higher energy events (K)$_2$VV and KVV + $\hbar\omega_p$ were said to occur in Be (at a primary electron beam energy of 2.5 keV) with about equal probability. The data were obtained in a standard Varian LEED system (base pressure <10^{-10} Torr) with the three-grid LEED optics as retarding potential analyzer, and identical results were obtained from both a bulk specimen with no surface contaminants detectible by Auger spectroscopy and a film evaporated in situ [2]. The general features of this spectrum agreed with the data of Suleman, Pattinson [3] who attributed the two peaks at 104 and 94 eV to clean Be and an oxidized Be surface, respectively; see, however [5].

The two main peaks at 92 and 104 eV together with lowered losses at 84, 75, 66, 53, and 45 eV are clearly shown in the Auger spectrum of a thermally cleaned Be surface by Maguire [4]. The lowered losses originate from the individual peaks in the Auger spectrum and represent multiple losses suffered by some of the emerging electrons to collective bulk oscillations ($\hbar\omega_p \approx 19$ eV plasmons) in the solid. On the high-energy side of the main spectrum two small peaks were observed at 112 and 124 eV. A separation in energy of these satellites from the main Auger peaks coupled with their similar behavior to oxygen exposure strongly suggested that the 112 and 124 eV peaks reflect plasmon-gain processes associated with the 92 and 104 eV peaks, respectively, i.e., stimulated de-excitation of some of the emerging Auger electrons by energy absorption of a bulk plasmon before emission from the solid. Another Auger emission spectrum was obtained from a freshly evaporated Be film at room temperature onto a clean Be surface originally characterized by the Auger spectrum just discussed. This second spectrum showed only one major peak at 103 eV of comparable width to the two major peaks observed in the first spectrum, with multiple lowered losses (n$\hbar\omega_p$) at 84, 64, and 37 eV. Also observable is a single high-energy satellite of ~1% relative intensity and separated in energy from the Auger peak by an amount equivalent to a bulk plasma vibration. The first spectrum could be regenerated from the second spectrum by annealing for 15 min at 750°C. The second spectrum could be obtained from the first spectrum, i.e., one

Auger peak and one satellite from originally two Auger peaks and two satellites, by 1 kV Ar ion bombardment. Subsequent annealing as mentioned reproduced the original two peak Auger spectrum including the two high-energy satellites. The differences in the Auger spectra have been previously explained [5] as originating from a well-defined separation of the 2s and 2p energy distributions to give two distinct peaks for a thermally cleaned surface, but to give only one peak after evaporation onto a cold surface or Ar ion bombardment due to a disordered phase produced. The different Auger spectra result from differing electron energy distributions which do not affect the volume electron density and hence the bulk plasma vibrations. The high-energy satellites observed in the Be KVV Auger spectra uniquely reflect differing electron energy distributions in the valence bands, due to stimulated de-excitation of the emerging Auger electrons by energy absorption from collective bulk plasma oscillations within the solid. The valence band structural characteristics arise from an ordered and disordered phase resulting from different surface preparation procedures. The high-energy satellite observed at 143 eV [2] was suggested to be due to an unidentified impurity [4]. Discrepancies in the Auger emission spectra [3] and [6] (erroneously assumed BeO formation at the surface) were also explained in [5].

A comparison of the results [2] with X-ray emission spectra of Be (Li and Na) showed that the satellite observed 18 eV above the K Auger peak in Be and ascribed to Auger electrons gaining 18 eV by volume-plasmon decay in bulk Be was not found in the X-ray spectrum. However, the probability of volume-plasmon gain by Auger electrons passing through the metal was considered to be much greater than the probability of creating a high-energy X-ray photon by the simultaneous decay of a core level electron and a volume plasmon. The satellite 38 eV above the K Auger peak was attributed to double-plasmon gain or to double ionization of the K level in Be [7].

Auger electron spectroscopy (AES) was applied to study the slow oxidation of Be (and Mg) surfaces from residual oxygen. The fine structure in the Auger spectra involving the valence band was interpreted in terms of the density of states of the valence band. Additional fine structures in the Auger spectra of clean Be and Mg were attributed to the plasmon-loss and ionization-loss mechanisms [8].

The Be Auger spectrum (from K-shell excitations by slow heavy-ion bombardment at 45° incidence) is composed of two main structures: a broad structure similar (not identical) to that observed under electron impact assigned to the KVV transition involving two valence electrons of the solid, due to atoms that decay in the bulk. The other structure is a sharper line originating from sputtered excited atoms which survive the transition to the surface and decay outside the solid after neutralization by resonant electron transfer from the Be valence band. The Auger yields indicated that excitations occur mainly from symmetric collisions between target atoms. This is supported by Monte Carlo simulations of the atomic collision cascade in the solid [9].

Auger electron spectroscopy (AES) was developed as a quantitative method for solid surface analysis (to calculate the AES signal vs. film thickness in the film monolayer overgrowth) and applied to Be on Cu and other metals. The calculated results agreed well with experimental data. Details of the method are described by Tokutaka et al. [10]. Auger electron spectroscopy was used to study the surface chemical composition of Be cathodes which show significant differences (especially contamination by C) depending on the time of operation as well as on variations in the manufacturing procedures [11].

As the high surface sensitivity of AES (and XPS) is directly related to the escape probability of the signal electrons emitted within a solid, a Monte Carlo simulation was used to study systematically the depth distribution function (DDF) which describes the attenuation of signal electrons in solids with thin overlayers. Extensive model calculations performed for several substrate-overlayer combinations of Be, Al, Si, In, Ag, and Au for signal electrons over the range of energies from 250 to 1000 eV revealed no significant effect of backscattering from the interface on the DDF. The shape of the DDF at the surface showed a strong deviation from exponential behavior; the extent of this effect depended on the material and the energy. However, the DDF of electrons originating from the substrate exhibits a significant dependence on the overlayer material. Results for the most interesting combinations Be/Au and Au/Be are given in figures in the paper [12].

Calculations with multiconfiguration interaction wave functions (including relativistic and mass polarization corrections) were performed for Be I (and B II) to obtain the energies of Auger electrons in decay processes. The results were used to identify the recalibrated high-resolution spectra. For Be, previously unidentified lines in the observed spectrum were identified. For both Be I and B II, agreement between the results from calculation and experiment was excellent [13].

References:

[1] Watts, C. M. K. (J. Phys. F **2** [1972] 574/83).
[2] Jenkins, L. H.; Zehner, D. M. (Solid State Commun. **12** [1973] 1149/51), Jenkins, L. H.; Zehner, D. M.; Chung, M. F. (Surf. Sci. **38** [1973] 327/40), Zehner, D. M.; Barbulesco, N.; Jenkins, L. H. (Surf. Sci. **34** [1973] 385/93).
[3] Suleman, M.; Pattinson, E. B. (J. Phys. F **3** [1973] 497/504, **1** [1971] L24/L27).
[4] Maguire, H. G. (Solid State Commun. **45** [1983] 71/3).
[5] Maguire, H. G.; Augustus, P. D. (Phil. Mag. [8] **30** [1974] 95/103).
[6] Musket, R. G.; Fortner, R. J. (Phys. Rev. Letters **26** [1971] 80/2).
[7] Arakawa, E. T.; Williams, M. W. (Phys. Rev. [3] B **8** [1973] 4075/8).
[8] Fadavi, M. (Diss. Univ. Keele, Staffs., Engl., 1981 from Phys. Abstr. **85** [1982] No. 1777, p. 737 entry 9230).
[9] Grizzi, O.; Baragiola, R. A. (Phys. Rev. [3] A **30** [1984] 2297/303).
[10] Tokutaka, H.; Nishimori, K.; Takashima, K.; Ichinokawa, T. (Surf. Sci. **133** [1983] 547/79, 574).

[11] Ritz, V. H.; Thomas, R. E.; Gibson, J. W.; Klebanoff, J. (SIA Surf. Interface Anal. **11** [1988] 389/97 from C.A. **109** [1988] No. 129933).
[12] Werner, W. S. M. (Surf. Sci. **257** [1991] 319/27).
[13] Chung, K. T. (J. Phys. B **23** [1990] 2929/43, 2939).

15.8.5 Impact-Induced Radiation

Transition radiation, the production of photons when charged particles cross the interface between two media, was applied to high-energy particle detection because the total X-ray output is directly proportional to the energy of the particle. The absolute differential production efficiencies (in photons emitted per unit solid angle, see equation 9 in the paper) for X-rays emitted from each of three transition radiators were measured for incident electron-beam energies of 17.2, 25, and 54 MeV. The radiators were made of stacks of 1.0-μm thin foils: 18 foils of Be, 18 foils of C, and 30 foils of Al. The radiation spectra were most intense between 0.5 and 2.5 keV, peaking at 0.8, 1.3, and 1.3 keV, respectively. The angular distribution of the transition radiation from the Be-foil stack was measured for the three electron-beam energies and found to agree well with theoretical predictions. Theoretical calculations including both the two-surface interference and photon attenuation in the foil material agreed well with these data. A method of enhancing the output using a split-foil stack was considered. Results, especially for Be, are given in tables and figures in the paper [1].

The bremsstrahlung spectra in the forward direction produced by 8.5 MeV electrons were measured with a sodium iodide scintillation crystal and Be, Ni, and Pt targets of several thicknesses. The measured spectra (shown in figures in the paper) agreed rather well with semiempirical calculations [2]. The bremsstrahlung cross section, differential in photon energy and angle, was determined by measuring the X-rays emitted from thin targets of Be (Al and Au) for incident electron energies of 0.5 and 1.0 MeV at angles of 0°, 10°, 20°, 30°, 60°, 90°, and 120°. A comparison of the cross section for Be and Al gave evidence for electron-electron bremsstrahlung. For figures and theoretical considerations, see the paper [3]. For the angular distribution of bremsstrahlung (from 17 MeV electrons), see [4], and other studies [5]. The angular dependence of the linear polarization near the high-frequency limit of the bremsstrahlung spectrum for 500 keV electrons, measured on Be and Au with a polarimeter dependent on the polarization sensitivity of the Compton process, confirmed the polarization reversal predicted at large angles [6].

The angular distribution of bremsstrahlung from 15 MeV electrons incident on thick targets of Be (Al and Pb) was measured. Excellent agreement was found between the measured and calculated spectral shapes. There was no significant angular variation in the ratio of the measured and calculated yields. The predictions of the theories may be improved by including target-scattered photons [7].

An improved cross-section formula for thin-target electron bremsstrahlung for Monte Carlo transport simulation was derived. Predictions of the proposed formula gave better agreement with experimental data than existing bremsstrahlung cross sections over wide ranges of electron energies between 1 and 250 MeV and target atomic numbers 4 to 82 [8]. For a calculation of the degradation of the bremsstrahlung spectra of electrons in light-weight scattering materials such as Be, see [9].

References:

[1] Piestrup, M. A.; Kephart, J. O.; Park, H.; Klein, R. K.; Pantell, R. H.; Ebert, P. J.; Moran, M. J.; Dahling, B. A.; Berman, B. L. (Phys. Rev. [3] A **32** [1985] 917/27).
[2] Kimura, M.; Mutsuro, N.; Ohnuki, Y.; Shoda, K.; Sugawara, M.; Tohei, T.; Yuta, H. (J. Phys. Soc. Japan **14** [1959] 387/96).
[3] Motz, J. W. (Phys. Rev. [2] **100** [1955] 1560/71).
[4] Lanzl, L. H.; Hanson, A. O. (Phys. Rev. [2] **83** [1951] 959/74, 968).
[5] Blunck, O. (Ann. Physik [6] **9** [1951] 373/80).
[6] Motz, J. W.; Placious, R. C. (Phys. Rev. [2] **112** [1958] 1039/40).
[7] Faddegon, B. A.; Ross, C. K.; Rogers, D. W. O. (Med. Phys. **18** [1991] 727/39 from C. A. **115** [1991] No. 192 024).
[8] Al-Beteri, A. A.; Raeside, D. E. (Nucl. Instrum. Methods Phys. Res. B **44** [1989] 149/57 from C. A. **112** [1990] No. 65 126).
[9] Zavgorodnii, S. F. (Vestn. Akad. Nauk Kaz. SSR **1989** No. 8, pp. 83/5 from C. A. **111** [1989] No. 182 552).

15.8.6 Fluorescence Yield

Some information about elemental Be in its standard state (experimental data (3.6 or 3.04) ×10^{-4}) is given in "Beryllium" Suppl. Vol. A 1, 1986, p. 161.

Applying a general theory of autoionizing states (based on projected function spaces) the fluorescence yield of Be^+ $1s2s^2$ ^2S was calculated to be 1.2×10^{-4} [1]. This value, being closer to the available experimental values from the solid state (see p. 131) than the results of earlier many-electron calculations, implies that the atomic character of fluorescence in the solid is much larger than previously [2] predicted. Fluorescence yields for free Be^+ were mostly calculated for the pure initial state ($1s2s^2$ ^2S) giving smaller values than those observed in the solid state. Therefore, experimental and theoretical data were considered not to be useful to compare since the initial Be^+ states in the valence band of solid Be may include some p character. This suggestion was confirmed. The ΔSCF cluster theory (UHF calculations on Be_{13}) was combined with calculations of atomic electron correlation to predict the 1s binding energy of Be metal as 115.4 eV in agreement with experimental data (115.2 to 115.6 eV). Upon creation of the 1s hole, the p character of the valence band increased significantly. The discrepancy between the experimental solid state fluorescence yield and the theoretical prediction for the atomic state could be semiquantitatively explained by this significant contribution of p character in the valence band for the excited state [3].

References:

[1] Nicolaides, C. A.; Komninos, Y.; Beck, D. R. (Phys. Rev. [3] A **27** [1983] 3044/52).
[2] Kelly, H. P. (Phys. Rev. [3] A **9** [1974] 1582/5).
[3] Nicolaides, C. A.; Zdetsis, A. D.; Andriotis, A. N. (Solid State Commun. **50** [1984] 857/60).

15.9 Positron Annihilation

The positron lifetimes in bulk metal (τ_f), where f denotes free, untrapped positrons, and in monovacancies (τ_{1v}) of deformed samples of Be are reported to be $\tau_f = 137$ ps and $\tau_{1v} = 178$ ps [1] from [2]. The earlier value $\tau_1 = 213$ ps (main component of the time distribution curve) given by Weisberg, Berko [3] is much higher. The value found by Bisi et al. [4] related to Al is also too high.

Local-density calculations agree with the lower experimental value $\tau_{exp} = 142$ ps (from literature), $\tau_{calc} = 134$ ps (calculated using the LDA for all electrons in the metal), $\tau'_{calc} = 140$ ps (obtained by calculating the core electron contribution separately using the "independent particle" model) [5].

The angular correlation of positron annihilation radiation (ACPAR) in Be single crystals was measured and the data were used to deduce the shape (anisotropy) of the Fermi surface [6]. Further measurements of ACPAR from oriented Be (and Mg) crystals revealed marked anisotropy in the case of Be, see [7]. The measurement of the two-gamma angular distribution of positron annihilation radiation in metals such as Be shows up the band character of the electrons as well as many-body effects [8].

The positron band structure of Be (e^+ annihilating with e^- from the thermalized state $k^+ = 0$, i.e., from the absolute minimum of the lowest band) was calculated using the APW method. Qualitative agreement was found between theoretical and experimental ACPAR curves; agreement became better when the higher momentum components of the Bloch functions were taken into account [9]. For previous studies of positron annihilation, carried out on oriented Be single crystals and showing the anisotropy of the Fermi surface, see [10]; cf. [8].

Older experimental investigations of the time and angular distribution of positron annihilation radiation in Be are reported in [11 to 16]; for studies on oriented Be (and Mg) crystals, see

[17], and for an old review, see [18]. Three-quantum annihilation of positrons in solids including Be is described in [19].

The measured widths of annihilation lines in uni- and bivalent metals were analyzed. For the light metals Be, Li, Na, and Mg the positrons (e⁺) were found to annihilate almost exclusively with conduction electrons [20].

The radiations from high-energy positrons incident on a Be target were investigated. Yields of positrons and nearly monoenergetic photons in the range 2 to 14 MeV were measured. The bremsstrahlung spectra from 8.5 MeV positrons and electrons were compared, showing agreement within the experimental error for the yields and spectral distributions; for experimental details and figures, see the paper [21].

The differential cross section at 0°C for the annihilation in flight of 8.5 MeV positrons was measured. The positrons were directed onto a Be target where annihilation occurred, and the annihilation photons were measured [22]. The scattering of 200 MeV positrons by electrons was studied using a Be target [23].

In a study on thermalization and diffusion of positrons in solids, the critical positron energy, E_c, below which thermalization of positrons by phonon scattering occurs, was estimated to be $E_c = 0.38$ eV for Be [24].

For the use of positron annihilation data in deriving the electronic configurations of the ion core in metals such as Be where the calculated electron densities and the interstitial variation agreed well with experimental data, see [25].

References:

 [1] Kögel, G.; Triftshäuser, W. (in: Coleman, P. G.; Sharma, S. C.; Diana, L. M.; Positron Annihilation, North-Holland, Amsterdam 1982, p. 595).
 [2] Seeger, A.; Banhart, F. (Phys. Status Solidi A **102** [1987] 171/9).
 [3] Weisberg, H.; Berko, S. (Phys. Rev. [2] **154** [1967] 249/57).
 [4] Bisi, A.; Faini, G.; Gatti, E.; Zappa, L. (Phys. Rev. Letters **5** [1960] 59/60).
 [5] Jensen, K. O. (J. Phys. Condens. Matter **1** [1989] 10595/602).
 [6] Stewart, A. T.; Shand, J. B.; Donaghy, J. J.; Kusmiss, J. H. (Phys. Rev. [2] **128** [1962] 118/9).
 [7] Berko, S. (Phys. Rev. [2] **128** [1962] 2166/8).
 [8] Terrell, J. H. (Diss. Brandeis Univ. 1965, pp. 1/124 from Diss. Abstr. B **27** [1967] 4516).
 [9] Oriade, J. O. (Intern. J. Quantum Chem. **20** [1981] 891/6).
[10] Shand, J. B. (Diss. Univ. North Carolina, Chapel Hill 1965, pp. 1/96 from Diss. Abstr. **26** [1965/66] 4034), Shand, J. B.; Stewart, A. T. (AED-CONF-65-11-8 [1965] 9 pp. from C.A. **66** [1967] No. 40848).

[11] Bell, R. E.; Graham, R. L. (Phys. Rev. [2] **90** [1953] 644/54).
[12] Lang, G.; DeBenedetti, S.; Smoluchowski, R. (Phys. Rev. [2] **99** [1955] 596/8).
[13] Erdman, K. L. (Proc. Phys. Soc. [London] A **68** [1955] 304/11).
[14] Baskova, K. A.; Dzhelepov, B. S. (Izv. Akad. Nauk SSSR Ser. Fiz. **20** [1956] 951/5; C.A. **1957** 4168).
[15] Lang, G.; DeBenedetti, S. (Phys. Rev. [2] **108** [1957] 914/21).
[16] Green, R. E.; Stewart, A. T. (Phys. Rev. [2] **98** [1955] 486/91).
[17] Berko, S. (Bull. Am. Phys. Soc. [2] **8** [1963] 142).
[18] Berko, S.; Hereford, F. L. (Rev. Mod. Phys. **28** [1956] 299/307).
[19] Graham, R. L.; Stewart, A. T. (Can. J. Phys. **32** [1954] 678/9).
[20] Średniawa, B. (Acta Phys. Polon. **20** [1961] 235/41).

[21] Jupiter, C. P.; Hansen, N. E.; Shafer, R. E.; Fultz, S. C. (Phys. Rev. [2] **121** [1961] 866/70).
[22] Seward, F. D.; Hatcher, C. R.; Fultz, S. C. (Phys. Rev. [2] **121** [1961] 605/9).
[23] Poirier, J. A.; Bernstein, D. M.; Pine, J. (Bull. Am. Phys. Soc. [2] **3** [1958] 421).
[24] Brandt, W.; Arista, N. (Phys. Rev. [3] B **26** [1982] 4229/38).
[25] Johnson, O. (J. Phys. Chem. Solids **42** [1981] 65/76).

15.10 Meson and Muon Capture

The capture process of negative mesons in solids has been a matter for discussion. The meson is first bound to a particular nucleus in a high quantum state ($n \approx 15$) and then proceeds to cascade inwards toward the nucleus, transferring its energy by radiative and Auger transitions. At each intermediate level the meson can 1) be absorbed directly from the intermediate level by the nucleus, 2) make a radiative transition to a lower state and emit a quantum, or 3) make a radiationless transition to a lower state, transferring the energy difference to an atomic electron (mesonic Auger effect). Alternatively, if the energy difference is sufficiently large, it might create an electron pair or excite the nucleus. All three processes mentioned above have been observed for π mesons (pions). In particular, nuclear absorption is exceedingly strong for pions in states of low angular momentum. A π meson in any s state will almost certainly be absorbed. Absorption from the 2p state is practically complete for $Z \gtrsim 11$. The study of the cascade process in π-mesonic atoms is complicated, because little is known about the pion-nucleus interaction. This difficulty does not exist for μ mesons (muons). The muon-nucleus interaction is very weak; it is practically certain that in the course of the cascade all muons reach the 1s state from which they are captured or in which they decay. The cascade time is more than a million times shorter than the muon lifetime. Thus muons are useful in studying the cascade process, in particular the competition between radiative and Auger transitions [1].

The first experimental evidence that polarized positive muons are depolarized when coming to rest inside condensed matter was contained in one of the experiments which showed nonconservation of parity in the $\pi^+ - \mu^+ - e^+$ decay chain. With a given incident beam, the polarization of muons at the time of decay depended critically on the medium in which they decayed. Concurrently, such a process was expected from the behavior of positrons and protons, which are known to attach electrons during the slowing down process and, respectively, form positronium and atomic hydrogen. Therefore it was expected that positive muons would form muonium (hydrogen-like atoms $\mu^+ e^-$) while stopping and depolarize either by the hyperfine interaction or by precession of the muonium (Mu) in fields known to exist in condensed matter [2].

The formation of short-lived π-mesonic atoms as a preliminary step in the absorption of π mesons by nuclei is well established; K, L, and M emission lines from these atoms were studied in varying detail, for Be targets (only K emission) by [3 to 7]. The specific pion-nucleus interaction is an important motivation for investigations. The effect of this interaction on the energy shift of the $2p \rightarrow 1s$(Kα) lines was discussed in [4]: the measured energies of the $2p \rightarrow 1s$ X-rays from the π-mesonic atoms Li through F were lower than those calculated from the Klein-Gordon equation assuming a point-charge Coulomb field and corrected for finite size effects and vacuum polarization. For Be the Kα energy was measured as 42.09 ± 0.10 keV (in excellent agreement with the value 42.0 ± 0.1 keV from [6]) and calculated to be 43.95 keV. This indicated that the net nuclear interaction between a meson in the 1s orbit and a nucleus is repulsive [4]. A value of 43 ± 3.5 keV was measured by [5].

Of equal interest is this interaction effect on the nuclear absorption of mesons directly from their excited states. This absorption process increases strongly with increasing Z; since it competes with radiative transitions, it is exhibited by a decrease in the mesonic X-ray yield. The absolute radiative yields (X-rays per stopped meson) and relative intensities of the K series (2p→1s, 3p→1s, etc.) were measured. The total yield is a maximum in the case of Be and amounts to 21%, and then falls to about 3% at Na. The higher transitions (Kβ, etc.) constitute 26% of the total yield [7].

The radiative yields of the K and L series of μ-mesonic atoms were measured for Li through K. In the low Z region both yield curves exhibited a rapid decrease in yield with decreasing Z. This behavior was attributed to competition between Auger and radiative transitions. For Be the energy of 2p→1s was found to be 33.3 keV, the relative radiative yield 0.33±0.03, and the ratio of higher transitions to total 0.22 [1]. The rapid drop in the yield of mesonic K X-rays in the light elements was suggested to be associated with the capture of μ mesons into the metastable 2s state. Mechanisms for transitions from the 2s to the 1s state and from various p states into the 2s state were investigated in detail for Li, Be, and B [8].

Early investigations on the scattering of mesons by Be targets are reported in [9 to 12]. The scattering of muons in Be plates was studied, and the observed angular distribution agreed with the distribution expected from a Coulomb interaction between muons and atoms of the metal [13]. The mean lifetime of muons (μ⁻ mesons) stopped in Be was determined to be $\tau^- = 2.05 \pm 0.06$ μs [14].

Attenuation cross sections and cross sections for scattering into a fixed angular interval were measured for pions of 33, 46, and 68 MeV energy; target materials were Be, C, Al, and Cu [15]. Positive π meson (π⁺) interactions in Be were obtained for three bands of energy from 15 to 30, 25 to 40, and 35 to 50 MeV. The absorption cross section was found to be energy-dependent [16].

The effects of parity nonconservation in the π–μ–e decay were used to measure the depolarization of positive muons for various substances including Be, see table in the paper [2]. For experiments with a polarized muon beam, using Be as target, see [17].

Muon stopping in Be metal was used as one of the two methods to study μ decay asymmetry and to determine the integral asymmetry parameter in μ decay [18]. The influence of defects on muon diffusion in polycrystalline Be (and In) was studied (in the case of In trapping of positive muons at vacancies was indicated) [19].

Differential cross sections for the production of charged pions (π⁻ and π⁺) by 590 MeV protons in Be, C, and Ni at pion production angles of 22.5° and 90° for pion energies from 6 to 35 MeV were measured because cross sections for the production of low-energy pions are needed for the optimal design of new facilities such as surface muon beams and pionic X-ray detection systems. At the lowest pion energies the cross sections tended towards zero. For Be and C there were more slow π⁺ than π⁻, see table and figures in the paper [20].

The negative muonium ion (Mu⁻≡μ⁺e⁻e⁻) was produced by double electron pickup as a beam of positive muons passed through thin foils of various materials such as Be, Al, and Au. Yields of Mu⁻, Mu and μ⁺ as well are reported in the paper for μ⁺ with kinetic energies from 380 to 800 keV impinging on the thin foils. A Monte Carlo simulation gave good agreement with the experimental data [21].

For high-energy protons, pions, and muons (up to 30 TeV) the energy and angle of the final state particles in bremsstrahlung, direct pair production and, for muons, deep inelastic scattering were determined as a function of the fractional energy loss of the incident particle. The results (given in figures in the paper) were parametrized for convenient use in Monte Carlo simulations. One striking feature of the TeV energy regime, with respect to particle transport, is that with increasing energy bremsstrahlung begins to dominate multiple Coulomb scattering as a source of angular diffusion. Likewise even for heavy particles, direct pair production as well as bremsstrahlung become more important than ionization as a source of energy loss in transport [22].

The cross section of the Compton effect on π^- meson was measured. The reaction $\pi^-A \rightarrow A\pi^-\gamma$ at 40 GeV/c was studied with a Sigma spectrometer on six different nuclear targets (Be, C, Al, Pb, Fe, Cu) in the region of four-momentum transfer $|t| < 0.05$ (GeV/c)^2. The elastic scattering of pions on photons of the nuclear Coulomb field was observed and its total and differential cross sections were measured, the latter being in agreement with theoretical calculations [23]. For theoretical studies on Be, see [24], and for further experimental studies on π^--Be interactions at 40 and 43 GeV/c, see [25 to 27]. The charge-exchange reactions $\pi^+ \rightarrow \eta$ and $\pi^+ \rightarrow \pi^0\pi^0$ were studied at 10.5 GeV/c with Be, Pb, and Cu. The Pb glass shower hodoscope was used to detect meson decay. The measured cross sections were compared with theory [28].

The inclusive cross section for charged D* production by 205-GeV/c π^- incident on a Be target was measured [29]; for follow-up studies on the neutral D production, see [30].

Measurements of direct photon production at high transverse momenta for π^- and p interactions on Be at 500 GeV/c were reported; the yields as a function of incident momentum and rapidity were in good agreement with calculated expectations [31]. The production of vector mesons (cross-section measurements) for a 530-GeV/c π^- beam incident on Be and Cu targets is described in [32].

Results for the positive muon Knight shift in Be metal are given in "Beryllium" Suppl. Vol. A 2, 1991, pp. 268/9.

The success of μSR techniques (where μSR means muon spin rotation, relaxation, or resonance, depending on the precise variant of the technique used, or in general muon spin research, to cover all possibilities) relies on two circumstances: these are the intrinsic polarization of the muons during their production from pions, and the anisotropy of the positron emission in their subsequent decay; both are expressions of the violation of parity conservation in processes where particles susceptible to weak interactions are involved: $\pi^+ \rightarrow \mu^+ + \nu_\mu$ ($\tau_\pi + 26$ ns), $\mu^+ \rightarrow e^+ + \bar{\nu}_e + \nu_\mu$ ($\tau_\mu = 2.2$ μs). The broad range of application of implanted-muon studies and the extent of information that is available from such studies are discussed in the review [33].

References:

 [1] Stearns, M. B.; Stearns, M. (Phys. Rev. [2] **105** [1957] 1573/82).
 [2] Swanson, R. A. (Phys. Rev. [2] **112** [1958] 580/6).
 [3] Stearns, M.; Stearns, M. B.; DeBenedetti, S.; Leipuner, L. (Phys. Rev. [2] **97** [1955] 240/2, [2] **95** [1954] 625, 1353/4, [2] **93** [1954] 1123/4).
 [4] Stearns, M.; Stearns, M. B. (Phys. Rev. [2] **103** [1956] 1534/44).
 [5] Camac, M.; McGuire, A. D.; Platt, J. B.; Schulte, H. J. (Phys. Rev. [2] **99** [1955] 897/905, [2] **89** [1953] 905).
 [6] West, D.; Bradley, E. F. (Phil. Mag. [8] **1** [1956] 97/100, [8] **2** [1957] 957/76).
 [7] Stearns, M.; Stearns, M. B. (Phys. Rev. [2] **107** [1957] 1709/11).
 [8] Ruderman, M. A. (Phys. Rev. [2] **118** [1960] 1632/41).
 [9] Valley, G. E. (Phys. Rev. [2] **73** [1948] 1251, [2] **72** [1947] 772/83).
[10] Wilson, R.; Perry, J. P. (Phys. Rev. [2] **84** [1951] 163/4).

[11] Chedester, C.; Isaacs, P.; Sachs, A.; Steinberger, J. (Phys. Rev. [2] **82** [1951] 958/9).
[12] Sigurgeirsson, T.; Yamakawa, K. A. (Phys. Rev. [2] **71** [1947] 319/20; Rev. Mod. Phys. **21** [1949] 124/32).

[13] Kirillov-Ugryumov, V. G.; Moskvichev, A. M. (Zh. Eksperim. Teor. Fiz. **34** [1958] 322/6; Soviet Phys.-JETP **7** [1958] 224/7), Kirillov-Ugryumov, V. G.; Moskvichev, A. M.; Lomakin, S. S. (Nekotorye Vopr. Inzhener. Fiz. No. 1 [1957] 22/9 from C. A. **1960** 7354).

[14] Bell, W. E.; Hincks, E. P. (Phys. Rev. [2] **88** [1952] 1424/5).

[15] Stork, D. H. (Phys. Rev. [2] **93** [1954] 868/80).

[16] Tenney, F. H.; Tinlot, J. (Phys. Rev. [2] **92** [1953] 974/7).

[17] Cassels, J. M.; O'Keeffe, T. W.; Rigby, M.; Wetherell, A. M.; Wormald, J. R. (Proc. Phys. Soc. [London] A **70** [1957] 543/6).

[18] Dincklage, von, R. D. (AIP Conf. Proc. No. 150 [1986] 561/6 from C. A. **105** [1986] No. 215150).

[19] Gauster, W. B.; Fiory, A. T.; Lynn, K. G.; Kossler, W. J.; Parkin, D. M.; Stronach, C. E.; Lankford, W. F. (J. Nucl. Mater. **69/70** [1978] 147/56 from C. A. **89** [1978] No. 93609).

[20] Crawford, J. F.; Daum, M.; Eaton, G. H.; Frosch, R.; Garzon, J.; Hirschmann, H.; Kettle, P.-R.; McCulloch, J. W.; Steiner, E. (Helv. Phys. Acta **53** [1980] 497/505).

[21] Kuang, Y.; Arnold, K.-P.; Chmely, F.; Eckhause, M.; Hughes, V. W.; Kane, J. R.; Kettell, S.; Kim, D.-H.; Kumar, K.; Lu, D. C.; Matthias, B.; Ni, B.; Orth, H.; Putlitz, zu, G.; Schaefer, H. R.; Souder, P. A.; Woodle, K. (Phys. Rev. [3] A **39** [1989] 6109/23, 6114, 6117).

[22] Van Ginneken, A. (Nucl. Instrum. Methods Phys. Res. A **251** [1986] 21/39).

[23] Antipov, Yu. M.; Batarin, V. A.; Bessubov, V. A.; Budanov, N. P.; Gorin, Yu. P.; Denisov, S. P.; Klimenko, S. V.; Kotov, I. V.; Lebedev, A. A.; et al. (Z. Physik C **24** [1984] 39/44; C. A. **101** [1984] No. 79631).

[24] Antipov, Yu. M.; Bezzubov, V. A.; Budanov, N. P.; Gorin, Yu. P.; Denisov, S. P.; Ech, F. A.; Kartasheva, V. G.; Klimenko, S. V.; Lebedev, A. A.; et al. (Yadern. Fiz. **39** [1984] 1461/5 from C. A. **101** [1984] No. 44784).

[25] Antipov, Yu. M.; Bezzubov, V. A.; Vishnevskii, A. V.; Gorin, Yu. P.; Gornushkin, Yu. A.; Denisov, D. S.; Eroshin, O. V.; Kartasheva, V. G.; Kulinich, P. A.; et al. (Yadern. Fiz **53** [1991] 1314/23 from C. A. **115** [1991] No. 58842).

[26] Antipov, Yu. M.; Batarin, V. A.; Bezzubov, V. A.; Bilenkii, M. S.; Budanov, N. P.; Vishnevskii, A. V.; Gorin, Yu. P.; Gornushkin, Yu. A.; Denisov, D. S.; et al. (Yadern. Fiz. **53** [1991] 1324/35 from C. A. **115** [1991] No. 58843).

[27] Antipov, Yu. M.; Batarin, V. A.; Bezzubov, V. A.; Bilenkii, M. S.; Budanov, N. P.; Denisov, D. S.; Eroshin, O. V.; Frabetti, P. L.; Gorin, Yu. P.; et al. (Nucl. Phys. A **536** [1992] 637/47 from C. A. **116** [1992] No.137614).

[28] Akimenko, S. A.; Belousov, V. I.; Blik, A. M.; Kolosov, V. N.; Kut'in, V. M.; Pavlinov, A. I.; Romanovskii, V. I.; Solov'ev, A. S.; Shelikhov, V. I. (Yadern. Fiz. **51** [1990] 437/43 from C. A. **112** [1990] No. 146958).

[29] Mooney, P.; Sarmiento, M.; Bishop, J. M.; Biswas, N.; Cason, N. M.; Dauwe, L.; Godfrey, K.; Kenney, V. P.; Pemper, R.; et al. (Phys. Rev. [3] D **39** [1989] 2494/8 from C. A. **111** [1989] No. 13567).

[30] Sarmiento, M.; Mooney, P.; Bishop, J. M.; Biswas, N.; Cason, N. M.; Dauwe, L.; Godfrey, J.; Kenney, V. P.; Pemper, R.; et al. (Phys. Rev. [3] D **45** [1992] 2244/8 from C. A. **116** [1992] No. 182926).

[31] Alverson, G.; Baker, W. F.; Ballocchi, G.; Benson, R.; Berg, D.; Blusk, S.; Bromberg, C.; Brown, D.; Carey, D.; et al. (Phys. Rev. Letters **68** [1992] 2584/7).

[32] Zieminski, A.; et al. (AIP Conf. Proc. No. 243 [1992] 891/6 from C. A. **116** [1992] No. 160403).

[33] Cox, S. F. J. (J. Phys. C **20** [1987] 3187/319, 3194).

15.11   Atom and Ion Impact Phenomena

15.11.1   Sputtering. Secondary Ion Emission

Early studies are reviewed, e.g., by Behrisch [1]; for other general references, see [2 to 4] and those cited in Section 5.10.3 of "Molybdenum" Suppl. Vol. A 2b, 1988, pp. 112/24.

In the first study in 1952 the **sputtering** rate for Be was measured using krypton ions accelerated through a potential difference of 1500 V [5]. Helium, krypton, and xenon ions of 100, 200, 300, and 600 eV were used to measure the sputtering yields at low energy. The yields varied from 0.03 atoms/ion for 100-eV krypton ions to 0.61 atoms/ion for 600-eV krypton ions. Helium yields varied from 0.04 atoms/ion for 100-eV helium ions to 0.34 atoms/ion for 600-eV helium ions. Using xenon the values were 0.1 atoms/ion for 200-eV ions and 0.42 atoms/ion for 600-eV ions [6]. For sputtering yields using argon and neon ions with energies from 50 to 600 eV, see [7], for sputtering by argon and mercury ions, see [8]. Threshold energies for sputtering by inert gas ions (12 eV for Ne^+, 15 eV for Ar^+, Kr^+ and Xe^+) at normal incidence on Be are determined in [9]. Sputtering yields from high-energy hydrogen ions were measured; a Be surface bombarded with 7-keV hydrogen ions H_2^+ and H_3^+ gave yields of 0.032 atoms/ion for H_2^+ and 0.046 atoms/ion for H_3^+ [10].

If sputtering yields, measured for normal incidence on the solid, for different ion-target combinations are plotted as a function of the ion energy in a log-log scale, all yield functions have similar shape for energies below a few keV. Therefore the sputtering yield data can be characterized by a normalized energy function (Y_N) and two parameters (Q, K) for each ion-target combination; (1/K is suggested to be the threshold energy). Both parameters depend mainly on the ion and target mass $(M_1$ and $M_2)$, and on the surface binding energy (E_b). The analytic expression for the normalized functions and both parameters allows estimation of unknown sputtering data. Experimental values for 1/K for the ions of H (27.5 eV), D (24 eV), and ^{4}He (33 eV), incident on a Be target, are tabulated in the paper [11] and compared with calculations showing general agreement.

The sputtering yield of Be was measured for D^+ and $^4He^+$ ions in the energy range from 60 to 3000 eV and the target temperature varying between room temperature and 760°C [12]. At ion energies >1 keV, no dependence of the yield values on the target temperature was observed. At the highest temperature (760°C) in these experiments the weight change was mainly due to thermal evaporation and not to sputtering. At energies below 1 keV, a change in the sputtering yield with the target temperature was found. This behavior was explained by assuming an oxide layer at the Be target surface (the existence of such an oxide layer was deduced from sputtering yield measurements at room temperature [13]). At elevated temperatures Be diffuses through this oxide layer and is sputtered from the surface of the BeO. If enough Be is supplied to the surface by this diffusion, the sputtering yield becomes independent of the target temperature. The measured yield data can be well reproduced by a universal relation as shown in **Fig.** 15-8 (for the formula and tabulated constants, see the paper [12]). A comparison of the threshold energy values for 300°C and those found for room temperature [14] indicated a reduction of the threshold energy at elevated target temperature. While the room temperature data [13] agree with estimates for the threshold energy [11] using the surface binding energy of BeO, the data above 300°C agree with a threshold energy based on the surface binding energy of pure Be. This reduction of the threshold energy also explains the increase of the sputtering yield with increasing temperature at low ion energies [12].

For early measurements of the **energy distribution** of secondary positive ions emitted from Be targets under ion bombardment, see [15, 16]. The ions H^+, H_2^+, He^+, O^+, O_2^+, N^+, N_2^+, CO^+, Ar^+, and Ar^{2+} with energies between 7.5 and 80 keV were used for the Be (and Mo) target by [16]; calculations of the detailed structure of the found energy spectra are given in the paper.

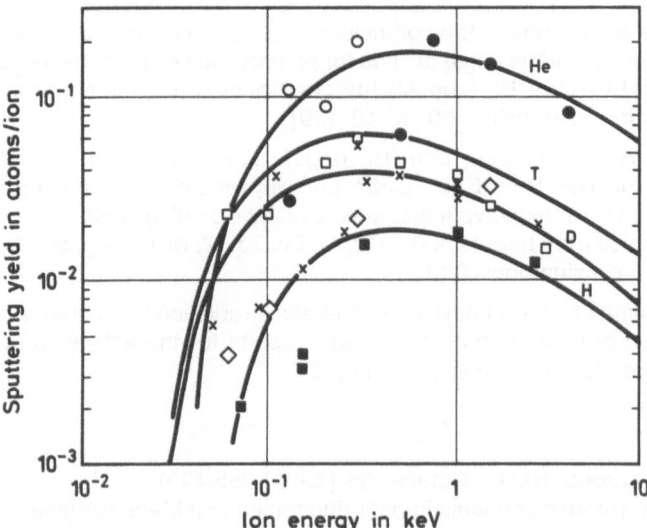

Fig. 15-8. Energy dependence of the sputtering yield of Be with H, D, ⁴He at different temperatures. Data for tritium, T, are estimated. The universal relationship is given by the solid lines; ◇ measurements at room temperature; □ ○ measurements at 800°C; ■ × ● measurements from [14].

In later studies devoted to clean, oxygen- and hydrogen-covered Be surfaces, 3 keV Ar⁺ ions were used incident at 45° to the sample normal. The secondary ion-energy distribution $N^+(E)$, obtained by sputtering Be, was analyzed in terms of competing processes, and was separated into an ionization coefficient $R^+(E)$ and a total energy distribution, $N(E)$; i.e., $N^+(E) = R^+(E) \cdot N(E)$. Experimentally, the dependence of $R^+(E)$ on both the energy and oxygen coverage indicated a linear superposition of adiabatic tunneling and resonance ionization processes from clean and oxygen-covered portions of the surface without contributions to the secondary ion yield from regions of intermediate coverage. The total energy distributions were deduced, and the principal features agreed with predictions of the collision cascade sputtering model; for details and figures, see the papers [17].

Simultaneous measurements of the energy distribution of secondary ions and atoms under identical ion-bombardment conditions (5.5 keV Ar⁺ beam) were performed, making it possible to determine how the degree of ionization of the sputtered particles (β^+) depends on their velocity (v). For this purpose, the sputtered atoms were ionized by electron impact (not accompanied by energy transfer from the electron to the atom). The results for the energy spectra of Be (and Cu) ions and atoms are shown in figures in the paper. The spectrum of the Be atoms can be described by an expression which follows from Sigmund's cascade theory (see [18]) of sputtering (with a surface binding energy $E_b = 2.8$ eV). This behavior is also typical for the energy distributions of other elements. The degree of ionization, β^+, as a function of the reciprocal velocity, $v_\perp^{-1}$, of the sputtered particles (also shown in a figure in the paper) is an exponential function for Be (W, Cu, and Ag) at velocities corresponding to energies $E > 3$ eV of the secondary particles, in agreement with the electron-tunneling model. The deviation from exponential behavior at $E < 3$ eV was ascribed to an effect of a surface potential barrier on the low-energy ions. An estimate of the degree of ionization averaged over the energy with the assumption of identical angular distributions of the ions and atoms gave $\bar{\beta}^+ = 8 \times 10^{-4}$ for Be [19].

In early studies the change in the coefficient of secondary emission of negative ions was determined as a function of the angle of slide for primary (50 keV) protons on targets of Be, Al, stainless steel, and Cu. For Be (and Al) the coefficient was negative at large angles and became positive for angles below 30° to 40° [20].

In the secondary ion emission from Be targets bombarded with 12 keV Ar^+ ions (45° incidence) the species Be^+, Be^{2+}, BeH^+, BeO^+, and Be_2^+ were detected (by mass spectrometry) [21]. High yields (>1%) of negative metal ions, usually including clusters of metal atoms and oxides, were observed from targets of Be, Cd, Al, Sn, Ta, W, Ni, Cu, Ag, and Au bombarded by a beam of positive caesium ions [22].

In the development of a target device for ion-beam irradiation in vacuum, which can be used for ion implantation or in sputtering technology, results for the activation of metallic Be by beams of $^{14}N^{3+}$ and $^{12}C^{3+}$ were presented in [23].

References:

[1] Behrisch, R. (Ergeb. Exakt. Naturw. **35** [1964] 295/443).
[2] Kaminsky, M. (Atomic and Ionic Impact Phenomena on Metal Surfaces, Springer, Berlin 1965, pp. 142/211).
[3] Behrisch, R. (Top. Appl. Phys. **47** [1981], **52** [1983]).
[4] Winograd, N. (Progr. Solid State Chem. **13** [1982] 285/375).
[5] Fisher, T. F.; Weber, C. E. (J. Appl. Phys. **23** [1952] 181/3).
[6] Rosenberg, D.; Wehner, G. K. (J. Appl. Phys. **33** [1962] 1842/5).
[7] Wehner, G. K.; Laegreid, N. (J. Appl. Phys. **32** [1961] 365/9).
[8] Strachan, J. F.; Harris, N. L. (Proc. Phys. Soc. [London] B **69** [1956] 1148/61).
[9] Stuart, R. V.; Wehner, G. K. (J. Appl. Phys. **33** [1962] 2345/52).
[10] KenKnight, C. E.; Wehner, G. K. (J. Appl. Phys. **35** [1964] 322/6).

[11] Bohdansky, J.; Roth, J.; Bay, H. L. (J. Appl. Phys. **51** [1980] 2861/5).
[12] Bohdansky, J.; Roth, J. (J. Nucl. Mater. **122/123** [1984] 1417/24).
[13] Roth, J.; Bohdansky, J.; Blewer, R. S.; Ottenberger, W. (J. Nucl. Mater. **85/86** [1979] 1077/9).
[14] Roth, J.; Bohdansky, J.; Ottenberger, W. (IPP-Rept. 9/26 [1979], Max-Planck-Institut für Plasmaphysik, Garching), cited in [12].
[15] Stanton, H. E. (J. Appl. Phys. **31** [1960] 678/83).
[16] Panin, B. V. (Zh. Eksperim. Teor. Fiz. **42** [1962] 313/24; Soviet Phys.-JETP **15** [1962] 215/21).
[17] Krauss, A. R.; Gruen, D. M. (Surf. Sci. **92** [1980] 14/28, **90** [1979] 564/78).
[18] Sigmund, P. (Phys. Rev. [2] **184** [1969] 383/416).
[19] Makarenko, B. N.; Popov, A. B.; Shaporenko, A. A.; Shergin, A. P. (Pis'ma Zh. Tekhn. Fiz. **14** [1988] 609/13; Soviet Tech. Phys. Letters **14** No. 4 [1988] 271/2).
[20] Mitropan, I. M.; Gumenyuk, V. S. (Zh. Eksperim. Teor. Fiz. **34** [1958] 235/6 from C. A. **1958** 14337).

[21] Beske, H. E. (Z. Naturforsch. **22a** [1967] 459/67).
[22] Krohn, V. E., Jr. (J. Appl. Phys. **33** [1962] 3523/5).
[23] Ryzhkov, V. A. (Prib. Tekhn. Eksperim. **1989** No. 1, pp. 48/50 from C. A. **110** [1989] No. 200831).

15.11.2 Stopping Power (Energy Loss of Ions)

When well-collimated particles, e.g. protons, pass through a thin target of random matter, they will lose their energy according to the thickness of the target and their direction will diverge due to multiple scattering with atomic nuclei in the target. The energy losses of 7-MeV protons in thin metallic (Be, et al.) and organic foils were obtained as functions of the emergence angles from 0° to 1.7°. Angular distributions due to multiple scattering were also measured. The energy loss was found to increase for all targets with increasing emergence angles. Because this increase in energy loss could not be explained by several effects considered in the paper, it was concluded to be due to a hitherto unknown effect, very likely the dependence of the energy loss on the average impact parameter with the atomic nucleus. Results are given in figures in the paper [1]. The predictions of multiple scattering theory with a reasonable impact-parameter-dependent inelastic energy loss function were at least a factor of 3 below the in [1] reported experimental data [2].

Extensive tables were prepared of the mean energy loss, path length, range, multiple scattering, path length straggling, time-of-flight, and nonelastic collision probability for protons of the energy range 1 keV through 10 GeV in all elements with atomic numbers 1 through 92 and in many compounds. Emphasis was placed on obtaining accurate results, particularly in the difficult low-energy and high-atomic-number regions. The accuracies of all proton range-energy parameters are tabulated. The energy loss below 20 keV was found by normalizing theoretical expressions for the ionization, excitation, and nuclear contributions to known values at 20 keV. The energy loss between 20 keV and 1.0 MeV was obtained by statistically evaluating the accuracy of the available experimental information and fitting least-squares curves. Above 1.0 MeV, the Bethe equation with corrections was used. The polarization effect was calculated in detail for each material. Path lengths were calculated by integrating the reciprocal of the energy losses. Ranges were obtained by use of the multiple Coulomb scattering theory. The straggling calculations included electronic, elastic nuclear, and charge exchange effects. A comparison of the calculation methods used in this compilation and in other proton range-energy tables is given in the paper [3].

In early studies a general expression was derived for the range-energy relation $R(E_p)$ for protons (where E_p is the proton kinetic energy) as a function of the mean excitation potential which enters into the Bethe-Bloch formula for the ionization loss. The expression for $R(E_p)$ was obtained by interpolation of previously calculated range-energy relations for Be, Al, Pb, and Cu [4]. Other range-energy data for pions, protons, deuterons, tritons, and ^{3}He particles in Be and other materials are tabulated and shown in graphs in the earlier paper [5].

Studies concerning the energy loss of protons impacting Be foils are reported in [6 to 12]; for theory, see [4]. Works on impacting deuterons, see [13, 14], on α-particle impact [15, 16]; for theory, see [17].

The formation of negative hydrogen ions during the passage of protons through thin metallic foils such as Be, Al, and Cu is described in [18]; about 10% of the protons (with energies in the range 11.5 to 28 keV) which fell on a Be foil were changed to negative hydrogen ions.

In a study of valence bond effects on mean excitation energies for stopping power in metals, the mean electronic excitation energies per cell (as related to the stopping of charged particles such as electrons, positrons, fast ions) were calculated for Be, Li, and Al by the local plasma approximation, using a valence bond model of the metallic state and Pauling's ion core charge ($p \approx 0.72$); the calculations were in agreement with experimental data [19].

References:

[1] Ishiwari, R.; Shiomi, N.; Sakamoto, N. (Phys. Rev. [3] A **25** [1982] 2524/8).

[2] Gras-Marti, A. (Nucl. Instrum. Methods Phys. Res. B **9** [1985] 1/5 from C. A. **102** [1985] No. 227 913).

[3] Janni, J. F. (At. Data Nucl. Data Tables **27** [1982] 341/529 from C. A. **97** [1982] No. 190 054).

[4] Sternheimer, R. M. (Phys. Rev. [2] **118** [1960] 1045/8, [2] **115** [1959] 137/42).

[5] Rich, M.; Madey, R. (UCRL-2301 [1954] 433 pp. from C. A. **1955** 737).

[6] Kahn, D. (Phys. Rev. [2] **90** [1953] 503/9).

[7] Madsen, C. B. (Kgl. Danske Videnskab. Selskab Mat.-Fys. Medd. 27 No. 13 [1953] 21 pp. from C. A. **1954** 460), Madsen, C. B.; Venkateswarlu, P. (Phys. Rev. [2] **74** [1948] 648/9).

[8] Neufeld, J. (Proc. Phys. Soc. [London] A **66** [1953] 590/6).

[9] Bakker, C. J.; Segrè, E. (Phys. Rev. [2] **81** [1951] 489/92).

[10] Hall, T. (Phys. Rev. [2] **79** [1950] 504/12), Hall, T.; Warshaw, S. D. (Phys. Rev. [2] **77** [1950] 754), Warshaw, S. D. (Phys. Rev. [2] **76** [1949] 1759/65).

[11] Aron, W. A.; Hoffmann, B. G.; Williams, F. C. (AECU-103 (UCRL-121 Rev. 2nd Ed.) [1948] 1/61 from N.S.A. **2** [1949] No. 1165), Aron, W. A. (UCRL-1325 [1951] 1/47 from N.S.A. **5** [1951] No. 4890).

[12] Zrelov, V. P.; Stoletov, G. D. (Zh. Eksperim. Teor. Fiz. **36** [1959] 658/68; Soviet Phys.-JETP **9** [1959] 461/7).

[13] Wilkinson, G. (J. Chem. Soc. Suppl. No. 2 [1949] S 360/S 363 from C. A. **1950** 9263).

[14] Huus, T. (Kgl. Danske Videnskab. Selskab Mat.-Fys. Medd. **26** No. 4 [1951] 3/16 from C. A. **1951** 8361).

[15] Barber, W. C. (Rev. Sci. Instrum. **24** [1953] 469/70).

[16] Dissanaike, G. A. (Phil. Mag. [7] **44** [1953] 1051/63).

[17] Lindhard, J.; Scharff, M. (Phys. Rev. [2] **85** [1952] 1058/9).

[18] Fogel, Ya. M.; Safronov, B. G.; Krupnik, L. I. (Zh. Eksperim. Teor. Fiz. **28** [1955] 711/8; Soviet Phys.-JETP **1** [1955] 546/51).

[19] Kamaratos, E.; Chang, C. K.; Wilson, J. W.; Xu, Y. J. (Phys. Letters A **92** [1982] 363/5).

15.11.3 Ion-Induced Electron Emission

Ion-induced electron emission (IIEE) experiments with relatively low projectile energies, Ar^+ and Ar^0 from 1 to 20 keV, on clean Be (Al, CuBe, and Cu) targets were performed; results are represented in figures in the paper [1]. For further experiments with Ar^+ ions of 5 to 40 keV and targets of Be and other metals, see [2], also [3]; with Ar^+ ions of 80 keV, see [4].

The SEE yield from mechanically cleaned metal foils of Be (Al, Ta, W, CuBe, Ag, AgMg, Au, and Pt) bombarded by beams of 20 to 60 keV Li^+ and Li^0 was measured [5].

Studies of SEE after proton bombardment with energies between 48 and 212 keV gave a yield of ~7.5 electrons per proton at normal incidence on an outgassed Be target [6]. For other SEE studies during bombardment of Be (Al, Zr, Cu, Ag, or Au) with 60 to 300 keV D^+ and D_2^+ ions, see [7]. Experiments of IIEE with projectile energies exceeding 250 keV, including those of [8], using H^+, H_2^+, H_3^+, He^+, Ne^+, Ar^+ ions in the range 0.075 to 0.9 MeV and targets of Be and other substances, are compiled in the review [9]; for follow-up studies, see [18]. Older studies with 1 MeV H^+ ions on Be and other metal targets are reported in [10]; see also [11], and later studies [12]; for theory, see [13]. Experiments with He^+ and He^{2+} ions with velocities corresponding to 1.1 MeV/amu on Be, Al, Cu, Ag, and Au targets were described in [14].

For IIEE studies after impact of positive ions of ^{6}Li, ^{7}Li, ^{20}Ne, ^{22}Ne, ^{39}K, and ^{41}K on Be, Mo, Cu, and Pt, see [15]. The spectra and yields of electrons emitted when heavy ions pass through thin films of Be, Ni, and Au were investigated by [16].

Theoretically, different production mechanisms for IIEE have been visualized. If the neutralization energy at the target surface for the impinging positive ion is more than a factor of 2 larger than the work function of the target material, an electron may be ejected from the target at the same time as another target electron neutralizes the projectile. This is essentially an Auger transition, and is referred to as potential SEE (for a review, see [17]). Being exothermic, it can occur at all projectile energies and is generally assumed to be a steadily decreasing function of the projectile energy. Potential ejection is the primary source for IIEE only at projectile velocities below approximately 10^7 cm/s. At higher projectile velocities, electrons may also be emitted by the kinetic mechanism (for a review, see [9]), i.e., they may result from atom-atom and/or atom-electron collisions in the solid. At an projectile energy of 80 keV as used in the work of [4], the kinetic ejection will dominate. The potential IIEE will cause less than 5% of the total IIEE. The measured total IIEE yield (defined as the average number of electrons emitted per incoming projectile) for the 80 keV Ar$^+$ ions obtained on Be at normal incidence ($\varphi = 0°$) is $\gamma = 2.3 \pm 0.1$. The measured angular dependence of γ follows a $\cos^{-1} \varphi$ law [4].

The intensity, energy distribution, and dependence of IIEE from 1 MeV protons on the emission angle were measured by [11], and from the energy spectrum as a function of the emission angle a modification of the Sternglass formula [13] for the SE coefficient of kinetic IIEE was proposed [12]. The emission coefficient depended on the nature and state of the target material and increased with the kinetic energy of the incident projectiles [2, 3]. A significant dependence on Z (= atomic number of the target) of the peak energy E_p of the energy distribution (measured from 100 to 1000 eV) of the loss electrons backscattered from metal targets for impacts of 1.1 MeV/amu He$^+$ and He^{2+} ions was found with the following order of E_p for the targets Au > Cu > Ag > Al > Be ($E_p = 260$ eV for Be). This dependence of E_p was explained in terms of energy loss for loss electrons in their transportation to the surface [14].

References:

[1] Schackert, P. (Z. Physik **197** [1966] 374/8).

[2] Cousinié, P.; Colombié, N.; Fert, C.; Simon, R. (Compt. Rend. **249** [1959] 387/9).

[3] Slodzian, G. (Compt. Rend. **246** [1958] 3631/4).

[4] Veje, E. (Nucl. Instrum. Methods Phys. Res. B **2** [1984] 536/9, B **33** [1988] 497/501; C.A. **100** [1984] No. 220065, **109** [1988] No. 220798).

[5] Bethge, K.; Lexa, P. (Brit. J. Appl. Phys. **17** [1966] 181/6).

[6] Allen, J. S. (Phys. Rev. [2] **55** [1939] 336/9).

[7] Leroy, J.; Prelec, K. (Comm. Energ. At. [France] Rappt. No. 1445 [1960] 11 pp. from C.A. **1960** 23783).

[8] Hasselkamp, D.; Hippler, S.; Scharmann, A.; Schartner, K.-H. (Z. Physik D **6** [1987] 269/74; Nucl. Instrum. Methods Phys. Res. B **18** [1987] 561/5; C.A. **106** [1987] No. 167239).

[9] Hasselkamp, D. (Springer Tracts Mod. Phys. **123** [1992] 1/95, 41).

[10] Gorodetzky, S.; Bergdolt, A. M.; Chevalier, A.; Bres, M.; Armbruster, R. (J. Phys. [Paris] [8] **24** [1963] 374/8).

[11] Kronenberg, S.; Nilson, K.; Basso, M. (Phys. Rev. [2] **124** [1961] 1709/12).

[12] Batrakin, E. N.; Zalyubovskii, I. I.; Karas', V. I.; Kononenko, S. I.; Mel'nik, V. N.; Moiseev, S. S.; Muratov, V. I. (Zh. Eksperim. Teor. Fiz. **89** [1985] 1098/101; Soviet Phys.-JETP **62** [1985] 633/4).

[13] Sternglass, E. J. (Phys. Rev. [2] **108** [1957] 1/12).

[14] Koyama, A.; Benka, O.; Sasa, Y.; Ishikawa, H.; Uda, M. (Phys. Rev. [3] A **36** [1987] 4535/8).

[15] Ploch, W. (Z. Physik **130** [1951] 174/95, 182; Z. Naturforsch. **5a** [1950] 570/1).

[16] Haines, E. L.; Whitehead, A. B.; Parker, R. H. (Beam-Foil Spectrosc. Proc. Conf., Tucson, Ariz., 1967 [1968], Vol. 1, pp. 177/92 from C.A. **70** [1969] No. 91623).

[17] Varga, P.; Winter, H. (Springer Tracts Mod. Phys. **123** [1992] 149/213).

[18] Hasselkamp, D.; Hippler, S.; Scharmann, A.; Schmehl, T. (Ann. Physik [Leipzig] **47** [1990] 555/67).

15.11.4 Impact-Induced Photon Emission

Upon bombardment of solid surfaces by energetic heavy ions, a certain fraction of sputtered particles leaves the surface in electronically excited states. The subsequent decay of these excited states gives rise to the photon emission. This phenomenon was the subject of many investigations in the last decade. For investigations on Be metal targets, see for instance [1 to 5]. The problem of how the sputtered atoms come into the excited states, has not been solved as of yet. Several models were proposed to explain the secondary ion emission in sputtering, but no satisfactory mechanism has been provided to cover all events related to the photon emission. The most predominant among the proposed models is that one in which the excited atoms originate from the atoms sputtered from the surface during collision cascade processes rather than from high-energy recoil particles. If the excited-atom formation in the sputtering is assumed to originate from the collision cascade process following primary ion bombardment on the target surface, then the excitation can be expected to be strongly affected by the properties of the surface atoms and to be insensitive to the nature of the primary ions. A quasi-molecular collision model was proposed by Blaise [6] and Thomas [7] which seemed to have promise to elucidate the excitation mechanism because of its similarity to the Fano-Lichten theory which was established for gas phase collisions, see, e.g. [5].

Excitation in the sputtering from a Be metal surface was considered to provide useful information for investigating the quasi-molecular model, because the photon emission from the sputtered Be atoms involves both the transitions in the singlet and triplet systems. There is an interest to determine the relative population distribution in the excited states and to reveal the electron-transfer processes leading to the formation of these singlet and triplet excited atoms. Excitational sputtering of Be atoms also involves two-electron transitions, e.g., formation of $Be(2p^2 \, ^3P)$ and presumably $Be^+(2p \, ^2P)$. Knowledge is important about how these transitions proceed during sputtering. As an extension of a study of inelastic collision processes in the gas phase, Inouye and Tanji-Noda [5] investigated the photon emission at a Be surface bombarded by a variety of ions with low kinetic energies such as Ne^+, Ar^+, Xe^+, C^+, CO^+, N^+, NO^+, D_2^+, BF_2^+, and CF_3^+ at a collision energy of 3.0 keV (photon emission due to He^+ and D^+ bombardment was too faint to be observed). The relative intensities of the emission lines (tabulated in the paper) scarcely depended on the ion species as expected. The intensity of Be^+ (3131 Å, $2p^2P-2s^2S$ transition) was about 60% that of Be (3321 Å, $3s^3S-2p^3P$). The intensity ratio of these lines is considered to be not very different from the true value; the result, however, is different from the value of Veje [8] who used 50 keV Ar ions, and slightly different from the data of Wright and Gruen [9]. Most interesting, however, is that the photon emission in the sputtering of Be atoms by bombardment of BF_2^+ and CF_3^+ ions gave different results from those of some other incident ions studied to date. The reason for this selective excitation is not yet understood, but the fluorine atom of the incident molecular ion seems to play an important role in the formation of electronically excited particles in the sputtering processes [5].

While Veje [3, 8] in his studies bombarded various targets at normal incidence, Larsen and Veje [4] reported and discussed results obtained by bombarding targets of solid elemental Be, B, Mg, and Au at different angles of incidence of the projectile (80 keV Ar$^+$ ions). Generally, the excitation intensities increased with increasing angle of incidence, but a few levels showed stronger angular variations than the majority of levels did. It was concluded that whereas the common increase in excitation is caused by the general increase in the number of fast, sputtered atoms, a few levels are in addition selectively excited through molecular-orbital electron-promotion processes taking place in binary collisions [4].

The probability of excitation to various states was measured for Be (and Al) atoms sputtered from clean and oxygen-covered Be (and Al) surfaces bombarded by 12 keV K$^+$ ions at normal incidence. The absolute values of excitation probabilities can not be described by the Boltzmann distribution. In all cases studied, the excitation probabilities (W) exhibited a power law dependence on the effective principal quantum number n*. For clean Be (and Al) surfaces the dependence can be approximated by $W = 1.3 \times 10^{-3}(n^*)^{-8.8}$. Both the thermodynamic models and the models based on the kinetic (collision) mechanisms of secondary atom excitation failed to describe the light emission from excited atoms [1]. Another test for the thermal origin of the excited neutrals sputtered from bombarded solids including Be was performed; in the case of thermal origin, the photon yield emitted by the neutrals is governed by a Boltzmann factor (as found in fact) [10]. For relative level populations for Be excited by 80 keV Ar$^+$ ions (several strong Be I and Be II emission lines were observed), see [3]. The results agree with an electron-pickup model for atomic excitation in sputtering [3], and with a quantum mechanical model based on a break-off mechanism of the emission of excited atoms [1].

Visible and UV radiation observed during ion bombardment (keV range) of solids was interpreted as deexcitation radiation from molecules built up by atoms from the incoming beam (40 keV Ar$^+$ ions), from the target material (Be, etc.), and from impurities (H, O). For Be, the emission from Be I, Be II, and BeH was observed [11].

For a review on bombardment-induced light emission (BLE), see Thomas [12].

References:

[1] Pop, S. S.; Drobnich, V. G.; Bandurin, Yu. A.; Evdokimov, S. A. (Phys. Chem. Mech. Surf. **3** [1985] 2280/99, 2286; Poverkhnost **1984** No. 8, pp. 39/46 from C. A. **101** [1984] No. 182077).

[2] Veje, E. (Nucl. Instrum. Methods Phys. Res. B **2** [1984] 520/4; C. A. **100** [1984] No. 218291).

[3] Veje, E. (Phys. Rev. [3] B **28** [1983] 88/94).

[4] Larsen, P.; Veje, E. (Phys. Rev. [3] B **28** [1983] 5011/8).

[5] Inouye, H.; Tanji-Noda, K. (J. Appl. Phys. **54** [1983] 6792/4).

[6] Blaise, G. (Surf. Sci. **60** [1976] 65/75).

[7] Thomas, G. E. (Radiat. Eff. **31** [1977] 185/6).

[8] Veje, E. (Surf. Sci. **110** [1981] 533/42).

[9] Wright, R. B.; Gruen, D. M. (J. Chem. Phys. **73** [1980] 664/72).

[10] Kelly, R.; Good-Zamin, C. J.; Shehata, M. T.; Squires, D. B. (Nucl. Instrum. Methods **149** [1978] 563/6 from C. A. **88** [1978] No. 145014).

[11] Braun, M.; Emmoth, B. (Nucl. Instrum. Methods **170** [1980] 585/9 from C. A. **92** [1980] No. 223754; Annu. Rept. Res. Inst. Phys. [Swed.] **1979** 34/9 from C. A. **92** [1980] No. 118944).

[12] Thomas, G. E. (Surf. Sci. **90** [1979] 381/416, 385, 402).

15.11.5   Impact-Induced Inner-Shell Ionization and Bremsstrahlung

During bombardment of a solid with energetic atomic particles, a variety of secondary processes can take place with the ejection of particles originally present in the solid from or through its surface and, in addition, various transport and excitation phenomena within the solid.

The processes leading to K X-ray production by slow ion-atom collision can in principle be described by direct Coulomb excitation $Z_1 \ll Z_2$ (with Z_1 and Z_2 being the atomic number of the projectile and target, respectively) and by molecular orbital (MO) interaction ($Z_1 \approx Z_2$). The latter process is usually discussed as a three-step mechanism: vacancy production in the $2p\pi$ MO leading to a vacancy occupation number N_π in the entrance channel (generally, N_π is a function of the projectile velocity v: $N_\pi = N_0 + N_v$), vacancy transfer by rotational coupling between the $2p\pi$ and $2p\sigma$ MO (σ_{rot}) to the K shell of the lighter collision partner, and vacancy sharing in the outgoing part of the collision resulting in a K vacancy of the heavier collision partner. This concept was successfully applied to a variety of collision systems [1] with excellent agreement between theory and experiment (especially for the $Ne^+ + O$ system), see [2].

Measurements of the X-ray production cross section, σ_x, for thin Be foils with 80 to 100 keV H^+, D^+, He^+, Li^+, Be^+, and B^+ ions cover the region of both the abovementioned excitation processes, giving data for testing these models in the low-energy and low-Z region. For Li^+, Be^+, and B^+ Coulomb interaction is negligible, but in the MO excitation model σ_{rot} seems overestimated by theory [3]. Among the results reported in previous studies [4], those for H^+ and D^+ were in good agreement with calculations for direct Coulomb ionization, whereas in the case of He^+ the σ_x values were considerably higher than expected for Coulomb excitation, thus indicating that in this low-energy region He^+ can no longer be regarded as a structureless particle, and MO interaction becomes dominant [3, 4].

Measurements with 0.3 to 1.3 MeV ions of Ne, O, N, and C as projectiles on Be were performed by Ootuka at al. [5]. Cross sections for single and double K-shell ionization of Be were obtained from the $K\alpha$ diagram and hypersatellite X-ray yields produced by the bombardment. The data compared with theoretical predictions [2] showed good agreement except for the collision system $Ne + Be$ [5]. The main peak at 108.5 eV is the $K\alpha$ diagram line due to single K-shell ionization by 1.1 MeV O^+ ion bombardment (the X-ray spectrum is given in a figure in the paper). The hypersatellite line, $K\overset{h}{\alpha}$, due to double K-shell ionization is observed at 146.1 eV [5]. These transition energies agree well with results of other experiments of Be bombardment with 1.2 MeV He^+ ions [6] and with theoretical calculations [7]. For 0.3 MeV proton bombardment a peak of double K-shell ionization was not observed [6].

A survey of XUV radiation from different targets excited by bombardment with 4.5 MeV He^+, Ne^+, and Ar^+ and 3.0 MeV Kr^+ ions is reported in [7]. The spectra were generally similar to electron-excited spectra except for the presence of strong K hypersatellites which were measured for elements in the Be-Ne range [7].

The K-shell ionization cross section, $\sigma_i(E)$, is related to the true X-ray production cross section, $\sigma_x(E)$, by $\sigma_x(E) = \omega \sigma_i(E)$ where ω is the K-shell fluorescence yield and E = incident energy, see, e.g. [5]; see also "Beryllium" Suppl. Vol. A 1, 1986, p. 142.

Continuum X-rays produced by bombardment of gaseous or solid targets with heavy charged particles or heavy ions were interpreted in terms of molecular-orbital (MO) X-rays, radiative electron capture (REC), radiative ionization (RI), and secondary-electron bremsstrahlung (SEB). Direct processes such as MO and REC play an important role in heavy-ion impact, while multistep processes such as RI and SEB play a dominant role in light-ion impact. A new continuum X-ray component coming from a kind of RI process was observed by Yamadera et al. [8] in bombardments of Be, Al, and C targets with 6 to 40 MeV protons with an Si(Li) detector. The high-energy end of this component changes with the proton energy and has a $\frac{1}{2}(m_e v_p^2)$ dependence, where m_e is the electron mass and v_p is the projectile velocity. This energy dependence of the continuum X-rays was explained in terms of bremsstrahlung produced by orbital electrons scattered in the projectile-Coulomb field. Satisfactory agreement was obtained between the experimental results and the bremsstrahlung calculated in the projectile frame assuming that the orbital electrons are free and at rest, i.e., the quasi-free electron bremsstrahlung (QFEB). Since the QFEB is a process occurring in the projectile frame, the Doppler effect is expected to appear in the spectrum [8].

Energy spectra and angular distributions of QFEB induced by 20 MeV protons bombarding a Be target were measured. The Doppler shift of QFEB was clearly observed in the energy spectra and depended on the observation angle. The production cross section calculated from a free-electron approximation was compared with the experimental result; agreement was quite satisfactory. The spectral shape near the high-energy point of the QFEB definitely reflects the velocity distribution of the orbital electrons of the target atom. Angular distributions of secondary-electron bremsstrahlung were also measured and compared with a calculation which included relativistic retardation effects; see figures in the paper [9].

In order to investigate the projectile-charge dependence of QFEB, continuum X-rays induced by bombardment of a Be target with 20.14 MeV/amu protons and $^3\mathrm{He}^{2+}$ ions were measured by [10] in the 90° direction with respect to the incident beam. Differences in the X-ray production cross sections multiplied by the X-ray energy and the ratio of the cross sections for the proton and $^3\mathrm{He}^{2+}$ ion impact were obtained as a function of the X-ray energy. Both the difference and the ratio show peaks in the region where the X-ray energy is equal to the relative kinetic energy between the projectile and the inner-shell electron to be scattered by the projectile. From the comparison with a theory of quasi-free electron bremsstrahlung based on the impulse approximation [11], it was found that this peak corresponds to the maximum of the velocity-distribution function of the inner-shell electron. Furthermore, the contribution of the radiative electron-capture process in the case of proton and $^3\mathrm{He}^{2+}$ impact was clearly found. The main part of the continuum X-rays in the region of $\hbar\omega \leqq E_r \, (\equiv \frac{1}{2}(m_e v_p^2))$ is due to the QFEB process, while the SEB is predominant in the region of $\hbar\omega \geqq E_r$. Radiative electron capture (REC) is predominant in the region of $\hbar\omega \approx E_r - I_i$ (where I_i is the ionization energy for the i-shell electrons) [10]. For a brief review on REC studies, see [12].

In a study on bremsstrahlung characteristics, a formula was deduced for calculation of the power of bremsstrahlung induced by a pulsed ion beam in the matter as a function of the coordinate, and some results are given for the interaction of a deuteron beam with a Be target [13].

References:

[1] Schneider, D.; Stolterfoht, N. (Phys. Rev. [3] A **19** [1979] 55/64).

[2] Taulbjerg, K.; Briggs, J. S.; Vaaben, J. (J. Phys. B **9** [1976] 1351/71).

[3] Brunner, K.; Hink, W.; Scharnagl, T. (Inn.-Shell X-Ray Phys. At. Solids Proc. Intern. Conf. X-Ray Processes Inn.-Shell Ioniz., Stirling, Scot., 1980 [1981], pp. 71/4; C. A. **96** [1982] No. 60429).

[4] Scharnagl, T.; Hink, W. (J. Phys. F **13** [1980] 4021/30).

[5] Ootuka, A.; Kawatsura, K.; Fujimoto, F.; Komaki, K.; Ozawa, K.; Terasawa, M. (J. Phys. Soc. Japan **53** [1984] 1001/5).

[6] Fujimoto, F.; Kawatsura, K.; Ozawa, K.; Terasawa, M. (Phys. Letters A **57** [1976] 263/4).

[7] Nagel, D. J.; Knudsen, A. R.; Burkhalter, P. G. (J. Phys. B **8** [1975] 2779/86).

[8] Yamadera, A.; Ishii, K.; Sera, K.; Sebata, M.; Morita, S. (Phys. Rev. [3] A **23** [1981] 24/33).

[9] Chu, T. C.; Ishii, K.; Yamadera, A.; Sebata, M.; Morita, S. (Phys. Rev. [3] A **24** [1981] 1720/5).

[10] Ishii, K.; Sera, K.; Arai, H.; Morita, S.; Tokuda, K. (Phys. Rev. [3] A **27** [1983] 2225/8).

[11] Jakubassa, D. H.; Kleber, M. (Z. Physik A **273** [1975] 29/35).

[12] Spindler, E.; Betz, H.-D.; Bell, F. (J. Phys. B **10** [1977] L561/L564).

[13] Boiko, V. I.; Evstigneev, V. V.; Shamanin, I. V. (Izv. Vyssh. Uchebn. Zaved. Fiz. **31** No. 4 [1988] 110/1 from C.A. **109** [1988] No. 138166).

15.12 Neutron Optics. Generation of Neutrons

In comparison with other commonly used monochromator (Si, Nb, Cu) materials, Be single crystals give a considerably higher intensity of the monochromatic beam for neutrons with a wavelength between 0.07 and 0.3 nm. This is due to its very low total attenuation coefficient, μ, for this wavelength range, the highest reflectivity, and the low penetration depth of the primary irradiated neutrons [1]; cf. "Beryllium" Suppl. Vol. A 1, 1986, pp. 124/5.

The attenuation of a neutron beam in a crystal is given by the sum of nuclear neutron-capture cross section, inelastic diffuse scattering, and multiple Bragg diffraction. **Fig.** 15-**9** from [2] reproduced in [1] shows values of the total attenuation coefficient for several monochromator materials as a function of the neutron wavelength, λ. For $\lambda > 0.07$ nm only Si has more favorable values than Be. However, the reflectivity derived from these data proved Be to be the best material, see **Fig.** 15-**10** from [3] reproduced in [1]. This is due to its high scattering power per volume unit. Another advantage of Be for obtaining optimum reflectivity is its low penetration depth of the primary incident neutron beam, see **Fig.** 15-**11** from [3] given in [1]. A higher penetration depth would cause a wider spread of the reflected beam and thus a further intensity loss. The values represented in Fig. 15-10 and Fig. 15-11 were calculated for a mean halfwidth of about 20' of the neutron reflection curve ($\lambda = 0.1$ nm). For practical use, the monochromator should be a single crystal with a low defect concentration and without subgrains [1].

Early investigations were performed on the behavior of slow monochromatic neutrons (with energies in the range 0.004 to 0.20 eV) impacting monocrystalline Be; the transmission results agreed with the theory of elastic scattering from crystals [4]. Further experiments involved the interference of slow neutrons in order to determine the phase of the scattered neutrons with respect to the primary neutron wave. The limiting angle for total reflection of neutrons of 1.873 Å from a Be mirror was found to be 12.0' and calculated as 11.1' [5]; see also previous studies of neutron reflection from mirrors such as Be [6].

Slow neutron transmission of Be metal samples of different thickness was measured using a neutron-beam spectrometer in the energy range 0.004 to 50 eV with particular emphasis on the very low energy region from 0.004 to 0.2 eV or from 0.6 to 5 Å. The effective slow neutron cross section was determined for several energy values (in units of 10^{-24} cm²/atom): 6.1 at 0.2 Å, 3.5 at 1.5 Å, 2.3 at 2.5 Å, and 0.6 at 4.6 Å [7]. For slowing down of neutrons in Be metal, elastic and inelastic cross sections are given in [8]. The attenuation of thermal neutrons by successive thicknesses of Be targets was measured by [9]. The design of cold neutron filters and tests with Be (and other) filter materials are described in [10].

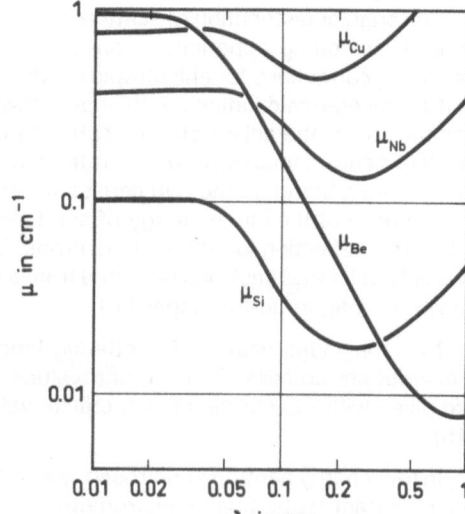

Fig. 15-9.   Total attenuation coefficient, μ, as a function of neutron wavelength, λ, for various monochromator materials at room temperature.

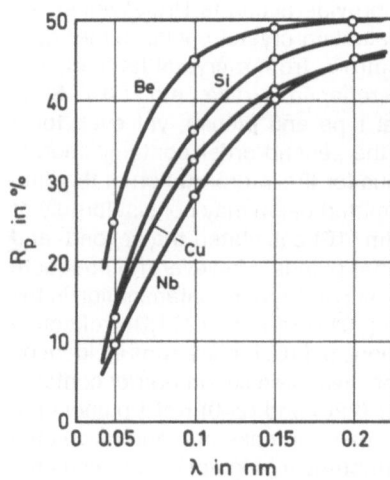

Fig. 15-10.   Maximum possible peak reflectivity, R_p, of various monochromator crystals as a function of neutron wavelength, λ, in transmission arrangement for a mosaic angle of about 4.4 mrad.

Fig. 15-11.   Optimum thickness, d_o, of various mono-chromator crystals as a function of neutron wave-length, λ.

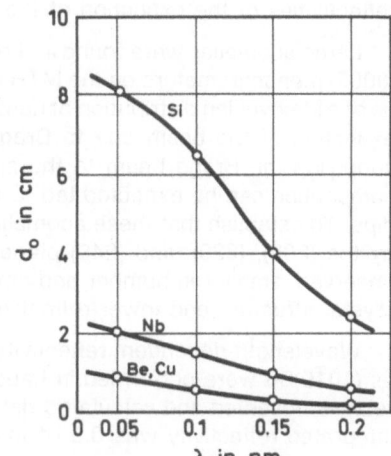

The angular distribution of scattered neutrons as a function of energy was determined from neutron scattering experiments on Be. From these data revised transport cross sections (including corrections for anisotropic scattering in the center of mass system) were calculated, and Fermi-age calculations with these new data agreed with experimental values [11]. A comparison of the anisotropy of scattering of cold neutrons from Be, Mg, and Pb at a given incident neutron temperature illustrated a change in the cause of anisotropy (an approximate method for separating the two causes for anisotropy in scattering is given in the paper). The dependence of the mean energy of scattered neutrons on the Debye temperature is verified [12]. The reflection of thermal neutrons from magnetized mirrors was studied, and the intensity of filtered neutrons reflected from a Be mirror as a function of the angle of incidence is shown in a figure in the paper [13].

Recommended values of scattering lengths and free and incoherent cross sections for elements are compiled in [14], and values of the scattering potential for various materials are given with the corresponding critical velocities for total reflection at normal incidence in [15].

In low-energy neutron spectroscopy at a nuclear reactor, the crystal spectrometer is one of the important tools. Such spectrometers are particularly effective in the region below 10 eV where high-resolution total cross section measurements provide accurate Breit-Wigner parameters for neutron resonances. A problem arises in evaluating crystal spectrometer data because the beam obtained in the Bragg reflection of neutrons from a crystal lattice is not monoenergetic owing to the presence of higher order reflections, see, e.g. [16]. Such reflections may be minimized by proper choice of crystal type and planes; yet even for a preferred choice of $(12\bar{3}1)$Be in the region of a few eV, the second-order contamination is 0.6%. Higher order contamination can be particularly serious for the interpretation of the data in studying the peak of a resonance. In this case the transmitted beam may consist largely of neutrons which are not from the first-order reflection. Holm [16] calculated the second- and third-order reflectivity expected from a number of Be crystal planes. However, no adequate experimental verification has been published. On this basis second-order contamination in the BNL neutron crystal spectrometer was measured for the $(10\bar{1}1)$Be and $(12\bar{3}1)$Be reflection over the energy range 0.08 to 5 eV. The reflectivity calculations of [16] for neutrons incident on thin Be (and NaCl) crystals were experimentally verified. The measured second-order contaminations as a function of energy for the Be planes (and for (220) and (240) NaCl planes) are constant for energies above 0.8 eV. The discrepancy between the measured and calculated values found for the planes with lower Miller indices was attributed to neglect in the calculated reflectivities of the extinction of the primary beam [17].

Large anomalies were found in the Bragg beam produced by Be $(10\bar{1}1)$, $(10\bar{1}3)$, $(10\bar{1}0)$, and (0002) monochromators on the MTR crystal spectrometer. Instead of a spectrum characteristic of a Maxwellian distribution of neutron velocities, many large dips appeared to be caused by extinction of the beam due to Bragg reflection by planes in the crystal other than those supplying the Bragg beam to the spectrometer. Calculations of the angles at which such competition can be expected led to the identification of planes responsible for the principal dips. To establish that these anomalies are due to crystal properties, other spectra produced by the (200), (220), and (240) planes of NaCl were examined. A few extinction dips were observed, smaller in number and amplitude than those found in Be, because of the simpler crystal structure and lower reflectivity of NaCl [18].

Wavelength-dependent reflectivity measurements of a few neutron monochromators such as $(10\bar{1}0)$Be were performed in Laue (transmission) geometry. Fair agreement between the relative observed and calculated data was found for $\lambda \gtrsim 1.27$ Å. For $(10\bar{1}0)$Be the measured integrated reflectivity was 0.9 of the calculated value [19].

The technique of (thermal) neutron pulse propagation was used to study four Be assemblies of different size in order to confirm experimentally the continuum contamination predicted for small assemblies. Continuum contamination was observed for all assemblies studied; it was associated with the sub-Bragg neutrons and appeared as oscillations in real and imaginary parts of the complex inverse relaxation lengths as a function of frequency (in the range of 0 to 1100 Hz) [20].

Calculations of steady-state space- and angle-dependent thermal neutron spectra in small Be assemblies were extended to assemblies of much greater transverse dimensions as well as neutron diffusion to much greater distances from the source plane (looking for a discrete mode of decay). In the forward direction, neutron distribution failed to attain equilibrium inside 140-cm thick assemblies with 150×150 cm transverse dimensions, whereas in the backward direction equilibrium was reached even inside an assembly of 80×80 cm transverse dimensions. In the forward direction, equilibrium was delayed by the presence of a penetrating beam of uncollided sub-Bragg neutrons of the source, such that attainment of equilibrium in the forward direction is difficult. The calculated value of the diffusion length was in excellent agreement with observed as well as with earlier theoretical values [21].

For measurements of phonon dispersion relations by means of slow neutron scattering, see "Beryllium" Suppl. A 2, 1991, pp. 127ff.

Thick-target neutron yield from the Be(p,n) reaction, for use in neutron cancer therapy machine design, was measured by several laboratories. It was observed that there is a significant discrepancy in shape and overall magnitude between (neutron energy) yield spectra measured (at 0°) especially for 35 MeV protons [22] and those measured by Waterman et al. [23] and Graves et al. [24] using 41 MeV protons. In the study [22] it was tried to resolve the difference in the high-energy end of the spectrum at 35 MeV incident proton energy by remeasuring the (0°) yield spectrum using a different experimental method (proton recoil spectroscopy) instead of neutron time-of-flight techniques. The integral neutron yield above 15 MeV was found to be 16.6×10^9 n/sr-µC; a figure comparing the new and earlier results is shown in the paper [22].

Attenuation and high-energy neutron production measurements for 190 MeV deuterons in Be (C and U) are presented. The high-energy neutron production cross section for Be is ~90 mb [25].

The possibility of producing pulsed photoneutron fluxes of high spatial-time and spectral density in a (synchrotron) radiation field of relativistic electrons channeled in a crystal is discussed in [26]; see also [27]. Calculated spectral and integral parameters of the "neutron focus" in Be for axial electron channeling in a W crystal are tabulated in [26].

Experimental data of the photoneutron yield produced in Be, Bi, and water targets under irradiation by a γ-ray beam (produced from a 900 MeV electron beam channeled in a 0.35 mm thick ⟨110⟩ diamond crystal) revealed the strong orientation dependence of the photoneutron production on the tilting angle of the diamond radiator relative to the electron beam. Calculated values of the total neutron yield from the targets irradiated by the γ-quanta beam from the 900 MeV axially channeled electrons are listed in the paper. For obtaining maximum neutron yield, it was necessary to use Be (or ^2H_2O) targets. The relatively narrow spectrum of channeling radiation should produce a powerful neutron source with a well-defined energy distribution [28].

The Be nucleus has an anomalously low threshold for the γ,n reaction [26]. The threshold for the $^9Be(\gamma,n)$ reaction at 1.665 MeV is the lowest for all stable nuclei; for measurements of the $^9Be(\gamma,n)^8Be$ threshold photoneutral cross section, see the paper [29].

References:

[1] Jönsson, S. (Naturwissenschaften **69** [1982] 483/90).

[2] Freund, A. (Hahn-Meitner Inst. Kernforsch. Berlin Ber. No. 273 [1978] from [1]).

[3] Aldinger, F.; Freund, A. (Beryllium 1977 Conf. Prepr. 4th Intern. Conf. Beryllium, London 1977 from [1]).

[4] Fermi, E.; Sturm, W. J.; Sachs, R. G. (Phys. Rev. [2] **71** [1947] 589/94).

[5] Fermi, E.; Marshall, L. (Phys. Rev. [2] **71** [1947] 666/77).

[6] Fermi, E.; Zinn, W. H. (Phys. Rev. [2] **70** [1946] 103).

[7] Rainwater, L. J.; Havens, W. W. (CP-2269 [1955] 18 pp. from N.S.A. **10** [1956] No. 5345).

[8] Gurney, R. (AECD-2449 (ANL-HDY-562) [1949] 11 pp. from N.S.A. **2** [1949] No. 935).

[9] Johnson, E. B. (ORNL-3193 [1961] 1/368, 4/9 from N.S.A. **16** [1962] No. 3838).

[10] Egelstaff, P. A.; Pease, R. S. (J. Sci. Instrum. **31** [1954] 207/12).

[11] Bulmer, J. J. (KAPL-M-JJB-1 [1955] 18 pp. from N.S.A. **10** [1956] No. 10916).

[12] Egelstaff, P. A. (AERE N/R-1165 [1957] 25 pp. from C.A. **1958** 8718).

[13] Hughes, D. J.; Brugy, M. T. (Phys. Rev. [2] **81** [1951] 498/506).

[14] Koester, L. (Springer Tracts Mod. Phys. **80** [1977] 1/55, 36).

[15] Steyerl, A. (Springer Tracts Mod. Phys. **80** [1977] 57/130, 61).

[16] Holm, M. W. (IDO-16115 (AT(10-1)-205) [1953] 39 pp. from N.S.A. **8** [1954] No. 1186) cited in [17].

[17] Haas, R.; Shore, F. J. (Rev. Sci. Instrum. **30** [1959] 17/21).

[18] Spencer, R. R.; Smith, J. R. (Nucl. Sci. Eng. **8** [1960] 393/9 from N.S.A. **15** [1961] No. 5522).

[19] Malik, S. S. (J. Appl. Cryst. **9** [1976] 273/8).

[20] Miles, R. E. (Diss. Univ. Missouri, Columbia 1972, pp. 1/166 from Diss. Abstr. Intern. B **34** [1973] 1126).

[21] Garg, S.; Ahmed, F.; Kothary, L. S. (Nucl. Sci. Eng. **63** [1977] 500/4 from C.A. **87** [1977] No. 124368).

[22] Ullmann, J. L.; Peek, N.; Johnson, S. W.; Raventos, A.; Heintz, P. (Med. Phys. **8** [1981] 396/7).

[23] Waterman, F. M.; Kuchnir, F. T.; Skaggs, L. S.; Kouzes, R. T.; Moore, W. H. (Med. Phys. **6** [1979] 160 from [22]).

[24] Graves, R. G.; Smathers, J. B.; Almond, P. R.; Grant, W. H.; Otte, V. A. (Med. Phys. **6** [1979] 123 from [22]).

[25] Knox, W. J. (UCRL-1427 [1951] 17 pp. from C.A. **1955** 7995).

[26] Eremeev, I. P.; Kumakhov, M. A. (Phys. Letters A **72** [1979] 359/60).

[27] Eremeev, I. P. (Pis'ma Zh. Eksperim. Teor. Fiz. **27** [1978] 13/7; JETP Letters **27** [1978] 10/4).

[28] Golovkov, V. M.; Zabaev, V. N.; Kalinin, B. N.; Kurkov, A. A.; Lunev, V. I.; Adishchev, Yu. N.; Potylitsin, A. P.; Vorobiev, S. A. (Nucl. Instr. Methods **212** [1983] 167/72; C.A. **99** [1983] No. 78623).

[29] Berman, B. L.; van Hemert, R. L.; Bowman, C. D. (Phys. Rev. [2] **163** [1967] 958/63).

16   Electrochemical Behavior

General References:

Chauvin, G.; Corou, H.; in: Bard, A. J.; Encyclopedia of the Electrochemistry of the Elements, Vol. 5, 1976, pp. 227/60.

Panzer, R. E.; in: Bard, A. J.; Parsons, R.; Jordan, J.; Standard Potentials in Aqueous Solution, New York – Basel 1985, pp. 675/86.

Piontelli, R.; Considérations sur l'Électrochimie des Métaux, J. Chim. Phys. **45** [1948] 115/22.

Shapovalov, E. T.; Gerasimov, V. V.; The Electrochemical Behavior of Beryllium in Aqueous Solutions, At. Energiya SSSR **26** [1969] 498/502, 578/81.

16.1   Potentials

16.1.1   Standard Potentials in Aqueous Solutions

References for this section are given on pp. 155/6.

Conventions

The standard potential at 25°C, symbolized by E°, is the potential of a half reaction with all participating species present in their standard state at unit activity. The standard states are defined as pure crystalline solids, pure liquids, ideal gases at unit fugacity ($f = 1\,atm = 1.01325 \times 10^5$ Pa), and ideal solutes at unit molality. In acid solution E° is referred to the standard hydrogen electrode with the half reaction $2\,H^+(aq) + 2e^- \rightleftharpoons H_2(gas)$, its E° value being fixed at zero Volt at all temperatures. In alkaline solutions (pH = 13.996) the standard potentials are referred to $H_2 + 2\,OH^- \rightleftharpoons 2\,H_2O + 2e^-$ with $E° = -0.8280$ V. The signs of the standard potentials are defined by the Stockholm Convention, i.e., E° is positive if the oxide species at unit molal activity is a better oxidant than $H^+(aq)$, a = 1 m, and it is negative if the reduced form is a better reducing agent than $H_2(gas)$, f = 1 atm.

The isothermal temperature coefficient, $(dE°/dT)_{isoth}$, shows the dependence of the isothermal cell Be|electrolyte|SHE or inert electrode|Be redox system-electrolyte|SHE on temperature; it is positive if the potential increases with increasing temperature. The thermal temperature coefficient, $(dE°/dT)_{th}$, gives the temperature dependence of a Be electrode or a Be redox system measured vs. SHE of fixed temperature. It is composed of the temperature effects at the electrode|electrolyte interface and at the liquid junctions. It is positive if the hot electrode is the positive pole.

General Remarks

A recent review concerned with the calculated standard potentials of various elements, including Be, in aqueous solution of pH = 1 or 13.996 at 25°C and of the isothermal temperature coefficients is given by Bratsch [1]. The thermodynamic data used for the calculations are taken from [2, 3]. The classic reference of standard potentials is Latimer [4], based on the thermodynamic data of [5], which are, however, obsolete. Potentials of Be couples in aqueous solutions, obtained by measurements, are rather reproducible. Thus, standard potentials derived from measured values are less reliable [6, p. 677].

The Be^I|Be Couple

For the reaction $Be^+ + e^- \rightleftharpoons Be$ is $E° = -1.13$ V. The value is based on measurements of the cell $Be|x M\ Be(ClO_4)_2|sat.\ KCl|SCE$ with $x = 0.00022$ to 0.215 [7]. According to [8] it was thought that Be passes into solution as univalent Be^+ which is subsequently oxidized to Be^{2+}: $Be^+ + H^+ \rightarrow Be^{2+} + \frac{1}{2}\ H_2$; details are given in the paper [7]; cf. p. 158.

The Be^{II}|Be Couple

In **acid solutions** the value $E° = -1.97$ V was calculated for the reaction $Be^{2+} + 2e^- \rightleftharpoons Be$ from $\Delta G° = -379.7$ kJ/mol [6, p. 677]. Another value $E° = -1.968$ V was calculated [1] based on the thermodynamic data of [2, 3]. This value is uncertain by less than ten units in the last digit; $(dE°/dT)_{isoth}$ was estimated to be 0.60 mV/K [1]. Previously $E° = -1.847$ V was calculated from $\Delta G° = -85.2$ kcal/mol [9], see also [10, 11]. The thermal temperature coefficient of this $E°$ value was calculated to be $(dE°/dT)_{th} = 1.44$ mV/K; the method of calculation is given in the paper [11]. The isothermal temperature coefficient $(dE°/dT)_{isoth} = 0.56_5$ mV/K was obtained from $(dE°/dT)_{th}$ by subtracting the calculated value of the thermal temperature coefficient of SHE (about 0.871 mV/K) [11]; cf. [10].

Other values (in V) are $E° = -1.85$ [4], calculated from the thermodynamic data [5]; $E° = -1.734$ calculated from a cyclic process [12]; $E° = -1.69$ extrapolated from measurements [12].

Referred to the standard state of the gaseous electron at 25°C, $E° = 2.72$ V was estimated by an empirical method. For the standard potential of a single hydrogen (H^+) electrode vs. the standard state of a gaseous electron at 25°C the estimated value $E° = 4.42$ V was used. Details are given in the paper [13]; see also [14].

For the reaction $Be^{2+} + H_2(gas) + 2e^- \rightleftharpoons BeH_2(cryst)$ the values $E° = -2.26$ V and $(dE°/dT)_{isoth} = -0.05$ mV/K were estimated [1].

Using the thermodynamic data [2, 3], $E° = -1.808$ V was calculated and $(dE°/dT)_{isoth} = 0.3$ mV/K was estimated for $BeOH^+ + H^+ + 2e^- \rightleftharpoons Be + H_2O$, and $E° = -1.880$ V was calculated for $Be_3(OH)_3^{3+} + 3H^+ + 6e^- \rightleftharpoons 3Be + 3H_2O$ [1].

For the following reactions $E°$ values were calculated from $\Delta G°$ [9, 10]:

$BeO + 2H^+ + 2e^- \rightleftharpoons Be + H_2O$: $E° = -1.785$ V for BeO from $\Delta G° = -139.0$ kcal/mol for BeO;
$BeO_2^{2-} + 4H^+ + 2e^- \rightleftharpoons Be + 2H_2O$: $E° = -0.909$ V from $\Delta G° = -155.3$ kcal/mol for BeO_2^{2-};
$Be_2O_3^- + 6H^+ + 4e^- \rightleftharpoons 2\ Be + 3H_2O$: $E° = -1.387$ V from $\Delta G° = -298.0$ kcal/mol for $Be_2O_3^-$.

For Be^{2+}, $Be_2O_3^{2-}|Be$ the value $E° = -0.935$ V was calculated from $\Delta G° = -85.2$ and -298.0 kcal/mol for Be^{2+} and $Be_2O_3^{2-}$, respectively, and for $Be_2O_3^{2-}$, $BeO_2^{2-}|Be$ the value $E° = -1.873$ V was calculated from $\Delta G° = -298.0$ and -155.3 kcal/mol for $Be_2O_3^{2-}$ and BeO_2^{2-}, respectively [9, p. 133].

In **alkaline solutions** the values $E° = -2.58$ V and $(dE°/dT)_{isoth} = -1.05$ mV/K were estimated for precipitated $Be(OH)_2$ in the reaction $Be(OH)_2 + 2e^- \rightleftharpoons Be + 2OH^-$ [1]. For the crystalline $\alpha(\beta)$ modifications of $Be(OH)_2$ the values $E° = -2.598$ (-2.609) V and $(dE°/dT)_{isoth} = -1.022$ (-1.001) mV/K, respectively, were calculated [1] from the thermodynamic data [2, 3].

For the reaction $BeO + H_2O + 2e^- \rightleftharpoons Be + 2OH^-$ the thermodynamic data [2, 3] were also used to calculate $E° = -2.606$ V and $(dE°/dT)_{isoth} = -1.174$ mV/K for crystalline BeO [1]. Other calculated values are $E° = -2.613$ V and $(dE°/dT)_{isoth} = -1.172$ mV/K [10].

For $Be(OH)_3^- + 2e^- \rightleftharpoons Be + 3OH^-$ $E° = -2.52$ V, and for $Be(OH)_4^{2-} + 2e^- \rightleftharpoons Be + 4OH^-$ $E° = -2.517$ V and $(dE°/dT)_{isoth} = -0.751$ mV/K were calculated [1] from thermodynamic data.

The value $E° = -2.62$ V was calculated from $\Delta G° = -298$ kcal/mol for $Be_2O_3^{2-}$ in $Be_2O_3^{2-} + 3H_2O + 4e^- \rightleftharpoons 2Be + 6OH^-$ [4], and $E° = -2.63$ V [10].

Correlations between Standard Potentials and Other Parameters

Relationships between $E°$ and the position in the periodic table are given by Deasy [15]. Elements such as Be which belong to an a-family of the periodic table show a greater value of $|E°|$ than any member of the corresponding b-family. In each a-family of the first three groups $|E°|$ increases with increasing atomic number (except Li) [15].

The standard potential depends on the sublimation energy, ΔH_s, the ionization potential, E_i, and the ionic radius, r_i, according to the relation $nFE° = \Delta H_s + E_i(1 - k/r) - B'$ (where k and B' are coefficients depending on the valence of the metal; for details, see the paper), which also reflects the variation of $E°$ of the metals in relation to their position in the periodic table. The negative values of $E°$ for $Be^{2+}|Be$ (and also $Mg^{2+}|Mg$), though Be and Mg have a high ionization energy, are due to their very small ionic radius [16]. Diagrams of $E°$ vs. the metals (Li to Au), arranged with increasing position in the electrochemical series, and of $(r_a - r_i)$, where $r_a =$ atomic radius and $r_i =$ ionic radius, vs. the same abscissa show practically the same shape (with exception of the third group metals), see the figure in the paper [17]. Standard potentials of alkali and alkaline earth metals as functions of their hydration enthalpies and reciprocal atomic radii are given in [18]. There is no definite relationship between $E°$ and the work function, Φ, of a metal. Approximately, the equation $E° = -4.25 + 0.204 e^{0.67\Phi}$ V is valid [19, p. 883]. For a relationship between the standard potential and the potential of zero charge, see [24].

The heats of reduction of various oxides such as Na_2O, MgO, Al_2O_3 by Be and further elements and the standard potentials of the reducing elements show parallelism [25].

Absolute Potential

The absolute potential E_{abs} is the single Galvani potential difference at the phase interface metal|electrolyte.

For standard $Be^{2+}|Be$ the absolute potential was calculated by the method of Latimer et al. [20], which was generalized for this purpose. The calculation involves a sequence of processes and always includes another metal. Several quantities are used for the calculation such as the free energy of sublimation, ionization potential, free energy of hydration of the ion, electronic work functions, and the standard electrode potential of the metal vs. SHE. For $Be^{2+}|Be$ (averaged over 16 self-consistent second metals) is $E_{abs} = -0.91$ V at 25°C with the Gibbs'-Stockholm sign convention [21, 22]. Parameters for the calculation of E_{abs} according to [20] and a relationship between the absolute potential and the potential of zero charge are given in [23]. An empirical approach to calculate the real free energy of solvation and the absolute potential for aqueous redox systems containing a monoatomic cation, with $Be^{2+}|Be$ among others, is presented in [26].

References:

[1] Bratsch, G. (J. Phys. Chem. Ref. Data **18** [1989] 1/21).

[2] Wagman, D. D.; Evans, W. H.; Parker, V. B.; Schumm, R. H.; Halow, I.; Bailey, S. M.; Churney, K. L.; Nuttall, R. L. (NBS-TN 270-3/270-8 [1968/81]).

[3] Wagman, D. D.; Evans, W. H.; Parker, V. B.; Schumm, R. H.; Halow, I.; Bailey, S. M.; Churney, K. L.; Nuttall, R. L. (J. Phys. Chem. Ref. Data **11** Suppl. 2 [1982]).

[4] Latimer, W. M. (The Oxidation States of the Elements and Their Potentials in Aqueous Solutions, 2nd Ed., Prentice Hall, New York 1961, pp. 313/5).

[5] Rossini, F. D.; Wagman, D. D.; Evans, W. H.; Levine, S.; Jaffe, I. (NBS-C-500 [1952]).
[6] Panzer, R. E. (in: Bard, A. J.; Parsons, R.; Jordan, J.; Standard Potentials in Aqueous Solution, New York–Basel 1985, pp. 675/86).
[7] Getman, F. H. (Trans. Electrochem. Soc. **66** [1934] 143/52).
[8] Bodforss, S. (Z. Physik. Chem. **124** [1926] 66/82).
[9] Van Muylder, J.; Pourbaix, M. (in: Pourbaix, M.; de Zoubov, N.; Van Muylder, J.; Deltombe, E.; Schmets, J.; Vanlengenhaghe, C.; Atlas d'Équilibres Électrochimiques, Paris 1963, pp. 132/8).
[10] Milazzo, G.; Caroli, S.; Sharma, V. K. (Tables of Standard Electrode Potentials, Wiley, New York, 1978, pp. 62/5).

[11] De Bethune, A. J.; Licht, T. S.; Swendeman, N. (J. Electrochem. Soc. **106** [1959] 616/25).
[12] Makishima, S. (Z. Elektrochem. **41** [1935] 697/712).
[13] Matsuda, A. (J. Res. Inst. Catal. Hokkaido Univ. **27** [1979] 167/9; C.A. **93** [1980] No. 32592).
[14] Matsuda, A.; Notoya, R. (J. Res. Inst. Catal. Hokkaido Univ. **28** [1980] 67/71; C.A. **93** [1980] No. 226617).
[15] Deasy, C. L. (J. Chem. Educ. **18** [1941] 514).
[16] Korovin, N. V. (Izv. Vyssh. Uchebn. Zaved. Khim. Khim. Tekhnol. **3** [1960] 844/7; C.A. **1961** 8112).
[17] Rajagopalan, K. S.; Mathur, P. B. (Naturwissenschaften **47** [1960] 375).
[18] Weissner, H. (Z. Elektrochem. **62** [1958] 445/57).
[19] Vasenin, R. M. (Zh. Fiz. Khim. **27** [1953] 878/88; C.A. **1954** 57).
[20] Latimer, W. M.; Pitzer, K. S.; Slansky, C. M. (J. Chem. Phys. **7** [1939] 108/11).

[21] De Bethune, A. J. (J. Chem. Phys. **29** [1958] 616/25).
[22] De Bethune, A. J. (J. Chem. Phys. **31** [1959] 847/8).
[23] Jakuszewski, B. (J. Chem. Phys. **31** [1959] 846/7).
[24] Koshurnikov, G. S. (Zh. Prikl. Khim. **25** [1952] 562; J. Appl. Chem. [USSR] **25** [1952] 635/6).
[25] Iskol'dskii, I. I.; Shokor, T. G. (Zh. Prikl. Khim. **19** [1946] 693/704; C.A. **1947** 4423).
[26] Notoya, R.; Matsuda, A. (Nippon Kagaku Kaishi No. 6 [1982] 990/6; C.A. **97** [1982] No. 79937).

16.1.2 Potentials in Aqueous Solutions

References for this section are given on pp. 161/2.

Potential – pH Diagram

The E – pH diagram, which reflects the electrochemical behavior of Be in aqueous solutions, was studied by Van Muylder and Pourbaix [1]. It is based on a sequence of equations which give the equilibrium potential vs. SHE at 25°C as a function of the pH value and the activity of the ions participating in the electrochemical process. The diagram takes into account the ions Be^{2+} and $Be_2O_3^{2-}$ and the solid species Be and crystalline $Be(OH)_2$ in the β-form. The electrolyte must not contain substances which form soluble Be complexes or insoluble Be salts. A simplified diagram is given in **Fig. 16-1**. The stability range of water is limited by the dotted lines a and b which denote the reactions $2H^+ + 2e^- \rightleftharpoons H_2$ and $2H_2O \rightleftharpoons O_2 + 4H^+ + 4e^-$, respectively. A concentration of $10^{-6}M$ for Be^{2+} and $Be_2O_3^{2-}$ was used for the solubility limits which define the regions of corrosion, passivation, and immunity. The stability range of Be metal lies far below that of water indicating that Be is a nonnoble, very unstable metal and a strong reducing agent. In acid solutions Be dissolves with the formation of Be^{2+} ions and H_2 evolution. In strong alkaline solutions Be also dissolves with H_2 liberation and formation of $Be_2O_3^{2-}$ and BeO_2^{2-} ions. In the intermediate pH range the metal becomes covered with a layer of BeO or $Be(OH)_2$ and thus shows passive behavior. Strong acids such as solutions of HCl and H_2SO_4 dissolve the metal with H_2 liberation; weak acids, e.g. citric or tartaric acid, attack the metal only at the beginning until a protective layer has formed. Cathodic protection of Be metal probably can not be achieved because of the very low potential value of its protection (about -2 V vs. SHE). For this reason a deposition of Be from aqueous solutions can not be realized [1]; see also [2, 3].

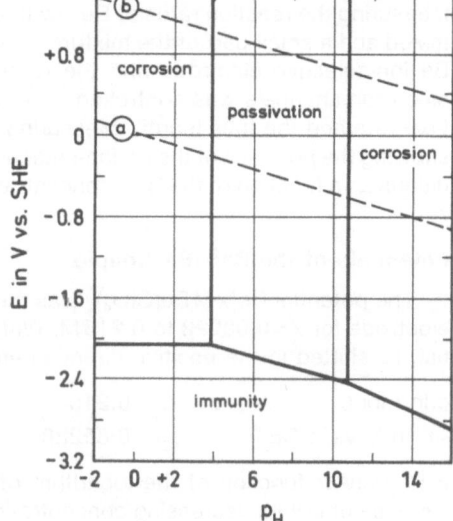

Fig. 16-1. Simplified E–pH diagram for Be in aqueous solution at 25°C.

An investigation of the corrosion of Ta, Nb, Zr, Ti, and Be showed that Be is corroded in all really corrosive media; but none of these metals is corroded under oxidizing conditions, e.g. in the presence of HNO_3. This behavior is due to the protective influence of a surface oxide film which is readily produced in the presence of an oxidizing agent [4].

Beryllium Ion-Selective Electrodes

An ion-selective electrode is a membrane electrode which responds selectively towards one (or several) ion species in the presence of others. "Membrane" denotes a thin section of an electrically conducting material that separates two solutions across which a potential is set up. A PVC-matrix membrane Be ion-selective electrode was prepared in the following way: at first the Be ion exchanger was formed by conversion of the Ca-form of (Orion 92-20-02) resin by reaction with $BeSO_4$ to the Be-form (type A) or by a similar conversion of Ca-bis-di[4-(1,1,3,3-tetramethylbutyl)phenyl]phosphate to the corresponding Be-form (type B). For membrane fabrication PVC was added to a solution of the respective Be ion exchanger in tetrahydrofuran. Then disks were cut from the master membrane for the electrode assembly, the internal solution of which was 0.1M KCl–0.1M $BeSO_4$. Both types of electrodes (prepared with Be ion exchanger A or B) were calibrated in 10^{-6} to 10^{-1} M $BeSO_4$ solutions vs. a calomel reference electrode. The detection limit was 4.3×10^{-5} M Be^{2+} and $\sim 1.1 \times 10^{-5}$ M Be^{2+} and the slope of the potential vs. log $a_{Be^{2+}}$ plot between $a_{Be^{2+}} = -4$ and -1 was 30 and 31 mV/decade for the type A and B electrodes, respectively. The response time was less than 2 min in both cases. Studies of pH influence were performed with both types of electrodes in 10^{-2} M $BeSO_4$ solution of pH varying in the range 1 to 6. There was no region in which the electrode response is independent of pH. Especially the type B electrodes were very sensitive to pH changes over the range studied [5].

Previously a rapid-mixing continuous-flow system, which utilizes liquid membrane ion-selective electrodes as sensors, was presented. Using an ion-selective Be resin in conjunction with the flow system, the rate of complex formation of Be^{2+} with biologically important ligands could be measured. A schematic diagram of the flow assembly is given in the paper [6]. A liquid divalent ion resin (Orion type 92-32) was converted into the Be^{2+} form. For this ion-selective electrode the internal reference solution consisted of 10^{-3} M Be_2SO_4 in 10^{-1} M KCl. For measuring the reaction rates of Be^{2+} with a special ligand, the solutions of the reactants were mixed and a small part of the mixture was led along a short flow path through the measuring Be ion-selective electrode and the reference electrode. The volume flowing through the electrode channels was controlled. The time of measurement was determined by the linear flow rate and the tube length. Measuring times as low as 10 to 15 ms were attained. A plot showing the potential of the Be ion-selective electrode vs. a flow-through SCE under flow conditions as a function of the Be^{2+} concentration of the reacting mixture is given in the paper [6].

Potentials of the Be²⁺|Be Couple

The potential $Be|x$ M $Be(ClO_4)_2$ was measured at 25°C with a saturated calomel reference electrode for $x = 0.00022$ to 0.215 M. With decreasing $Be(ClO_4)_2$ concentration, c, the potential, E, shifted in the positive direction (increased). Selected values are:

c in mol/L	0.215	0.0860	0.0215	0.0022	0.0002
−E in V vs. SCE	0.85238	0.83050	0.78723	0.72582	0.66421

E is a linear function of the logarithm of concentration. The anomalous behavior, i.e., the increase of E with decreasing concentration, is explained by the assumption that metallic Be goes into solution as Be^+ [7]. But the postulation that Be passes into solution as a monovalent ion is experimentally not justified [8, p. 378]. In the anodic dissolution, however, the effective valence lies between $n = 1$ and $n = 2$ [9]; see also pp. 194/6.

The potential $Be|1N$ $BeX_2|X_2$, i.e., the decomposition potential, E_{dec} (in V), of 1N BeX_2 is given by the equation $E_{dec} = 22.93\ (r_c/r_a) - 1.266$ with r_c = radius of the cation and r_a = radius of the anion. Corresponding relationships are given for Mg, Ca, Sr, Ba, and Ra halogenides. The plots E_{dec} vs. r_c/r_a extended to $r_c/r_a \rightarrow 0$ meet in the origin of the coordinates for the Ca, Sr, Ba, and Ra halogenides, but not so for Be and Mg halogenides [10].

Potentials in Solutions Initially Not Containing Be Species

The potential of a metal in a solution that does not contain the respective metal ions is not an equilibrium potential. It depends on numerous variables such as the time of immersion, temperature, concentration of the electrolyte, formation of complex ions, impurities, pretreatment of the electrode surface, formation of surface films, and the magnitude of exchange currents of the possible electrode reactions. In most cases it is referred to as dissolution, corrosion, or simply irreversible potential.

The dissolution potentials of Be in HF, HCl, $HClO_4$, and H_2SO_4 solutions were determined at 30°C as functions of time, acid concentration, and current density (for the latter, see the section "Behavior as Anode" pp. 192 ff.). For this purpose the emf of the cell $Be|acid|sat.$

KCl|1N KCl, Hg$_2$Cl$_2$|Hg was measured in air and then converted to SHE by adding 0.279 V. Two samples of Be were used: A (vacuum arc-cast of 99.48% purity) and B (hot pressed from 98.5% powder). The exposed surface of the Be electrode was ground and polished. **Fig.** 16-2 shows the potential −E vs. SHE as a function of time for various acid concentrations. In H$_2$SO$_4$ solutions E increases with time. A stronger increase is observed in HF solutions, at least up to about 70 min. In HCl and HClO$_4$ solutions the increase is rather small; in 0.05N HClO$_4$ solution a decrease is found. Generally, the purer sample A, which contains less oxygen, shows less noble potentials than sample B under the same conditions. At a fixed time of measurement (not given) E decreased for HCl, HClO$_4$, and H$_2$SO$_4$ solutions with increasing acid concentration in the range 0.01 to 0.8N. An increase with concentration was only observed in HF solutions. Again sample A showed more negative E values than sample B [8]. The potential-time curves were interpreted in the following way: the Be metal electrode was assumed to contain BeO, finely divided or as solid solution. In HF solution the acid attacks first the Be metal while the

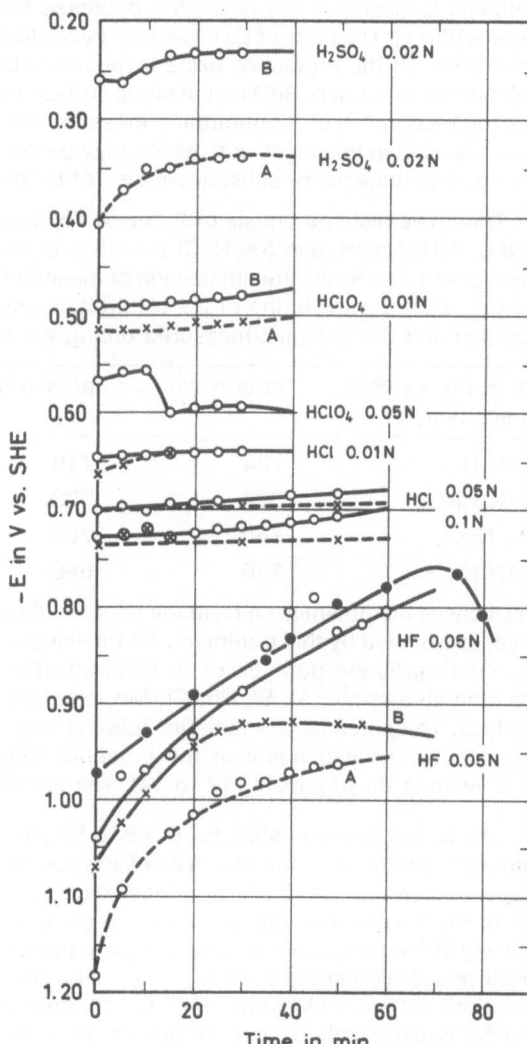

Fig. 16-2. Potential −E of Be in various acids as a function of time. Dashed curves: Be sample A of 99.48% purity; solid curves: Be sample B of 98.5% purity.

less soluble BeO accumulates on the surface. Additionally an adherent film of Be(OH)$_2$ or basic salts may be formed by direct reaction of Be with dissolved oxygen or with water. This layer has a more noble potential than Be. If the oxide layer dissolves partially or breaks down mechanically, a drop of potential follows, as can be observed in Fig. 16-2, p. 159. A decrease in oxygen content of Be will entail a less noble potential, as can be seen from the values of the samples A and B (the latter has a higher oxygen content and a more noble potential). A similar behavior as in HF was observed in H$_2$SO$_4$ solutions; in this case the potential-time curves lie in a more noble range than in HF solutions thus indicating a denser and more adherent protective layer. For HCl and HClO$_4$ solutions for which the increase of E with time is rather small or even absent (sample A), it is assumed that BeO does not accumulate on the surface. This may be indicative of BeO solubility in the acids or of rapid breakdown of the protective layer. During dissolution of Be in HCl and HClO$_4$ solutions, disintegration of Be was observed. Possibly the breakdown of the oxide layer is accompanied by the formation of metallic particles due to disintegration. Thus fresh metal is exposed to the action of the acids and the increase in potential is slowed down or even a decrease takes place. Further details are given in the papers [8, 11]. The plots of E vs. acid concentration for HCl, HClO$_4$, and H$_2$SO$_4$ indicate a rapid breakdown of the protective oxide layer with increasing acid concentration, whereas in HF solution the amount of BeO on the metal surface rapidly increases with the acid concentration. For the recognition of disintegration these potential measurements are not useful because a potential shift in the negative direction may be due to the breakdown of protective layers and also to simultaneous or subsequent loss of Be particles from the dissolving surface [8, 11].

The irreversible potentials of 22 metals including Be were determined in 0.1N HCl, 0.1N HNO$_3$, 0.1N NaOH, and 3% NaCl solutions at room temperature. At first the potential was measured 1 and 5 min after immersion of the metal electrode into the electrolyte. Then a value was measured with stirring of the electrolyte, and then the electrode was rubbed under the solution and the potential measured during the rubbing process:

$-E$ in mV vs. SHE in solutions of	after 1 min	after 5 min	with stirring	rubbing under solution
0.1N HCl	724	710	712	717
0.1N HNO$_3$	397	322	273	341
3% NaCl	760	764	753	1016
0.1N NaOH	746	666	664	1216

In all cases the potential shifts in the negative direction during rubbing as the protective oxide layer is removed by this treatment. All the initial potentials (before rubbing) are more positive than the equilibrium potential of the electrode (Be^{2+}|Be). The results show a great tendency of Be (and also of Mg, Al, Si, Nb, Cr, Mo, and Mn) to form protective layers on the electrode surface. The potential change after rubbing was similar for metals of the same group of the periodic table independent of the potential before rubbing. The influence of pH was not discussed in these papers [12 to 14]; see also [15].

The dissolution potential, E$_d$, of 99 to 99.5% Be was measured at 20°C in acid chloride, bromide, and iodide solutions, free of oxygen and saturated with hydrogen, at fixed pH; E$_d$ depended on the nature of the halogenide (X) and on its concentration. The range of activity $a = 10^{-2}$ to 1M was investigated. In all cases E$_d$ became more negative by about 70 mV when the activity of the halogenide increased by one power of ten. At constant activity, E$_d$ shifted in the positive direction in the order Cl$^-$ → Br$^-$ → I$^-$. The values of E$_d$ (± 10 mV) vs. SHE at 20°C for solutions with $a_{x^-} = 1$ M were: -787 mV for chloride, -627 mV for bromide, and -405 mV for iodide. When a cathodic current density–potential curve of Be in acid halogenide solution is extended to the dissolution potential, the current density of corrosion can be obtained, as represented by curve 1 in Fig. 16-4 on p. 180, in 0.1M (Na,H)Cl solutions as a function of pH in the range 0 to 6 at 20°C. At fixed pH, the current density of corrosion increases only slightly with increasing halogenide concentration. The rate of corrosion increases by one power of ten when a_{x-} is increased by 4.6 powers of ten. This can be calculated from the dependence of the dissolution potential on the halogenide concentration and the slope of the cathodic current density–potential curve [9].

The corrosion potentials of a 99.7% Be electrode were measured at 80°C after the usual pretreatment (cleaning, degreasing, washing) in various 0.1N electrolytes. The measurements were performed immediately after immersion (A) and after standing for 20 h (B) in the electrolyte:

electrolyte	$NaNO_3$	Na_2SO_4	$CuSO_4$	$Cu(NO_3)_2$
E in V vs. SHE (A)	−0.14 to −0.01	−0.41 to −0.34	−0.045	+0.36
E in V vs. SHE (B)	+0.24	−0.35 to −0.33	−0.05	+0.56

electrolyte	KOH	HNO_3	H_2SO_4
E in V vs. SHE (A)	+0.02 to +0.05	+0.19	−0.57 to −0.48
E in V vs. SHE (B)	+0.14 to −0.13	+0.06	−0.52

Based on these results, the activating influence of sulfate ions is discussed. In media containing sulfate, the potential does not shift in the positive direction during standing; Cu^{2+} ions, on the other hand, have a passivating effect as the comparison of $NaNO_3$ and $Cu(NO_3)_2$ shows. In sulfate solution the presence of Cu^{2+} entails intensified pitting corrosion compared with Na_2SO_4. In 0.5N NaCl solution the corrosion rate of Be, determined from the weight loss in 120 h, is 0.012 $g \cdot m^{-2} \cdot h^{-1}$ and the corrosion current density is 7.18 $\mu A/cm^2$ [16].

The potential of a Be electrode in aqueous alkaline solutions is still fairly negative (about −1.48 V vs. SHE) in the first moments after immersion. Then the potential shifts in the positive direction up to about −1.1 to −1.05 V. The rate of this shifting process decreases as the potential approaches its stationary value. This potential drift to more noble values is indicative of self-passivation. The latter effect is caused by a chemisorbed layer of oxygen and probably not by an oxide film which possibly would dissolve spontaneously [17, p. 581].

References:

[1] Van Muylder, J.; Pourbaix, M. (in: Pourbaix, M.; de Zoubov, N.; Van Muylder, J.; Deltombe, E.; Schmets, J.; Vanlengenhaghe, C.; Atlas d'Équilibres Électrochimiques, Paris 1963, pp. 132/8).

[2] Delahay, P.; Pourbaix, M.; Van Rysselberghe, P. (Compt. Rend. Reunion Comite Thermodynam. Cinet. Electrochim. 1951 [1952], pp. 15/29; C.A. **1954** 1119).

[3] Levy, D. J. (NP-11 154 [1961] 1/18; N.S.A. **16** [1962] No. 4540).

[4] Inglis, N. P.; Cotton, J. B. (Eng. Mater. Design **2** [1959] 78/80; N.S.A. **13** [1959] No. 10083).

[5] Moody, G. J.; Thomas, J. D. R.; Yarmo, M. A. (Lab. Pract. **30** [1981] 1111/2; C.A. **96** [1982] No. 59777).

[6] Fleet, B.; Rechnitz, G. (Anal. Chem. **42** [1970] 690/3).

[7] Getman, F. H. (Trans. Electrochem. Soc. **66** [1934] 143/52).

[8] Straumanis, M. E.; Gnanamuthu, D. S. (Corros. Sci. **4** [1964] 377/86).

[9] Heusler, K. E. (Z. Elektrochem. **65** [1961] 192/7).

[10] Weissner, H. (Z. Elektrochem. **62** [1958] 445/57).

[11] Straumanis, M. E.; Gnanamuthu, D. S. (AD-299972 [1963] 1/20; N.S.A. **17** [1963] No. 19989).

[12] Akimov, G. V.; Clark, G. B. (Trans. Faraday Soc. **43** [1947] 679/97).

[13] Clark, G. B.; Akimov, G. V. (Trans. 2nd Russ. Conf. Corrosion Metals **2** [1943] 33/51).

[14] Clark, G. B.; Akimov, G. V. (Compt. Rend. Acad. Sci. [URSS] **30** [1941] 805/9).

[15] Akimov, G. V. (Corrosion **11** [1955] 477t/486t, 515t/534t).

[16] Gerasimov, V. V.; Shapovalov, E. T. (Zh. Prikl. Khim. **46** [1973] 953/7; J. Appl. Chem. [USSR] **46** [1973] 1017/20).

[17] Shapovalov, E. T.; Gerasimov, V. V. (At. Energiya SSSR **26** [1969] 498/502; Soviet At. Energy **26** [1969] 578/81).

16.1.3 Potentials in Nonaqueous Solutions

The standard potential of $Be^{2+}|Be$ in liquid NH_3 of 1 m acid (i.e., 1 m NH_4^+) at 25°C was calculated from the free energy of formation of solvated Be^{2+} cations: $E = -1.5$ V referred to the standard hydrogen electrode in NH_3 ($E_H = 0$) [1].

The standard potentials of $Be^{2+}|Be$ were also calculated for various other nonaqueous electrolytes from the standard molal free energy of solvation. They are referred to the standard state of electrons in the gaseous state at 25°C [2]:

solvent	CH_3OH	C_2H_5OH	HCOOH	CH_3CN
E in V (25°C)	2.80	2.71	3.26	3.05

solvent	$HCONH_2$	N_2H_4	DMSO	quinoline
E in V (25°C)	2.58	1.37	2.78	1.89

Panzer [3] measured the potentials of Be electrodes in various nonaqueous solvents: hexamethylphosphoramide (HMPA), NH_3(liquid), acetonitrile (AN), dimethylformamide (DMF), 4-butyrolactone (4-BL), methylpyrolidone (M-pyrol), tetrahydrofuran (THF), propylene carbonate (PC), and $POCl_3$. The solvents did not contain Be species initially. The electrodes were of 98.3% Be or of SR-grade Be with 0.6% BeO and 0.01% C. They were chemically pretreated (for details see the paper) so that even traces of oxides were stripped from the surface. Smooth uniform surfaces were obtained. In most cases an aqueous $Ag|0.1M\ Ag^+$ or sat. Ag^+ electrode was used as reference with the exception of the solvents NH_3, THF, PC, and $POCl_3$ in which cases $Pb|Pb(NO_3)_2$ sat. $LiNO_3$ sat. in NH_3, $Ag|0.1M\ LiClO_4$, $Li|1M\ LiClO_4$, and $Mo|POCl_3$ were used, respectively. Potentials E converted to SHE_{aq} (probably at 25°C):

solvent	HMPA	NH_3	AN	DMF	4-BL
supporting electrolyte	0.1M $LiClO_4$	—	0.1M $LiClO_4$	0.1M KPF_6	0.1M KPF_6
E in V	+0.50	−0.52	+0.56	+0.66	+0.77

solvent	M-pyrol	THF	PC	$POCl_3$
supporting electrolyte	0.1M $LiClO_4$	0.1M $LiClO_4$	—	—
E in V	+0.79	+0.30	−2.68	+1.18

The results show that the potentials in the nonaqueous solvents are considerably more positive than the thermodynamic values because of interfering reactions. The formation of

BeO was observed in the nonaqueous solutions just as in aqueous solutions. This is likely the result of the disproportionation of a solvo-Be^I species (probably Be_2^{2+}) which results in powdery Be metal. In neutral nonaqueous solutions the potentials are comparable to the values observed in aqueous solutions of the Be^{II}|Be couple depending on the anion of the system [3].

References:

[1] Bratsch, S. G.; Lagowski, J. J. (J. Solution Chem. **16** [1987] 583/601, 596).
[2] Matsuda, A.; Notoya, R. (J. Res. Inst. Catal. Hokkaido Univ. **28** [1980] 67/71; C.A. **93** [1980] No. 226617).
[3] Panzer, R. E. (NASA-CR-86272 [1969] 1/71; C.A. **73** [1970] No. 30962).

16.1.4 Standard Potentials and Potentials in Melts

References for this section are given on pp. 171/2.

A review of electrode potentials in fused electrolytes including the couple Be^{II}|Be is given by Bard [1] and by Delimarskii and Markov [2, 3]. For an older survey up to 1956, see [4].

Pure Beryllium Halogenide Melts

The reversible potential E° of Be|BeCl$_2$ referred to a standard chlorine electrode (unit fugacity) was calculated from thermodynamic data in the temperature range T = 25 to 550°C (the accuracy of the quantities used for calculation was not evaluated):

T in °C	25	100	200	300	350	400	450	500	550
–E° in V	2.435	2.382	2.315	2.252	2.222	2.192	2.167	2.144	2.122

The values from 25 to ~400°C refer to the solid electrolyte, those above 400 to 550°C to the molten electrolyte. The standard potential in aqueous solution Be|BeCl$_2$, at 25°C and unit activity, vs. the standard chlorine electrode, $E_{aq}^° = -3.21$ V, is given for comparison [5].

The potential Be|BeCl$_2$ referred to a Cl$_2$ (1 atm),C electrode, i.e., the decomposition potential of BeCl$_2$, was measured in a cell in an atmosphere of purified Ar. Between 419 and 448°C the decomposition potential is given by the equation $E = 2.789 - 11.19 \times 10^{-4}$ T (± 0.0006) V, the Be electrode being the negative pole. The change in the chlorine partial pressure, caused by the change in the BeCl$_2$ vapor pressure, was taken into account [6]. For the free energies of formation of BeF$_2$ and BeCl$_2$ used for the calculation of the standard potentials Be|BeF$_2$ and Be|BeCl$_2$, see [7].

The following potentials of Be|BeCl$_2$ at 700°C are given in [2]: $E = -0.90$ V vs. a hydrogen reference electrode ($E_{H^+|H_2} = 0$); $E = +1.47$ V vs. an Na reference electrode ($E_{Na^+|Na} = 0$); $E = +1.292$ V referred to $E_{Na^+|Na} = 0$, calculated from thermodynamic data. At the melting point of BeCl$_2$ (399°C) a value of $E = +1.17$ V vs. $E_{Na^+|Na}$ was obtained. The decomposition potentials of BeCl$_2$ at its melting point and at 700°C are given as 2.08 and 1.92 V, respectively [2]; see also [8].

LiF–BeF$_2$ Melts

A beryllium reference electrode compatible with molten fluorides was developed in the following way: a tubular piece of Be metal was brought onto an Ni support. The contact between Ni and the melt was prevented by a boron nitride sleeve. Helium bubbled through the centered hole of the Be and protected it from the attack of any residual HF. Measurements with two such Be electrodes vs. each other in an Li_2BeF_4 melt indicated them to be completely stable, reproducible, and reversible [9].

The potential of Be|BeF_2–LiF was measured as a function of melt composition in the range $X_{BeF_2} = 0.30$ to 0.90 and of temperature, T, between 500 and 900°C; Pt, H_2, HF in the same melt was the reference electrode. The assumed cell reaction is Be(s)+2HF(gas)$\rightleftharpoons$$BeF_2$(melt)+ H_2(gas). Accurate measurements with pure BeF_2 as the electrolyte were not possible presumably due to its high viscosity and/or high electrical resistivity. Selected potential values are:

X_{BeF_2}	T in °C	E in V	X_{BeF_2}	T in °C	E in V
0.30	585.1	1.8864	0.60	521.0	1.8160
0.30	681.0	1.8213	0.60	710.0	1.6760
0.40	597.5	1.8069	0.80	611.5	1.7484
0.40	706.0	1.7316	0.80	800.4	1.6027
0.50	503.3	1.8467	0.90	702.0	1.6620
0.50	728.0	1.6765	0.90	866.0	1.5307

By relating these results to BeF_2 liquidus data (for details, see the paper [10]), the standard potential Be|BeF_2 (X=1) was calculated using values a_{BeF_2} derived from the phase diagram: $E° = 2.4430 - 0.0007952\,T$. The free energy and the heat of formation of liquid BeF_2 between 298 and 1000 K were obtained when the results of these measurements were combined with available thermochemical data of HF. The Be|BeF_2–LiF electrode proved to be stable and reversible and thus is suited as a reference electrode [10, 11].

The potential of a Be|LiF(66 mol%)–BeF_2 (34 mol%) electrode, isolated in a boron nitride compartment, was measured vs. Ni|Ni^{II} in LiF (65.6 mol%)–BeF_2 (29.4 mol%)–ZrF_4 (5 mol%) at 500°C as a function of the Ni^{II} concentration. The measured values were extrapolated using the Nernst equation to a potential E which is referred to Ni|Ni^{II} (unit mole fraction): E = −2.120 V [12].

The potential of Be|LiF–BeF_2 was determined with reference to an Li-Bi alloy electrode in the same melt at 819 K as a function of alloy concentration and of electrolyte composition. The alloy concentration varied between ~10^{-4} and 3×10^{-3} mole fraction of Li at a fixed electrolyte composition ($X_{BeF_2} = 0.3275$) and the melt composition between $X_{BeF_2} = 0.2$ and about 0.35 (values taken from figures 2 and 4 in the paper [13]). The plot of the potential E vs. X_{Li} fits the Nernst equation thus indicating that the activity coefficient in the Li-Bi alloy is constant in the concentration range studied and that the Li-Bi alloy follows Henry's law. The plots E vs. X_{BeF_2} of this study [13] are compared with the measurements of cells with transference of [10, 11] and [14].

Concentration cells with transference Be|LiF–$BeF_2(X_2)$ (cell II)‖LiF–$BeF_2(X_1)$|Be (cell I) were investigated at 500 and 610°C. The liquid junction between the indicator half cell II and the reference half cell I was made from coarse-grade fritted silica disks. At 500 (610)°C the fixed BeF_2 concentration in the reference half cell I was $X_1 = 0.3400$ (0.7012), respectively. Selected values of E_t (emf of the cell with transference):

500°C	X_2	0.3338	0.3478	0.4040	0.4485	0.5096
	E_t in mV	−12.3	5.4	71.9	127.6	167.9

610°C	X_2	0.6211	0.4992	0.3842	0.3268	0.2831
	E_t in mV	17.33	72.73	186.23	263.61	321.41

The emf of the same cell without transference can be calculated from these results if the transference number of Li^+ in the melt is known. Details for the calculation are given in the paper [14].

$BeCl_2$-NaCl and $BeCl_2$-KCl Melts

The potential $Be|BeCl_2$ (58.7 wt%)–NaCl (41.3 wt%) vs. a Cl_2,C electrode in the same melt was obtained by determining the decomposition potential of this melt with a carbon, Pt, or Mo cathode and a carbon anode. Measured values are $-E = 1.93$, 1.96, 1.99, 2.02, and 2.08 V at 700, 640, 600, 540, and 420°C, respectively [15]. For an investigation of the decomposition potential as a function of the $BeCl_2$/NaCl ratio in the melt, see [16].

The $Be|BeCl_2(X)$–NaCl(1−X) potential was investigated as a function of the mole fraction of $BeCl_2(X)$ in the range 0.25 to 0.80 and of the temperature between 300 and 700°C. The reference electrode was a chlorine or Pt electrode in the same melt. Selected values at 500°C are:

X	0.25	0.35	0.45	0.51	0.65	0.80
$-E$ vs. Cl_2 in V	2.14	2.03	2.00	1.93	1.92	1.92
$-E$ vs. Pt in V	1.770	1.738	1.726	1.718	1.712	1.706

As the potential values deviate widely from the Nernst equation, complex formation in the melt was assumed (for low values of X anionic complexes, for high values of X cationic complexes). The activity coefficients of $BeCl_2$ in the melt were calculated from the results [17].

Somewhat more negative values of E vs. Cl_2 reference electrode are given in [18], e.g., $E = -2.057$ V for $X = 0.51$ at 500°C. The authors obtained $E° = -1.9864$ V for $X = 1$ by extrapolation. The formation energy of $BeCl_2$,$2NaCl$ at 500°C was calculated to be ~7.5 kcal [18]. Further measurements of the potential $Be|BeCl_2(X)$–NaCl(1−X) vs. a chlorine electrode were performed for $X = 0.4$ to 0.7 and the temperature range 400 to 500°C. By extrapolation of the Nernst plot $E° = -1.924$, -1.890, and -1.839 V was found at 400, 440, and 500°C, respectively, for $X = 1$. From thermodynamic data $E° = -1.938$ V was calculated at 400°C. The cell reaction was considered to be $Be(s) + Cl_2(gas) \rightleftharpoons BeCl_2(l)$. A complex such as Na_2BeCl_4 was probably formed in the melt [19].

The potential $Be|BeCl_2(X)$–KCl(1−X) vs. a Cl_2,C electrode in the same melt was measured between 400 and 530°C for $X = 0.4$ to 0.7. Due to complex formation the melt remained stable beyond the boiling point of pure $BeCl_2$ (483°C). The potential of the Be electrode is given by the following equations (T in °C):

$X = 0.450$	$-E = 2.131 - 6.4 \times 10^{-4} (T - 400)$	$X = 0.600$	$-E = 2.057 - 7.1 \times 10^{-4} (T - 400)$
$X = 0.500$	$-E = 2.105 - 6.6 \times 10^{-4} (T - 400)$	$X = 0.700$	$-E = 2.018 - 7.5 \times 10^{-4} (T - 400)$

For $X = 1$ the value $-E° = 1.927 - 8.3 \times 10^{-4} (T - 400)$ was extrapolated which is in fair agreement with $-E° = 1.910 - 6.6 \times 10^{-4} (T - 400)$ calculated from thermodynamic data. Therefore, a two-electron reaction $Be(s) + Cl_2(gas) \rightleftharpoons BeCl_2(l)$ was considered as the cell reaction. The plot E vs. log 1/X shows large deviations from an ideal solution for the $BeCl_2$–KCl melt [20].

The Gibbs energy for the reactions $Be(s) + \frac{1}{2}Cl_2(gas) \rightleftharpoons BeCl(l)$ and $Be(s) + Cl_2(gas) \rightleftharpoons BeCl_2(l)$, which are assumed to proceed in the cell $Be|BeCl_x$–$KCl|Cl_2$, C, were calculated [21]. The products BeCl and $BeCl_2$ are formed as ions in the melt, not in the pure state. From the Gibbs energy the following apparent standard electrode potentials (in V) vs. a chlorine electrode were obtained and compared with experimental values of [22]:

T in K	1100	1200	1300	1400
$-E^{\circ}_{Be^+|Be}$ calc.	1.85	1.79	1.72	1.65
$-E^{\circ}_{Be^+|Be}$ exp.	1.81	1.77	1.73	1.67
$-E^{\circ}_{Be^{2+}|Be}$ calc.	2.31	2.21	2.12	2.02
$-E^{\circ}_{Be^{2+}|Be}$ exp.	2.19	2.14	2.10	2.10

The calculated $E^{\circ}_{Be^+|Be}$ values differ by only 5% from the experimental values, whereas the deviations for $E^{\circ}_{Be^{2+}|Be}$ amount to 15%. This result shows that the interaction between the strongly polarizing cations and the solvent must be taken into account [21]; cf. also p. 170.

$BeCl_2$ Melts in Eutectic LiCl–KCl

The system $Be|BeCl_2$–LiCl–KCl was intensively investigated. Yang and Hudson [23] used a W wire electrode on which Be was electrodeposited and a graphite,Cl_2 reference electrode, separated by a glass diaphragm for the Be electrode and its surrounding melt. Potentials were measured for mole fractions $X_{BeCl_2} = 0.0021$ to 0.1514 between 761 and 818 K and corrected for the thermal electromotive force between W and graphite. Potentials E at 817 K and dE/dT over the whole temperature range are:

X_{BeCl_2}	0.0021	0.0259	0.0416	0.1514
$-E$ in V (at 818 K)	2.4524	2.3434	2.3109	2.2175
dE/dT in 10^{-4}V/K	3.37	4.00	4.66	5.88

The standard potentials extrapolated from the measured values are $E^{\circ} = -2.0074$, -1.9926, and -1.9888 V for 713, 728, and 734 K, respectively. The calculated E° values of [5] are more negative by ~ 0.2 V whereas E° values obtained from the standard free energy change ΔG of the reaction $Be + Cl_2 \rightleftharpoons BeCl_2$ given in [24] are in agreement with the extrapolated data within 35 mV. This agreement indicated that probably a two-electron reaction takes place in the LiCl–KCl melt [23]. In nearly the same temperature range (767 to 825 K) measurements were performed for $X_{BeCl_2} = 0.0322$ and 0.0416 vs. an $Ag|AgCl(X_{AgCl})$ electrode. The correction term for conversion of the potential values vs. Ag|AgCl to a chlorine reference electrode was determined [25].

After a careful preelectrolysis of the LiCl–KCl base melt under Ar to remove contaminations, $BeCl_2$ (X = 0.2 to 0.6) was added and the surface of the carbon reference electrode was saturated with chlorine. Then the Be electrode was immersed. A contact potential was reached after about 10 min. In the temperature range studied (410 to 500°C) the following relationships were obtained (T in °C):

X_{BeCl_2}	0.188	0.410	0.613
$-E$ in V .	$2.165 - 8.1 \times 10^{-4}(T - 410)$	$2.103 - 8.7 \times 10^{-4}(T - 410)$	$2.045 - 9.2 \times 10^{-4}(T - 410)$

The standard potential ($X_{BeCl_2} = 1$) was extrapolated from these results: $-E^{\circ} = 1.950 - 9.9 \times 10^{-4}$ $(T - 410)$, and compared with the calculated value: $-E^{\circ} = 1.931 - 6.5 \times 10^{-4} (T - 410)$ [26, 27].

Plambeck [28] produced the Be^{2+} ions in the eutectic LiCl–KCl melt by anodizing Be; the Be^{2+} concentration was $\leqq 0.2$M. At concentrations below 0.006M Be^{2+}, constant potentials could not be obtained. Potential measurements were done at 450°C with $Pt|1M\ Pt^{II}$ in eutectic LiCl–KCl as the reference electrode. The plot E vs. log $c_{Be^{2+}}$ is linear in the range $c_{Be^{2+}} = 0.006$ to 0.2 M with a slope of 0.0723 ± 0.0017 V/log unit, that is close to the theoretical value of 0.0717 for a two-electron process. The experimental value of the number of electrons involved in the electrode process is $n = 1.9 \pm 0.2$. The standard potential obtained from the experimental results is $E^{\circ} = -2.039 \pm 0.013$ V on molarity scale [28]. This value is in poor agreement with that of [23] if the latter is converted to molarity scale, 450°C, and Pt|1M Pt^{II} reference electrode (−1.781 V). The deviation was attributed to the higher temperature and the more concentrated melts used by [23] the behavior of which is non-Henrian [28, p. 930].

The temperature dependence of the reversible potential of Be (99.6 or 99.84%) in molten eutectic LiCl–KCl containing 0.023, 0.169, and 1.13 wt% BeCl$_2$ was measured between 380 and 940°C with a chlorine electrode in the same base electrolyte as reference. The equations $E = -2.629 + 2.19 \times 10^{-4}$ T, $E = -2.650 + 3.39 \times 10^{-4}$ T, and $E = -2.695 + 4.8 \times 10^{-4}$ T, respectively, were obtained for the given BeCl$_2$ concentrations. From the slope of the isotherms $E = f(\log X_{Be})$, with X_{Be} = mole fraction of the Be species in the melt, the valence of the Be ions in contact with the Be metal electrode can be found. As the temperature dependence of the equilibrium constant of the redox reaction $Be^{2+}(melt) + Be(s) \rightleftharpoons 2 Be^{+}(melt)$ is known, the mole fractions of Be^{2+} and Be^{+} can be calculated at any temperature. If equilibrium is established, the sum of the potentials $Be^{2+}|Be$ and $Be^{2+}|Be^{+}$ is equal to the potential $Be^{+}|Be$. Thus, from the measured potential values, the following standard potentials (in V) were calculated for the temperature range 700 to 900 K: $E^{\circ}_{Be^{2+}|Be} = -2.702 + 6.4 \times 10^{-4}$ T, $E^{\circ}_{Be^{+}|Be} = -2.212 + 4.0 \times 10^{-4}$ T, and $E^{\circ}_{Be^{2+}|Be^{+}} = -3.192 + 8.8 \times 10^{-4}$ T. The decomposition potential of molten Be monochloride and the free energy of its formation were derived from the results [29]; cf. [30, p. 3].

Later measurements of the potential Be|BeCl$_2$ (0.1 to 0.6 mol%) in eutectic LiCl–KCl were performed between 400 and 600°C vs. an Ag|AgCl electrode and converted to a C,Cl$_2$ reference electrode. According to the BeCl$_2$ concentration and temperature, equilibrium was reached 5 to 30 min after immersion of the Be electrode. The isotherms E vs. log X_{BeCl_2} follow the Nernst equation for 400 to 450°C (deviating only by 0.5 mV). For higher temperatures there are considerable deviations from linearity, e.g., for 0.1 mol% BeCl$_2$ at 500 (600)°C: 3 (7) mV. It was assumed that the deviations are caused by the initial formation of Be^{+}; but the presence of Be^{+} could not be confirmed. For the standard potential Be^{2+}|Be (mole fraction scale) the value $E^{\circ} = -2.684 + 0.6 \times 10^{-3}$ T (± 0.003) V was obtained for the temperature range 400 to 600°C [31, 32].

An Mo and a Be electrode (the latter deposited on Mo) as well as a chlorine reference electrode were immersed in a crucible containing a eutectic mixture of LiCl–KCl melt with 10 wt% BeCl$_2$. The variations of the potentials of the Be and Mo electrodes vs. the chlorine electrode and vs. each other were measured as a function of time at a fixed temperature between 630 and 960°C. The potential of the Be electrode reached a constant value rather rapidly: -2.44 ± 0.01 V vs. the chlorine electrode at 770°C. The potential of the Mo electrode was at first much more positive and shifted in the negative direction: −1.13 V after 18 min, −1.82 V after 1480 min, finally approximating the potential of the Be electrode. The results show that the equilibrium $Be^{2+}(melt) + Be(s) \rightleftharpoons 2 Be^{+}(melt)$ is rapidly established near the surface of the Be electrode. Be^{+} ions diffuse into the electrolyte; their accumulation in the melt is limited by the rate of diffusion of the less mobile Be^{2+} ions to the electrode surfaces. The change of ΔE (potential difference between the Be and the Mo electrode) was attributed to a change of the redox potential $Be^{2+}|Be^{+}$ in the molten system as the concentration of Be^{+} near the Mo surface increased. The potential $Be^{2+}|Be$ at the Mo electrode shifts most quickly at the beginning of the redox reaction when the Be^{+} concentration is still rather small and the concentration ratio $c_{Be^{2+}}/c_{Be^{+}}$ decreases quickly with the accumulation of Be^{+}. More details are given in the paper [33].

BeCl$_2$ in NaCl–KCl Melts

The emf of the cell Be|equimolar NaCl–KCl–(1 to 3 wt%)BeCl$_2$|BeO,C was measured in an atmosphere of Ar for temperatures between 682 and 1040°C. The Be electrode was the negative pole, its potential was converted afterwards to a C,Cl$_2$ electrode. Selected values are:

T in °C	682	753	778	851	1040
−E vs. BeO,C in V	1.538	1.498	1.487	1.450	1.346
−E vs. Cl$_2$,C in V	2.240	2.274	2.323	2.389	—

The cell reaction with the BeO,C electrode is $BeO(s) + \frac{1}{2}C(graphite) - 2e^- \rightleftharpoons Be^{2+}(melt) + \frac{1}{2}CO_2(gas)$. The temperature dependence of the cell with regard to the thermal emf between the Mo lead of the Be electrode and the graphite lead of the BeO electrode is given by $-E = 2.036 - 5.16 \times 10^{-4}$ T (± 0.005) V. Thermodynamic data of the cell reaction were calculated [34].

The electrochemical behavior of high melting metals including Be, immersed in an NaCl–KCl melt, is characterized by three reversible processes which are for Be: $Be \rightleftharpoons Be^+ + e^-$, $Be \rightleftharpoons Be^{2+} + 2e^-$, and $Be^+ \rightleftharpoons Be^{2+} + e^-$. The comparison of the standard potentials of the three reactions (for unit activity of the participating ions) shows that $E^\circ_{Be^{2+}|Be}$ lies between the two other standard potentials; $E^\circ_{Be^+|Be}$ is more positive and $E^\circ_{Be^{2+}|Be^+}$ more negative. The corresponding behavior was found for the other high melting elements of the main groups [35].

Potentials of pure Be in equimolar NaCl–KCl with 4 wt% BeCl$_2$ were measured vs. an electrode of Be-Cu alloys as a function of alloy composition (2.1 to 99.5 at% Be) and temperature (710 to 835°C). The plots of the potentials vs. alloy composition show distinctly the changes in the phase composition of the alloy. Activity coefficients of Be and Cu as well as thermodynamic data of the alloy were calculated from the experimental results [36]. The potential Be|10 wt% BeCl$_2$ in equimolar NaCl–KCl was measured vs. a Be-Ni alloy of varying composition in the temperature range 963 to 1113 K by [37, 38].

For Be-Ni alloys with 3.49 to 50 at% Be the potential of the Be electrode shifted in the positive direction up to about 10 at% Be; this region corresponds to the α-phase of the alloy. Between 10 and 50 at% Be the potential remained constant indicating the coexistence of the α- and β-phase. At a fixed alloy composition the Be electrode potential increases linearly with the temperature [37]. Further studies of the Be potential vs. Be-Ni alloys with about 51 to 96 at% Be showed sharp shifts in the positive direction, the first one between 51.32 and 54.42 at% Be (corresponding to the β-phase) and the second one between 80.09 and 81.35 at% Be (corresponding to the γ-phase). In the two-phase regions 54.42 to 80.09 at% Be (β- + γ-phases) and 81.35 to 95.8 at% Be (γ- + δ-phases) the potential was constant. The temperature dependence was linear. Details are given in the paper [38]. The activities of Be and Ni in the alloys and thermodynamic data were calculated from the experimental results [37, 38].

BeCl$_2$ in Further Chloride and Chloride–Fluoride Melts

The potential of Be|BeCl$_2$ in NaCl–KCl–SrCl$_2$ vs. a chlorine electrode in the same melt, determined as the decomposition potential of the melt, is −2.27 V at the melting point (399°C) of BeCl$_2$ and −2.06 V at 700°C. The value of the potential at the melting point vs. an Na reference electrode is +0.98 V, at 700°C +1.33 V [2]. For the experimental procedure and composition of the melt, 28% NaCl–22% KCl–50% SrCl$_2$ (probably wt%), see previous studies [39]. The potential Be|BeCl$_2$–NaCl–AlCl$_3$ at the melting point of BeCl$_2$ is −2.26 V vs. a chlorine electrode and +0.99 V vs. an Na reference electrode [2]; see also [8].

Addition of fluoride ions to an equimolar NaCl–KCl melt shifted the potential of a Be electrode by 0.6 to 0.8 V in the negative direction. This was attributed to the formation of Be^{2+}–F$^-$ complex species, the Be^{2+}–F$^-$ bond being stronger than the Be^{2+}–Cl$^-$ bond. Measurements were performed of Be|BeCl$_2$–NaF–NaCl–KCl vs. a chlorine reference electrode between 953 and 1133 K with X_{BeCl_2} about 0.01 to 0.09 and X_{NaF} about 0.08 and 0.39. For a ratio

$X_{NaF}/X_{BeCl_2} > 4$ the complex BeF_4^{2-} is predominant, whereas for $3 < X_{NaF}/X_{BeCl_2} < 4$ a mixture of BeF_3^- and BeF_4^{2-} is present. Selected values for $X_{NaF}/X_{BeCl_2} > 4$ are:

T in K	$X_{Be^{2+}}$	X_{F^-}	$-E$ in V	$-E°$ in V
953	0.011500	0.0091997	3.177	2.192
991	0.064193	0.386048	3.279	2.174
1123	0.070177	0.338074	2.980	2.108

The temperature dependence of the calculated $E°$ values is given by the equation $E°_{Be^{2+}|Be} = -2.64 + 4.7 \times 10^{-4}\,T$. From the potential values the instability constant of the BeF_4^{2-} complex, K_1, was calculated. Selected values for $3 < X_{NaF}/X_{BeCl_2} < 4$ are:

T in K	$X_{Be^{2+}}$	X_{F^-}	$-E$ in V	$-E°$ in V
991	0.091078	0.351827	3.233	2.174
1073	0.030086	0.092283	2.804	2.107

From these potential values and K_1, the instability constant of BeF_3^- can be obtained. An equation for the equilibrium potential of $Be|BeCl_2$ in equimolar NaCl–KCl with relatively high fluoride concentration (>5 mol%) is also given in [40]. Butorov et al. [41] also established that the introduction of fluoride into the chloride melt caused a shift of the steady state potential of a Be electrode in the negative direction. The authors used melts with a concentration ratio fluoride to Be (in the melt) of 2 or 4 at 700°C. Increasing the c_{F^-}/c_{Be^+} ratio entails a further shift to still more negative values. On the addition of YCl_3 into the melt the potential became more positive. Empirical equations for the dependence of the Be potential on its mole fraction in the melt are presented in the paper [41].

Be^{2+} in Bromide and Iodide Melts

The potential $Be|Be^{2+}$ was measured at 700°C vs. an Na reference electrode for the following melts: NaBr–KBr ($E = 1.33$ V), NaBr–$AlBr_3$ ($E = 1.40$ V), and NaI ($E = 1.24$ V) [2, p. 184], cf. [8].

The following values were obtained for the potential $Be|BeI_2$ (5 mol%) – NaI (95 mol%) vs. I_2 in the same melt: $-E = 1.48$, 1.38, and 1.28 V at 600, 700, and 800°C, respectively. The potentials were determined as the decomposition potentials of the melts [42].

For **comparison of Be^{2+} in various melts** Delimarskii [43] determined the potentials of metals including Be in the melt of $BeCl_2$ and in the base melts NaCl–KCl–$SrCl_2$, NaCl–$AlCl_3$, NaBr–KBr, NaBr–$AlBr_3$, NaI, and NaI–AlI_3. The resulting values are given for 700°C and at the melting points of the respective halides and referred to an Na electrode. The electrochemical series in the various melts are established and the influence of the anions on the electrode potentials is discussed in the paper [43]; see also [2, 8, 44]. Graphs of the electrode potential of Be and of other metals vs. an Na electrode in fused fluorides, chlorides, bromides, and iodides as a function of the atomic number of the metal are given in [45].

In order to study the influence of the cations of the base melt on the potential of a Be electrode, the following systems were investigated [46]: $BeCl_2$ in LiCl, eutectic LiCl–KCl, KCl, and CsCl between ~700 to 1300 K. A chlorine electrode in the respective base melt was used as the reference, the potential differences of this electrode in the various base melts being close to zero (± 5 mV). Taking into account the thermal emf between the Mo lead of the Be electrode and the carbon lead of the chlorine electrode, the following equations were obtained:

melt: mol% BeCl$_2$		temperature range in °C	potential E in V
2.0	in LiCl	649 to 905	$-2.519 + 3.56 \times 10^{-4}$ T
0.46	in LiCl	620 to 880	$-2.514 + 2.77 \times 10^{-4}$ T
8.17	in LiCl–KCl	421 to 755	$-2.695 + 4.80 \times 10^{-4}$ T
1.2	in LiCl–KCl	393 to 940	$-2.650 + 3.39 \times 10^{-4}$ T
9.77	in KCl	779 to 1007	$-2.684 + 3.49 \times 10^{-4}$ T
0.60	in KCl	796 to 1023	$-2.629 + 1.69 \times 10^{-4}$ T
10.1	in CsCl	666 to 930	$-2.799 + 4.25 \times 10^{-4}$ T
0.47	in CsCl	695 to 950	$-2.782 + 2.74 \times 10^{-4}$ T

The slopes of the isotherms E vs. log $X_{Be^{2+}}$ give the value of the number of electrons participating in the electrode process, $n < 2$. In each base melt n decreases with increasing temperature: LiCl n = 1.92 (900 K), 1.76 (1200 K); LiCl–KCl n = 1.93 (700 K), 1.64 (1200 K); KCl n = 1.82 (1100 K), 1.73 (1300 K); CsCl n = 1.96 (1000 K), 1.82 (1200 K). The decrease of n is due to the formation of Be⁺ ions according to the equilibrium $Be^{2+}(melt) + Be \rightleftharpoons 2Be^+(melt)$. The standard potentials $E^{\circ}_{Be^{2+}|Be}$ (in V) for each melt can be calculated from the mean valence of the Be ions in the melt, the total Be concentration in the melt, and the empirical equation for E with its temperature dependence (see above). Thus, $E^{\circ}_{Be^{2+}|Be} = -2.538 + 5.7 \times 10^{-4}$ T in LiCl, $-2.702 + 6.4 \times 10^{-4}$ T in LiCl–KCl, $-2.727 + 5.3 \times 10^{-4}$ T in KCl, and $-2.834 + 5.9 \times 10^{-4}$ T in CsCl [46]; see also [47]. The considerable difference of $E^{\circ}_{Be^{2+}|Be}$ in the various base melts is caused by the nature of the cation. The relationship $E^{\circ}_{Be^{2+}|Be} = -3.089 + 5.83 \times 10^{-4}$ T $+ 0.43(e/r)$ with e = elementary charge of the cation and r = radius of the cation (in Å) is established. For mixtures of alkali chlorides the mean effective radius is used. Expressions for $E^{\circ}_{Be^{2+}|Be}$ in alkali chloride mixtures of any composition may be derived from this equation. From the difference of $E^{\circ}_{Be^{2+}|Be}$ in different melts the ΔG value for the reaction $M_2BeCl_4 + 2M'Cl \rightleftharpoons 2MCl + M'_2BeCl_4$ can be calculated [46]; see also [47].

The potential Be|Be^{n+} was measured for two ion fractions of Be in the melt and two temperature ranges in LiCl, eutectic LiCl–KCl, KCl, and CsCl as the base melts. From the measured values the standard potentials $E^{\circ}_{Be^{2+}|Be}$, $E^{\circ}_{Be^+|Be}$, and $E^{\circ}_{Be^{2+}|Be^+}$ (in V) vs. a chlorine electrode in the same base melt were calculated for each medium (details of the calculation are given in the paper) [48]:

base melt	$E^{\circ}_{Be^{2+}|Be}$	$E^{\circ}_{Be^+|Be}$	$E^{\circ}_{Be^{2+}|Be^+}$	temperature range in K
LiCl	$-2.526 + 5.40 \times 10^{-4}$ T	$-2.503 + 7.60 \times 10^{-4}$ T	$-2.549 + 3.20 \times 10^{-4}$ T	900 to 1300
eutectic LiCl–KCl	$-2.723 + 6.30 \times 10^{-4}$ T	$-2.488 + 6.53 \times 10^{-4}$ T	$-2.958 + 6.07 \times 10^{-4}$ T	800 to 1300
KCl	$-2.699 + 4.63 \times 10^{-4}$ T	$-2.374 + 5.0 \times 10^{-4}$ T	$-3.024 + 4.26 \times 10^{-4}$ T	1100 to 1400
CsCl	$-2.803 + 5.30 \times 10^{-4}$ T	$-2.545 + 6.95 \times 10^{-4}$ T	$-3.061 + 3.65 \times 10^{-4}$ T	1000 to 1300

From the results the equilibrium constant K of the reaction $Be^{2+}(melt) + Be \rightleftharpoons 2Be^+(melt)$ and its temperature coefficient were obtained. With an increasing radius of the cation of the base melt, $E^{\circ}_{Be^{2+}|Be}$ and $E^{\circ}_{Be^{2+}|Be^+}$ are shifted in the negative direction and the equilibrium of the aforementioned redox reaction is displaced to the side of the Be²⁺ ions [48].

The **absolute** potential and the real free energy of solvation of molten redox systems containing a monoatomic cation, including $Be^{2+}|Be$, were evaluated by an empirical approach assuming that the surface potentials of fused salts may be neglected. Values (referred to the standard state of the gaseous electron) are given for mixed melts of chlorides, nitrates, sulfates, thiocyanates, carbonates, metaphosphates, and acetates in the paper [49].

References:

[1] Bard, A. J. (Encyclopedia of Electrochemistry of the Elements, Vol. 10: Plambeck, J. A.; Fused Salt Systems, New York – Basel 1976, pp. 65, 114, 312/3, 366).

[2] Delimarskii, Yu. K.; Markov, B. F. (Electrochemistry of Fused Salts, Washington 1961, pp. 138/44, 176/92).

[3] Delimarskii, Yu. K. (Fiz. Khim. Elektrokhim. Rasplavl. Solei **1965** 3/35 from C.A. **64** [1966] 9239).

[4] Delimarskii, Yu. K. (Usp. Khim. **26** [1957] 494/515; C.A. **1957** 12703).

[5] Hamer, W. J.; Malmberg, M. S.; Rubin, B. (J. Electrochem. Soc. **103** [1956] 8/16).

[6] Chukreev, N. Ya.; Saparov, A. (Ukr. Khim. Zh. [Russ. Ed.] **39** [1973] 1108/12; C.A. **80** [1974] No. 43423).

[7] Glassner, A. (ANL-5750 [1957] 1/70; C.A. **1958** 4303).

[8] Delimarskii, Yu. K. (Zh. Fiz. Khim. **29** [1955] 28/38; C.A. **1955** 8713).

[9] Dirian, G.; Romberger, K. A. (ORNL-3789 [1965] 1/372, 76/7).

[10] Hitch, B. F.; Baes, C. F. (Inorg. Chem. **8** [1969] 201/7).

[11] Hitch, B. F.; Baes, C. F. (ORNL-4257 [1968] 1/46; C.A. **70** [1969] No. 25089).

[12] Jenkins, H. W.; Mamantov, G.; Manning, D. L. (J. Electrochem. Soc. **117** [1970] 183/5).

[13] Sood, D. D.; Braunstein, J. (J. Electrochem. Soc. **121** [1974] 247/9).

[14] Romberger, K. A.; Braunstein, J. (Inorg. Chem. **9** [1970] 1273/5).

[15] Delimarskii, Yu. K.; Skobets, E. M. (Zh. Fiz. Khim. **20** [1946] 1005/10; C.A. **1947** 2341).

[16] Abramov, G. A.; Vetinkov, M. M.; Gupalo, I. P.; Kostinkov, A. A.; Lozhkin, L. N. (Metallurgizdat **1953** 251/300 from [2]).

[17] Sheiko, I. N.; Delimarskii, Yu. K. (Ukr. Khim. Zh. **23** [1957] 713/20; C.A. **1958** 14384).

[18] Markov, B. F.; Delimarskii, Yu. K. (Zh. Fiz. Khim. **31** [1957] 2589/90; C.A. **1958** 8708).

[19] Kuroda, T.; Matsumoto, O. (J. Electrochem. Soc. Japan **33** No. 1 [1965] 29/34).

[20] Oyamada, R.; Kuroda, T. (Trans. Natl. Res. Inst. Metals [Tokyo] **9** [1967] 289/92; C.A. **69** [1968] No. 100201).

[21] Smirnov, M. V.; Khaimenov, A. P. (Dokl. Akad. Nauk SSSR **158** [1964] 1172/5; Dokl. Phys. Chem. Proc. Acad. Sci. USSR **154/159** [1964] 954/7).

[22] Smirnov, M. V.; Chukreev, N. Ya. (Tr. Inst. Elektrokhim. Akad. Nauk SSSR Ural'sk. Filial No. 3 [1962] 3/15; C.A. **59** [1963] 9398).

[23] Yang, L.; Hudson, R. G. (Trans. AIME **215** [1959] 589/601).

[24] Brewer, L.; Bromley, L. A.; Gilles, P. W.; Lofgren, N. L. (in: Quill, L. I.; Chemistry and Metallurgy of Miscellaneous Materials: Thermodynamics, Chapter 6, McGraw-Hill, New York 1950, p. 104).

[25] Yang, L.; Hudson, R. G. (J. Electrochem. Soc. **106** [1959] 986/90).

[26] Kuroda, T.; Matsumoto, O. (Bull. Tokyo Inst. Technol. No. 61 [1964] 29/41; C.A. **64** [1966] 1629).

[27] Kuroda, T.; Matsumoto, O. (J. Electrochem. Soc. Japan **30** No. 4 [1962] E193/E197).

[28] Plambeck, J. A. (Can. J. Chem. **46** [1968] 929/31).

[29] Smirnov, M. V.; Chukreev, N. Ya. (Dokl. Akad. Nauk SSSR **127** [1958] 1066/9; Proc. Acad. Sci. USSR Phys. Chem. Sect. **124/129** [1959] 707/10).

[30] Smirnov, M. V. (Electrochemistry of Molten and Solid Electrolytes, New York 1961, pp. 3/5).

[31] Shapoval, V. I.; Baranenko, V. M.; Barchuk, V. T.; Saparov, A. S. (Elektrokhim. Ionnykh Rasplavov **1979** 95/109; C. A. **93** [1980] No. 103 783).

[32] Shapoval, V. I.; Baranenko, V. M.; Barchuk, V. T. (Fiz. Khim. Elektrokhim. Rasplavl. Tverd. Elektrolitov Tezisy Dokl. 7th Vses. Konf. Fiz. Khim. Ionnykh Rasplavov Tverd. Elektrolitov, Sverdlovsk 1979, Vol. 1, pp. 67/8; C. A. **93** [1980] No. 156 578).

[33] Smirnov, M. V.; Chukreev, N. Ya. (Zh. Neorgan. Khim. **4** [1959] 2536/43; Russ. J. Inorg. Chem. **4** [1959] 1168/72).

[34] Smirnov, M. V.; Chukreev, N. Ya. (Zh. Neorgan. Khim. **3** [1958] 2445/9; Russ. J. Inorg. Chem. **3** No. 11 [1958] 18/24).

[35] Baimakov, Yu. V. (Freiberger Forschungsh. B No. 118 [1967] 43/71, 52).

[36] Anfinogenov, A. I.; Smirnov, M. V.; Ilushchenko, N. G.; Belyaeva, G. I. (Tr. Inst. Elektrokhim. Akad. Nauk SSSR Ural'sk. Filial No. 3 [1962] 83/100; C. A. **59** [1963] 8187).

[37] Kornilov, N. I.; Ilushchenko, N. G.; Rossokhin, B. G.; Belyaeva, G. I. (Electrochemistry of Molten and Solid Electrolytes, Vol. 7, New York 1969, pp. 74/81).

[38] Ilushchenko, N. G.; Kornilov, N. I.; Rossokhin, B. G. (Electrochemistry of Molten and Solid Electrolytes, Vol. 7, New York 1969, pp. 60/7).

[39] Delimarskii, Yu. K.; Ryabokon, V. D.; Kolotti, A. A. (Ukr. Khim. Zh. **15** [1949] 149/58).

[40] Smirnov, M. V.; Komarov, V. E.; Chukreev, N. Ya. (Zh. Neorgan. Khim. **10** [1965] 2001/5; Russ. J. Inorg. Chem. **10** [1965] 1089/92).

[41] Butorov, V. P.; Nichkov, I. F.; Novikov, E. A. (Izv. Vyssh. Uchebn. Zaved. Tsvetn. Metall. **19** No. 6 [1976] 50/6; C. A. **86** [1977] No. 147 726).

[42] Delimarskii, Yu. K.; Kolotti, A. A. (Zh. Fiz. Khim. **23** [1949] 97/100; C. A. **1949** 4581).

[43] Delimarskii, Yu. K. (Ukr. Khim. Zh. **16** [1950] 414/37; C. A. **1954** 473).

[44] Delimarskii, Yu. K. (Ukr. Khim. Zh. **21** [1955] 449/56; C. A. **1956** 6154).

[45] Delimarskii, Yu. K. (Zh. Obshch. Khim. **26** [1956] 2968/72; J. Gen. Chem. [USSR] **26** [1956] 3303/6).

[46] Smirnov, M. V.; Chukreev, N. Ya. (Zh. Neorgan. Khim. **6** [1961] 1361/8; Russ. J. Inorg. Chem. **6** [1961] 699/704).

[47] Smirnov, M. V.; Chukreev, N. Ya. (Fiz. Khim. Rasplavl. Solei Shlakov Tr. Vses. Soveshch., Sverdlovsk 1960 [1962], pp. 227/35; C. A. **58** [1963] 970).

[48] Smirnov, M. V.; Chukreev, N. Ya. (Tr. Inst. Elektrokhim. Akad. Nauk SSSR Ural'sk. Filial No. 3 [1962] 3/15; C. A. **59** [1963] 9398).

[49] Notoya, R.; Matsuda, A. (Nippon Kagaku Kaishi **1982** No. 6, pp. 990/6; C. A. **97** [1982] No. 79 937).

16.1.5 Potential of Zero Charge and Capacity

The potential of **zero charge**, E_2, of a Be electrode in an equimolar NaCl–KCl melt was obtained by capacity measurements, and $E_2 = -1.66$ V was found vs. Ag|0.05 M AgCl in the same base melt at 700°C [1]. The same value, $E_2 = -1.66$ V vs. Ag|10 wt% AgCl at 700°C in equimolar NaCl–KCl, was determined by measuring the differential electrode capacity of a single crystal Be electrode (a Pt electrode served as the second working electrode). Converted to a standard chlorine electrode in the same base melt, the value is $E_2 = -2.75$ V. According to their E_2 values in molten equimolar NaCl–KCl, a series of metals is given with the most negative components Mg (−1.97 V), Mn (−1.70 V), and Be (−1.66 V). This series is close to the electrochemical series based on standard potentials in chloride melts [2]; see also [3].

A relationship is presented between E_2 of liquid metals in equimolar NaCl–KCl at 700°C and the Gibbs free energy of formation, ΔG, of the respective chloride with Pb|2.5 wt% $PbCl_2$ in the same base melt as the reference electrode. In order to use this relationship for solid metals, e.g. Be, a correction term for the heat of fusion, ΔH, of the metal is added: $E_2 = 0.235 + (1/nF)\Delta G_{973} − (1/F)\Delta H$. From the latter equation $E_2 = −1.705$ V was calculated for a Be electrode [4]; cf. [3].

According to Vasenin [5], E_2 in aqueous solutions is a linear function of the electronic work function, Φ, obtained from the photoelectric effect. The best approximation is $E_2 = −4.25 + 0.86\,\Phi$ referred to SHE. For Be the value $E_2 = −0.88$ V was calculated with $\Phi = 3.92$ eV [5], and $E_2 = −1.06$ V vs. SHE was calculated with $\Phi = 3.7$ eV [2]; for details of the calculation, see the latter paper.

If E_2 values of metals are determined as the electrocapillary maximum in molten salts, the values are represented by $E_2 = −4.2 + 0.77\,\Phi$ (in V) vs. SHE [5]. The E_2 values determined in equimolar NaCl–KCl vs. Ag|10 wt% AgCl in the same base melt show a linear dependence on the work functions given by $E_2 = \Phi/e − 5.3$ (in V) where e is the electron charge. The constant in this linear relation (here 5.3) depends on the reference electrode, base melt, and temperature [2]. Relations between E_2, the electronic work function, Φ, the standard enthalpy, $\Delta H°$, and the standard Gibbs free energy, $\Delta G°$, of the crystal lattice of metals were investigated; the energy of the crystal lattice is connected with the sublimation of the metal and the subsequent ionization in electrons and positively charged ions. As the energy of the metal lattice is a more reliable quantity than Φ, the equation $E_2 = 0.305\,\Delta G_2/q + A$ was established with $G_2 =$ change of the Gibbs free energy accompanying sublimation and subsequent ionization to doubly charged ions, q = charge of one mol electrons in F (q=1), and A = −4.72, obtained by introducing the experimental values of Hg into the equation. For Be the value $E_2 = −0.07$ V was calculated with reference to $E_2(Hg) = −0.20$ V [6].

The **capacity** of a Be electrode in 5 N KOH at the oxygen evolution potential is 4.34 μF/cm² if the coefficient of roughness is assumed to be equal to three. In neutral solutions the capacity values are somewhat higher and do not differ much at the oxygen and hydrogen potentials: 7.05 and 7.47 μF/cm², respectively. These values are considerably lower than the capacity of a double layer and, therefore, they are considered to be an evidence for the presence of an oxygen film on the electrode [7]. A capacity of 77.5 μF/cm² in 0.5 N NaCl solution at the stationary potential of the Be electrode was obtained from the anodic charging curves. Taking the roughness factor into account, the capacity amounts to 23.6 to 25.8 μF/cm². A similar value was calculated from the cathodic charging curve. It was assumed that these values are close to the capacity of the double layer at the zero potential in uni-univalent electrolytes [8].

The relationship between resistive and capacitive components of the impedance of various metal electrodes including Be were investigated in equimolar NaCl–KCl melts at 700°C with Ag|10 wt% AgCl in the same base melt as reference electrode. For the potential of zero charge the capacity curve C vs. E (direct current polarization potential) passes through a minimum and the curve R (resistive component of impedance) vs. E through a maximum. The plot CR vs. C is linear, with the Be curve not being parallel to the linear plots of other metals (Zr, Ta, Mn, Cr, Fe, Co); the same result was found for the plot CR vs. fC (f = frequency of the alternating current in the range 0.3 to 2 kHz) [9].

The capacity of a single-crystal Be electrode (of at least 99.99% purity) was recorded as a function of the ac polarization potential (frequency 2 kHz) at 700°C in equimolar NaCl–KCl. A Pt sheet of 12 to 15 cm² served as the second working electrode and Ag|10 wt% AgCl in the same base melt as the reference electrode. The curves showed a minimum at -1.66 V which corresponds to the zero potential. The specific capacity of the Be single-crystal electrode at zero potential is 40 $\mu F/cm^2$ in good agreement with the values of liquid metals. This result indicates that the Be single crystal had no pores and thus the apparent surface area is equal to the true surface [2].

The impedance of a Be wire (0.1 mm in diameter) electrode in eutectic LiCl–KCl at 450°C was measured as a function of frequency with an alternating current bridge under cathodic polarization. For measurements under equilibrium conditions $BeCl_2$ was added to the melt. The double-layer capacity, C_d, was obtained from the experimental results by extrapolation (the procedure is described in the paper) and was compared with the high-frequency capacity in the base electrolyte. The latter gave a parabolic curve with a minimum of about 30 $\mu F/cm^2$ at -2.75 V vs. a chlorine electrode in the same melt when plotted vs. the polarizing potential E. The two C_d values were commensurable, thus demonstrating that the extrapolated value represents the double-layer capacity of the Be electrode in equilibrium. An equivalent-circuit of the impedance for the Be electrode in equilibrium was established; it is a double-layer capacitance shunted by a Faraday impedance without adsorption processes. The standard exchange-current density (exchange current at 1 M $BeCl_2$ in the melt) and the rate constant of the electrode process in equilibrium were also determined; see the paper [10].

References:

[1] Delimarskii, Yu. K.; Kikhno, V. S. (Ukr. Khim. Zh. **31** [1965] 872/3; C. A. **64** [1966] 3045).
[2] Delimarskii, Yu. K.; Kikhno, V. S. (Elektrokhimiya **5** [1969] 145/50; Soviet Electrochem. **5** [1959] 132/7).
[3] Ryabukhin, A. G. (Elektrokhimiya **15** [1979] 1817/9; Soviet Electrochem. **15** [1979] 1557/9).
[4] Ryabukhin, A. G. (Elektrokhimiya **17** [1981] 1513; Soviet Electrochem. **17** [1981] 1253).
[5] Vasenin, R. M. (Zh. Fiz. Khim. **27** [1953] 878/88; C. A. **1954** 57).
[6] Khomutov, N. E. (Zh. Fiz. Khim. **36** [1962] 2721/6; Russ. J. Phys. Chem. **36** [1962] 1475/8).
[7] Shapovalov, E. T.; Gerasimov, V. V. (At. Energiya SSSR **26** [1969] 498/502; Soviet At. Energy **26** [1969] 578/81).
[8] Gerasimov, V. V.; Shapovalov, E. T. (Zh. Prikl. Khim. **46** [1973] 953/7; J. Appl. Chem. [USSR] **46** [1973] 1017/20).
[9] Delimarskii, Yu. K.; Gorodyskii, A. V.; Kikhno, V. S. (Teor. Eksperim. Khim. **4** [1968] 554/6; Theor. Exptl. Chem. [USSR] **4** [1968] 355/7).
[10] Delimarskii, Yu. K.; Baranenko, V. M.; Barchuk, V. T. (Ukr. Khim. Zh. **45** [1979] 75/7; Soviet Progr. Chem. **45** [1979] 92/4).

16.2 Ionic Conductance

The limiting equivalent conductance of Be^{2+} ions at 18°C is $\Lambda_0 = 28$ $cm^2 \cdot \Omega^{-1} \cdot val^{-1}$, and the Stokes radius derived from this value is $r_s = 5.52$ Å, whereas the crystalline radius is only 0.37 Å. From this difference it was concluded that the Be^{2+} ion is surrounded by an inner sphere of 4 H_2O molecules and an outer sphere of 20 H_2O molecules [1].

The specific diffusion coefficient, D, of a cation in 0.1 M solution at 10°C was correlated with the ionic conductance at 25°C. The relation is given by $1/k = \Lambda_{25}/n \cdot D_{10}$ where n is the valence of the ions. Most of the cations, e.g., Ba^{2+}, Cd^{2+}, Mn^{2+}, Ni^{2+}, Fe^{2+}, and Cu^{2+}, showed a value of

$1/k \approx 62.8$. Solutions containing Be^{2+} (measurements in Be nitrate) showed a strong deviation from this average: $1/k = 50$. This was attributed to the presence of $(Be_2O)^{2+}$ ions in Be salt solutions containing less than 0.1M (or even 0.01M) free acid [2]. Sutra [3] found $\Lambda_0 \approx 40$ cm²·Ω^{-1}·val⁻¹ at 25°C. This value is rather inaccurate because of the hydrolysis of the Be salt solutions. From Λ_0 the Stokes radius $r_s = 4.58$ Å was calculated by means of the equation $r_s = 0.819 \, n/\eta\Lambda_0$, where $\eta = 0.00895$ is the viscosity of water.

The conductance of hydroxide solutions, $Be(OH)_2$ among others, was determined in very dilute solutions of $\sim 6.1 \times 10^{-5}$ to 1.2×10^{-7} M at 25°C. The ionic conductance of Be^{2+} ions was derived to be $\Lambda_0 = 42.6$ cm²·Ω^{-1}·val⁻¹. But even in such dilute solutions undissociated molecules were present. This followed from the conductance coefficient Λ/Λ_0 for 1×10^{-6} M $Be(OH)_2$ which was considerably smaller than that calculated from the Onsager equation [4].

Based on a theory which describes an electrolyte solution as a perturbation of the water lattice, an exponential relation was derived between the Jones and Dole parameter B (correlated with the fluidity of the solution, for details see the paper) and the limiting ionic conductance of various ions of crystal radius $\leqq 4$ Å, including Be^{2+}. The dependence of the limiting ionic conductance on the temperature, T, and the pressure, p, can be calculated from $\partial B/\partial T$ and $\partial B/\partial p$ [5].

References:

[1] Darmois, E. (J. Chim. Phys. **43** [1946] 1/20, 1/6).
[2] Jander, G.; Blohm, C.; Grüttner, B. (Z. Anorg. Allgem. Chem. **258** [1949] 205/20).
[3] Sutra, G. (J. Chim. Phys. **43** [1946] 189/204, 193/4).
[4] Hłasko, M.; Salitowna, I. A. (Roczniki Chem. **19** [1939] 396/408; [Ger. 407/8]).
[5] Podolsky, R. J. (J. Am. Chem. Soc. **80** [1958] 4442/51).

16.3 Electrokinetic Phenomena

The mobility of various ions, including Be^{2+} (in $BeCl_2$ solutions), across a membrane of cellulose acetate cured at 85°C was investigated. The mobility of Be^{2+} was very high compared with that of other doubly charged ions (such as of Mg, Ca, Sr, Ba). It reached approximately the mobility of singly charged ions (such as of Li, Na, K, Rb, and Cs). The potential difference across the membrane is due to the contributions of the diffusion potential of a thin rejecting layer, the streaming potential across the porous sublayer, and an asymmetric double layer on both sides of the membrane. In a $BeCl_2$ electrolyte the Be^{2+} ions are preferentially adsorbed on the membrane, whereas in the other chloride solutions this is true for the Cl^- ions [1].

Electrophoresis was performed in thin layers of SiO_2 or kieselgur gel plated on glass in order to separate various ions. The migration of Be^{2+} in 0.05 M lactic acid at a voltage of 46 V/cm, was 58 mm/5 min (for comparison: Sr^{2+} 20, Ba^{2+} 65, Mg^{2+} 70, and Cd^{2+} 75 mm/5 min) [2]. The migration of Be^{2+}, UO_2^{2+}, Tl^+, and $Fe(CN)_6^{4-}$ on filter paper was compared with various acids (HCl, $HClO_4$, HNO_3, H_2SO_4, acetic acid, citric acid, tartaric acid, and lactic acid) as background electrolytes. Plots of the mobility vs. the logarithm of the acid concentration are given in the paper [3].

The migration of Be^{2+} in 0.75 M formic acid solution (pH 2) on Whatman No. 3 MM paper was 16.5 cm towards the cathode for 100 V/cm at 11°C and 1.5 psi in 20 min (conversion factors are given on pp. 276/7). The relative migration rate related to K^+ under the same conditions was 0.432. In alkaline solution such as ~ 0.1 M $(NH_4)_2CO_3$ (pH 8.9) for 80 V/cm at 11°C and 1.5 psi no migration was observed in 20 min [4].

The relative electrophoretic mobility of various ions including Be^{II} was determined on Whatman paper in a high-potential apparatus with 1500 V and cooling water at 15°C for 30 min runs. All metal ion solutions were obtained by dissolving a small amount of nitrate salt in 0.1, 0.5, or 1N acetic acid–sodium acetate buffer. In 1N buffer the relative (dimensionless) movement of Be^{II} was −4, i.e. anionic (Al^{III} 0); in 0.5N buffer the Be^{II} movement was +4, i.e. cationic (Al^{III} 2), and in 0.1N buffer the Be^{II} movement was +36, i.e. cationic (Al^{III} 24). A comparison of the electrophoretic ion mobilities with ion-exchange data obtained on a WA-2 amberlite resin paper revealed that metal ions which move anionically in paper electrophoresis are also strongly adsorbed on the resin [5].

Previous studies on paper electrophoresis with 0.1N tartaric acid and 0.1N NH_4SCN as base electrolytes showed that Be^{2+} belongs to the fast moving ions together with, e.g., Zn, Cd, Pb, Ni, and Co, whereas Bi ions moved slowly [6]. The electrophoresis of Be^{2+} and other ions was compared on plain and on stannic antimonate papers. The background electrolyte was 0.1M NH_4Cl; 100 V were applied along the paper strip for 8 h. The Be^{2+} ions showed the usual behavior, i.e., higher migration rate on the plain than on the impregnated paper due to the ion-exchange effect of stannic antimonate [7].

References:

[1] Choi, K. W.; Bennion, D. N. (Ind. Eng. Chem. Fundam. **14** [1975] 296/305; C. A. **83** [1975] No. 148205).
[2] Moghissi, A. (Anal. Chim. Acta **30** [1964] 91/5).
[3] Pučar, Z.; Jakovac, Z. (J. Chromatog. **2** [1959] 320/1).
[4] Gross, D. (J. Chromatog. **10** [1963] 221/30).
[5] Mazzei, M.; Lederer, M. (Anal. Letters **1** [1967] 67/70).
[6] Mukerjee, H. G. (Z. Anal. Chem. **163** [1958] 408/11).
[7] Qureshi, M.; Varshney, K. G.; Rajput, R. P. S. (Ann. Chim. [Rome] **66** [1976] 337/43).

16.4   Behavior as Cathode

16.4.1   Beryllium Cathodes in Aqueous Solutions

16.4.1.1   Overvoltage of Hydrogen

In the following section η means the overvoltage of hydrogen and j the current density. Being representative of a cathodic process, η has negative values throughout, but the statement "η increases" means that the numerical value of η increases, i.e., the cathode potential shifts to more negative values. Conversely, "η decreases" means that it shifts to more positive values.

Experimental Results

The cathodic behavior of Be in aqueous solutions is complex and depends markedly on the purity of the Be and the electrolyte as well as on the atmosphere above the solution.

Very few measurements concerning the hydrogen overvoltage, η_H, of Be cathodes are available. Bockris [1] used a direct method to study η_H on Be cathodes (0.5% impurities, mainly Mg and Si) in 0.1N HCl solutions, presumably at room temperature. A stream of H_2, directed onto the cathode, minimized the concentration polarization. At a given constant current density, η rises rapidly at first and then more slowly to a maximum value. Potentials were measured when a steady value was reached; this was generally accomplished in 15 to 30 min. At current densities of 10^{-3}, 10^{-2}, and 5×10^{-2} A/cm² η was −0.63, −0.73, and −0.85 V, respectively. As Be tends to dissolve at the lowest current densities (10^{-3} A/cm²), even in 0.1N HCl solution, the reproducibility was somewhat worse than ±0.01 V. The η values obtained at 10^{-1} A/cm² were not reproducible; the Be surface darkened. The results with 0.1N HCl solutions yielded a Tafel constant b = −0.11 V/decade and an estimate of the Tafel constant a ≈ −1.0 V [1]. For 1N HCl solutions at 25°C, the exchange-current density, $j_o = 10^{-9}$ A/cm², and the transfer coefficient, α = 0.5, were derived from the results of Bockris [1]; see [2 to 4]. Again, from the latter data (Bockris [3]) a value η ≈ −0.69 V (see figure in the paper) was calculated for Be in 1N HCl at a current density of 10^{-3} A/cm² [8]. For the H_2 evolution on Be cathodes in aqueous 1.0N HCl at 20°C, the following values for the Tafel constants a and b in the range of −log j = 3.0 to 1.3 are given: −a = 1080 mV, −b = 120 mV/decade, and −log j_o ≈ 9 (j_o in A/cm²), without description of the experimental method [5]. From polarization measurements of Be cathodes (purity ≧ 99.0%) in 0.1M (Na, H) Cl solution of pH = 3.1 at 20°C, a Tafel slope of −320 (±10%) mV/decade is obtained [6]; cf. Fig. 16-3 on p. 180. Polarization of (98.6% pure) Be cathodes was investigated by the direct method at 22°C in the following electrolytes: 1N H_2CrO_4, 1N H_3PO_4, 1N CH_3COOH, 1N and 0.1N $NaNO_3$, 0.1N NaCl, 0.1N Na_2SO_4, 0.1N $KMnO_4$, 0.1N NaF, 1N CH_3COONa, 1N Na_2CrO_4, 1N Na_2CO_3, 1N Na_3PO_4, 1N and 0.1N NaOH, 1N $NaAlO_2$, 1N Na_2SiO_3. A reproducible sharp increase in cathodic polarization is observed at 15 to 40 μA/cm² for most of the electrolytes. This is attributed to a transition from polarization caused by films of $Be(OH)_2$ or adsorbed hydrogen gas to polarization caused primarily by hydrogen evolution. Above the transition zone, i.e., in the range of hydrogen evolution, the cathodic polarization curves exhibit a Tafel region common to all electrolytes, with a slope of −0.18 V/decade. Extrapolation of the Tafel region to the reversible hydrogen potential of −0.241 V vs. SCE indicates an exchange current for hydrogen evolution of only 10^{-11} A/cm², see [7] and pp. 179/80. In neutral and alkaline aqueous solutions the discharge of hydrogen ions on Be cathodes (99.7% pure) is stated to proceed with an exceedingly high overvoltage. It is suggested that chemisorbed oxygen forms on the Be in alkaline and neutral media; see [9] and pp. 181/2.

Reaction Mechanism and Relation of Parameters of H_2 Evolution to Other Quantities

The basic reaction paths possible for the cathodic hydrogen evolution are: the discharge (or "Vollmer" reaction) $H_3O^+ + Be(e^-) \rightarrow Be-H + H_2O$ (1a) in acid media, or $H_2O + Be(e^-) \rightarrow Be-H + OH^-$ (1b) in alkaline media, followed by either chemical desorption ("Tafel" reaction) Be-H + Be-H → 2Be + H_2 (2) or by electrochemical desorption ("Heyrovsky" reaction) Be-H + H_3O^+ + Be(e^-) → $H_2 + H_2O + 2Be$ (3) where Be-H = atomic H adsorbed on the Be metal. Each of these reaction steps may determine the rate of the overall reaction ($2H^+ \rightarrow H_2$), see, e.g., [24].

Plots of log j_o vs. Φ, the thermionic work function, yield a straight line for a group of metals including Be. Theoretical considerations suggest an electrochemical mechanism, i.e., either reaction (1a) or reaction (3) are rate-determining; for details, see [10]; cf. [4, 11, 22].

In a plot of η at 10^{-3} A/cm² vs. D_{M-H} (the metal-hydrogen bond strength) for various metals under comparable conditions, Be belongs to a group of metals for which η increases with increasing D_{M-H}, see figure in [12]; for this group of metals electrochemical desorption was considered to be the rate-determining step. Correlations are discussed in [8, 13]. According to theoretical considerations by Rüetschi and Delahay [14], the differences in η for various metals at given conditions of electrolysis are determined by differences in the heat of adsorption of atomic hydrogen on the metals and not by the corresponding work functions. A plot of experimentally determined values of $η_H$ vs. the heats of adsorption (calculated for a fraction of covered surface equal to zero, assuming the additivity of metal-metal and H-H bond energies) gave a straight line, as expected for metals including Be with $\eta_H > 0.1$ to 0.15, i.e., when the effect of the backward electrode reaction can be neglected. The slope $1/\alpha \cdot F$ yielded $\alpha = 0.45$ [14].

Based on a feedback theory involving the existence of a metallic-bonded H-metal complex, H(M), on the cathode surface, the reaction step $1/2 H_2(M) = H(M)$ was considered to be rate-determining for Be; details are given in [15]. In comparison to other metals, beryllium exhibits a considerable hydrogen overvoltage and a low thermionic work function. Various metals investigated were arranged according to increasing η_H values, determined in aqueous 1N HCl solutions at 10^{-3} A/cm^2, and η_H was plotted vs. the corresponding thermionic work function. Decreasing work function with increasing η_H values was considered evidence that the adsorption of H atoms is a predominating factor in the H_2 evolution reaction on such metals, see [16, 17]; cf. [11, 23]. Corresponding periodic variations, but in the opposite sense, were observed in plots of η_H and the thermionic work function vs. the atomic number [18]. For a discussion of the relations of η_H vs. the atomic radius and vs. the interatomic distance in a solid cathode metal, see [19, 20]. A plot of η_H (measured in 1N HCl solution at 10^{-3} A/cm^2 and room temperature) vs. the logarithm of the melting point of the metal (T in K) yielded a linear relationship for the value of $\eta = -0.782 \log T + 2.933 = -0.3396 \ln T + 2.933$ which also fits for Be with T=1553 K [21].

References:

[1] Bockris, J. O'M. (Trans. Faraday Soc. **43** [1947] 417/29).
[2] Conway, B. E. (Electrochemical Data, Amsterdam – Houston – London – New York 1952, p. 847).
[3] Bockris, J. O'M. (NBS-C-524 [1953] 243/62, 256; C.A. **1954** 4338).
[4] Bockris, J. O'M. (in: Bockris, J. O'M.; Conway, B. E.; Modern Aspects of Electrochemistry, No. 1, London 1954, pp. 180/276, 198/200).
[5] Parsons, R. (Handbook of Electrochemical Constants, London 1957, p. 95).
[6] Heusler, K. E. (Z. Elektrochem. **65** [1961] 192/7).
[7] Levy, D. J. (AD-267 976 [1961] 18 pp., pp. 3-3, 3-6; C.A. **58** [1963] 8639).
[8] Conway, B. E.; Bockris, J. O'M. (J. Chem. Phys. **26** [1957] 532/41).
[9] Shapovalov, E. T.; Gerasimov, V. V. (At. Energiya SSSR **26** [1969] 498/502; C.A. **71** [1969] No. 66578).
[10] Bockris, J. O'M.; Potter, E. C. (J. Electrochem. Soc. **99** [1952] 169/86).

[11] Bockris, J. O'M. (Z. Elektrochem. Angew. Physik. Chem. **55** [1951] 105/11).
[12] Bockris, J. O'M.; Srinivasan, S. (Proc. Ann. Power Sources Conf. **19** [1965] 4/8; C.A. **64** [1966] 3056).
[13] Bockris, J. O'M.; Wroblowa, H. (J. Electroanal. Chem. **7** [1964] 428/51, 435/6, 444/6).
[14] Rüetschi, P.; Delahay, P. (J. Chem. Phys. **23** [1955] 195/9).
[15] Chittum, J. F. (Nature **183** [1959] 589/90).
[16] Bockris, J. O'M. (Chem. Rev. **43** [1948] 525/77, 536/7, 540/1).
[17] Bockris, J. O'M.; Azzam, A. M. (Experientia **4** [1948] 220/1).
[18] Bockris, J. O'M. (Nature **159** [1947] 539/40).
[19] Khomutov, N. E. (Tr. 3rd Soveshch. Elektrokhim., Moscow 1950 [1953], pp. 97/104; C.A. **1955** 8010).
[20] Khomutov, N. E. (Zh. Fiz. Khim. **24** [1950] 1201/3; C.A. **1951** 5496).

[21] Swarup, J. (Indian J. Chem. **4** [1966] 145/6).
[22] Kortüm, G.; Bockris, J. O'M. (Textbook of Electrochemistry, Vol. II, Amsterdam 1951, p. 420).

[23] Vagramian, A. T.; Petrova, Yu. S. (Physico-Mechanical Properties of Electrolytic Precipitates, Moscow 1960, 206 pp., 56/62).

[24] Bockris, J. O'M.; Reddy, A. K. N. (Modern Electrochemistry, Vol. 2, Plenum, New York 1970, pp. 1231/50).

16.4.1.2 Polarization Curves

Acid Solutions

The cathodic polarization curve on Be (purity $\geqq 99.0\%$) in 0.1M (Na,H)Cl solution (pH = 3.1) at 20°C, see **Fig.** 16-3, p. 180, represents H_2 evolution. The limiting current density depends on the pH value of the solution, see **Fig.** 16-4, p. 180, curve 2, but the polarization curve is not much influenced by the kind of halide ion or its concentration. Extrapolation of the cathodic polarization curve to the potential, at which anodic dissolution sets in, yields the corrosion current density, which depends on the pH as shown in Fig. 16-4, curve 1 [1]; cf. p. 161. In 2N H_2SO_4 solution at 100°C and 20 mA/cm², a Be cathode shows similar or decreased activity in methanol oxidation as compared with Pt-black, see [2]. In various acid (and other) electrolytes cathodic (and anodic) polarization of Be (polished rods, purity 98.6% Be, 0.95% BeO, other impurities see table in the paper) was measured at room temperature (22 ± 1°C) by hand regulation of the current, the results being checked in part by use of a potentiostat. The polarizing current density (in $\mu A/cm^2$) for the potential E vs. SCE (in V) of the Be cathode is given in the table:

electrolyte	E =				
	0	−0.5	−1.0	−1.5	−2.0
1N CH_3COOH		40	400	1500	3000
1N H_3PO_4			2100	($>10^4$)	
1N H_2CrO_4	0.8	280	9000	($>10^4$)	

For the potential ranges in which the cathodic (and anodic) current densities were less than 0.1 $\mu A/cm^2$, see the table in the paper [3]. The polarization curve for 1N H_2CrO_4 solution is shown in **Fig.** 16-5, p. 180. No corrections were applied for the ohmic drop in the electrolyte and for the liquid junction potentials. Strong cathodic voltages were found to activate the Be surface; corrosion was observed in the solutions of CH_3COOH and H_3PO_4. Cathodic polarization was less pronounced than anodic polarization [3].

The polarization curves of Be measured with rotating disk cathodes (680 rev/min) in solutions of HNO_2 and KNO_2 in 2M KCl as supporting electrolyte at 25 ± 0.5°C do not exhibit a wave in the acid solutions; reduction waves of the nitrous acid reduction to NH_4^+ and NH_2OH were expected and found with other metal cathodes (see the paper). The plot of the cathode potential vs. log $j/(j_{lim}-j)$ is linear; the slope ($2 \times 3 \times RT/n\alpha F$, with n = number of electrons involved in the cathodic process, α = transfer coefficient) yields $\alpha \cdot n < 1$ (as for other metals), indicating concentration and electrochemical polarization [5].

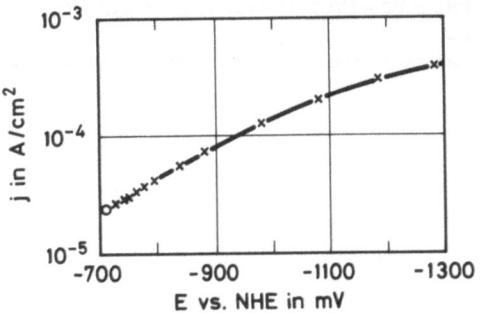

Fig. 16-3. Cathodic polarization of Be in 0.1M (Na,H)Cl solution at pH = 3.1 and 20°C, representing H_2 evolution; ○ extrapolated point at the potential of anodic Be dissolution.

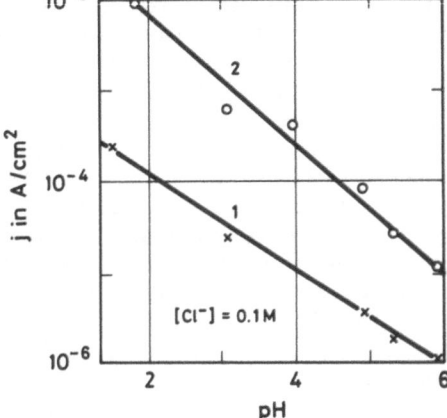

Fig. 16-4. Current density of corrosion (curve 1) and cathodic limiting current density (curve 2) on Be in 0.1M (Na,H)Cl solution at 20°C as functions of pH.

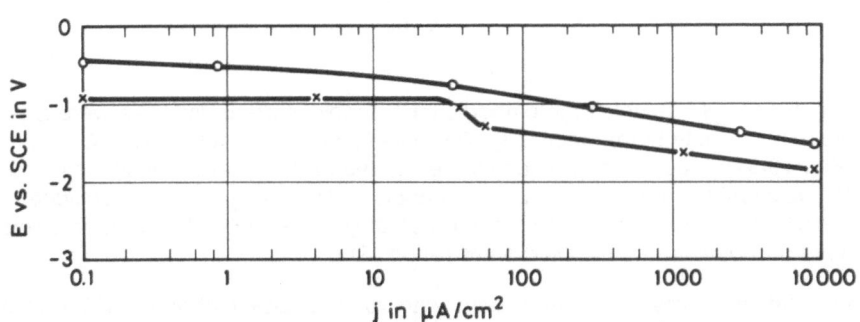

Fig. 16-5. Polarization curves of a Be cathode at 22°C in aqueous 1N H_2CrO_4 (○) and 0.1N NaCl (×) solution.

Alkaline Solutions

The following data were obtained for the Be cathodes in NaOH solutions by Levy [3] with the method described above for acid solutions; polarizing current density (in $\mu A/cm^2$) for the potential E vs. SCE (in V) of the Be cathode:

electrolyte	E =			
	−0.5	−1.0	−1.5	−2.0
1N NaOH		20	200	>10⁴
0.1N NaOH	0.1	36	100	>10⁴

For the ranges of potential observed at current densities $<0.1\,\mu A/cm^2$, see the table in the paper [3].

Cathodic (and anodic, cf. p. 192) polarization curves of hot-pressed 99.7% Be electrodes were measured by means of a potentiostatic apparatus over a wide range of potentials in order to elucidate passivation of Be in the absence of such depassivators as Cl^- and SO_4^{2-} ions. In unstirred 5N KOH solution, the cathodic polarization curve shows only a section representing the evolution of oxygen, see the potentiodynamic curve 1 in **Fig. 16-6**. Stirring shifts the stationary potential to more positive values. It apparently caused a considerable increase in the concentration of oxygen in the near-electrode layer and decreased the thickness of the diffusion layer. Sections corresponding to the ionization of oxygen and to the limiting diffusion current are evident on the cathodic polarization curves 3 and 5, see Fig. 16-6. Increasing the speed of stirring increased the rate of the cathode process with oxygen depolarization. The application of an oxide film onto Be by exposing the Be to water at 300°C shifted the stationary potential in the positive direction by nearly 1 V (see **Fig. 16-7**, p. 182), i.e., the cathode process was intensified and the hydrogen polarization changed to oxygen depolarization. The results and the absence of a visible oxide film on the untreated Be electrode indicated that the Be passivity has an adsorptive or chemisorptive character. The absence of lags on cathodic charging curves in alkaline (and neutral) solutions suggests that a layer of chemisorbed oxygen forms on Be when immersed in the alkaline (or in neutral) solutions, with subsequent or

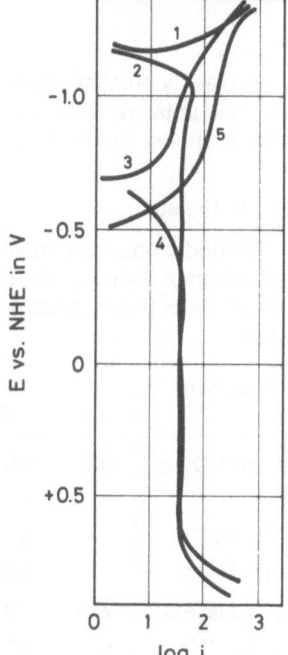

Fig. 16-6. Cathodic (1, 3, 5) and anodic (2, 4) polarization curves of Be in 5N KOH solution under potentiodynamic conditions ($\Delta E/\Delta t = 3.6$ V/h). Curves 1 and 2 without stirring, curves 3, 4, and 5 with stirring (the rate of stirring in 5 was higher than in 3 and 4). Current density, j, measured in $\mu A/cm^2$.

parallel adsorption of OH⁻ ions. Further evidence for the presence of an oxygen layer on the Be surface is given by the values for the capacitance, estimated from charging curves, viz. 4.34 μF/cm² at the O_2 evolution potential in 5N KOH solution, as well as 7.05 and 7.47 μF/cm² at the O_2 and H_2 evolution potentials, respectively, in neutral solution [4].

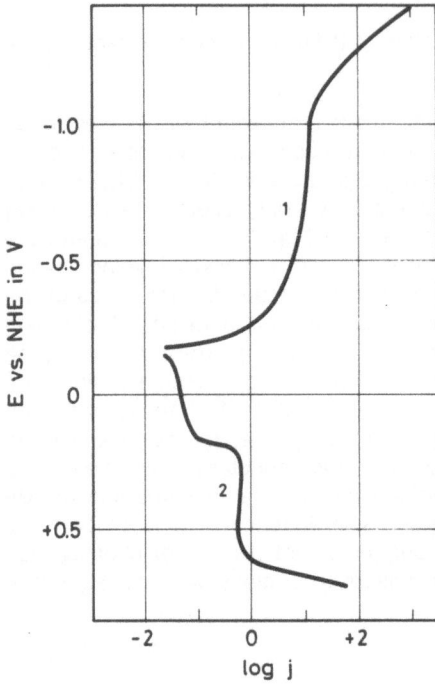

Fig. 16-7. Cathodic (1) and anodic (2) polarization curves of Be in 5N KOH solution under potentiodynamic conditions ($\Delta E/\Delta t = 0.225$ V/h); Be with an oxide film formed in H_2O at 300°C and then kept 4 h in 5N KOH solution. Current density, j, measured in μA/cm².

In weakly alkaline nitrite solutions Be rotating disk electrodes do not show a wave on the cathodic polarization curves, as could be expected because of nitrite reduction, see [5]; for corresponding results in acid solution, cf. p. 179.

Salt Solutions

Cathodic polarization was investigated for Be in various aqueous salt solutions by hand regulation of the currents necessary to maintain given potential values, see [3]. The polarizing current density (in μA/cm²) for the potential E vs. SCE (in V) of the Be cathode is given in the table:

electrolyte	E =				electrolyte	E =			
	−0.5	−1.0	−1.5	−2.0		−0.5	−1.0	−1.5	−2.0
1N NaNO₃	0.6	100		>10⁴	1N CH₃COONa	25	31	400	>10⁴
0.1N NaNO₃	12	61	1600	>10⁴	1N Na₂SiO₃		23	500	>10⁴
0.1N NaF		30	170		1N Na₃PO₄	0.1	30	200	>10⁴
0.1N NaCl		35	400	>10⁴	1N NaAlO₂		3	200	>10⁴
0.1N Na₂SO₄	7	40	250	>10⁴	1N Na₂CrO₄		0.6	130	>10⁴
1N Na₂CO₃	5	17	600	>10⁴	0.1N KMnO₄	0.2	500	1300	>10⁴

For the ranges of electrode potential at current densities $<0.1\,\mu A/cm^2$, see the table in the paper. A polarization curve of a Be cathode in aqueous 0.1N NaCl solution at $22\pm1°C$ is shown in Fig. 16-5 on p. 180. Most of the polarization curves show a reproducible marked increase in cathodic polarization at 15 to 40 $\mu A/cm^2$ [3]. In neutral air-saturated aqueous media the cathodic polarization curves display sections corresponding to oxygen ionization, limiting diffusion current, and hydrogen ion discharge. The discharge of hydrogen ions occurs with exceedingly high overvoltage; the reaction is appreciable only at $-1.1\,V$ (vs. NHE) with $\log j = 1.85$ (j in $\mu A/cm^2$). Depassivation, e.g. by Cl^- or SO_4^{2-} ions, immediately causes H_2 evolution; see above and [4]. In neutral nitrite solutions there is no reduction wave on the polarization curves of Be rotating disk cathodes according to [5]; cf. p. 182.

Galvanostatic polarization curves were obtained for (99.9% pure) Be and 20 other metal rotating disk electrodes in an aqueous solution of 0.01M Na m-nitrobenzenesulfonate – 0.5M K_2SO_4 (pH = 6.2) at 25°C. The polarization curve for Be, see **Fig. 16-8**, shows one wave with $E_{1/2} = -1.5\,V$ vs. SCE, due to the reduction of the nitro group in m-nitrobenzenesulfonate. E_{40}, the potential at 40 mA/cm², is $-1.75\,V$ vs. SCE for a Be rotating disk electrode. The value of E_{40} lies in the region of hydrogen evolution (see table in the paper [6]), and is very similar for most of the metals tested, indicating a common mechanism for the electrochemical process. The value of $E_{1/2}$ depends on the rate of rotation, that of E_{40} does not. The limiting current densities depend linearly on the square root of the rate of rotation, indicating diffusion control of the cathodic process. The value of $E_{1/2}$ for Be is the most negative for the metals tested. The values of the half-wave potentials and the shape of the polarization curves are taken to indicate correspondence between the reducing capacity of the cathodes and the reducing capacity of the cathode metals in chemical reactions, i.e., their position in the periodic system, see [6].

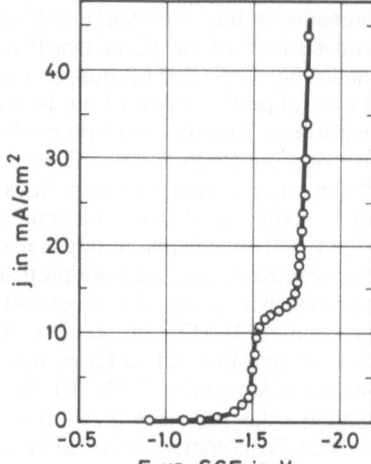

Fig. 16-8.   Polarization curve for Be in 0.01M Na m-nitrobenzenesulfonate – 0.5M K_2SO_4 solution (pH = 6.2) at 25°C, 460 rpm.

References:

[1] Heusler, K. E. (Z. Elektrochem. **65** [1961] 192/7).
[2] Bockris, J. O'M.; Wroblowa, H. (J. Electroanal. Chem. **7** [1964] 428/51, 435/6).
[3] Levy, D. J. (AD-267976 [1961] 18 pp.; C.A. **58** [1963] 8639).
[4] Shapovalov, E. T.; Gerasimov, V. V. (At. Energiya SSSR **26** [1969] 498/502; Soviet At. Energy **26** [1969] 578/81).

[5] Khomutov, N. E.; Stamkulov, U. S.; Vinogradov, V. A.; Kudryavtsev, N. T. (Tr. Inst. Mosk. Khim. Tekhnol. Inst. No. 65 [1970] 99/103; C. A. **76** [1972] No. 20673).

[6] Khomutov, N. E.; Skornyakova, T. N. (Zh. Fiz. Khim. **39** [1965] 195/200; Russ. J. Phys. Chem. **39** [1965] 101/5).

16.4.1.3 Cathodic Deposition (Plating) on Beryllium

The application of Be as a structural material (e.g., in aerospace and nuclear systems) requires suitable coatings of the Be objects in order to improve their resistance to corrosion and oxidation, to facilitate joining, and to improve the resistance to wear (e.g., under thermal shock) [1]. For general literature and surveys, see [2 to 4].

For the formation of anodic films on Be and the electropolishing of Be (as anode), see Sections 16.5.1.2 and 16.5.1.3, pp. 197 and 202, respectively.

A cleaned Be surface is quickly covered by a protective thin oxide film on exposure to air or water; a limiting thickness of ~100 Å is reached when Be is exposed to air at room temperature for 2 h [1, 5 to 7]. Beryllium resists further attack by dry H_2, N_2, air, and CO_2 at temperatures up to ~800°C, by pure water (free of halogen ions), and by oxygen-free liquid metals. However, Be is corroded by water containing small amounts of chloride (e.g., 0.5 ppm), by water at high temperatures, by oxygen in liquid metals (such as Na), by moist CO_2, and by fluorinated oils [1, 8]. Nonmetallic inclusions (such as chlorides) stimulate the corrosion of Be surfaces exposed to air [7]. Chromic acid solutions passivate Be [9].

The difficulties experienced in obtaining sound, adherent electroplates on Be are due to the presence of the oxide film and to contaminations and defects present on the Be surface [10]. The surface of Be parts which are mostly prepared by powder metallurgy techniques is inherently rough and porous, and machining may even increase the surface roughness. Hence a careful pretreatment of the Be surface prior to the electroplating process is of the utmost importance for obtaining high-quality adherent deposits. Careful mechanical treatment (including "descaling") should be followed by a cleaning process in which grease and oils are usually removed by organic solvents. Residual dirt may be removed by cathodic treatment in NaOH or KOH solutions [9, 11]. Subsequent chemical polishing in an H_2SO_4–H_3PO_4–CrO_3 solution [1, 11, 12] reveals and removes microcracks and twinning, but it may produce a pitted surface because oxide and impurity particles in the Be surface layer are removed as well [1, 4]. The semipassive surface film produced by the chemical polishing is removed by an acid activating pickle in an H_2SO_4 solution [11, 13, 14], and plating is performed directly afterwards. The method and the plating baths used depend on the metal to be plated onto Be. While some metals (such as Zn, Al, Sn, Ni, Fe, Cu, Ag) yield good adherent electroplates on Be by direct plating, other metals (e.g., Cr) need an intermediate bonding layer of one or more other metals such as Zn (commonly applied by immersion in a "zincating" solution) [10, 11], Cu (applied electrolytically, "Cu-strike") [6, 10], Ag ("Ag-strike") [9, 11], or Au (applied by sputtering) [15].

Detailed procedures for the pretreatment of the Be surface are given in [1, 2, 6, 10, 11, 16 to 19]. Missel [6] tabulated about 30 methods, including those of Beach and Faust [11] and of Kolodney [16], for the treatment of Be prior to Ni electroplating from a sulfamate bath. The table also contains the results of the solar furnace exposure (2 to 12 s) tests of the final Ni plates.

Recommended operation sequences are also given for Ni electroplating by Missel [6], by Beach and Faust [9, 11, 17], and by Rothschild and Nabors [10]. For the pickling of Be, see also [14]. Some details of the procedures and plating baths are given as follows.

For the pretreatment suitable for a good adherent Ag plate, see [20]. An excellent Ni plate is obtained by the pretreatment of [21].

Rothschild and Nabors [10] found a Zn immersion treatment with a $ZnSO_4–Na_2P_2O_7–NaF$ solution at pH 7.5 to 8 to yield the best Zn, Cr, Ni, Fe, and Cu plates on Be. Utz [22] reported successful Zn, Cr, Ni, Cu, and Ag plates after thorough cleaning of the Be surface by a ketone rinse followed by a Co-strike from a $CoSO_4–(NH_4)_2SO_4$ solution (at pH 4 to 6, 10 to 40°C, and a current density of 25 to 45 A/ft²).

Dini and Johnson [23 to 28] used ring shear and conical head tension measurements to test the adhesion of electroplates on Be (mostly Ni plates). A Zn immersion treatment was found to improve the adhesion of the electroplates considerably. Adhesion was further enhanced if the pH of the Zn immersion solution was in the range 3.0 to 7.7 and if the plating solution was a solution suitable for plating on Zn (i.e., low pH value, low free cyanide). The duration of the zincate treatment was found to be a critical factor [28]. Further studies see [42].

Atomic hydrogen, absorbed by Be during polishing, pickling, and/or plating, must be removed by heat treatment (outgassing) [11, 13, 20].

For a 5×10^{-4} in Cr plate on Be (preceded by Zn immersion and Cu-strike treatments), resistant to, e.g., liquid alkali metals at elevated temperatures, see [29].

Beryllium pretreated with a nitro-fluoro-phosphoric acid bath and plated with 15 to 20 µm Mg from an aqueous $Mg(NO_3)_2$ bath, resisted corrosion in moist CO_2 (several vpm H_2O) up to ~750°C [30].

At higher temperatures diffusion and alloying at the Be-electroplate interface may involve changes in the crystal lattice and volume and cause a deterioration of the adhesion. Diffusion at the Ni-Be and Fe-Be interfaces starts in the temperature ranges 350 to 400°C and 500 to 550°C, respectively. Thus Ni- and Fe-plated Be is not suitable for extended use at temperatures above 350 and 500°C, respectively. Chromium plated on Cu on immersion-zinc-coated Be showed good adhesion even at elevated temperatures [1, 5, 11, 13, 20, 31].

For the precleaning of Be prior to spray coating with platinum-black, see [32].

A pretreatment for badly oxidized Be prior to brazing is given by [33]. The oxide film and the impurities of Be powder produced by wet milling of electrolytic Be flake were removed by leaching in 75% HNO_3 or 10% oxalic acid solutions, improving the quality and stability of the Be powder [34]. A highly reflective, gray Be surface was obtained by treating an oxidized Be surface with an $H_2SO_4–H_3PO_4–CrO_3$ mixture at 110°C [12].

Plating Solutions

In the following section aqueous solutions are compiled in which metal deposits form on Be cathodes. For technical details as well as for the pretreatment of Be, see [1, 11, 13, 15, 16, 18, 19, 21, 23, 35 to 38]. For the formation of Al plates on Be cathodes from nonaqueous solutions, see p. 187, and from melts, see p. 191.

Nitrate Baths. Onto a Be cathode **magnesium** is electrodeposited from a solution of 100 g/L $Mg(NO_3)_2$ at room temperature and 1 A/dm² [30].

Sulfate Baths. From $CrO_3–H_2SO_4$ solutions **chromium** deposited on Be cathodes (at 45 or 54°C and 110 or 220 A/ft²); for details, see [11, 13, 17, 29]. Solutions of $MnSO_4–(NH_4)_2SO_4–Na_2SO_3$ of pH 7.5 at ~38°C and 40 A/ft² gave **manganese** deposits [17]. **Cobalt** platings were obtained on Be in $CoSO_4–(NH_4)_2SO_4$ baths of pH 4 to 6 at 10 to 40°C and 25 to 45 A/ft² [22]. Solutions of $CuSO_4–H_2SO_4$ did not yield **copper** deposits on Be directly [38]; cf. [21, 23]. 10 to

15 μm thick Cu deposits were obtained on a rotating Be cathode from a bath containing $CuSO_4$, the sodium salt of EDTA, and Na_2SO_4 at pH 8.5, 40°C, and 5 mA/cm² [41].

Baths Based on Sulfate and Chloride. From $ZnSO_4$–NH_4Cl–$MgSO_4$–$Al_2(SO_4)_3$ baths at pH 4.0 **zinc** was electrodeposited on Be cathodes at ~27°C and 25 A/ft², see [1, 11, 13], also from $ZnSO_4$–NH_4Cl–NaF solutions containing a so-called "joiner's glue" at room temperature and pH 5.9 to 6.2. The NaF content in the solution is reported to render the potential of Be more negative than that of Zn by 100 to 150 mV; details are given in [39]. **Nickel** was plated on Be from $NiSO_4$–NH_4Cl–$MgSO_4$–H_3BO_3 solutions in the pH range 5 to 6 at 32 to 38°C and 15 A/ft²; for details, see [1, 11, 13, 17]; cf. [21]. **Iron** deposited on Be cathodes from $FeSO_4$–$FeCl_2$–$(NH_4)_2SO_4$–NaCOOH–H_3BO_3 solutions of pH 4 at 60°C and 40 A/ft² [1, 11, 17]; cf. [13].

Cyanide, Pyrophosphate, Acetate, and Sulfamate Baths. From CuCN–NaCN–Na_2CO_3–$Na_2S_2O_3$–tartaric acid solutions of pH 9 at 49°C and 25 A/ft² **copper** plates formed on Be [1, 11, 13, 17], also from an alkaline $K_2Cu(CN)_3$–NaCl solution, see [38]. **Silver** is deposited onto Be cathodes first from an AgCN–NaCN solution (at 25°C and 7.5 A/ft²), and then from an AgCN–KCN–K_2CO_3–KOH solution (at pH 13.0, 49°C, and 25 A/ft²); for details, see [1, 11, 13, 17]. **Copper** is also reported to be directly plated on Be cathodes from mildly alkaline pyrophosphate baths of pH 8.5 at 60°C and 40 A/ft² [11, 13]; cf. [19]. In Na_2SnO_3–NaOH–Na acetate solutions at 66°C and 25 A/ft² Be cathodes become covered with **tin** [1, 11, 13, 17]. The same is the case with Na stannate–NaOH–Na acetate–H_2O_2 solutions at 60 to 90°C and 0.15 A/in² [16]. Ni sulfamate–boric acid baths (of pH 3.0 to 3.5) gave **nickel** platings on Be at 53.4 to 55.1°C and 50 A/ft² [1, 6]; cf. [21, 23 to 27, 40].

References:

[1] Beach, J. G. (DMIC-Memo-197 [1964] 8 pp.; N.S.A. **19** [1965] No. 9598).

[2] Stacy, J. T. (in: White, D. W., Jr.; Burke, J. E.; The Metal Beryllium, ASM, Cleveland, Ohio 1955, pp. 295/303).

[3] Gex, R. C. (SB-61-4 [1961] 32 pp.; N.S.A. **15** [1961] No. 14681).

[4] Mueller, J. J.; Adolphson, D. R. (in: Floyd, D. R.; Lowe, J. N.; Beryllium Science and Technology, Vol. 2, Plenum, New York – London 1979, pp. 417/33, 426/33).

[5] Beach, J. G.; Faust, C. L. (Chem. Eng. Progr. Symp. Ser. **50** No. 11 [1954] 31/8; C.A. **1954** 9302).

[6] Missel, L. (Metal Finishing **58** No. 3 [1960] 53/7).

[7] Whitby, L.; Gowen, E.; Levy, D. J. (Tech. Proc. Am. Electroplaters' Soc. **48** [1961] 106/8; C.A. **57** [1962] 10906).

[8] Vachon, L. J. (J. Nucl. Mater. **6** [1962] 139/41; C.A. **58** [1963] 1121).

[9] Beach, J. G.; Faust, C. L. (Metal Finishing **51** No. 10 [1953] 68/70, 73).

[10] Rothschild, B. F.; Nabors, R. G. (Metal Finishing **63** No. 7 [1965] 49/53).

[11] Beach, J. G.; Faust, C. L. (J. Electrochem. Soc. **100** [1953] 276/9).

[12] Wikle, K. G. (Metal Ind. [London] **93** [1958] 529/33).

[13] Faust, C. L.; Beach, J. G. (Plating **43** [1956] 1134/42).

[14] Gurklis, J. A.; McGraw, L. D. (PB-161235 [1961] 19 pp.; N.S.A. **15** [1961] No. 11450).

[15] Dini, J. W. (Proc. 11th World Congr. Met. Finish., Jerusalem 1984, pp. 212/20; C.A. **103** [1985] No. 185834).

[16] Kolodney, M. (U.S. 2588734 [1952]; C.A. **1952** 4936).

[17] Beach, J. G.; Faust, C. L. (U.S. 2729601 [1956]; C.A. **1956** 5430).

[18] Reinsch, H. H. (Angew. Mess-Regeltech. **24** No. 6 [1966] A237/A238; C.A. **66** [1967] No. 91111).

[19] Missel, L.; Titus, R. K. (Metal Finishing 65 No. 10 [1967] 59/61, 66).
[20] Chauvin, G.; Coriou, H. (Corrosion Anticorrosion 12 No. 4 [1964] 153/65).

[21] Pilite, S.; Molchadskii, A. M.; Roseniene, R. K.; Karpavicius, A.; Matulis, I.; Altovskii, R. M.; Urazbaev, M. I. (U.S.S.R. 850753 [1978/81]; C.A. 95 [1981] No. 177752).
[22] Utz, J. J. (U.S. 2798036 [1957]; C.A. 1957 12710).
[23] Dini, J. W.; Johnson, H. R. (Plating Surf. Finish. 63 No. 6 [1976] 41/6).
[24] Dini, J. W.; Johnson, H. R. (ASTM Spec. Tech. Publ. No. 640 [1976] 305/26; C.A. 89 [1978] No. 33116).
[25] Dini, J. W.; Johnson, H. R. (AES Symp. Plat. Difficult-Plate Met. Proc., New Orleans 1980, Paper 5, 17 pp.; C.A. 95 [1981] No. 15008).
[26] Dini, J. W.; Johnson, H. R. (Plating Surf. Finish. 68 No. 10 [1981] 64/9; C.A. 96 [1982] No. 13063).
[27] Dini, J. W.; Johnson, H. R. (Galvanotechnik 75 [1984] 1116/23).
[28] Dini, J. W.; Kelley, W. K.; Johnson, H. R. (ASTM Spec. Tech. Publ. No. 947 [1986/87] 320/8; C.A. 107 [1987] No. 159490).
[29] Townsend, R. G. (U.S. 2901408 [1959]; C.A. 1959 21291).
[30] Darras, R.; Dewanckel, B.; Leclercq, D. (Fr. 1457463 [1965/66] 3 pp.; C.A. 67 [1967] No. 17370).

[31] Chauvin, G.; Coriou, H.; Hardy, J.; Mallen, J.; Weisz, M. (Mem. Sci. Rev. Met. 66 No. 1 [1969] 67/73).
[32] Missel, L.; Greear, G. R. (Metal Finishing 61 No. 8 [1963] 46/8, 54; C.A. 59 [1963] 14972).
[33] Zunick, M. J.; Illingworth, J. E. (Mater. Methods 39 No. 3 [1954] 95/7).
[34] Williams, J.; Munro, W.; Jones, J. W. S. (AERE-M-R-1679 [1955/58] 27 pp.; N.S.A. 12 [1958] No. 16386).
[35] Beach, J. G.; Faust, C. L. (BMI-732 [1952] 46 pp. from C.A. 1956 10568).
[36] Beach, J. G.; Faust, C. L. (Ind. Finishing [London] 9 [1956] 173/4 from C.A. 1957 890).
[37] Taylor, R. D.; Steyert, W. A.; Fox, A. G. (Rev. Sci. Instrum. 36 [1965] 563).
[38] Gel, R. P.; Drobotenko, G. A.; Kolosov, V. N. (Khim. Khim. Tekhnol. Metall. Redk. Elem. 1982 138/41; C.A. 98 [1983] No. 115771).
[39] Plaskeev, E. V.; Azhogin, F. F.; Batrakov, V. P. (U.S.S.R. 313907 [1966/71]; C.A. 76 [1972] No. 9849).
[40] Dini, J. W.; Johnson, H. R. (SAND-79-8069 [1980] 25 pp.; C.A. 93 [1980] No. 122466).

[41] Bettelheim, A.; Raveh, A.; Mor, U.; Ydgar, R.; Segal, B. (J. Electrochem. Soc. 137 [1990] 3151/3).
[42] Il'yasov, R. S.; Rudoi, V. M.; Levin, A. I.; Menshchikov, G. P. (SPSTL-85 Khp-D80 [1980] 12 pp.; C.A. 96 [1982] No. 93972).

16.4.2 Beryllium Cathodes in Nonaqueous Solvents

Beryllium cathodes in a diethyl ether solution of $AlCl_3$ (400 g/L) and LiH (6 g/L) were satisfactorily plated with Al reaching 0.001 in thickness after 120 min at room temperature and 10 A/ft² [1]. Similarly, adherent Al platings formed on Be wire cathodes in a diethyl ether solution of $AlCl_3$ (3.4 M) and $LiAlH_4$ (0.45 M) under a dry N_2 atmosphere at 215 to 403 A/m² [2].

References:

[1] Faust, C. L.; Beach, J. G. (Plating **43** [1956] 1134/42).
[2] Clay, F. A. (BDX-613-865 [1973] 19 pp.; C.A. **80** [1974] No. 33176).

16.4.3 Beryllium Cathodes in Melts

Alkali Halide Melts

The polarization of a Be cathode in molten eutectic **LiCl–KCl** at 500°C is shown in **Fig.** 16-**9** [1]. Below current densities of 10^{-2} A/cm², the potential of the Be cathode changes little. Comparing the polarization curve with that of an Mo cathode in the same melt at the same temperature indicated that at current densities $>\sim10^{-2}$ A/cm² the Li–K alloy formation at the surface of the Be cathode was accompanied by a clearly expressed depolarization of the Be cathode and also by alloy formation between Li and Be [1].

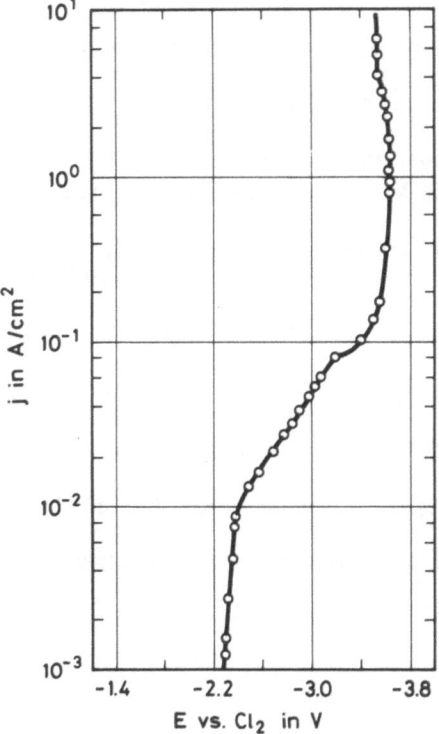

Fig. 16-9. Cathodic polarization of Be in molten eutectic LiCl–KCl at 500°C.

A cathodic (and anodic) polarization curve (j vs. E) of Be in molten (eutectic) LiCl–KCl containing 10 wt% BeCl₂ at 450 to 500°C indicated an extremely fast electrochemical system. A zero current potential of about −1.175 V vs. Ag|AgCl–eutectic LiCl–KCl (mole fraction of AgCl 10^{-2}) was obtained and the polarization curve gave a straight line up to about 5.5 A/dm², see figure in the papers [2, 3].

Wohlfarth [4] determined the j vs. η relation for a Be cathode during Be deposition at 750 K from molten eutectic LiCl–KCl with $BeCl_2$ added (N_{BeCl_2} = 0.1; anode: Be), see **Fig.** 16-**10**. The calculated curve was obtained based on the assumption that only the transfer reaction and ionic diffusion are reaction steps (i.e., crystallization, nucleation, and surface diffusion are neglected) using j_o = 165 A/cm² and j_{lim} = -4×10^{-3} A/cm² from electrode kinetic data for $a_{Be^{2+}}$ = 10^{-6} mol/cm³. Good agreement was obtained between experimental data and the calculated curve for small values of η. The divergence at j > 10 mA/cm² was attributed to an increase in $a_{Be^{2+}}$ due to anodic dissolution; for details, see [4] and p. 260.

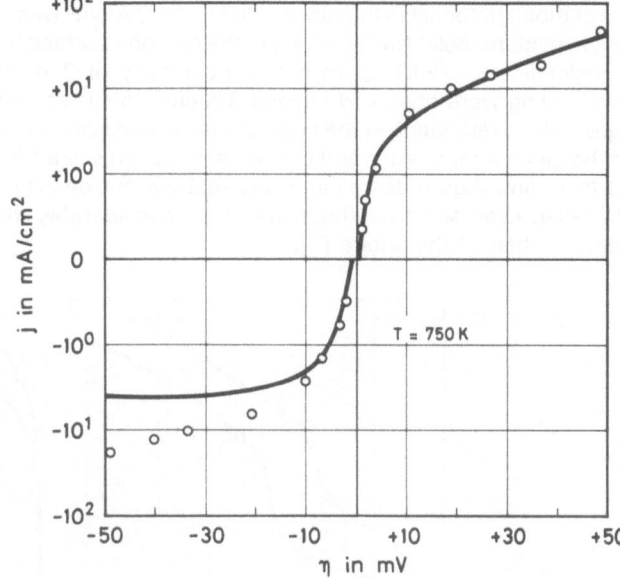

Fig. 16-10. Current density vs. overvoltage, η, for a Be cathode in a $BeCl_2$–LiCl–KCl melt (N_{BeCl_2} = 0.1) at 750 K (anode: Be); solid line = calculated curve, ○ = experimental values.

Investigations concerning nucleation and crystallization of Be on Be cathodes from eutectic LiCl–KCl melts with added $BeCl_2$ are treated on pp. 255/8.

The impedance of a Be wire electrode (~0.1 mm diameter, polished to mirror luster) was measured at 450°C as a function of frequency in the range 1 to 70 kHz, 1) in high-purity eutectic LiCl–KCl melt under cathodic polarization and 2) with additions of $BeCl_2$ under conditions of equilibrium. The capacitance of the Be electrode at high frequency, e.g. 20 kHz, in eutectic LiCl–KCl shows a parabolic dependence on the potential with a minimum of ~30 μF/cm² at 2.75 V vs. a chlorine electrode, see figure in the paper [5]. The double-layer capacitance of the Be electrode at equilibrium was determined in the 2 to 20 kHz range by an extrapolation method for $BeCl_2$ concentrations of c_{BeCl_2} = (2.94 to 7.35)×10^{-5} mol/cm³ (see table in the paper). The values for the exchange current density were found to be j_o = $J_o c_{BeCl_2}^{(1-\alpha)}$, where J_o = 42.7 A/cm² and (1−α) = 0.79 (α = transfer coefficient) [5].

In molten equimolar **NaCl–KCl** the differential capacitance of a Be cathode (single crystal of ≧99.99% purity, investigated surface not specified) and of other metal cathodes was measured at 700 ± 2°C as a function of the potential and ac frequency. The "null-point" of Be was determined to be −2.75 V vs. a chlorine electrode from the minimum of the capacitance vs. the potential plot for an ac frequency of 2 kHz; the specific capacitance is 40 μF/cm² for the Be electrode. Comparison with data of the other metal cathodes indicated the absence of pores in the Be single crystal. At lower frequencies (<~1 kHz) the capacitance depended strongly on the frequency (dispersion of the capacitance). The null-points of the various metals investigated yield a series similar to the electrochemical series of the metals in molten chlorides. For details and discussion with respect to the electronic work function, see [6].

The polarization of a Be cathode in a **BeCl$_2$–NaCl** melt (41 mol% BeCl$_2$) was measured in the temperature range 325 to 420°C. The overheating of the near-electrode layers (at current densities $\gtrsim$1 A/cm^2) was recorded simultaneously, see **Fig. 16-11**. The surface of the Be cathode was polished to a mirror shine. Polarizations for 10 s at a given current density were sufficient to reach constant potentials and overheating temperatures in the near-electrode layer. Significant polarization of the Be cathode sets in at ~0.1 A/cm^2. A certain stabilization of the cathode potential in the range ~0.3 to ~1.0 A/cm^2 was attributed to the existence of a two-phase mixture (solid NaCl + liquid) in the cathode surface layer of the electrolyte. Overheating is insignificant ($\leqq$2°C) up to a current density of 2 A/cm^2. Considerable polarization and overheating were observed beyond 3 A/cm^2; this is attributed to a depletion of BeCl$_2$ in the liquid. The overheating in the near-electrode layer of a Be cathode is less than for a Be anode at the same temperature and current density. This is attributed to an increase in the cathode surface, and thus a decrease in the real current density in the course of the experiments. However, even small overheatings cause considerably more polarization in the case of the cathode than of the anode [7].

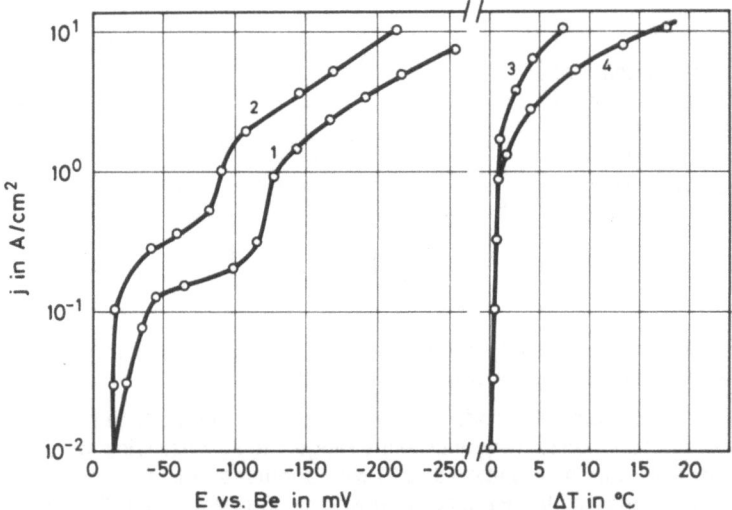

Fig. 16-11.   Polarization (curves 1 and 2) and overheating, ΔT, in the near-electrode layer (curves 3 and 4) for a Be cathode in a BeCl$_2$–NaCl melt (41 mol% BeCl$_2$) at the temperatures: 1 = 325°C, 2 = 420°C, 3 = 400°C, 4 = 350°C.

Other Melts

The polarization of a Be cathode was measured in an **MgCl$_2$–KCl** melt (40 and 60 mol%, respectively) at 510 $\pm$ 2°C in a closed quartz apparatus with an atmosphere of purified Ar. The cathodic potentials were measured in the current density range of 5×10^{-4} to 8 A/cm^2. Below 3×10^{-3} A/cm^2, the cathodic potential (~ −2.4 V vs Cl$_2$) corresponds to the discharge of Be ions, present in the near-cathode layer due to corrosion. A subsequent gradual shift of the potential in the negative direction was attributed to the discharge of Mg^{2+} with the formation of a Be-Mg alloy. As the diffusion of Mg into Be is very slow, the Be surface layers become saturated with Mg and the cathodic potential reaches the potential of metallic Mg (~ -2.8 V vs. Cl_2) at a current density of ~ 0.3 A/cm², see [8].

Beryllium cathodes (rods of 2 to 20 mm diameter) were investigated in dehydrated and oxygen-free molten eutectic $CaCl_2$–CaF_2 (80 and 20 mol%, respectively) at 800 to 1100°C, under Ar atmosphere at current densities of 25 to 60 A/dm². Micrographic and electron-microscopic observations as well as X-ray analysis showed the formation of a continuous layer of the alloy $Be_{13}Ca$ (~ 25 wt% Ca) with a hardness ~ 1350 VPN (30 g load). The thickness of the layer varied from 5 to 100 µm, depending on the experimental conditions, especially the temperature. A second (thinner) metallic layer was observed between $Be_{13}Ca$ and the electrolyte, paradoxically containing only ~ 0.8 wt% Ca. Its hardness was ~ 180 VPN (30 g load), which is close to that of the Be (~ 250 VPN) [9], cf. [10].

In an $AlCl_3$–$NaCl$ melt (80 and 20%, respectively) at 177°C and 18 A/ft² adherent plates of Al (thickness 0.001 in after 75 min) were deposited on Be cathodes, see [11].

Melts Containing Organic Compounds. Plates of Al were obtained on Be cathodes from the molten eutectic of ethyl pyridinium bromide–$AlCl_3$, see [12]. Homogeneous Al plates (thickness 30 µm after 3 h) were also electrodeposited on Be from the electrolyte $Na[(C_2H_5)_3AlFAl(C_2H_5)_3] \cdot 3.4\,C_6H_5CH_3$ at 95 to 100°C and 1.0 A/dm² [13].

References:

[1] Smirnov, M. V.; Chukreev, N. Ya. (Zh. Fiz. Khim. **32** [1958] 2165/73; C.A. **1959** 6833).

[2] Boisdé, G.; Broc, M.; Chauvin, G.; Coriou, H.; Hardy, L.; Jarny, P. (Bull. Inform. Sci. Tech. [Paris] No. 62 [1962] 29/38; C.A. **58** [1963] 12172).

[3] Boisdé, G.; Broc, M.; Chauvin, G.; Coriou, H.; Hure, J.; Jarny, P. (J. Nucl. Mater. **6** [1962] 256/64; C.A. **58** [1963] 8644).

[4] Wohlfarth, H. (Diss. Univ. Stuttgart 1981, pp. 1/119, 63/5).

[5] Delimarskii, Yu. K.; Baranenko, V. M.; Barchuk, V. T. (Ukr. Khim. Zh. [Russ. Ed.] **45** No. 1 [1979] 75/7; Soviet Progr. Chem. **45** No. 1 [1979] 92/4).

[6] Delimarskii, Yu. K.; Kikhno, V. S. (Elektrokhimiya **5** [1969] 145/50; Soviet Electrochem. **5** [1969] 132/7).

[7] Lebedev, V. A.; Pyatkov, V. I.; Titov, G. N.; Agapitov, V. A.; Klimovskikh, N. M. (Izv. Vysshikh Uchebn. Zaved. Tsvetn. Metall. **1976** No. 4, pp. 82/6; C.A. **85** [1976] No. 113743).

[8] Mordovin, A. E.; Nichkov, I. F.; Raspopin, S. P.; Sunegin, G. P. (Fiz. Khim. Elektrokhim. Rasplavl. Solei Shlakov **1969** No. 2, pp. 121/3; C.A. **74** [1971] No. 134079).

[9] Boisdé, G.; Broc, M.; Chauvin, G.; Schaub, B. (Metaux Corros. Ind. **47** [1972] 205/9; C.A. **78** [1973] No. 37093).

[10] Broc, M. (CEA-R-4921 [1978] 103 pp., pp. 89/90; C.A. **91** [1979] No. 201108).

[11] Faust, C. L.; Beach, J. G. (Plating **43** [1956] 1134/42).

[12] Beach, J. G.; Faust, C. L. (J. Electrochem. Soc. **100** [1953] 276/9).

[13] Doetzer, R.; Stoeger, K. (Ger. Offen. 2166843 [1971/76] 1/20; C.A. **85** [1976] No. 168867).

16.5   Behavior as Anode

16.5.1   Beryllium Anodes in Aqueous Solutions

16.5.1.1   Polarization and Dissolution

Polarization

Beryllium anodes in aqueous solutions of inorganic acids, bases, and salts show two types of behavior: considerable anodic polarization (e.g., in solutions of OH^-, NO_3^-, PO_4^{3-}, AlO_2^-, CrO_4^{2-}) or activation (e.g., in solutions of Cl^-, ClO_4^-, SO_4^{2-}). The observed polarizations may be due to the formation of insoluble BeO or salt films, adsorbed films of H_2 or O_2, or slow chemical reactions. Measurements were carried out by Levy [1] at 22°C with 98.6% pure Be anodes in unstirred solutions in air with hand regulation of the voltage, the potential being measured vs. SCE over a current density range of 10^{-7} to 10^{-2} A/cm². Starting at strong cathodic potentials (about −2 V vs. SCE) activated the test electrodes and otherwise did not interfere with the anodic polarization measurements. In the voltage range −2 to +4 V vs. SCE the polarizing current densities were determined for 1N solutions of CH_3COOH, H_3PO_4, NaOH, $NaNO_3$, Na_2CO_3, CH_3COONa, Na_2SiO_3, Na_3PO_4, $NaAlO_2$, Na_2CrO_4, and H_2CrO_4, and for 0.1N solutions of NaOH, $NaNO_3$, NaF, NaCl, Na_2SO_4, and $KMnO_4$ (see table in the paper). Typical polarization curves are given for 0.1N NaCl and 1N H_2CrO_4 solutions. The anodic polarization curve for Be in the chromic acid solution shows a Tafel region at potentials more noble than +2.5 V vs. SCE and oxygen is evolved. Maximum anodic polarization was observed in the Na_3PO_4 solution. Dark anodic films indicative of BeO were produced in solutions of $NaNO_3$, $NaAlO_2$, or Na_2CrO_4. With the H_2CrO_4 solution somewhat less anodic polarization was produced than with the Na_2CrO_4 solution. With the chromic acid solution there was no visual evidence of an anodic film. Carbonate and acetate produced some polarization (see table in the paper), but acetic acid as well as sodium acetate created unstable, fluctuating anodic (and cathodic) potentials. The 0.1N solutions of NaCl, NaF, $NaClO_4$, and Na_2SO_4 caused activation of the Be surface. The dilute (0.1N) NaOH solution produced significantly more anodic polarization than the more concentrated 1N NaOH solution, while an opposite effect was observed with $NaNO_3$ [1].

The activation of Be by Cl^- and SO_4^{2-} ions was studied with anodes of hot-pressed 99.9% pure Be. Cl^- ions exert a distinct activation effect upon the anodic dissolution of Be. This is illustrated by the anodic charging curve for Be in 0.5N NaCl solution, see figure 3 in the paper [2]. The potentiostatic anodic polarization curves for 0.01N $NaNO_2$ solutions additionally containing 0.001 and 0.0001N NaCl (see figure 1 in the paper) further demonstrate the activating effect of Cl^- ions. The potentiostatic curve (also given in the figure) for an aqueous H_3BO_3 solution (40 g/L) is a straight line in the potential vs. log j plot and positioned at more negative potentials than the curves for the $NaNO_2$–NaCl solutions. The potentiodynamic (3.6 V/h) anodic curves for a 0.01N KOH solution with various NaCl contents, see **Fig. 16-12**, show that hydroxyl ions inhibit the activating action of the chloride ions considerably. When the hydroxyl ion concentration is more than 10 times the chloride ion concentration, the latter has no longer an influence on the anodic curve. To a less pronounced degree than Cl^-, SO_4^{2-} also suppresses the shift of the Be anode potential to positive values and enhances corrosion, see the potentiostatic anodic curve for 1N Na_2SO_4 (figure 1) in the paper. The anodic charging curve in the 1N Na_2SO_4 solution shows the activation of the Be to proceed in two stages, at first slowly and then rapidly. While in media containing sulfate ions (0.1N Na_2SO_4 or H_2SO_4) the potential does not shift in the positive direction with increasing time of immersion at 80°C, in the presence of copper ions, i.e. in 0.1N $CuSO_4$ solution, intensified corrosion occurs. This was explained by a passivating action of copper, because Be is passivated in a 0.1N $Cu(NO_3)_2$ solution according to its potential values; see table 2 in the paper [2].

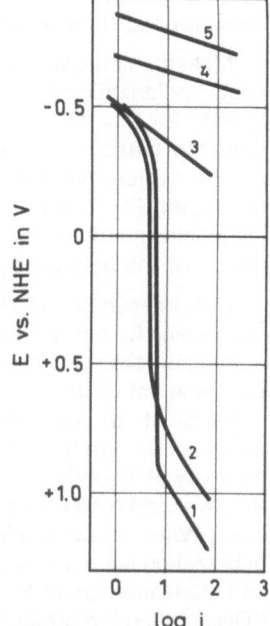

Fig. 16-12. Potentiodynamic anodic curves ($\Delta E/\Delta t = 3.6$ V/h) for Be in 0.01N KOH solution with various NaCl contents (N): 1) 0; 2) 0.0001; 3) 0.01; 4) 0.1; 5) 1. The current density, j, is measured in $\mu A/cm^2$.

Polarization was further studied in the course of an investigation of the anodic dissolution mechanism of 98.6 and 99.18% pure Be in solutions of $K_2SO_4 + KF$, $K_2SO_4 + KCl$, $K_2SO_4 + KBr$, $K_2SO_4 + KI$, $K_2SO_4 + KNO_3$, $K_2SO_4 + K_3PO_4$, $KCl + KNO_3$, $LiCl + K_2SO_4$, $NaCl + Na_2SO_4$, and $MgCl_2 + MgSO_4$ at current densities from 2.0 to 80 mA/cm². The ionic strengths were held constant at 1.00 and the temperature was 25°C. From these investigations the Tafel curves for 0.33 M K_2SO_4 solutions containing KCl, KBr, or KI (each 0.01M) are shown in **Fig.** 16-13. Tafel curves for 0.333 M K_2SO_4 solution and for solutions of 0.33 M K_2SO_4 containing 0.01 M KNO_3 or K_3PO_4 are given in the paper [3]. Linear sections over 1.0 to 1.5 decades are found in all the curves; for polarization values, see the table in the paper. Since local corrosion occurs simulta-

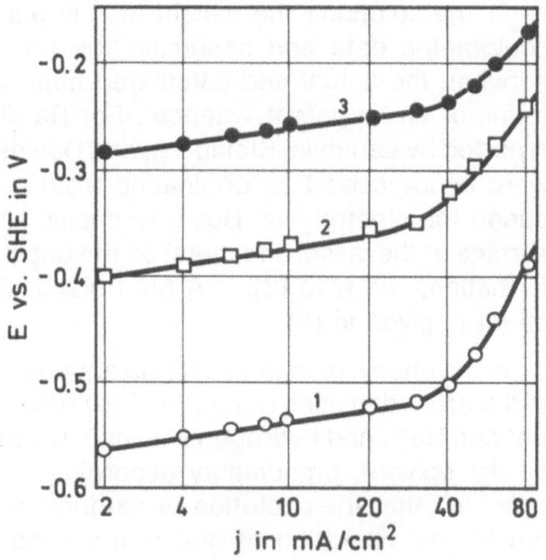

Fig. 16-13. Tafel curves for the anodic dissolution of Be in solutions of (0.01M KCl + 0.33 M K_2SO_4) = 1, (0.01M KBr + 0.33 M K_2SO_4) = 2, and (0.01M KI + 0.33 M K_2SO_4) = 3 at 25°C.

neously with the anodic dissolution, it is questionable if the Tafel slopes can be used for any straightforward mechanistic considerations [3].

Measurements with a rotating (5000 rpm) Be disk electrode with potentiodynamic (2 V/min) anodic polarization have been carried out in aqueous 4 N solutions of NaCl, KBr, KI, NaClO$_4$, NaNO$_3$, 2 N Na$_2$SO$_4$, and a saturated NaF solution (polarization curves are given in the paper). With the latter three electrolytes, considerable current densities were found to begin at potential values (vs. NHE) of at least ~1.4, 4.3, and 1.8 V, respectively (values estimated from the figure) [4]. It has been stated that the anodic polarization curves of Be in aqueous halogenide solutions (of NaCl, KCl, bromide, and iodide) depend only on the type and concentration of the halogenide ion; they do not depend on the pH [5].

Hot-pressed Be (99.7% pure) after traditional preparation (cleaning, washing, and degreasing) retains its passive state on anodic polarization in 5 N KOH solution over a wide range of potentials. Current densities calculated from measured weight losses essentially fit the experimental values of the anodic potentiostatic curve, see figure 1 in the paper [6]. For the influence of stirring on the polarization curve, see Fig. 16-6, p. 181. Anodic oxygen evolution occurs at 0.65 to 0.75 V vs. NHE and is characterized by an overvoltage of about 0.2 V. The passivity of Be under these conditions is not due to the presence of an oxide film; it is assumed that in alkaline solutions passivity of Be stems from an adsorptive or chemisorptive mechanism. When an oxide film is formed by the exposure of Be to water at 300°C prior to the polarization experiments, the stationary potential is "nobilized" by nearly 1 V and the currents on polarization are reduced. With longer holding times of the oxidized Be (up to 432 h) in the KOH solution the protective properties of the oxide film become weaker, apparently because of its dissolution in the electrolyte. With such passivating oxide films, still unexplained steps in the potentiodynamic polarization curves (i.e., rapid increases of the current density on increasing anodic polarization) are observed at 0.15 to 0.2 V vs. NHE (see figure 3 in the paper). The polarization of a Be anode covered with a friable white film formed during accelerated corrosion in water at 300°C is also described. In a 1 N NaNO$_3$ solution the anodic potentiostatic curve (see figure 4 in the paper) does not show a clearly marked overpassivation section; oxygen in this case is visibly evolved at about 5 V vs. NHE accompanied by darkening of the surface, which finally becomes black [6].

Dissolution

A fundamental aspect of the anodic dissolution of metals is the valence with which the ions enter the solution. The weight loss in many instances is greater than that calculated using coulometric data and assuming the normal valence of the metal ions. This discrepancy between the actual and calculated quantities (difference effect, D.E.) may be expressed in terms of an "apparent valence". For Be dissolving anodically, such a discrepancy was first reported by Laughlin, Kleinberg, and Davidson [7]. Small metallic particles found in the anolyte were hypothesized as originating from a disproportionation reaction of Be$^+$ ions produced during the electrolysis. But Straumanis and coworkers, see, e.g. [8 to 12], established the surface of the dissolving metal as the origin of the particles and disclaimed any monovalent ion formation, see also [3]. – A brief discussion of the methods for determining the difference effect is given in [2].

In the anodic oxidation of Be in aqueous NaCl solutions (0.2 and 5.0 M) at room temperature and current densities of 0.03 to 0.20 A/cm² for 5 to 140 min Be was converted to a mixture of Be$^+$ and Be^{2+}, and hydrogen was evolved at the anode from the oxidation of the unipositive Be by the solvent, presumably according to the equation $2 Be^+ + 2 H_2O = 2 BeOH^+ + H_2$. It was observed that the evolution of cathodic hydrogen ceased as soon as the current was interrupted, but liberation of gas at the anode continued for an additional five to ten minutes.

Furthermore, during approximately the first 15 min of electrolysis the anolyte remained clear and transparent. Beyond this time, however, a white gelatinous material ($Be(OH)_2$) and a black substance were slowly formed throughout the anolyte; the latter substance was shown to consist of Be and beryllium oxide. It was concluded that metallic Be is formed by disproportionation according to $2Be^+ = Be^{2+} + Be$ [7].

In alkaline solutions the potential at which anodic Be dissolution occurs is more noble than the potential of the reversible hydrogen electrode. It was suggested, in accordance with the above results, that the hydrogen formed at a Be anode (proportional to the current) must be due to the reaction of water with an unstable product of the Be dissolution. From experiments with a Be disk electrode the lifetime of Be^+ was estimated to be of the order of 10^{-3} to 10^{-1} s [13]; for details of the ring-disk electrode system, see [5]. Later detailed experiments with a ring-disk electrode system in aqueous KCl solutions (10^{-3} to 1 mol/L) at 25°C clearly showed that the anodic dissolution of a rotating Be disk electrode includes both the reactions $Be \rightarrow Be^{2+}_{aq} + 2e^-$ and $Be \rightarrow Be^+_{aq} + e^-$. Based only on this method, other causes for the anomalies of anodic Be dissolution (low apparent valence, negative difference effect) can not be excluded. With the ring electrode (amalgamated Pt) being kept at potential values of -0.2 to $+0.6$ V vs. SCE, anodic currents were measured which first increased when the potential of the rotating disk electrode increased and then decreased again rapidly. The heights of the current maxima thus observed increased with the ring electrode potential, see figure 1 in the paper [14]. With increasing KCl concentration in the electrolyte, the current at the ring electrode increased more slowly (approximately with the square root of the concentration) than the current of the disk electrode. The intermediate product, which is oxidized at the ring electrode, is concluded to be Be^+ alone. The quantity of Be^+ which reaches the ring electrode apparently depends on the Be disk potential. An analysis of the results indicated that the Be^+ ions are also electrochemically oxidized at the disk electrode ($Be^+ \rightarrow Be^{2+} + e^-$, with an estimated rate constant of $\sim10^{-2}$ cm/s). The amount of Be^+ formed (up to 37%) and the stability of the Be^+ ion in the electrolyte depended not only on the anode potential, but also on the type and concentration of the electrolyte. The Be^+ lifetime was estimated to be $\sim10^{-2}$ to 10^{-1} s [14].

In halide solutions (NaCl, KCl, bromide, iodide) the apparent valence for the anodic dissolution of Be was found to have values between 1 and 2. It was higher for more noble dissolution potential values and independent of the pH of the solution [5]. Also, the following values for the apparent valence were determined in aqueous solutions: 1.07 to 1.22 (depending on the current density) in 1M KCl, 1.41 in 1M KI, 1.82 in 0.33 M K_2SO_4, and 1.21 in 0.01 M LiCl + 0.33 M K_2SO_4 solution [15]. The effect of various anions and cations on the anodic dissolution was further studied with 99.18 and 98.62% pure Be in aqueous solutions of mixtures of K_2SO_4 with LiCl, KF, KCl, KBr, KI, KNO_3, or K_3PO_4, of KCl with KNO_3, of NaCl with Na_2SO_4, and of $MgCl_2$ with $MgSO_4$ (ionic strength 1.00; temperature 25°C; current densities 2.0 to 80 mA/cm²). The amount of Be dissolved was determined from the weight loss in order to calculate the apparent valence. It was found that the apparent valence depended on the anions, but only slightly on the cations (if at all), and was relatively independent of the current density in the range studied. In most cases the apparent valence values ranged from 1.07 to 1.82 (see tables in the paper [3]); only in solutions containing 0.99 to 1.00 mol/L of nitrate, where the electrodes were passivated and oxygen was evolved, the apparent valence reached up to 10.9. Otherwise a slight hydrogen evolution was usually noted, indicating that local corrosion occured simultaneously with the anodic dissolution. The halide anions were found to cause the greatest decrease in the apparent valence. An anodic dissolution mechanism, involving local corrosion, disintegration, and the effect of ions, was proposed to explain the results. Microscopic observations of the Be anode surfaces showed relations to the apparent valence values and supported this view. It seems that for solutions which do not contain highly adsorbing anions (e.g., Cl^- and Br^-), disintegration of Be is not significant, because large numbers of attack sites do not become interconnected causing "chunks" of the metal to be dislodged. The effect of the anions on the apparent valence seems to be related to the anion size, especially in the case of Cl⁻, Br⁻, and I⁻; for details and discussion, see [3].

When Be is dissolved anodically in an NaCl solution, after some time the solution becomes dark; the black deposit obtained consists of metallic Be [7]. It was pointed out by Straumanis and coworkers [8 to 10] that the solution blackens as a result of a partial disintegration of the Be anode. When Be is anodized in neutral NaCl solutions, Be needles and particles are found to separate from the dissolving compact metal. The same is observed when Be is dissolved in HCl ($\leqq 0.5$ N) [9, 16], in HBr [8], or in HClO$_4$ (~ 0.3 N), while on dissolution in solutions of HF or H$_2$SO$_4$ the metal surface remains smooth and no particels separate [16]. Dissolution potentials of 98.5 and 99.48% pure Be anodes were measured at $\sim 30°$C in dilute HF, HCl, HClO$_4$, and H$_2$SO$_4$ solutions as functions of time, acid concentration, and current density; the values were found to be unreliable for the recognition of disintegration during the Be dissolution [10]. For Be dissolution in dilute HCl, see also [17, 18].

In certain cases of electropolishing, a layer is formed on the anode which is almost free from H$_2$O and contains many ions capable of being adsorbed on the metal surface. Therefore an electric field builds up sufficiently high to transfer the metal ions through the adsorbed layer into the solution. The high anodic potential often observed, especially in solutions containing ClO$_4^-$, will be induced by this layer and not by an oxide film. The electric field in the layer is thought also to accelerate the diffusion of metal ions with lower than the normal valence from the metal surface into the electrolyte. Thus it is found that Be$^+$ ions go into solution in the case of electropolishing in aqueous baths containing ClO$_4^-$, while only Be^{2+} ions do so in baths containing phosphoric acid, chloride, fluoride, etc. [19]. When Be$^+$ is oxidized in the solution by ClO$_4^-$, chloride is also formed [20], cf. [21, 22]; for further details and references on this subject with regard to nonaqueous solvents, see p. 203.

References:

[1] Levy, D. J. (AD-267976 [1961] 18 pp.; C.A. **58** [1963] 8639).
[2] Gerasimov, V. V.; Shapovalov, E. T. (Zh. Prikl. Khim. **46** [1973] 953/7; J. Appl. Chem. [USSR] **46** [1973] 1017/20).
[3] Sheth, K. G.; Johnson, J. W.; James, W. J. (Corros. Sci. **9** [1969] 135/44).
[4] Davydov, A. D.; Kashcheev, V. D.; Kozlov, M. V. (Zashch. Met. **9** [1973] 436; Protect. Metals [USSR] **9** [1973] 398/9).
[5] Heusler, K. E. (Z. Elektrochem. **65** [1961] 192/7).
[6] Shapovalov, E. T.; Gerasimov, V. V. (At. Energiya SSSR **26** [1969] 498/502; Soviet At. Energy **26** [1969] 578/81).
[7] Laughlin, B. D.; Kleinberg, J.; Davidson, A. W. (J. Am. Chem. Soc. **78** [1956] 559/61).
[8] Straumanis, M. E. (Metall. Soc. Conf. Proc. **24** [1963] 645/55).
[9] Straumanis, M. E.; Mathis, D. L. (J. Less-Common Metals **4** [1962] 213/5).
[10] Straumanis, M. E.; Gnanamuthu, D. S. (Corros. Sci. **4** [1964] 377/86).

[11] Straumanis, M. E. (J. Electrochem. Soc. **105** [1958] 284/6, **106** [1959] 535/6).
[12] Straumanis, M. E. (J. Electrochem. Soc. **108** [1961] 1087/92).
[13] Heusler, K. E. (Z. Elektrochem. **64** [1960] 1112/3).
[14] Eckert, J.; Forker, W. (Z. Phys. Chem. [Leipzig] **253** [1973] 153/60).
[15] Vaidyanathan, H.; Straumanis, M. E.; James, W. J. (J. Electrochem. Soc. **121** [1974] 7/12).
[16] Straumanis, M. E.; Mathis, D. L. (J. Electrochem. Soc. **109** [1962] 434/6).
[17] Straumanis, M. E. (Metall [Berlin] **16** [1962] 102/7).

[18] Straumanis, M. E. (Latv. PSR Zinat. Akad. Vestis Khim. Ser. **1967** 643/50; C.A. **68** [1968] No. 74581).

[19] Epelboin, I. (Z. Elektrochem. **62** [1958] 813/8).

[20] Epelboin, I. (Z. Phys. Chem. [Leipzig] **215** [1960] 380/7).

[21] Brouillet, P.; Epelboin, I.; Froment, M. (Compt. Rend. **239** [1954] 1795/7).

[22] Epelboin, I. (Z. Elektrochem. **59** [1955] 689/92).

16.5.1.2 Protective Anodic Coatings

General Remarks

In spite of its character as a very unnoble metal, beryllium is quite resistant to atmospheric corrosion because of the thin but dense oxide film formed on its surface. This film can be reinforced by anodic oxidation [1], which has been achieved with good results mostly in chromic acid or chromate solutions, see, e.g. [1 to 4]; other electrolytes were also used, such as solutions of boric acid [5 to 7], ammonium molybdate [8], sodium carbonate [9, 10], etc. It was shown that anodic films produced on Be in a number of aqueous electrolytes are characterized by one or more inopportune properties, such as porosity, the tendency to extensive dissolution, and crystallinity. This includes films produced in aqueous solutions of KOH, Na_2CO_3, $(C_2H_5)_2HPO_4$, $(C_2H_5)_2SO_4$–H_2SO_4 mixture, CrO_3–HNO_3 mixture, and acetic acid [9].

Formation of Anodic Coatings

When foils of 98.8% Be were anodized in electrolytes containing **chromic acid**, e.g., 20% CrO_3 solutions at 20°C with 20 A/dm², the anodic films obtained consisted of a thin barrier layer next to the metal and a thicker, porous outer layer. The thickness of the barrier layer was independent of the anodizing time; this was deduced from capacitance measurements. For the porous layer the breakdown voltage increased from 400 to more than 1000 V as the formation time increased from 30 to 240 min [2]. Up to 30 μm thick anodic films may be produced on 99.7% pure Be in aqueous chromic acid solutions (20% wt/vol CrO_3) at 0, 10, or 18°C with current densities of 5, 10, or 15 A/dm² (anodizing time up to 4 h). The film formation is accompanied by dissolution of Be and by oxygen evolution. The thickest films were obtained at 10°C. Changes in the CrO_3 concentration (2 to 20%) had no effect on the film thickness. For a constant quantity of electricity passed, the thickness and the porosity of the film were independent of the current density in the range 5 to 20 A/dm². Porosity first increased rapidly and then linearly with time up to about 50% (see figure 3 in the paper [3]). The applied voltage increased rapidly at first and then remained practically constant at somewhat below 4 V (see figure 4 in the paper [3]). The behavior described is probably related to the initial formation of a nonporous layer, on which in subsequent stages a porous anodic film grows [3, 11].

In anodizations of 98.5% pure Be in aqueous chromic acid (1 to 35% wt/vol) at 25 ± 3°C the film thickness depended on the electrolyte concentration up to ~10% CrO_3 [12]. Anodic films with thicknesses of 25, 150, and 3500 Å were formed after 15, 90, and 600 s, respectively, on vacuum cast and hot-compacted Be in 10% aqueous HNO_3 containing 200 g CrO_3/L (actual current density on the Be anode 30 mA/cm²). The anodic oxidation of Be is considered to be ionic in character and to be controlled by the slow diffusion of Be ions through the anodic film [13], see also [14]. Significant Be dissolution occurred during the anodization of 99.5% pure Be foils in the above HNO_3–CrO_3 electrolyte at 22°C [9]. For the formation of anodic films on Be in Na chromate solutions, see [8, 15]; cf. p. 198.

According to a patent specification, a protective coating of amorphous, hydrated BeO is formed on Be by anodic treatment in aqueous solutions containing 0.4 to 10 wt% Na chromate and sufficient chromic acid to give a pH 5.5 to 7.5. The electrolyte must be free of Cl^- and SO_4^{2-} ions. Anodization is carried out at 20 to 40°C with 10.7 to 21.5 A/dm², and the 0.0025 to 0.6 mm thick coatings may be converted to a crystalline form by heating to 800°C [4]. Black, 20 to 30 µm thick films produced by this method were reported to have up to 40% porosity; for electron microscopic and electron-diffraction studies, see [16].

Ammonium molybdate solutions were used with the aim of achieving thicker, high-quality films on Be than those obtained in chromic acid baths (30 to 40 µm). A comparative investigation was performed in aqueous solutions of 1) ammonium molybdate (5 to 20%, pH 5 to 7.5), 2) chromium trioxide, and 3) sodium chromate. Curves of the growth of film thickness with time (see figure 1 in the paper [8]) at current densities of 5 to 20 A/dm² show that with the ammonium molybdate solution (as with the others) thickness increases almost linearly. This is explained by the constant thickness of a barrier layer through which the beryllium ion must diffuse in anodization, the diffusion being constant with time. This corresponds to results with chromic acid (see [2]) and is proved by experiments with the isotope ^{99}Mo incorporated in the first anodic layer (for details, see the paper [8]). The porosity of the anodic films obtained in the chromic acid and sodium chromate baths, as shown in **Fig.** 16-**14**, is much higher than that obtained in the ammonium molybdate bath. The specific porosity (the derivative of porosity with respect to thickness) decreases with an increase in thickness to 25 to 30 µm, and then remains almost unchanged for all the above electrolytes at a current density of 10 A/dm². With increasing current density the film thickness and the porosity increase, but this only corresponds to the larger amount of electricity that has passed. Electrolyte temperatures of about 0 up to 70°C were investigated with the above electrolytes at a 30 A/dm² current density. For a given amount of electricity passed, maxima of the anodic film weight and minima of porosity were observed at 10, 20, and 40°C for 10% solutions of chromic acid, sodium chromate, and ammonium molybdate, respectively (see figure 4 in the paper). In an ammonium molybdate solution below 40°C only barrier anodizing occurs and an outer porous layer does not form. In this electrolyte, coatings of good quality may be obtained with a thickness of more than 100 µm. Suitable pH ranges are 5 to 7.5 for the ammonium molybdate and 2 to 9 for the sodium chromate solutions. For the influence of chloride and sulfate ions present in the electrolyte on the anodic film, see the paper [8].

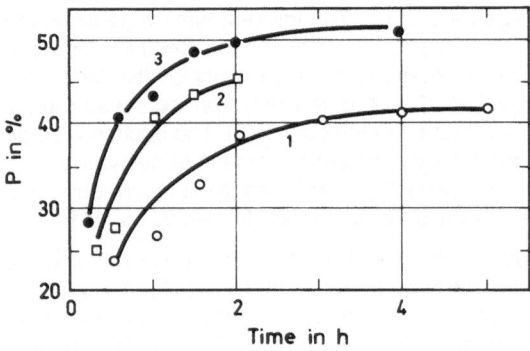

Fig. 16-14. Change in the total porosity, P, of anodic films in the course of Be anodizing in ammonium molybdate 1, sodium chromate 2, and chromic acid 3. Current density 10 A/dm². The change in the specific porosity of the films on Be anodizing in the same media is shown (one curve) in the original figure in the paper [8].

Barrier-type anodic films are produced in aqueous electrolytes containing **H_3BO_3, NH_4OH**, and **ethylene glycol**; they are characterized by an amorphous structure and minimal porosity. With a thickness of 0.5 to 1 µm they have optimal properties with regard to corrosion. The kinetic characteristics of the anodic oxidation in such electrolytes were investigated under optimal operating conditions, i.e., with current densities of 12, 60, and 60 A/m² at temperatures of 10, 35, and 40°C, and anodizing times of 45, 25, or 5 min, respectively; the respective electrolyte compositions were 40, 145, or 160 g/L of H_3BO_3, 150, 140, or 150 g/L of ethylene glycol, and 40, 145, or 160 mL of a 25% NH_4OH solution per liter of the electrolyte. At current densities >100 A/m² crystalline oxide films are formed [5]. For a patent specification concerning the above electrolytes and procedure, see [6]. A threshold current density (5.0 mA/cm² or higher) is observed for the patented electrolyte compositions below which the voltage builds up very slowly (see figure 1 in the paper [7]). The limiting film formation voltage is 120 to 150 V and the thickness of the resulting films is 250 to 350 nm or more. The films are translucent and slightly tinted from yellow to blue, as a result of interference. Anodization at a constant terminal voltage is accompanied by a comparatively fast decrease of the current. For the results of electron microscopic studies, showing both polycrystalline and amorphous structures of the films, see the paper [7].

A blue, 1500 Å thick film may be produced on Be which is as corrosion-resistant to aqueous NaCl as are much thicker films formed by anodizing in chromic acid baths. This blue film was obtained by anodization with a current density of 50 mA/cm² at 30°C in an electrolyte of 100 g $Na_2B_4O_7 \cdot 10H_2O$ per liter. The current density was kept constant during the anodization time (total 20 min) by raising the applied voltage (to 100 V maximum). Also aqueous solutions of 100 g $NaBO_4 \cdot 4H_2O$ or 100 g $Na_2HPO_4 \cdot 2H_2O$ per liter can be used as the electrolyte. For details and cleaning of the Be before anodizing, see the patent specification [17]. Black films on Be were obtained according to a patent by anodizing in a bath containing Na borate and NaOH in a ratio of 2.69 to 0.77 mol, respectively; these quantities were the minimum per liter H_2O; details are given in [18]. The kinetics of Be anodization were studied in a 0.5M Na_2CO_3 or K_2CO_3 solution which was saturated with oxygen. The metal was covered with an oxide film of increasing thickness and oxygen gas was liberated. The oxide layers were reported to be practically free of pores [10]; cf. [9].

An aqueous electrolyte for the preparation of anodic coatings on Be or its alloys, which are improved with respect to throwing power and corrosion resistance, is described in a patent specification as consisting of **Na acetate** (10 to 20 wt%), **ammonium acetate** (2 to 5 wt%) and distilled water [19].

Properties of Anodic Coatings

Anodic films produced on Be of 98.8 to 99.0% purity in chromic acid solutions (e.g., 20% CrO_3, cf. [3]) at 5 to 15 V were black and 20 to 30 μm thick. They had a porosity of up to 40%, the porosity being visible on electron micrographs. The diameter of the pores reached 200 to 400 Å. On annealing, the black oxide lost 5 to 10% of its weight (by heating up to 100°C) due to the elimination of moisture. On further heating up to 800°C, its weight remained practically unchanged. Since Be hydroxide decomposes at 300°C with weight losses up to 40%, the black anodic films can not contain appreciable amounts of hydroxide. The black films consisted of amorphous Be oxide; upon annealing in air at 750 to 800°C for 6 h they changed their color to gray and became crystalline as shown by X-ray patterns. It was concluded that the black films consisted of nonstoichiometric, O-deficient BeO. Films obtained on Be by anodizing at 60 V in solutions used for anodizing Al (solutions not specified in the paper) were 10 μm thick, light gray, and had a porosity of up to 15%. According to electron-diffraction patterns they consisted of crystalline Be oxide with a crystal size close to 100 Å [16].

Coatings obtained by anodic oxidation of hot-pressed Be in aqueous solution of chromium oxide (20%, stated to be Cr_2O_3) consist of fine-crystalline Be oxide according electron-diffraction and X-ray studies [20]. Anodic oxide films produced on electropolished polycrystalline Be in nitric acid (1 or 10%) or in a mixture of nitric and chromic acid (10% HNO_3 and 200 g CrO_3 per liter) were always crystalline with the oxide layer next to the substrate showing a weak preferred orientation with the {001} planes parallel to the substrate [14]. Anodic BeO layers grown at a constant rate up to a thickness of several thousand Å in a nitric and chromic acid solution initially were amorphous, i.e., as long as they were very thin, but on reaching a thickness of 100 to 150 Å and thereafter, the surface consisted of randomly disposed BeO crystals, 60 Å or more in diameter. It was suggested that the surface temperature must have exceeded 300°C for the recrystallization of BeO to take place [13]. Beryllium foils (0.1 mm thick and 99.9% pure) on an Mo substrate were anodized in an electrolyte containing H_3PO_4, H_2SO_4, glycol, and water with an anode current density of 4 A/cm². With respect to corrosion processes, secondary ion-ionic emission was studied on the composition of the anodically produced films [21].

When Be was anodized in an aqueous solution of 20% H_2CrO_4 at 10°C with 10 A/dm² for 1 h, an amorphous 15 μm thick oxide film formed, which prevented the interaction of the Be metal with stainless steel (type 1Kh 18N9T) at 700°C in vacuum for one year. Changes in the chromic acid concentration (concentrations investigated 2 to 20%) had no substantial effect on the quality of the films. The porosity of the outer layer of 15 to 20 μm thick films was not more than 40%; this layer contains randomly arranged pores, 200 to 400 Å in diameter. By changing current density, temperature, and anodization time the thickness and the porosity of the films can be varied within wide limits [11]. In investigations of the protection of Be metal by anodic films against reaction with UO_2 at temperatures of 700 or 800°C, films produced in three different ways were compared: the anodization was carried out either in an aqueous solution of HNO_3 (10%) and 200 g CrO_3/L, or in pure H_2CrO_4 solution. Further, films obtained from the first electrolyte were additionally heated in oxygen. In contrast to untreated Be metal, which reacted extensively with UO_2 pellets at 800°C (100 h) in a vacuum or with flowing dry CO_2 at 700°C (1600 h), all anodized Be samples were corrosion-protected except for the Be samples treated in the first procedure which showed some spalling. For the behavior of anodized Be in the presence of CO_2 containing 3 to 4% H_2O by volume, see the paper [22].

An anodizing treatment was applied to dense and to porous Be (purity of both ≧98.5%) for 5 min at 22 mA/cm² in diluted nitric acid (50% by volume), and the protective effect of the films formed by this procedure was tested in comparison to other protective methods in the presence of salt (marine), air, and moisture [23]. The corrosion resistance of Be and anodized Be in demineralized and tap water solutions was tested and compared by [24]. For the corrosion resistance of anodic films on Be towards 3% NaCl solutions, see [17, 19]. Beryllium anodized in Na dichromate–chromic acid baths (pH 7.0) at 24°C with 17.2 A/dm² showed good corrosion resistance towards dry air at 800 to 900°C, moist air at 800°C, CO_2-free dry air at 850 to 1100°C, dry O_2 at 700 to 1030°C, NaCl solution sprays, 0.1N NaCl solution, deionized water at 43°C and ordinary water; see tables in [4].

According to a patent, a corrosion-resistant surface was obtained on Be by anodizing in a chromate solution with 5 to 40 A/dm² after immersing the metal in a solution containing H_3PO_4, H_2SO_4, and H_2CrO_4 [15]. Another patent specification describes coatings resistant to a corrosive salt spray environment; they can be applied to metals and their alloys (which have the property of electrolytic rectification) by an anodic treatment in a strongly alkaline aqueous solution containing a fluoride and an Fe salt. A sufficiently high voltage must be applied to obtain spark discharge at the surface to be anodized [25].

Beryllium (99%) may be colored (yellow, dark green) by stepwise anodization at 25°C in a 2% $H_2C_2O_4$ bath; for details, see [26].

An apparatus for chromic acid anodizing of Be, including operating conditions, dimensional changes of the Be specimens, density of the coatings, and process efficiency, is described in [12]. A slight loss in weight and dimensions of the articles anodized was observed. Elevating the temperature to $50 \pm 3°C$ considerably increases the coating thickness and the amount of reacting Be. However, in this case the adherent glossy black films of BeO tend to become quite porous, especially with increasing CrO_3 concentration [12].

Luminescence was observed during the formation of anodic oxide films on Be in borate baths. It can be characterized as a prebreakdown electroluminescence, such as observed, for instance, in the case of Al anodization; details are given in [27].

References

[1] Von Krusenstjern, A.; Stiehr, W. (Metall [Berlin] 21 [1967] 692/5).

[2] Vol'fson, A. I.; Markova, N. E.; Chernyshev, V. V.; Lebedev, V. N.; Vavakin, V. V. (Zashch. Met. 9 [1973] 346/7; Protect. Metals [USSR] 9 [1973] 321/2).

[3] Al'tovskii, R. M.; Fedotova, A. G.; Urazbaev, M. I.; Korolev, S. I. (Zashch. Met. 5 [1969] 206/9; Protect. Metals [USSR] 5 [1969] 172/4).

[4] Csontos, L. J.; Stonehouse, A. J. (Fr. 1446592 [1964/66] 4 pp. from C.A. 66 [1967] No. 51675).

[5] Chernykh, M. A.; Rossina, N. G.; Lazarev, V. F.; Chernyshev, V. V. (Fiz.-Khim. Geterog. Sist. 1984 120/7 from C.A. 103 [1985] No. 61449).

[6] Vol'fson, A. I.; Umov, V. S.; Polonskii, E. L.; Markova, N. E.; Chernyshev, V. V.; Lebedev, V. N. (U.S.S.R. 305210 [1970/71]; C.A. 75 [1971] No. 104504).

[7] Chernyshev, V. V.; Vol'fson, A. I. (Zashch. Met. 13 [1977] 245/7; Protect. Metals [USSR] 13 [1977] 209/11).

[8] Al'tovskii, R. M.; Urazbaev, M. I. (Zashch. Met. 22 [1986] 447/50; Protect. Metals [USSR] 22 [1986] 364/7).

[9] Shehata, M. T.; Kelly, R. (J. Electrochem. Soc. 122 [1975] 1359/65).

[10] Heusler, K. E. (Ber. Bunsen-Ges. Phys. Chem. 67 [1963] 943/9).

[11] Al'tovskii, R. M.; Vasina, E. A. (At. Energiya SSSR 38 [1975] 333/4; Soviet At. Energy 38 [1975] 424/6).

[12] Whitby, L.; Gowen, E.; Levy, D. J. (Tech. Proc. Am. Electroplat. Soc. 1961 106/8; N.S.A. 17 [1963] No. 36237).

[13] Kerr, I. S.; Wilman, H. (J. Inst. Metals 84 [1956] 379/85).

[14] Levin, M. L. (Trans. Faraday Soc. 54 [1958] 935/40).

[15] Kawanishi, M. (Jpn. 48-27188 [73-27188] [1968/73] 3 pp. from C.A. 80 [1974] No. 66165).

[16] Al'tovskii, R. M.; Gornyi, D. S.; Momin, E. Yu.; Urazbaev, M. I. (Zashch. Met. 11 [1975] 89/91; Protect Metals [USSR] 11 [1975] 87/9).

[17] Imperial Metal Industries (Kynoch) Ltd. (Fr. 1558159 [1967/69] 3 pp.; C.A. 72 [1970] No. 27769).

[18] Pittman, R. D. (Br. 1078165 [1964/67] 10 pp.; C.A. 67 [1967] No. 104603).

[19] Konchus, A. D.; Syzranova, Zh. V.; Maslova, L. N. (U.S.S.R. 521361 [1973/76]; C.A. 85 [1976] No. 132830).

[20] Smirnova, A. I.; Kancheev, O. D.; Politko, S. K.; Konchus, A. D.; Medvedev, A. A. (Mater. 9th Vses. Konf. Elektron. Mikrosk., Tiflis 1973, pp. 297/8; C.A. 85 [1976] No. 150806).

[21] Kolot, V. Ya.; Rybalko, V. F.; Fogel, Ya. M.; Tikhinskii, G. F. (Zashch. Met. 3 [1967] 723/30; Protect. Metals [USSR] 3 [1967] 631/6).

[22] Vachon, L. J. (J. Nucl. Mater. 6 [1962] 139/41).

[23] Weatherill, B. T.; Loasby, R. G. (Beryllium 1977 Conf. Prepr. 4th Int. Conf. Beryllium, London 1977, Paper No. 42, 12 pp.; C.A. 92 [1980] No. 219538).

[24] Schmitt, C. R. (Y-1397-Rev. 1 [1975] 24 pp. from C.A. **83** [1975] No. 123011).

[25] Habermann, C. E.; Garrett, D. S. (U.S. 466834 7A [1985/87] 6 pp. from C.A. **107** [1987] No. 86092).

[26] Iwasaki, K. (Japan. Kokai 52-45548 [77-45548] [1975/77] 2 pp. from C.A. **87** [1977] No. 59964).

[27] Chernyshev, V. V.; Pershina, E. N. (Elektrokhimiya **22** [1986] 418/9; Soviet Electrochem. **22** [1986] 390/1).

16.5.1.3 Corrosion, Electropolishing, and Machining

Beryllium becomes passivated in acid, alkaline, and neutral aqueous media over a wide range of potentials provided that depassivators, such as Cl^- and SO_4^{2-} ions, are absent. With passivated Be in neutral aqueous solutions, it takes a long time to achieve weight losses exceeding the error in the weighing methods [1]. For surveys on the **corrosion** of Be, see [2, 13].

Failure of Be by stress corrosion in aerated synthetic sea water (pH 8.2, chloride content 19500 ppm) appears to be closely related with random pitting attack. Certain pits remain active and cause severe localized attack and deep penetration along a plane nearly perpendicular to the applied stress. Experiments with Be sheet material of commercial (98.3%) purity showed that applied anodic currents at a tensile stress of 30000 psi (for conversion factors, see pp. 276/7) at 22°C reduced the time to failure considerably (see figure in the paper), while cathodic currents prevented pitting and the subsequent stress corrosion [3].

Data on **electropolishing** of Be are given in [4], on electropolishing and etching in [5]. For some aspects of the electrochemical mechanism of Be electropolishing in ClO_4^- solutions, see p. 196.

Electrochemical **machining** is possible at practical metal-removal rates with aqueous $NaNO_3$ or NaCl electrolytes. Thus smooth surfaces (roughness about 10 to 50 μin) were obtained with a block-pressed and forged extra-fine-grain Be material containing 3.2% BeO (Type II A beryllium) and an electrolyte of 480 g $NaNO_3$/L at 100°F with feed rates of 0.02 to about 0.1 in/min. Chemical milling of Be materials is generally carried out in aqueous solutions of, e.g., H_2SO_4, NH_4HF, or HNO_3–HF mixtures, see [12].

A patent specification [6] reports an electrochemical procedure for obtaining smooth U-shaped cuts in beryllium (e.g., high-purity Be sheet) without cracks and tears as sometimes result from other cutting methods. In this method an aqueous nitrate solution (e.g., 500 g NH_4NO_3/L at 42°C) is applied as a pressure jet from a Pt nozzle (inside diameter 0.02 to 0.04 in) serving as cathode to the Be specimen as anode. The distance between cathode and anode should be less than 0.040 in and the current should exceed 0.5 A at a potential difference of 60 to 120 V between the Be and the nozzle; for details, see the paper [6].

Electrolytical local polishing for metallography is described in [7], and the metallographic preparation of Be by electrolytic lapping in [8]. For an investigation to compare the technique of electropolishing of Be with that of chemical milling to remove machine damage, see [9].

Successive thin layers (2.6 to 17 nm thick) may be removed from the surface of macroscopic Be specimens by anodic sectioning. To do this an amorphous anodic film is produced on the Be in an ethylene glycol-based bath containing sulfate and phosphate ions; this film is then dissolved in a 10% KOH solution [10].

Goodman [11] described a method and equipment to remove an Ni coating from a Be wire (diameter 0.005 in) and to obtain a smooth finish. The method consists of an electrolysis carried out at room temperature in $2N$ HNO_3 solution, containing $0.2N$ KCl, with 19.6 A/in² for 8 s; the process is further enhanced by application of ultrasonic energy.

References:

[1] Shapovalov, E. T.; Gerasimov, V. V. (At. Energiya SSSR **26** [1969] 498/502; Soviet At. Energy **26** [1969] 578/81).

[2] English, J. L. (in: White, D. W.; Burke, J. E.; The Metal Beryllium, Vol. 2, ASM, Cleveland, Ohio, 1955, pp. 530/48).

[3] King, T. T.; Myers, J. R. (Corrosion [Houston] **25** [1969] 349/51).

[4] Gurklis, J. A.; McGraw, L. D.; Faust, C. L. (DMIC-Memo-98 [1961] 16 pp. from N.S.A. **15** [1961] No. 16009).

[5] Udy, M. C. (in: White, D. W.; Burke, J. E.; The Metal Beryllium, Vol. 2, ASM, Cleveland, Ohio, 1955, pp. 505/29, 507/10).

[6] Distler, W. B.; Wiesner, H. J. (U.S. 3556963 [1968/71] 2 pp. from C.A. **74** [1971] No. 134183).

[7] Jaquet, A. (J. Less-Common Metals **1** [1959] 439/55).

[8] Price, C. W. (Metallography **1** [1968] 5/18).

[9] Helms, J. R. (SAND-75-8052 [1975] 16 pp.; C.A. **85** [1976] No. 165300).

[10] Shehata, M. T.; Kelly, R. (J. Electrochem. Soc. **127** [1980] 579/84).

[11] Goodman, E. (Natl. SAMPE Symp. Exhib. Proc. **12** [1967] 9 pp. from C.A. **70** [1969] No. 83502).

[12] Gurklis, J. A. (MCIC-72-03 [1972] 84 pp.; C.A. **77** [1972] No. 38365).

[13] Miller, P. D.; Boyd, W. K. (AD-824446 [1967] 37 pp.; C.A. **70** [1969] No. 60126).

16.5.2 Beryllium Anodes in Nonaqueous Solvents

When an ethanol solution containing (50 g/L) $LiClO_4$ and (25 g/L) water is used as the electrolyte, a Be anode dissolves with an apparent valence (cf. p. 194) n_a of 1.00 to 1.13 depending on the applied potential (1 to 25 V; Pt cathode) and temperature, see figures in the papers [1, 2]. The minimum valence value of 1.00 is observed at ~6 and ~9 V for the electrolyte at 25 and 50°C, respectively. With only 0.1 g H_2O/L in the electrolyte at 60°C, the apparent valence values range from about 0.78 at low potentials (< about 1 V) through a broad maximum of 1.02 at about 8 V to a constant value of 1.00 at 13 to 25 V. For the voltage range with an apparent valence $n_a = 1$ for the Be dissolution, the metal surface appeared electropolished (SEM observations), while for $n_a > 1$ granular aspects were observed and for $n_a < 1$ the anode surface appeared rugged. In the case of electrolytes containing 25 g H_2O/L, the anode surface testified very heterogeneous Be dissolution. It is thought that in the case of the anhydrous solution the Be goes into solution as the univalent ion (cf. pp. 194/6). The formation of the transient Be^+ in the initial dissolution step is attributed to a deficiency of solvating molecules in the immediate neighbourhood of the anode, causing a local inversion of the relative thermodynamic stabilities of the Be^+ and Be^{2+} ions; Be^{2+} is then formed in the solution by a chemical oxidation. In the presence of water the mechanism is more complex. For surface investigations of the anode with a scanning electron microscope and for theoretical considerations in favor of the above explanation, see [1], cf. [2 to 5]; for discussions, see [6, 7], and for microscopic investigations [8]. Solutions of $LiClO_4$ in propylene carbonate (0.1 g H_2O/L, 60°C) behaved similarly to the

$LiClO_4$ solutions in C_2H_5OH, with the apparent valences somewhat lower (always <1) [5]. For some comparable results with $Mg(ClO_4)_2$ solutions in ethanol, also with varying amounts of water added, see [3, 4, 9]. Other investigations on anodic Be dissolution in nonaqueous electrolytes (e.g., 0.2 M LiCl in methanol, 0.2 M KI in dimethylformamide, 0.2 M $LiClO_4$ in ethanol, average water content at the end of a valence determination 0.30 to 0.68%) at $24 \pm 1°C$ gave apparent valence values between 0.34 and 1.75, which were attributed to anodic disintegration of the Be rather than to the occurence of monovalent Be. Also Be is passivated in nonaqueous media containing NO_3^-. In a 0.2 M $NaNO_3$ solution in methanol its surface appeared dark with shiny Be crystals on it. Anodic disintegration occurred in all these media, and cyclic voltammograms did not show peaks corresponding to two oxidation states of Be. The anions seemed to exert a greater influence on the anodic dissolution than the medium; for details, see [10]; a discussion of these results is reported in [11].

Further studies concerning the influence of the water content in the solutions were carried out with Cl^- and ClO_4^- solutions (LiCl, NaCl, and $XClO_4$, where X = Li, Na, K, NH_4) in methanol, ethanol, or N,N-dimethylformamide (DMF) at 25°C, with the water contents kept below 0.1 g/L. Apparent valences of the Be ions going into solution were 0.37 to 1.8 with the lower values obtained in the alcohol solutions with Cl^-. Considerable gas evolution from the Be surface was observed. Experimental details and results are tabulated in the paper [12]. The polarization curves for the solutions studied (see figures in the paper) do not indicate any passivation, but to some extent relate to the experimentally found apparent valences. There are no peaks or discontinuities on the cyclic voltammograms (see figures in the paper) for the anodic dissolution of Be in the methanol and ethanol solutions. A current peak is observed in the scan for 0.2 M $NaClO_4$ solution in DMF. The cyclic studies indicate the presence of surface films on the Be and large differences in the total dissolution rates in the different solvents. The low water contents seem to hinder the reforming of an oxide or hydroxide film which could protect the metal from attack. Beryllium will chemically react once a current disrupts the surface film. From the results (together with product analyses) the authors [12] inferred possible cell reactions in the ethanol electrolytes. The low experimental values for the apparent valences in the alcohol solutions were attributed to chemical dissolution of the Be rather than to the formation of Be^+ [12].

When Be (or Mg, Cd, Al) is anodized in pyridine solutions containing certain organic additives which are potential electron acceptors, the metal enters solution with a mean valence number significantly lower than its normal valence of 2. Pyridine solutions of 2 m LiCl with one of the following additives were tried (air excluded, temperature 40°C, current density 0.004 to 0.007 A/cm², applied voltage 130 to 190 V): nitrobenzene, benzaldehyde, benzophenone, bromobenzene, benzonitrile, ethyl benzoate, trans-stilbene, and 1-heptyne. The low valence numbers of 1.63 and 1.78 were obtained with nitrobenzene and benzaldehyde, respectively, and values near 2 with the other additives [13]; from the results, the metals could be arranged according to their activity as reducing agents when used as anodes in electrolyses carried out in pyridine solutions: Mg > Al > Be > Cd.

A solution in ethylene glycol saturated with both Na_2HPO_4 and Na_2SO_4 is used for anodizing Be (final potential 50 V, current density increasing from 3 to 10.35 mA/cm², anodizing time 1.5 min, temperature ~22°C) to produce amorphous BeO films formed with negligible dissolution in the electrolyte. Scanning electron micrographs suggest that the films are significantly free from porosity. Also saturated solutions of Na_2HPO_4 or KH_2PO_4 in ethylene glycol gave good films. Experiments to anodize Be in diethyl phosphate or in a mixture of diethyl sulfate and H_2SO_4 gave less promising results [14].

References:

[1] Aida, H.; Epelboin, I.; Gareau, M. (J. Electrochem. Soc. **118** [1971] 243/8).
[2] Garreau, M. (Compt. Rend. C **270** [1970] 16/9).
[3] Garreau, M. (Met. Corros. Ind. **45** [1970] 425/47, 439).
[4] Epelboin, I.; Froment, M. (Met. Corros. Ind. **32** [1957] 55/72, 67/8).
[5] Epelboin, I.; Froment, M.; Garreau, M. (Corros. Trait. Prot. Finition **18** [1970] 433/40).
[6] Aida, H.; Epelboin, I.; Gareau, M. (J. Electrochem. Soc. **118** [1971] 1961).
[7] James, W. J.; Straumanis, M. E. (J. Electrochem. Soc. **118** [1971] 1960/1).
[8] Epelboin, I.; Froment, M.; Garreau, M.; Aida, H. (J. Microsc. [Paris] **15** [1972] 313/22).
[9] Froment, M. (Corros. Anticorros. **7** [1959] 46/56, 98/109, 134/45; 108/9; C.A. **1959** 14771).
[10] Vaidyanathan, H.; Straumanis, M. E.; James, W. J. (J. Electrochem. Soc. **121** [1974] 7/12).

[11] Epelboin, I.; Froment, M.; Garreau, M. (J. Electrochem. Soc. **121** [1974] 1604/6).
[12] Johnson, J. W.; Chen, S. C.; Chang, J. S.; James, W. J. (Corros. Sci. **17** [1977] 813/31).
[13] Rausch, M. D.; McEwen, W. E.; Kleinberg, J. (J. Am. Chem. Soc. **77** [1955] 2093/6).
[14] Shehata, M. T.; Kelly, R. (J. Electrochem. Soc. **122** [1975] 1359/65).

16.5.3 Beryllium Anodes in Melts

Alkali Halide Melts

In chloride melts Be dissolves anodically in the monovalent as well as the divalent state; e.g., in the eutectic LiCl–KCl at 500°C about a third of the Be passes over into the electrolyte as univalent ions in the region of low current densities of the order of 10^{-3} A/cm² [1]; cf. [2, 3]. The dissolution and the polarization of a Be anode in eutectic LiCl–KCl was studied in detail at 400, 500, 600, and 800°C at $j = 10^{-3}$ to 5 A/cm² [1]. Below a current density of 10^{-2} A/cm² the potential of Be is close to its rest potential in the melt, especially at the lower temperatures (see diagram in the paper [1]). Above 10^{-2} A/cm² the Be potential begins to increase with the current density j, up to 10^{-1} A/cm² $\sim (RT/2F)\log j$. On a further increase of j the anode potential increases abruptly and approaches the equilibrium potential E of Be in a melt of its chloride ($E = -2.59 + 1.32 \times 10^{-3}$ T in V vs. Cl_2, according to calculations from data in the literature). Addition of fluoride to the melt shifts the Be electrode potential strongly to the negative direction (by 0.57 and 0.84 V for 25 and 50 mol% F^-, respectively). In mixed fluoride–chloride melts the range of potential change of a Be anode in the above given range of current densities becomes broader than in the chloride melts and reaches 1.2 to 1.4 V [1].

A molten mixture of the LiCl–KCl eutectic with 10 wt% $BeCl_2$ reacts with metallic Be forming univalent Be ions. The reaction is followed by measuring the change in the redox potential of the Be^+–Be^{2+} system in the chloride melt with time. Furthermore, when the solidified melt (after the experiments) is dissolved in water, hydrogen is liberated according to the reaction $Be^+ + H_2O = Be(OH)^+ + \frac{1}{2}H_2$. In experiments with a high-purity Be anode (cast material, 0.11% metallic impurities, including 0.05% Al, 0.03% Fe, 0.01% Mn) and an Mo cathode in an LiCl–KCl eutectic melt at temperatures from 351 to 600°C and a current density of 10^{-2} A/cm², the anodic current efficiency for the Be dissolution was determined (the rate of corrosion, i.e., spontaneous dissolution of Be in the melt being subtracted). The anodization was carried out until about 1 wt% $BeCl_2$ was in the melt. Under these conditions between

0.1736 and 0.2169 g Be/Ah were dissolved anodically from the Be electrode in the above temperature range (see table in the paper [2]), compared to 0.1682 g/Ah if Be^{2+} alone would have been formed. From these results the equilibrium constant for the reaction Be^{2+} (melt) + Be = 2 Be^+ (melt) was evaluated for the above temperature range (see also the table in the paper). It was concluded that under industrial conditions for the electrolysis of a molten mixture of 50% $BeCl_2$ and 50% NaCl at 350°C less than 0.5% of the total Be in the electrolyte would be in the monovalent state. In mixed fluoride–chloride melts (about 12 wt% KF were added to the eutectic LiCl–KCl) the fluoride ions were found to accelerate the spontaneous dissolution of the Be and to displace the above equilibrium to the left (monovalent Be could not be detected under the experimental conditions). This is attributed to the formation of BeF_4^{2-} ions [2]; cf. [3].

Lebedev and coworkers [4] studied the polarization of a Be anode (and a Be cathode, cf. p. 190) in a melt of NaCl containing 43 mol% $BeCl_2$ in the temperature range 300 to 600°C, determining simultaneously the temperature in the near-electrode layer of the electrolyte; for details and discussion, see [4].

Other Melts

According to a patent specification Be was anodized at 450 to 550°C in a molten eutectic mixture of $Na_2Cr_2O_7$ and $K_2Cr_2O_7$ with a current density of 30 to 60 A/dm² in order to increase its corrosion resistance in the atmosphere and at high temperatures [5]. The anodic coatings obtained on Be in Na and K dichromate melts consist, according to electron diffraction and X-ray studies, of crystalline Be oxide. The growth of the oxide crystals was observed to be fast and uniform on top of the fine-crystalline oxide film formed during the first stage of the anodization [6].

In a melt of $KF \cdot 2 Be(C_2H_5)_2$ at about 80°C a Be anode is dissolved (Be and Be_2C being simultaneously deposited at the cathode) and $Be(C_2H_5)_2$ forms. The anode mechanism consists of the discharge of the ion $Be(C_2H_5)_2F^-$, the disproportionation $[Be(C_2H_5)_2F] \rightarrow FBe(C_2H_5) + \cdot C_2H_5$, and the quantitative reaction of the radical $\cdot C_2H_5$ with Be [7].

References:

[1] Smirnov, M. V.; Chukreev, N. Ya. (Zh. Fiz. Khim. **32** [1958] 2165/72; C.A. **1959** 6833).
[2] Smirnov, M. V.; Chukreev, N. Ya. (Zh. Neorg. Khim. **4** [1959] 2536/44; Russ. J. Inorg. Chem. **4** [1959] 1168/72).
[3] Smirnov, M. V.; Chukreev, N. Ya.; Yushina, L. D. (Tr. Inst. Khim. Akad. Nauk SSSR Ural'sk. Filial **1958** No. 2, pp. 171/6 from C.A. **1960** 9555).
[4] Lebedev, V. A.; Pyatkov, V. I.; Titov, G. N.; Agapitov, V. A.; Klimovskikh, N. M. (Izv. Vyssh. Uchebn. Zaved. Tsvetn. Metall. **1976** No. 4, pp. 82/6; Soviet Non-Ferrous Metals Res. **1976** 159/61).
[5] Konchus, A. D.; Maslova, L. N.; Syzranova, Zh. V.; Mamontova, T. E. (U.S.S.R. 356426 [1970/72] from C.A. **78** [1973] No. 127790).
[6] Smirnova, A. I.; Kancheev, O. D.; Politko, S. K.; Konchus, A. D.; Medvedev, A. A. (Mater. 9th Vses. Konf. Elektron. Mikrosk., Tiflis 1973, pp. 297/8; C.A. **85** [1976] No. 150806).
[7] Strohmeier, W.; Popp, G. (Z. Naturforsch. **23b** [1968] 38/41).

16.6 Polarographic and Voltammetric Characteristics

16.6.1 Survey

Quantitative polarographic determination of beryllium is possible [1]; however, difficulties are evidenced by the divergence of the results and interpretations [1, 2]. The electrolysis of beryllium salts in aqueous solutions is complicated by strong hydrolysis; the electrochemical reduction of Be^{2+} can only take place in acidic solutions ($pH \leq 4.14$ [3]), as $Be(OH)_2$ starts to form at $pH \approx 5$ and precipitates at $pH = 6$ [4]. Beryllium polarography is further complicated by the reduction of H^+ ions at -1.1 to -1.4 V vs. NCE and H_2 evolution in acidic aqueous solutions [5 to 8]. The deposition of Al at -1.6 to -1.7 V vs. NCE may mask the deposition of Be, because Al is often contained in Be compounds as an impurity [5, 9]. Hence a strict control of the pH value of the solutions containing Be^{2+} is necessary for polarographic studies [3, 10]. Results from Al and Th polarography indicated that Be polarographic waves may result from the reduction of H^+ from the Be aquo complexes rather than from the reduction of the Be^{2+} ion [11 to 13].

The polarographic studies of aqueous Be^{2+} solutions were carried out with solutions of LiCl, KCl, $MgCl_2$, tetramethyl or tetraethyl iodides as supporting electrolytes, with or without additions of, e.g., dioxane, oxalate, or other complexing agents. In alkaline aqueous media Be(II) complexes with a number of organic complexing agents give rise to polarographic waves, which can be used for the polarographic determination of beryllium.

Polarographic and voltammetric studies in nonaqueous solutions of Be salts (with solvents such as, e.g., alcohols, diethyl ether, propylene carbonate, acetonitrile) were carried out with a variety of supporting electrolytes, such as LiCl, $LiClO_4$, and tetraalkyl-ammonium halides. Alkali chloride melts were used mainly for studies with molten electrolytes.

Unless otherwise stated, a dropping mercury electrode (DME) was used in the following descriptions.

References:

[1] Panzer, R. E. (in: Bard, A. J.; Parsons, R.; Jordan, J.; Standard Potentials in Aqueous Solutions, New York – Basel 1985, pp. 675/86, 677, 686).

[2] Novoselova, A. V.; Batsonova, L. R. (Analytical Chemistry of Beryllium, Jerusalem 1968, pp. 78/9, 201/2).

[3] Venkataratnam, G.; Raghava Rao, B. S. V. (J. Sci. Ind. Res. [India] B **17** [1958] 360/2; Anal. Abstr. **6** [1959] No. 2875).

[4] Shirvington, P. J.; Florence, T. M.; Harle, A. J. (Austral. J. Chem. **17** [1964] 1072/84).

[5] Heyrovsky, J.; Berezicky, S. (Collect. Czech. Chem. Commun. **1** [1929] 19/46, 35).

[6] Fornasari, E.; Gagliardo, E. (Atti Ist. Veneto Sci. Lettere Arti Classe Sci. Mat. Nat. [2] **115** [1957] 217/32; C.A. **1958** 12652).

[7] Kryukova, T. A.; Sinyakova, S. I.; Aref'eva, T. V. (Polarographische Analyse, Leipzig 1964, pp. 1/711, 214, 239).

[8] Kolthoff, I. M.; Coetzee, J. F. (J. Am. Chem. Soc. **79** [1957] 1852/8).

[9] Heyrovsky, J. (Acta Chim. Acad. Sci. Hung. **9** [1956] 3/16, 73/91).

[10] Everest, D. (The Chemistry of Beryllium, Amsterdam – London – New York 1964, pp. 114/7, 130/3).

[11] Stradyn, Ya. P.; Lepin, L. K. (Zh. Fiz. Khim. **32** [1958] 196/200).

[12] Mašek, J. (Zeitschrift für Physikalische Chemie, Sonderheft, Leipzig 1958, pp. 108/18; C.A. **1959** 5919).

[13] Mašek, J. (Collect. Czech. Chem. Commun. **24** [1959] 159/69).

16.6.2 Acid Aqueous Solutions

References for this section are given on p. 217.

Solutions of Beryllium Salts

Kemula and Michalski [1] electrolyzed aqueous 0.1 and 0.01N $BeCl_2$ and 0.005, 0.01, and 0.1N $BeSO_4$ solutions in the usual polarographic apparatus (under H_2 atmosphere). The 0.005N $BeSO_4$ solutions contained either 0.03N CH_3COONa and 0.005N KCl or 0.05N CH_3COOK to suppress the concentration of H^+. Two waves were obtained: first, a wave at −1.64, −1.61, and −1.41 V vs. NCE for 0.005, 0.01, and 0.1N solutions of Be^{2+}, respectively (sensitivity of the galvanometer: 2.7×10^{-8} A/mm), attributed to the reduction of H^+ ions; and second, a wave at −1.83, −1.82, and −1.79 V vs. NCE for 0.005, 0.01, and 0.1N solutions of Be^{2+}, respectively (sensitivity of the galvanometer: 1.1×10^{-6} A/mm). The latter wave was attributed to the reduction of the Be^{2+} ion; this was confirmed by the agreement of the experimental and the calculated shifts of the potential due to changes in the Be^{2+} ion concentration. The Be wave became indistinct for higher concentrated Be^{2+} solutions and disappeared completely in alkaline solution. A value of −1.70 V vs. NCE was found for the so-called normal deposition potential of Be by extrapolation and correction to a galvanometer sensitivity of 10^{-8} A/mm, in the absence of an H_2 interference. Polarographic estimation of Be was impossible in the presence of Al salts, even in the presence of complexing salts, because of the proximity of the deposition potentials (Al: −1.66 V vs. NCE). The presence of ions of nobler metals did not interfere with the deposition of Be [1]; cf. [2 to 4].

Solutions of Beryllium Salts with LiCl as Supporting Electrolyte

A single reduction wave was obtained by Ralea and Saviuc [5] for solutions of Be^{2+} (no concentration range or pH value is given in the paper) in 0.1M LiCl at $E_{1/2} = -1.3$ V vs. SCE, using an H-type Kalousek cell. The wave height was not found to be proportional to the Be^{2+} ion concentration. Two polarographic waves were observed for 2.5×10^{-4} to 1×10^{-3}M Be^{2+} solutions in 0.1M LiCl (or Li_2SO_4), pH 3.5 to 3.8, in the presence of 0.006% agar-agar. The first wave at −1.4 V vs. SCE is attributed to the discharge of hydrogen, the second, well-developed wave at $E_{1/2} = -1.85$ V to the reduction of the Be^{2+} ion [6]. Two well-defined Be waves were obtained for 1.25 to 12.50 mM $BeCl_2$ solutions in 0.5M LiCl: e.g., for 5×10^{-3}M $BeCl_2$ at pH 2.38, $E^I_{1/2} = -1.79$ V vs. SCE and $E^{II}_{1/2} = -2.00$ V. The pH was adjusted carefully by means of dilute HCl and LiOH solutions; pure N_2 was bubbled through the solution prior to measurements. The waves were poorly defined in 0.1M LiCl, but a "hydrogen-wave" always preceded the Be waves. For the dependence of the half-wave potentials and the limiting diffusion currents of the two Be waves on the pH (range 2.38 to 4.14), and on the $BeCl_2$ concentration (range 1.25×10^{-3} to 12.50×10^{-3}M) at pH 2.5 and 3.33, see the tables in the paper [7]; the half-wave potentials are shifted to more negative values with decreasing pH. The diffusion current of the first reduction wave is constant for a given $BeCl_2$ concentration in the pH range examined (2.38 to 4.14) and is strictly proportional to the Be^{2+} ion concentration up to concentrations of 8×10^{-3}M for pH values below 3.3. This wave can therefore be used for the polarographic estimation of Be^{2+} in the above concentration and pH ranges. The diffusion current of the second Be wave is independent of the pH for values below 2.66 and above 3.33, and is proportional to the Be^{2+} ion concentration up to 7.5×10^{-3}M Be^{2+} for pH values below 2.66. In the pH range 2.66 to 3.33, the diffusion current increases at the steady rate of 1.1 µA/0.1 unit of pH. A diffusion current constant of 1.66 was obtained for the normal diffusion wave for both reductions; the value calculated from capillary characteristics and conductance data was 1.467. The authors attributed the two waves to the reduction steps $Be^{2+} \xrightarrow{+e^-} Be^+ \xrightarrow{+e^-} Be^0$. The reciprocal slopes of the log $I/(I_d - I)$ vs. E plots (where I = current at potential E,

I_d = limiting current) were 0.052 and 0.050 for the first and second reductions, respectively, and supported the suggested two-stage reduction of the Be^{2+} ion (theoretical value for a one-electron reduction: 0.059). For further details, see the paper [7].

Fig. 16-15 from [8] shows the polarogram for Be^{2+} in 4×10^{-4} M $BeCl_2$ in 0.1M LiCl solution, obtained with a Kalousek cell with a separated SCE (oxygen was removed from the cell by passage of N_2). The surface of the dropping mercury electrode was observed microscopically. The shape of the polarographic curve is explained (by analogy to results for Al) by a stepwise reduction of hydrogen ions from Be aquo complexes at the surface of the electrode. The first maximum is associated with the direct reduction of Be^{2+} ions from the partly hydrolyzed complex to the metallic state. This is accompanied by the liberation of OH^- ions and hence by the precipitation of $Be(OH)_2$, which covers the mercury drop. The second maximum may be due to a catalytic effect of Be ions occluded in the hydroxide layer. The complicated reaction was studied in detail (including current-time curves) for solutions of Al^{3+} [8, 9].

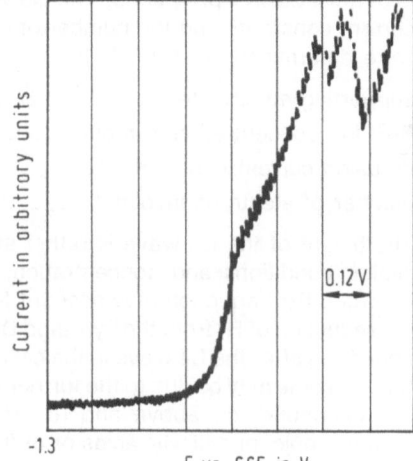

Fig. 16-15. Polarographic curve for Be^{2+} in 4×10^{-4} M $BeCl_2$ in 0.1M LiCl solution (sensitivity 1/20).

Györbiró [10, 11] also attributed the two waves on the polarograms of Be^{2+} ion solutions, see **Fig. 16-16**, to the reduction of H^+ from the H_3O^+ ion and from a Be aquo complex. He used pure $BeSO_4$ and $Be(ClO_4)_2$ dissolved in 0.1M and 0.2M solutions of LiCl or $N(CH_3)_4I$ or in a

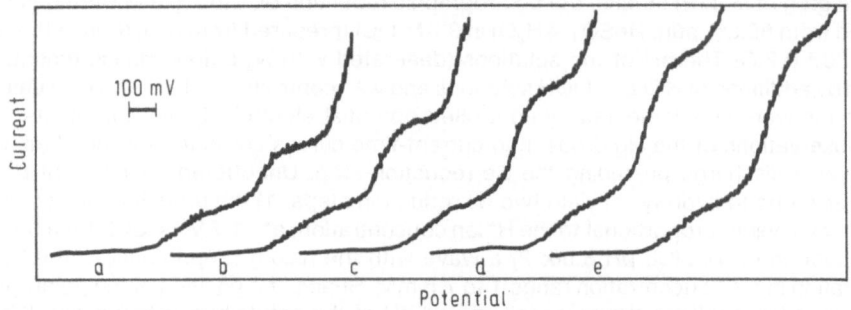

Fig. 16-16. Polarograms of solutions of Be^{2+} in 0.1M LiCl (sensitivity 1/1000) with Be^{2+} ion concentrations: a = 0.00182M, b = 0.00246M, c = 0.00334M, d = 0.00413M, and e = 0.00461M.

0.02M solution of $MgCl_2$. The pH of the solution was adjusted by means of dilute HCl, $[N(CH_3)_4]OH$, or LiOH solution. The polarograms show two waves for pH values of 3.5 to 4.4 (the "natural" pH value of the Be salt solution at the concentration concerned should not be exceeded), e.g., at −1.62 and −1.85 V vs. NCE for Be^{2+} ion solutions in 0.1M LiCl and pH 4.5. The first wave is attributed to the reduction of H^+ from H_3O^+, as its height is a function of the H^+ ion concentration; it disappears at higher pH values. The height of the second wave is proportional to the Be^{2+} ion concentration and can be observed even for Be^{2+} concentrations as low as 4×10^{-4} M (in 0.1M LiCl or $N(CH_3)_4I$). This wave is suitable for quantitative polarographic determination of Be in the concentration range 0.8×10^{-3} to 6×10^{-3}M. At low Be^{2+} concentrations (< 0.001M), the Be wave is improved by the presence of 25% p-dioxane. At Be^{2+} concentrations $> 5 \times 10^{-3}$M, the "H^+ wave" splits, especially in the presence of dioxane; 0.01% gelatin prevents the splitting of the "H^+ wave", but blurs the "Be wave". With decreasing pH the $E_{1/2}$ value of the first wave is shifted somewhat to more positive values, while that of the second wave becomes more negative by ~ 0.1 V per pH unit; thus the separation of the two waves is better at lower pH values. The Be wave is shown to be diffusion-controlled; the diffusion current constants and the number of electrons involved in the reaction (at −1.95 V vs. SCE) were determined at $25 \pm 0.1°C$:

supporting electrolyte	0.1M LiCl	0.2M $N(CH_3)_4I$	0.1M $N(CH_3)_4I$
Be^{2+} ion concentration in mol/L	1.82×10^{-3}	4.00×10^{-3}	1.00×10^{-3}
diffusion current constant	2.36	2.47	2.34
number of electrons involved	1.95	2.04	1.94

The height of the Be wave is rather small compared to the waves of other bivalent ions at similar conditions and concentrations, but the lower diffusion coefficient of the strongly hydrated Be^{2+} aquo ion may account for this. $[Be(H_2O)_4]^{2+} + 2e^- \rightarrow Be(OH)_2 + 2H_2O + 2H$, i.e., the reduction of H^+ from the hydrated Be^{2+} ion is considered the probable process. A maximum appearing after the Be wave in the case of a 5×10^{-4}M Be^{2+} solution in 0.015M LiCl containing 25% dioxane may be due to the further reduction of the $Be(OH)_2$ formed, or of some other basic Be compound, cf. above and [8, 9]. The addition of 0.04 to 0.1M solutions of oxalic, chromotropic, or salicylic acids or of fluorides does not shift the Be wave. As the Be wave is considered to correspond to the H^+ reduction of a weak acid ($Be(H_2O)_4^{2+}$), buffer systems of weak acids can not be used in the polarographic determination of Be. The Al^{3+} and Ba^{2+} can not be separated from Be^{2+} polarographically, as they have similar half-wave potentials. Na^+ and K^+ do not interfere unless their concentrations exceed that of Be^{2+} by one or more orders of magnitude. More details are given in [10, 11].

Shirvington et al. [12] studied the polarographic reduction of 1 to 7.5 mM $BeSO_4$ solutions (prepared from 99.5% pure $BeSO_4 \cdot 4H_2O$) in 0.5M LiCl (prepared from anhydrous 98.5% pure LiCl) at $30 \pm 0.2°C$. The pH of the solutions (deaerated with N_2 before measurements) was adjusted by additions of HCl and LiOH solutions and was controlled within ± 0.01 pH units. The polarograms were supplemented by controlled potential electrolysis measurements, microscopic observations of the Hg drops, and current-time curves corrected for the blank due to the hydrogen discharge preceding the Be reduction step. Unbuffered solutions of Be salts (slightly acid due to hydrolysis) yield two dc reduction steps: 1) the reduction of free protons with the wave height proportional to the H^+ ion concentration, at −1.7 V vs. SCE for a 5×10^{-3}M Be^{2+} solution in 0.5M LiCl, pH 3.90; 2) a wave with the height proportional to the Be^{2+} ion concentration in the concentration range 1 to 7.5 mM. Shape and position of the polarographic waves depend strongly on the composition and pH of the solution; see figure 1 in [12]. The Be^{2+} reduction step can only be studied in a narrow pH range, viz. 2.5 to 5.0, as the H^+ reduction interferes at pH < 2.5, and $Be(OH)_2$ formation and precipitation at pH 5 to 6. There is no plateau between the Be^{2+} reduction step and the decomposition of the supporting electrolyte, cf. [7, 10], but small and erratic drop times in this potential region prevented a detailed study. Size and shape of the Be step do not change for $2.5 \leq pH \leq 3.9$. It consists of two sections for which logarithmic analysis yields two straight lines (slopes 94 and 56 mV/decade) intersecting at the half-wave potential, $E_{1/2}$, thus indicating two steps of approximately equal height. For $3.9 \leq pH \leq 5.0$ the shape of the wave changes considerably. **Fig. 16-17** shows that with increasing pH the limiting current decreases and $E_{1/2}$ becomes more positive, corresponding to an increase in complex formation between Be^{2+} ions and OH^- ions; see the diagram in the paper [12], which gives the concentrations of the various Be species present in solutions of pH 2.5 to 5.2. $[Be(H_2O)_4]^{2+}$ and $[Be_3(OH)_3(H_2O)_6]^{3+}$ are considered the main two Be species involved in aqueous solution. Diffusion control of the limiting current of the Be wave was ascertained by its dependence on the mercury head. Irreversibility of the electrode reaction is indicated by the temperature coefficients for the diffusion current and $E_{1/2}$, viz. $+0.8 \pm 0.2\%$/deg and $+2.7 \pm 0.3$ mV/deg, respectively, and by the small wave obtained in ac and streaming mercury electrode polarography. An insoluble product $(Be(OH)_2)$ is formed and H_2 is evolved at Hg and Pt macroelectrodes during exhaustive electrolysis at -2.02 V vs. SCE of 5×10^{-3} M Be^{2+} in 0.5M LiCl solutions of pH 3.9; no Be was found in the Hg cathode.

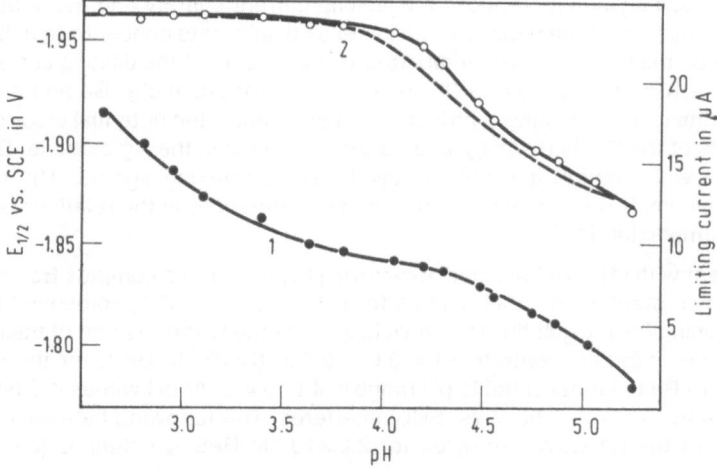

Fig. 16-17. The effect of pH on the characteristics of the Be wave. Curve 1: half-wave potential (left scale); curve 2: limiting current (right scale); o experimental values, --- theoretical values (predicted).

Conductivity measurements yielded a diffusion coefficient of $(6.2 \pm 0.3) \times 10^{-6}$ cm²/s at 30°C, a diffusion current constant of 3.1 ± 0.2, and a value of 2.0 ± 0.1 for n, the number of electrons involved in the electrode reaction, for pH values between 2.5 and 3.9; n decreases for pH >3.9. The presence of 0.002% gelatin splits the Be wave into two equal steps ($E_{1/2}$ values -1.757 and -1.89 V vs. SCE at pH 2.8) with the total height about half the original limiting current; the addition of 0.001% polyacrylamide shows hardly any effect. The authors [12] considered the experimental facts to be inconsistent with a mechanism such as $Be^{2+} + 2e^- \rightleftharpoons Be^0$, $Be^0 + 2 H_2O \rightleftharpoons Be(OH)_2 + H_2$, but suggested the following mechanism, involving the reduction of protons from the coordinated H_2O molecules in the hydrated Be^{2+} ion and the formation of a soluble intermediate: first, $3[Be(H_2O)_4]^{2+} \rightleftharpoons [Be_3(OH)_3(H_2O)_6]^{3+} + 3 H^+ + 3 H_2O$, $3H^+ + 3e^- \rightleftharpoons 1.5 H_2$; and second, $[Be_3(OH)_3(H_2O)_6]^{3+} \rightleftharpoons 3 Be(OH)_2 + 3 H_2O + 3 H^+$, $3 H^+ + 3e^- \rightleftharpoons 1.5 H_2$;

the first step is considered to be rate determining. Further details are given in the paper [12]. The increasing dependence of $E_{1/2}$ on the pH for values >3.9, see Fig. 16-17, is attributed to the decreasing concentration of $Be[(H_2O)_4]^{2+}$, causing the current-potential characteristics to be controlled more by the more reversible second step. The effect of complexing agents (at concentrations up to twice that of Be^{2+}) on the Be^{2+} reduction step is complicated, the most satisfactory pH range for their use being 3.1 to 3.9. The effect of oxalate (Ox) additions (2.5 to 10.0 mM, pH 3.1, in the presence of 0.001% polyacrylamide to ensure a more clearly defined base line) is shown in figure 5 in the paper [12]. With increasing oxalate concentration the diffusion current decreases, the polarogram becomes more negative, and the two sections of the Be reduction step become less distinct. Surface films are observed on the Hg drops. Logarithmic analysis of the reduction step yields a straight line with a slope of 63 mV/decade; the temperature coefficient of $E_{1/2}$ for a $5 \times 10^{-3}M$ Be^{2+}–$5 \times 10^{-3}M$ oxalate solution is $+2.0 \pm 0.2$ mV/K; $E_{1/2}$ becomes more positive with increasing pH. The results indicate that there is no radical change in the reaction mechanism postulated for uncomplexed Be, i.e., reduction of protons from coordinated H_2O molecules in the Be^{2+} complex. The drop in the diffusion current is attributed to the increased formation of $[Be(Ox)_2]^{2-}$ and to differences in the diffusion coefficients of $[Be(Ox)(H_2O)_2]^0$ and $[Be(H_2O)_4]^{2+}$. The addition of salicylate (3.75 to 7.5 mM, pH 2.9, 0.0005% polyacrylamide) reduces the limiting current sharply and the polarograms (see figure 6 in [12]) become more negative with increasing salicylate concentration, but there is no film formation on the Hg drops. The temperature coefficients of the limiting current and of $E_{1/2}$ are $+2.0 \pm 0.2\%/K$ and $+4.0 \pm 0.4$ mV/K, respectively, for equimolar Be and salicylate solutions, $E_{1/2}$ becoming more positive with increasing pH. Controlled potential electrolysis precipitated only 40% of the Be as $Be(OH)_2$, and no Be was found in the Hg cathode. Similar results were obtained with fluoride and sulfosalicylate as complexing agents. The differences in behavior of the complexing agents are attributed to differences in the relative stabilities of the reduction intermediates [12].

In agreement with [12], Rekalič and Jovanovič [13] reported a complex Be reduction wave consisting of two steps with equal heights for pH >3.8 for $BeSO_4$ solutions in a 0.2M LiCl solution containing 0.01% gelatin. The wave heights of the two parts and of their sum depend linearly on the Be^{2+} ion concentration for 0.6×10^{-3} to $6 \times 10^{-3}M$ Be^{2+}; for the polarographic determination of Be the most suitable pH range is 4.0 to 4.5. At pH values <3.8 the wave due to the H^+ discharge ($E_{1/2} = -1.53$ V vs. SCE) interferes. The following half-wave potentials of the two parts of the Be wave are given for $2.0 \times 10^{-3}M$ $BeSO_4$ solutions (current 10 μA):

pH of solution	2.6	2.8	3.0	3.5	3.8	4.0	4.2
$E_{1/2}^{I}$ in V vs. SCE	−1.81	−1.75	−1.75	−1.62	−1.63	−1.63	−1.62
$E_{1/2}^{II}$ in V vs. SCE	−1.95	−1.94	−1.92	−1.85	−1.85	−1.84	−1.83

The polarographic behavior of Be^{2+} ion solutions in 0.2M LiCl in the presence of oxalyl dihydrazide is similar to that in 0.2M KCl and $(C_2H_5)_4NI$ solutions, see p. 214, and p. 216, respectively. There are two waves; the height of the first one with $E_{1/2} = -1.46$ to -1.49 V vs. SCE (~ 0.25 to 0.29 V more positive than the second one) depends linearly on the Be^{2+} ion concentration for 0.6×10^{-3} to $4.0 \times 10^{-3}M$ Be^{2+} in the pH range 4.0 to 4.5. The linear dependence of the heights of the Be waves on the square root of the height of the Hg column showed the waves to be diffusion-controlled; for further details, see [13].

For the polarographic behavior of Be^{2+} in 0.2×10^{-3} to $1.4 \times 10^{-3}M$ $BeSO_4$ solutions in 0.2M LiCl (containing 0.01% gelatin) in the presence of oxamide and rubeanic acid, see pp. 214/5 and [14].

Medyntsev and Dyatlova [15, 16] studied the polarographic behavior of Be in 10^{-4} to $10^{-2}M$ $Be(NO_3)_2$ (using $Be(NO_3)_2 \cdot 4H_2O$ of AR grade) in 0.1M LiCl solution in order to clarify the reduction mechanism taking place at the DME and to develop a method for the polarographic determination of Be (details for the determination of beryllium in 0.1M LiCl solution of pH 3.5 are given in [15]). The pH values of the solutions ranged from 2.68 to the "pH of hydrolysis", pH_{hydr}, i.e., the "natural" pH of the Be salt solution; the authors determined pH_{hydr} to be 2.29 − 0.696 log c_{Be}, where c_{Be} is the Be concentration in mol/L. The pH was adjusted with an accuracy of ±0.02 pH units with the help of HCl. The temperature was maintained at 25.0±0.1°C, except for the measurements of the temperature dependences of the limiting current and the half-wave potential of the second wave (see below). Experimental details are given in the paper [15]. There are two waves in the polarograms of solutions having a pH < pH_{hydr}; the first, i.e., the more positive, wave disappears at pH values near pH_{hydr}, while both waves are absent at pH > pH_{hydr}. The first wave ($E^I_{1/2}$ in the range −1.5 to −1.6 V vs. SCE, estimated from a diagram in [15], no values are given in the papers [15, 16]) corresponds to the discharge of protons either formed by hydrolysis of the Be aquo ion according to $Be(H_2O)_4^{2+} + H_2O \rightleftharpoons Be(OH)(H_2O)_3^+ + H_3O^+$, etc., or present on account of the acidification of the solution with HCl. The second wave ($E^{II}_{1/2}$ in the range −1.7 to −2.0 V vs. SCE, from diagrams in [15]) is attributed to the depolarizing effect of the Be aquo complex. The diffusion character and the irreversibility of this wave were established in the usual way. In the temperature range 15 to 50°C, the temperature coefficient of $E^{II}_{1/2}$ was +3.50 mV/K, and the following temperature coefficients of the limiting current (in %/K) were obtained for the given pH values and Be^{2+} concentrations, c:

pH	c_{Be} in mol/L					
	1.3×10^{-3}	2.6×10^{-3}	3.9×10^{-3}	5.2×10^{-3}	7.8×10^{-3}	10.2×10^{-3}
2.69	4.45	2.07	1.63	1.43	1.31	1.15
2.90	3.87	1.77	1.70	1.44	1.22	1.07
3.39	2.54	1.79	1.67	1.44	—	—

The limiting current depended linearly on the Be^{2+} ion concentration up to 7.8×10^{-3} M; it increased slightly with increasing pH ≤ pH_{hydr}. The Ilkovič equation applies. The number of electrons involved in the electrode process, n=2.03, and the diffusion current constant of the value 2.47 (theorectical value 2.43) were obtained using a diffusion coefficient of 0.4×10^{-5} cm²/s. For solutions with pH > pH_{hydr} the limiting current drops sharply with increasing pH, see **Fig. 16-18**, curve 6, the wave disappearing at pH 4.8 to 5.0. This corresponds to Be^{2+} concentrations of 10^{-4} M and represents the limits of measurements; the polarograms at lower Be^{2+} ion concentrations or with pH ≥ pH_{hydr} were poorly reproducible. $E^{II}_{1/2}$ was found to shift in

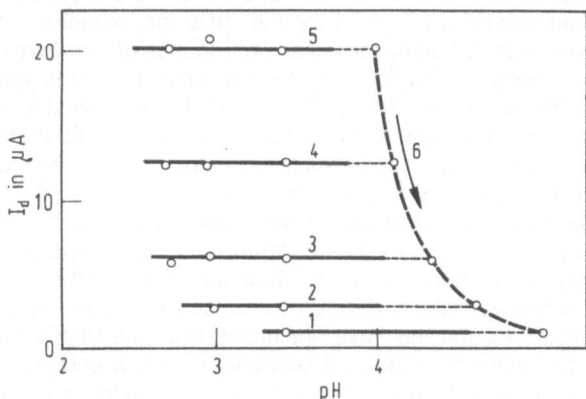

Fig. 16-18. Diffusion current, I_d, of the second wave as a function of pH of the solution for the following Be salt concentrations: $1 = 1.3 \times 10^{-4}$ M, $2 = 5.2 \times 10^{-4}$ M, $3 = 1.3 \times 10^{-3}$ M, $4 = 2.6 \times 10^{-3}$ M, $5 = 3.9 \times 10^{-3}$ M (curve 6 = limiting current).

the positive direction by 58 mV for an increase in the Be^{2+} ion concentration by a factor of 10 at constant pH and by 180 mV/pH unit with increasing pH at constant Be concentration, provided $pH < pH_{hydr}$. At $pH = pH_{hydr}$, $E^{II}_{1/2}$ shifts in the positive direction by 110 mV/pH unit with increasing pH_{hydr}. These results together with detailed theoretical considerations support a mechanism involving the discharge of Be ions with the formation of an amalgam rather than a mechanism involving the discharge of protons from a Be aquo complex. Considering the hydration and hydrolysis of Be ions in aqueous solution, the authors [15, 16] propose that the dissociation of hydroxy and aquo complexes of Be (the latter dissociate more slowly) precedes the discharge of Be ions with the formation of Be amalgam, which is decomposed by water yielding $Be(OH)_2$ and setting free H_2: $Be(OH)_i(H_2O)_j \rightleftharpoons [Be]^* + i(OH) + jH_2O$ (I), $[Be]^* + 2e^- \rightarrow Be(Hg)$ (II), $Be(Hg) + 2H_2O \rightarrow Be(OH)_2 + H_2 + (Hg)$ (III), where steps I and II are considered rate-determining. The authors derived theoretical expressions for the mean limiting currents for the two mechanisms and found that the experimental results agreed much better with the "amalgam" mechanism. Similarly, a shift of $E^{II}_{1/2}$ to more negative values with increasing height of the Hg column supports this mechanism; for further details, see [15, 16]. The addition of complexing agents, such as tartrate, nitrilo triacetic acid, ethylene diamine tetraacetic acid, pyrophosphate, and ethylene diamine bisalkylphosphonic acid (i.e., weak acids), causes a drop in pH; the Be^{2+} in the solution is present mainly as the hydrated form, and the reduction wave of free (i.e., uncomplexed) Be is shifted to the negative side. In addition, the wave due to the discharge of protons interferes and waves due to complexed Be can not be observed. The authors attribute the waves (assigned to the reduction of Be complexes in [12]) to the reduction of free Be, see [16].

Solutions of Beryllium Salts with Other Supporting Electrolytes

The polarographic behavior of Be^{2+} ions in aqueous **KCl** solutions in the absence or presence of oxalyl dihydrazide (ODH) was found to be similar to that of Al^{3+} ions [13]. A double wave is found at $E^{I}_{1/2} = -1.66$ and $E^{II}_{1/2} = -1.86$ V vs. SCE for 1×10^{-3} M $BeSO_4$ in 0.2 M KCl solution containing 0.01% gelatin. Polarographic determination of Be is possible for concentrations of 0.6×10^{-3} to 6×10^{-3} M in the pH range 4.0 to 4.5 (cf. p. 212). Oxygen in the apparatus was removed by the passage of nitrogen for 5 min before measurements; the temperature was 25°C, and the pH was adjusted with HCl or NaOH solutions. The polarograms of $BeSO_4$ in 0.2 M KCl – 4×10^{-3} M ODH solutions containing 0.01% gelatin show the reduction wave due to ODH ($E_{1/2} = -1.72$ V vs. SCE for a 4×10^{-3} M solution of ODH in 0.2 M KCl, pH range 4 to 9) to split into a double wave. The height of the more positive wave ($E_{1/2} = -1.51$ V vs. SCE) depends linearly on the Be^{2+} concentration for 0.6×10^{-3} to 6.0×10^{-3} M Be^{2+} and 4×10^{-3} M ODH in the pH range 4.0 to 4.5 (in 0.2 M KCl solution with 0.01% gelatin) and can be used for the polarographic determination of Be under these conditions. A wave due to the discharge of H^+ ions appears at $E_{1/2} = -1.29$ V vs. SCE for solutions with pH < 4. Diffusion control of all waves was established by the linear dependence of the wave heights on the square root of the height of the Hg column, see [13]. A behavior similar to the above was found for solutions of $BeSO_4$ in 0.2 M solutions of KCl, LiCl, or $(C_2H_5)_4NI$ containing oxamide (in the presence of 0.01% gelatin). For 0.2×10^{-3} to 1.4×10^{-3} M Be^{2+} in 0.92×10^{-3} M oxamide – 0.2 M KCl solutions, pH range 4.4 to 5.3, the oxamide wave splits (oxamide solutions show only one wave at $pH \gtrsim 4.4$). The height of the more positive wave at $E_{1/2} = -1.56 \pm 0.02$ V vs. SCE depends linearly on the Be^{2+} ion concentration (diffusion control was ascertained experimentally). The wave is attributed to the formation of a beryllium–oxamide compound (Be–Ox), for which a ratio $Be:Ox \approx 2:1$ was estimated from the experimental results. If the concentration of Be^{2+} ions was in excess of the above ratio, a clearly pronounced wave due to the reduction of free Be^{2+} could not be observed and the wave originating from the Be–Ox compound was flat. Beryllium solutions containing rubeanic acid also give rise to a polarographic wave due to a Be–rubeanic acid compound at a potential more positive than the reduction wave of rubeanic acid, but the wave is poorly defined in the pH range 3.5 to 5.0 and can not be used for the polarographic determination of Be; for details, see the paper [14].

According to Rekalič and Jovanovič [17], the hydrolyzed Be^{2+} ions act as proton donors for the electrochemical reduction of the organic depolarizers oxamide (Ox), oxalyl dihydrazide (ODH), benzil (Bz), benzoin (Bzn), rubeanic acid, and azobenzene in unbuffered aqueous solutions, pH > 4, with 0.2M KCl as supporting electrolyte. If the depolarizer is in excess with regard to the Be–depolarizer compound, the single polarographic waves of the above depolarizers are split into a more positive part, corresponding to the reduction involving protons from the hydrolyzed Be ion, and a more negative part, corresponding to the reduction of the depolarizer with water as the proton donor. For example, for 0.60×10^{-3}M $BeSO_4$ in 0.2M KCl -0.92×10^{-3}M oxamide solution, pH 4.8, the respective half-wave potentials are -1.54 and -1.66 V vs. SCE. The stoichiometric ratio Be : depolarizer in the Be–depolarizer compound and the number of electrons participating in the reaction were obtained by analysis of the polarograms. Two waves are also observed if the Be^{2+} ions are in excess: the more positive wave corresponds to the reduction of the depolarizer and the more negative wave to the reduction of protons from the hydrated Be^{2+} ion. In the paper [17], further details are given only for Al solutions, but the reaction steps are formulated for the complex electrode reactions involving Be^{2+} ions and some organic depolarizers, viz. oxamide (Ox): $Ox + 2 Be^{2+} + H_2O + 2e^-$ $\rightleftharpoons Ox \cdot 2H + Be_2O^{2+}$, Ox : Be = 1 : 2, n = 2; oxalyl dihydrazide (ODH): $ODH + 2 Be^{2+} + H_2O + 2e^- \rightleftharpoons$ $ODH \cdot 2H + Be_2O^{2+}$, ODH : Be = 1 : 2, n = 2; benzil (Bz): $2Bz + 2Be^{2+} + H_2O + 2e^- \rightarrow 2$ $Bz \cdot H$ (dimer) $+ Be_2O^{2+}$, Bz : Be = 1 : 1, n = 1; benzoin (Bzn): $Bzn + Be^{2+} + H_2O + 2 e^- \rightarrow Bzn \cdot 2H + BeO$, Bzn : Be = 1 : 1, n = 2. The reductions of benzil and benzoin are irreversible, the others are reversible [17].

Blasius et al. [22, 23] studied the dc polarographic reduction wave of the Be(II) complex with o-(2-hydroxy-5-methyl-phenylazo)-benzoic acid (HMPB) in aqueous acetate buffer solutions (pH ≈ 5.0) containing KCl as supporting electrolyte. The reduction reaction at $E_{1/2} =$ -0.42 V vs. SCE (aqueous buffer solution [23]), $= -0.45$ V (aqueous buffer solution containing 31.6% by wt of CH_3CN [23]), and $= -0.52$ V (aqueous buffer solution containing CH_3OH [22]) was found to be diffusion-controlled. It could be used for the polarographic determination of Be (at 20°C, under N_2) in the concentration range 10^{-6} to 10^{-9} mol/mL. The most suitable methanolic solution contained 0.1M total acetate, 0.1M KCl, 50% (by volume) of CH_3OH, and 0.32 μmol HMPB/mL; for further details, see [22]. The concentration of CH_3CN (~3 to 39% by wt) hardly affected the limiting diffusion current. Interfering cations, e.g., Al^{3+}, can be masked with Solochrome Violet RS. Inverse voltammograms (with C electrodes and without the addition of organic solvents to the aqueous acetate buffer) could be used to determine Be(II) concentrations as low as 10^{-9} to 10^{-10} mol/mL [23].

Fogg et al. [24] studied the polarographic behaviour of the Be(II) complexes with o-hydroxy-o'-carboxyazo dyes in aqueous acetate buffer solutions, containing KCl as supporting electrolyte. The Be(II) complex with Mordant red 60 in acetate buffer solutions of pH 4 to 6 (0.1M KCl) showed a polarographic reduction wave at ~-0.45 V vs. Hg; the position was independent of the pH in the above range. The Be(II) complex with Mordant yellow 8 in an acetate buffer at pH 4.8 did not cause a reduction wave. A reduction wave at -0.53 V vs. Hg was observed for aqueous methanolic solutions (acetate buffer, KCl as supporting electrolyte) of the 1:1 Be(II) complex with Mordant red 74 in the pH range 4.5 to 6.5. The height of the diffusion-controlled reduction step depended on the concentrations of the methanol and the buffer (see diagrams in the paper); 0.1M KCl, 0.1M acetate buffer, and 50% (by volume) of CH_3OH were optimum concentrations for the polarographic Be determination. Be(II) concentrations of 10^{-7}M could be detected and minimum Be(II) concentrations of 10^{-6}M could be determined provided the concentration of the dye was more than tenfold in excess of the Be concentration. For interfering cations and anions and further details, see [24].

For the polarography of $BeSO_4$ and $Be(ClO_4)_2$ solutions in 0.02 M **MgCl$_2$**, see [10, 11] and pp. 209/10; for the polarographic behavior of 2.5×10^{-4} to 1×10^{-3} M Be^{2+} solutions in 0.1 M **Li$_2$SO$_4$** of pH 3.5 to 3.8, see [6] and p. 208. $BeSO_4$ (~ 0.003 M) in dilute **Li$_2$SO$_4$–H$_2$SO$_4$** solutions (~ 0.2 and ~ 0.005 N, respectively), containing $\sim 0.025\%$ gelatin, yielded a poorly defined polarographic wave. Similar results were obtained in dilute solutions of **acetic** and **tartaric** acids. The wave is intensified by the presence of Al ions, as the Be and Al reduction waves coincide. The Be wave is interfered with by high concentrations of gelatin, see [18].

Kemula and Michalski [1] reported two waves for 0.005 N $BeSO_4$ solutions in 0.03 N **CH$_3$COONa** – 0.005 N **KCl** or 0.05 N **CH$_3$COOK** solutions: the more positive wave at −1.64 vs. NCE was attributed to the discharge of H^+ ions, the more negative one at −1.83 V vs. NCE to the discharge of Be ions [1].

Clearly defined polarographic waves are obtained with solutions of Be salts using **tetramethyl** or **tetraethyl ammonium** salts as supporting electrolytes. For the results of $BeSO_4$ and $Be(ClO_4)_2$ solutions in 0.1 and 0.2 M **(CH$_3$)$_4$NI**, see [10, 11] and pp. 209/10. As **(C$_2$H$_5$)$_4$NI** suppresses the current maxima observed on polarograms of Be salts, 0.1 M (C$_2$H$_5$)$_4$NI solutions were used as supporting electrolyte in polarographic studies with 5×10^{-4} to 2×10^{-3} M $BeSO_4$ solutions at 19°C. The diffusion current of the wave at about −1.9 V vs. SCE (from the diagram in the paper [21]) depended linearly on the $BeSO_4$ concentration, provided the pH was $\leqq 2.2$ to 2.4. At higher pH values, solid $Be(OH)_2$ formed, the "limiting" pH value depending on the concentration of the Be salt; e.g., the solid phase begins to form at pH values of 2.35, 2.5, and 2.65 for $BeSO_4$ concentrations of 2×10^{-3}, 1×10^{-3}, and 5×10^{-4} mol/L, respectively [21]. The Be wave for $BeSO_4$ solutions in 0.1 and 0.2 M (C$_2$H$_5$)$_4$NI (containing 0.01% gelatin) consists of two very close waves. The value of $E_{1/2}$ for the combined wave is −1.81 V vs. SCE for 2×10^{-3} M Be^{2+}, 0.1 M (C$_2$H$_5$)$_4$NI, 0.01% gelatin, pH 4.1, at 25°C. In the pH range 4.0 to 4.5 the sum of the two wave heights depends linearly on the Be^{2+} ion concentration in the range 0.6×10^{-3} to 6×10^{-3} M and can be used for the polarographic determination of Be. At pH values < 2.6 the wave height depends on the pH of the solution. The wave due to the H^+ discharge ($E_{1/2}$ values range from −1.60 V vs. SCE at pH 2.40 to −1.53 V vs. SCE at pH 3.75) interferes in the pH range 3 to 4, but the wave height of the Be wave is independent of the pH in the range 3.75 to 4.80. The value of the $E_{1/2}$ of the Be wave becomes more positive with increasing pH; thus for 1×10^{-3} M $BeSO_4$ in 0.1 M (C$_2$H$_5$)$_4$NI [13]:

pH	2.40	2.60	2.95	3.00	3.75	4.10	4.80
$E_{1/2}$ vs. SCE in V	−1.96	−1.90	−1.88	−1.88	−1.81	−1.73	−1.74

A double wave appears in the presence of oxalyl dihydrazide (ODH), similar to the case of 0.2 M KCl solutions, cf. pp. 214/5; the height of the more positive part ($E_{1/2}^{i} = -1.58$ to −1.65 V vs. SCE for 0.6×10^{-3} to 4.0×10^{-3} M Be^{2+} in 0.2 M (C$_2$H$_5$)$_4$NI) depends linearly on the Be concentration in the above range for pH values 4.0 to 4.5. It is attributed to the reduction of a "compound" between Be^{2+} and ODH. The wave due to the reduction of ODH is more negative than the first wave by ~ 0.25 to 0.29 V, cf. p. 214. The Be wave was shown to be diffusion-controlled experimentally in the usual way. For further details, see [13]. For polarographic studies of Be^{2+} solutions in 0.2 M (C$_2$H$_5$)$_4$NI in the presence of oxamide and rubeanic acid, see [14] and pp. 214/5.

The single-sweep polarographic reduction peak at −1.33 V vs. SCE for Be(II) solutions in 0.025 M sodium citrate – 0.025 M HCl solution (pH 3.5) containing 2.33×10^{-4} M DBC-chlorophosphonazo can be used for the determination of Be(II) in the range 2.2×10^{-8} to 4.0×10^{-7} mol Be(II)/L (detection limit 1.1×10^{-8} mol Be(II)/L) [25].

Solutions of Be^{2+} in **buffered acetate** solutions (pH 2.5 to 12.4), containing a small amount of gelatin, showed no reduction wave in the absence or presence of the complex-forming hexamethylene diamine tetraacetic acid [19]. Similarly, no reduction wave was observed in the case of 1×10^{-3} M Be^{2+} solutions in 1M solutions of **sodium lactate, sodium malate,** and **sodium salicylate** at pH 6 [20].

References:

[1] Kemula, W.; Michalski, M. (Collect. Czech. Chem. Commun. **5** [1933] 436/42).
[2] Kolthoff, I. M.; Lingane, J. J. (Polarography, 2nd Ed., Vol. 2, New York 1952, p. 430).
[3] Kryukova, T. A.; Sinyakova, S. I.; Arefeva, T. V. (Polarographische Analyse, Leipzig 1964, 711 pp., 214, 239).
[4] Meites, L. (Polarographic Techniques, Interscience, New York, N.Y., 2nd Ed., 1965, 752 pp., 615/8).
[5] Ralea, R.; Saviuc, E. (Analele Stiint. Univ. Al. I. Cuza Iasi 1 c [2] **14** [1968] 135/8; C. A. **71** [1969] No. 66698).
[6] Banerjee, T.; Bhattacharya, H. (J. Sci. Ind. Res. [India] B **16** [1957] 377; C. A. **1958** 1816).
[7] Venkataratnam, G.; Raghava Rao, B. S. V. (J. Sci. Ind. Res. [India] B **17** [1958] 360/2; Anal. Abstr. **6** [1959] No. 2875).
[8] Heyrovsky M. (Advan. Polarog. Proc. 2nd Intern. Congr., Cambridge, Engl., 1959 [1960], Vol. 3, pp. 854/60; C. A. **57** [1962] 10920).
[9] Heyrovsky, M. (Collect. Czech. Chem. Commun. **25** [1960] 3120/36).
[10] Györbíró, K. (Magy. Kem. Foly. **65** No. 9 [1959] 354/7; C. A. **1960** 7376).

[11] Györbíró, K. (Acta Chim. Acad. Sci. Hung. **22** [1960] 225/33; C. A. **1961** 230).
[12] Shirvington, P. J.; Florence, T. M.; Harle, A. J. (Austral. J. Chem. **17** [1964] 1072/84).
[13] Rekalič, V. J.; Jovanovič, M. M. (Glasnik Hem. Drustva Beograd **37** No. 3/4 [1972] 165/72; C. A. **79** [1973] No. 13220).
[14] Jovanovič, M. M.; Rekalič, V. J. (J. Electroanal. Chem. Interfacial Electrochem. **43** [1973] 135/43).
[15] Medyntsev, V. V.; Dyatlova, N. M. (Tr. Vses. Nauchn. Issled. Inst. Khim. Reaktivov Osobo Chist. Khim Veshchestv No. 28 [1966] 258/69; C. A. **68** [1968] No. 26566).
[16] Medyntsev, V. V.; Dyatlova, N. M. (Tr. Vses. Nauchn. Issled. Inst. Khim. Reaktivov Osobo Chist. Khim Veshchestv No. 30 [1967] 318/28; C. A. **68** [1968] No. 35280).
[17] Rekalič, V. J.; Jovanovič, M. M. (Glasnik Hem. Drustva Beograd **49** No. 3 [1984] 105/12; C. A. **101** [1984] No. 62512).
[18] Zuman, P. (Chem. Listy **46** [1952] 326/8; C. A. **1953** 417).
[19] Lastovskii, R. P.; Vainshtein, Yu. I.; Dyatlova, N. M.; Temkina, V. Ya.; Kolpakova, I. D. (Zh. Anal. Khim. **10** [1955] 128/31; J. Anal. Chem. [USSR] **10** [1955] 117/20).
[20] Adam, I.; Dolezal, J.; Zyka, J. (Zh. Anal. Khim. **16** [1961] 395/8; J. Anal. Chem. [USSR] **16** [1961] 405/8).

[21] Kovalenko, P. N.; Geiderovich, O. I. (Zh. Anal. Khim. **14** [1959] 634/5; J. Anal. Chem. [USSR] **14** [1959] 699/701).
[22] Blasius, E.; Janzen, K.-P.; Fallot-Burghardt, W. (Talanta **18** [1971] 273/8).
[23] Blasius, E.; Janzen, K.-P. (Fresenius Z. Anal. Chem. **258** [1972] 257/63).
[24] Fogg, A. G.; Kumar, J. L.; Burns, D. T. (Analyst **94** [1969] 262/8).
[25] Chen, Y.; Sheng, D. (Fenxi Shiyanshi **10** No. 5 [1991] 4/7 from C. A. **116** [1992] No. 186931).

16.6.3 Neutral Aqueous Solutions

Solutions of 1×10^{-3} M Be^{2+} in neutral (pH 7) 1M solutions of sodium lactate, sodium malate, and sodium salicylate showed no Be reduction wave.

Adam, I.; Dolezal, J.; Zyka, J. (Zh. Anal. Khim. **16** [1961] 395/8; J. Anal. Chem. [USSR] **16** [1961] 405/8).

16.6.4 Alkaline Aqueous Solutions

Reduction waves due to Be^{2+} ions could not be observed in strongly alkaline solutions [1], in 2N NaOH or KOH solutions [2], or in 1×10^{-3} M Be^{2+} solutions in 1M sodium lactate, malate, or salicylate (pH 9 or 12) solutions [3]. Similarly, there were no Be reduction waves in polarograms of Be^{2+} solutions in a 0.1M NH_4OH–0.1M NH_4Cl buffer (pH 9.35) or in 0.1M NaOH (pH 12.4) in the absence or presence of the complex-forming hexamethylene diamine tetraacetic acid [4], nor was a polarographic wave observed for a Be(II)-Mordant yellow 8 complex in an ammonium buffer solution (pH 9.2), containing KCl as supporting electrolyte [5]. An oxidation peak due to a Be(II) complex with adrenalin (AD) was observed at +0.35 V vs. SCE on linear sweep voltammograms (with a carbon paste electrode) for Be(II) solutions at $22 \pm 2°C$ in ammonium buffer solutions (pH 8.5). The peak height was proportional to the Be(II) concentration in the range 1×10^{-5} to 6×10^{-5} mol Be(II)/L; poorly developed reduction peaks on cyclic voltammograms indicate irreversibility of the electrochemical process; for details, see [6]. No peaks were observed on cyclic voltammograms of the above solutions if the AD was replaced by dopamine [6]. Beryllium could be detected at concentrations of 1.0×10^{-7} mol Be(II)/L in aqueous solutions of 0.2M KCl–0.4M ethylene diamine (pH 11.3) containing Alizarin Blue S (ABS) by the reduction wave at -0.75 to -0.76 V vs. SCE, attributed to a Be(II)-ABS complex [7]. The peak height of the polarographic wave of the Be–morin complex in aqueous 0.1M EDTA–0.1M ethylene diamine hydrochloride solutions (pH 10.8) (see diagrams in the paper) was proportional to the Be(II) concentration in the range 2.5×10^{-7} to 1.0×10^{-4} mol Be(II)/L [8].

The polarographic behavior of alkaline solutions (ammonium buffers) of the Be(II)-Beryllon III complex was studied by [9 to 11]. Beryllium(II) may be determined oscillopolarographically in aqueous NH_3–NH_4Cl–EDTA–Beryllon III-ascorbic acid solutions in the range 0.2 to 100 ppb Be(II) (detection limit 2ppt Be(II)) [10]. A solution containing 0.8N NH_4OH, 0.5N NH_4Cl, 4% EDTA, and 0.002% Beryllon III was found to be the most suitable for the determination of Be(II) by derivative polarography. The peak height at -0.80 V vs. SCE was proportional to the Be(II) concentration in the range 1.5×10^{-8} to 1.0×10^{-5} mol Be(II)/L [9]. Detailed voltammetric studies with a hanging mercury drop electrode were carried out with Be(II) solutions in NH_3–NH_4Cl–EDTA–Beryllon III solutions by Sun et al. [11]. The cell was deaerated with pure N_2 and the solution was kept at a "preconcentration potential", E_a, for t_a s prior to measurements. The peak current of the reduction wave due to the Be(II)–Beryllon III complex at -0.78 V vs. SCE (at optimum conditions, see below) depended on the scan rate, on the temperature (the temperature coefficients of the peak current were 2.6, 0, and -1.0%/K for the temperature ranges 5 to 13°C, 13 to 20°C, and 20 to 60°C, respectively), on the presence of surfactants, on E_a, and on the pH of the solution for pH < 9.7. The peak potential depended on the scan rate and on the pH of the solution. The optimum conditions for the Be(II) determination in the concentration range 1×10^{-9} to 8×10^{-8} mol Be(II)/L (detection limit 5×10^{-10} mol Be(II)/L) were: 0.5 M NH_4OH–NH_4Cl (pH 9.7), 0.12% EDTA, 2.5×10^{-7} mol Beryllon III/L, scan rate 100 mV/s, $t_a = 90$ s, $E_a = -0.30$ V vs. SCE. For further details, see [11].

The polarographic wave at −0.70 V vs. SCE of a Be(II)–Thoron(II) complex in an NH_4OH–NH_4Cl buffer (pH 9.5), containing EDTA and diphenyl guanidine, could be used for the determination of Be(II) in the concentration range 0.0005 to 0.1 µg/mL (detection limit 0.2 ng Be(II)/mL) [12]; cf. [13].

References:

[1] Banerjee, T.; Bhattacharya, H. (J. Sci. Ind. Res. [India] B **16** [1957] 377; C. A. **1958** 1816).
[2] Fornasari, E.; Gagliardo, E. (Atti Ist. Veneto Sci. Lettere Arti Classe Sci. Mat. Nat. **115** [1957] 217/32; C. A. **1958** 12652).
[3] Adam, I.; Dolezal, J.; Zyka, J. (Zh. Anal. Khim. **16** [1961] 395/8; J. Anal. Chem. [USSR] **16** [1961] 405/8).
[4] Lastovskii, R. P.; Vainshtein, Yu. I.; Dyatlova, N. M.; Temkina, V. Ya.; Kolpakova, I. D. (Zh. Anal. Khim. **10** [1955] 128/31; J. Anal. Chem. [USSR] **10** [1955] 117/20).
[5] Fogg, A. G.; Kumar, J. L.; Burns, D. T. (Analyst **94** [1969] 262/8).
[6] Ye, X. Z.; Hou, J. M.; Gao, S.; Gao, X. X.; Wang, D. M. (Chinese Chem. Lett. **2** [1991] 731/4; C. A. **116** [1992] No. 161045).
[7] He, W.; Li, N. (Fenxi Huaxue [Analytical Chemistry] **19** No. 10 [1991] 1189/91 from C. A. **116** [1992] No. 186907).
[8] Xu, Y.; Li, N. (Beijing Daxue Xuebao, Ziran Kexueban **27** No. 4 [1991] 430/4 from C. A. **116** [1992] No. 227139).
[9] Tan, Z. (Fenxi Huaxue [Analytical Chemistry] **9** No. 4 [1981] 401/5 from C. A. **96** [1982] No. 192479).
[10] Luo, Y.; He, L.; Yang, C. (Huaxi Yike Daxue Xuebao **20** No. 2 [1989] 227/30 from C. A. **112** [1990] No. 25277).

[11] Sun, C.; Wang, J.; Hu, W.; Xie, T.; Jin, W. (Anal. Chim. Acta **259** No. 2 [1992] 319/23).
[12] Wang, C.; Zhang, Y. (Fenxi Huaxue **18** [1990] 536/40 from C. A. **113** [1990] No. 204094).
[13] Luo, W.; Yu, X. (Xinan Shifan Daxue Xuebao, Ziran Kexueban No. 1 [1988] 46/9 from C. A. **111** [1989] No. 49512).

16.6.5 Aqueous Alcoholic Solutions

By analogy to thorium polarography, Mašek [1 to 3] suggested that the depolarizing effect of Be^{2+} ions in aqueous and aqueous alcoholic media may be due to the reduction of protons from Be aquo and Be alcohol complexes, but no experimental data are given.

References:

[1] Mašek, J. (Chem. Listy **52** [1958] 7/15; C. A. **1958** 13479).
[2] Mašek, J. (Zeitschrift für physikalische Chemie, Sonderheft, Leipzig 1958, pp. 108/18; C. A. **1959** 5919).
[3] Mašek, J. (Collect. Czech. Chem. Commun. **24** [1959] 159/69).

16.6.6 Influence of Be²⁺ Ions on the Reduction Waves of Other Metal Ions in Aqueous Solutions

The decrease of the catalytic hydrogen reduction wave by Be^{2+} ions was observed during the electrolytic reduction of Co^{2+} in an NH_4OH-NH_4Cl buffer ($pH \approx 9$) in the presence of 8-hydroxyquinoline. The decrease in the catalytic current is shown to be proportional to the Be^{2+} ion concentration in the range 1 to 8 µg Be/L for solutions containing 3×10^{-3} g-ion Co^{2+}/L and 2×10^{-6} M 8-hydroxyquinoline in a 0.4 M NH_4OH–0.4 M NH_4Cl buffer. Beryllium can be determined in the above concentration range with an error of −5.2 to +13.0%, even in the presence of Al [1]. Addition of 10^{-3} mol/L Be^{2+} to a 1×10^{-4} M Ni^{2+}–1×10^{-3} M oxybis-(ethylene amine isopropylphosphonic acid) solution shifts the Ni^{2+}-reduction wave sharply to more negative potentials [2].

References:

[1] Toropova, V. F.; Elizarova, G. L. (Zh. Anal. Khim. **19** [1964] 174/7; J. Anal. Chem. [USSR] **19** [1964] 158/61).

[2] Kabachnik, M. I.; Dyatlova, N. M.; Medved', T. Ya.; Medyntsev, V. V.; Rudomino, M. V. (Dokl. Akad. Nauk SSSR **164** [1965] 1311/4; Dokl. Chem. Proc. Acad. Sci. USSR **160/165** [1965] 1023/5).

16.6.7 Nonaqueous Solutions of Beryllium Salts

References for this section are given on pp. 226/7.

The polarographic reduction of Be^{2+} was studied in nonaqueous mixtures of **amyl** and **ethyl alcohols** (mostly in the ratio 1:4, respectively) with LiCl (concentration 1M) as supporting electrolyte by Ralea and Saviuc [1]. Advantages of the presence of ethyl alcohol are, e.g., better solubilization of the LiCl and an improved conductivity of the medium. An H-type Kalousek cell was connected to the SCE reference electrode via an agar-agar bridge saturated with KCl. **Fig. 16-19** shows the comparative polarograms obtained by the authors for solutions of Be^{2+} (concentration not given in the paper) in water, in pure ethyl alcohol, and in a mixture of amyl and ethyl alcohols. In the aqueous solution of Be^{2+} there is only one wave ($E_{1/2} = -1.3$ V vs. SCE), the height of which is not proportional to the Be^{2+} concentration, whereas there are two waves ($E_{1/2}^I = -1.2$ V, $E_{1/2}^{II} = -1.6$ V vs. SCE) in the alcoholic solutions. The height of the first wave (studied in amyl alcohol–ethyl alcohol) was found to be proportional to the concentration of Be^{2+} (and thus suitable for analytical purposes), while that of the second wave was independent of it; see figure in the paper [1]. The authors attribute the first wave (represented by a diffusion current) to the reduction of free Be^{2+} ions, and the second wave (represented by an adsorption current) to the reduction of adsorbed Be^{2+} ions. The more negative potential of the second wave indicates that the Be^{2+} ions are the adsorbed species. Confirmation of the diffusive and adsorptive characters of the first and second waves, respectively, are supplied by experimental data. The wave heights of the first and second waves were found to be proportional to $\sqrt{h}$ and h, respectively, where h = height of the Hg reservoir (see figure in the paper). Also, polarograms with a given Be^{2+} ion concentration for amyl alcohol–ethyl alcohol mixtures in the ratios 1:11 to 1:1, respectively, showed that the height of the second wave did not change, while that of the first wave decreased slightly with an increasing amount of amyl alcohol. This decrease can be accounted for by the increased viscosity and the change in the diffusion coefficient of the Be species for the different alcohol mixtures [1].

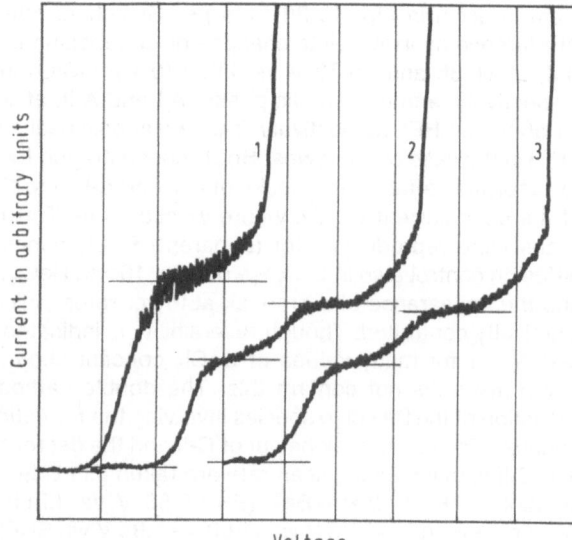

Fig. 16-19. Polarograms for the reduction of Be^{2+} in 1=water, 2= C$_2$H$_5$OH, 3=amyl alcohol – C$_2$H$_5$OH (1:4); supporting electrolyte LiCl, initial potential –1 V vs. SCE.

Single-sweep derivative voltammograms (sweep rate 1 V/s) of 0.0945 M beryllium acetate in an Al plating solution (3.73 M **AlCl$_3$** – 0.33 M **LiAlH$_4$** in **diethyl ether)** show a strong reduction peak at –0.93 V vs. Pt on the forward part of the curve (in addition to the Al reduction peak) and a relatively large residual voltage on the reverse part. The experiments were conducted in a glove box with freshly cleaned Pt working electrodes in an all-Pt electrode system. The derivative curve for BeCl$_2$ (0.122 M) in 3.55 M AlCl$_3$ – 0.31 M LiAlH$_4$–diethyl ether shows a strong peak at –2.40 V vs. Pt, but the peak can not be assigned to a Be reduction, as no Be was found in the Al deposit [2].

Voltammetric studies (scan rates: 14 to 276 mV/s) of solutions of BeCl$_2$ (concentration ≦0.01M) in various solvent systems (cf. p. 226) showed that only solutions in **propylene carbonate** (PC) provided reproducible evidence of reversible electrode reactions of the "solvo Be" species present in such solutions. Details of the purification of the substances used, apparatus and experimental procedure (exclusion of water and oxygen) are given in the paper [3]; Pt, Be, and Hg-pool electrodes were used as indicator electrodes, the latter yielded the best data. The background voltammograms depended on the supporting electrolyte (0.1 M KPF$_6$ or 0.1M LiClO$_4$) and on the indicator electrode (Hg-pool or Pt). The peaks associated with Li-Pt alloy formation lie in the same region as those due to the reduction of the Be species and may therefore interfere. Voltammetric scans of BeCl$_2$ solutions (concentration ranging from traces of BeCl$_2$ to saturation) in KPF$_6$–PC indicated that some side reaction produced anomalous results. The peak heights of the initially symmetrical curves were time-dependent, and eventually a light, gelatinous precipitate of a mixture of solvated Be(PF$_6$)$_2$ and KPF$_6$ appeared. However, there are double peaks on both cathodic and anodic scans. The cathodic peaks are separated by 90 to 120 mV, and the peak height difference decreases with increasing scan rate (up to 113 mV/s); the peak heights tended towards the same value at higher scan rates. In BeCl$_2$ solutions in 0.1M LiClO$_4$–PC a different Be solvo species is involved. Successive scans with a freshly cleaned Be indicator electrode (98.3% Be) in 1×10^{-3} M BeCl$_2$ in 0.1M LiClO$_4$–PC indicated a gradual covering of the Be surface, i.e., a reduction of the electroactive area. The effects of BeCl$_2$ concentration (0.5, 1.0, 2.0, 5.0, and 10 mM) and of scan rate were studied with Hg-pool electrodes in 0.1M LiClO$_4$–PC solutions. The above concentrations are apparent concentrations as the dissociation of the Be solvo complex is not known. A typical voltammogram is given in **Fig.** 16-**20** from [3]. Switching potentials were chosen carefully to avoid interference by irreversible changes or Li deposition. There are two cathodic peaks, C-I and C-II, at +0.86 and +0.73 V vs. Li|Li$^+$ (0.1M LiClO$_4$ in PC), or −1.82 and −1.95 V vs. SHE, respectively, and two anodic peaks, A-I and A-II, at +1.03 and +1.12 V vs. Li|Li$^+$ (−1.65 and −1.56 V vs. SHE), respectively. The two anodic peaks merge at +0.78 V vs. Li|Li$^+$ (−1.90 V vs. SHE) in the case of the lowest BeCl$_2$ concentration for scan rates >56 mV/s; this is attributed to adsorption effects. The peaks of C-II and A-I were shown to be diffusion-controlled by plots of the peak current vs. $\sqrt{v}$, where v = scan rate. This did not quite apply to C-I and A-II. The scans were reproducible for (apparent) BeCl$_2$ concentrations $\geqq$1×10^{-3}M, but C-II showed diffusion control also in the case of 0.5±10^{-3} M BeCl$_2$. Plots of the peak current vs. $\sqrt{v}$ for C-I and the appearance of a prepeak at lower rates indicate that C-I is partially adsorption- and kinetically controlled. Though reversibility is indicated by the ratio of the peak heights of C-II and A-I≈1 for many values of BeCl$_2$ concentration and scan rate, the peak separation of ~300 mV does not confirm this. The double cathodic peaks are attributed to a two-step reduction of the Be solvo species involving the formation of Be$^+$ (or [Be-Be]$^{2+}$) which disproportionates. The lower peak height of C-II and the decreasing difference in the peak heights of C-I and C-II with increased scan rate are taken as evidence for the following reactions in the PC solutions: $2\,Be^{2+} + 2\,e^- \rightarrow Be_2^{2+}$ (E = +0.86 V vs. Li|Li$^+$ = −1.82 V vs. SHE); $Be_2^{2+} \rightarrow Be + Be^{2+}$; $Be^+ + e^- \rightarrow Be$ (E = +0.73 V vs. Li|Li$^+$ = −1.95 V vs. SHE). The voltammetric data at, e.g., a scan rate of 113 mV/s yield a half-time of 1.3 s for the relatively slow disproportionation reaction and a rate constant of K = 0.53/s, if the reaction is taken as being of first order [3]; cf. pp. 203/4.

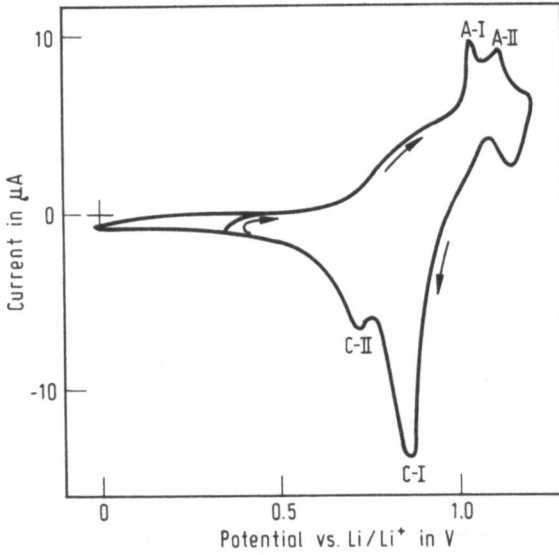

Fig. 16-20. Voltammogram of a 2× 10^{-3} M BeCl$_2$ solution in 0.1M LiClO$_4$ solution in propylene carbonate (scan rate 113 mV/s, Hg-pool electrode).

A well-defined, diffusion-limited reduction wave (E$_{1/2}$ = −1.60 V vs. SCE) was obtained for Be^{2+} in **1,2-propanediol carbonate** (PDC) at the mercury dropping electrode. The measurements were carried out under anhydrous conditions at 25.0±0.1°C, using solutions of Be(ClO$_4$)$_2$ in 0.1M (C$_2$H$_5$)$_4$NClO$_4$ in PDC (potential range: +0.2 to −2.3 V vs. SCE). A special diaphragm separated the aqueous and nonaqueous phases of the system; the values of potential were corrected for damping and ohmic losses. Experimental details are given in the paper [4]. The flat Be wave exhibits a maximum of the second kind at −2.0 V vs. SCE; it increases with increasing height of the Hg vessel, but can be suppressed by addition of 0.2% gelatin. Potentiostatic micro-coulometric measurements established that two electrons are involved in the electrode process.

The slope of the potential vs. log $I/(I_d-I)$ plot, viz. 0.260 V/decade, and the temperature coefficient of $E_{1/2}$, viz. 4.6 mV/K, showed the electrode process to be irreversible. The Il'kovič equation was found to apply for Be^{2+} concentrations of 1×10^{-4} to 1.5×10^{-3} M, based on measurements of the dependences of the diffusion current on the Be^{2+} concentration, the temperature, and the square root of the height of the Hg column. The diffusion current had a temperature coefficient of 0.4%/K. The diffusion current constant of 1.21 and the diffusion coefficient, $D=1.00\times10^{-6}$ cm²/s, were calculated with the help of the Il'kovič equation. The presence of water in concentrations up to 1% had no effect on the polarograms [4].

A special polarographic cell was used for studies in anhydrous **ethylene diamine**, see [5]. Solutions of Be^{2+} in 0.1M $NaNO_3$ in ethylene diamine at 23°C did not yield a polarographic wave prior to the wave due to the supporting electrolyte [5, 6].

In **acetonitrile** (AN) the Be^{2+} ion is less solvolyzed than in water and the solvated H^+ ion is reduced at a much more positive potential than in water. Solutions of Be^{2+} ions in acetonitrile give a well-defined, though incompletely reversible wave at the dropping mercury electrode. Coetzee et al. [7 to 9] studied solutions made up of $Be(ClO_4)_2 \cdot 4 H_2O$ (the anhydrous $Be(ClO_4)_2$ decomposed at $\sim110°C$) in 0.1M $(C_2H_5)_4NClO_4$ solution in acetonitrile at $25\pm0.2°C$. For experimental details, see [7, 8]. All diffusion currents were corrected for the residual current, and all potentials for the ohmic drop. The solutions were deaerated with nitrogen. The electrolytic solution contained a certain amount of water because of the hydrated $Be(ClO_4)_2$ used and because of the residual water content in the AN, viz. $(1$ to $2)\times10^{-3}$ M. However, water concentrations of <0.1M were found to have little effect on the values of $E_{1/2}$ and the diffusion current. The diffusion current is proportional to the Be^{2+} concentration over a wide range $(\sim10^{-4}$ to $\sim10^{-2}$ M$)$. For 1×10^{-3} M Be^{2+} in 0.1M $(C_2H_5)_4NClO_4$ at 25°C, the diffusion current constant in acetonitrile is 4.22, and $E_{1/2}^{AN}=-1.6$ V vs. SCE (for comparison, $E_{1/2}^{water}=-1.8$ V; the slightly more positive $E_{1/2}$ value in AN is attributed to a lower solvation energy in AN than in water). The waves are rather drawn out. Polarograms with Be^{2+} concentrations $>1\times10^{-3}$ M exhibit an irregular section of ~0.2 V in the rising part of the curve, i.e., the current increase with the growth of the Hg drop and the current-voltage plot are erratic. Beyond a certain potential, the interference stops abruptly and a normal diffusion plateau is reached (see figure in the paper) [7]. This phenomenon was attributed to H_2 evolution and the formation of an insoluble film on the surface of the Hg drop as a result of a solvolysis reaction, such as $Be(SH)_n^{2+} \rightleftharpoons BeS(SH)_{n-1}^+ + H_{solvated}^+$, where SH denotes a solvent molecule, i.e., a molecule of acetonitrile or water. The $H_{solvated}^+$ is then reduced at the electrode, while the $BeS(SH)_{n-1}^+$ forms an insoluble product with the anions present in the solution. Microscopic observation of gas bubbles at the capillary in the critical potential range support the above explanation. Further evidence is supplied by controlled potential electrolysis at the Hg-pool electrode at potentials corresponding to the erratic section on the polarogram using 4.4×10^{-3} M $Be(ClO_4)_2$ solutions in 0.2M $(C_2H_5)_4NClO_4$–AN (N_2 was passed through the apparatus to effect deaeration). Unless the Hg-pool surface is agitated, the current decays to almost zero within a few seconds. With a stirred Hg–solution interface, electrolysis proceeds with H_2 evolution until all the Be present is precipitated as a white compound (solvolysis product). Electrolysis at potentials corresponding to the diffusion plateau yields black Be amalgam. The addition of up to 1% water or a few tenths of a percent of acetic acid to the solution eliminates the solvolysis effect, possibly because of dissolution of the solvolysis product. The addition of larger amounts of water ($>1\%$) shifts the polarographic wave to more negative potentials without affecting the value of the diffusion current; further details are given in [7, 8]. The problem of liquid junction potentials was avoided by using the reversible $E_{1/2}$ of the Rb ion as a reference; for solutions of Be^{2+} in 0.10M $(C_2H_5)_4NClO_4$–AN, $E_{1/2}=-1.6$ V vs. SCE $=+0.4$ V vs. the reversible $E_{1/2}$ of Rb^+ under comparable conditions. For a comparison of half-wave potentials in nitriles (AN in the case of $Be(ClO_4)_2 \cdot aq$) and in water, see [9].

Shirvington et al. [10] studied the polarographic behavior of $Be(ClO_4)_2 \cdot 4H_2O$ and anhydrous $BeCl_2$ in acetonitrile (AN) solutions at $30 \pm 0.2°C$, using 0.1M solutions of $(n\text{-}C_3H_7)_4NX$, where $X = ClO_4$, Cl, Br, or I, as supporting electrolytes. The water content of the AN was 2 to 7 mM, the residual currents were < 0.5 µA at -2.0 V vs. SCE in 0.1M $(C_3H_7)_4NClO_4$. The all-glass polarographic cell was deaerated with dry, oxygen-free argon (for further details, see the paper). Mass spectrometric analyses of the gaseous products formed during controlled potential electrolysis of Be^{2+} solutions in 0.1M solutions of the supporting electrolytes indicated that H_2 gas was a reduction product of all the Be steps observed, while exhaustive electrolysis precipitated all Be present as $Be(OH)_2$, the number of electrons consumed per initial Be^{2+} ion being 2. Dc polarograms of 5×10^{-4}M $Be(ClO_4)_2$ solutions (supporting electrolyte 0.1M $(C_3H_7)_4NClO_4$) show one wave at ~ -1.5 V vs. SCE, see Fig. 16-21, curve A. In the presence of halide ions there are two polarographic waves with $E_{1/2}$ of the first step at values more positive than -0.8 V vs. SCE, see Fig. 16-21, curves B, C, D. Similarly, two waves were obtained with 5×10^{-4}M $BeCl_2$ solutions. In the presence of two halide ions, e.g., Cl^- and Br^-, the "prestep" consisted of two waves. As the "presteps" correspond exactly to the wave of the halogen acid concerned in AN, the authors concluded that the Be halides interact with water or AN in the bulk of the solution to form the strongly associated halogen acids and various solvolysis products, e.g., $BeCl_2(SH)_n \rightleftharpoons Be(S)_2(SH)_{n-2} + 2HCl$, where $SH = AN$ or H_2O. The degree of interaction increases from iodide to chloride. The critical effect of water is demonstrated by the decrease of the height of the prestep, and by the precipitation of a white compound of the composition $Be_4Cl_3(OH)_5 \cdot x\, CH_3CN$ from $BeCl_2$ solutions in 0.1M $(n\text{-}C_3H_7)_4N\text{-}ClO_4\text{-}CH_3CN$. As water is considered to affect the above solvolysis equilibrium and because the water content of the $(n\text{-}C_3H_7)_4NCl$ solutions could not be reduced below 7×10^{-3}M, polarographic measurements were carried out with 3×10^{-3}M $BeCl_2$ solutions and various water : beryllium ratios. The degree of solvolysis, α, was calculated from the diffusion currents. A plot of α vs. $\log[H_2O]$ (see figure in the paper) shows that maximum solvolysis (99%) occurred in $BeCl_2$–0.1M $(n\text{-}C_3H_7)_4NCl$–CH_3CN solutions at a water : beryllium ratio of 4:1. However, the evidence is insufficient to prove water to be the solvolytic agent, see [10].

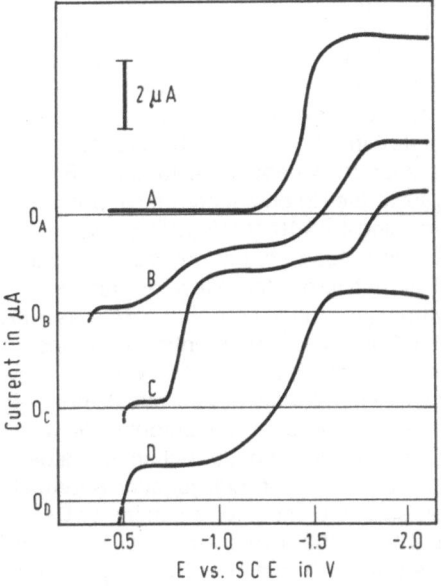

Fig. 16-21. Dc polarograms of 5×10^{-4}M Be-$(ClO_4)_2$ solutions in acetonitrile with the supporting electrolytes A = 0.1M tetrapropylammonium perchlorate, B = the same as in A, additionally 3×10^{-3}M Br^-, C = 0.1M tetrapropylammonium bromide, and D = 0.1M tetrapropylammonium iodide.

well as for an anode design and an optimum operating system, see [6]; cf. "Beryllium" Suppl. Vol. A 1, 1986, pp. 112/3.

The **forms of deposited Be**, i.e., the sizes and crystalline shapes of the Be deposits from $BeCl_2$–NaCl melts range from powder to needles, dendrites, and flakes. The dimensions of the $BeCl_2$ crystals were reduced considerably by decreasing the $BeCl_2$ concentration in the $BeCl_2$–NaCl melt from 50 to 30 wt%, and even spongy Be metal appeared at a cathode consisting of preberyllized Ni. Low $BeCl_2$ concentration also leads to an increase in the impurity content of the Be and a decrease in current yield, but good results were obtained with 40 to 50 wt% $BeCl_2$. Well-shaped, bright Be crystals were obtained at an optimum temperature of 340°C with cathodic current densities of 4 to 12 A/dm², while at 16 A/dm² also small amounts of Be sponge formed; see [6]. High current densities yield fine Be powder; however, if the electrolyte is very pure, large Be crystals may be obtained even at high current densities [2, 9]. An increase in the size of the Be crystals in the course of the electrolysis may be due to a decrease in the real current density, as the cathode surface area grows with the Be deposit [13]. As powdery, partially adherent Be deposits are expected to form by a secondary reduction ($2\,Na + Be^{2+} \rightarrow 2\,Na^+ + Be$, cf. p. 248), larger Be flakes produced during the later stages of an electrolysis may indicate that a secondary reduction is diminished [2, 13]. Jaeger [28] maintains that the $BeCl_2$ concentration and the impurities present in the electrolyte are the main factors determining the crystal shape of the Be (see photos in the paper). The leaflike, dendritic crystals are mostly found to be twins; the platelike crystals have a layered structure with terraces and grow oriented on the substrate; the Be needles show prismatic or obelisk-like structures. The determination of the ratio of the lattice parameters (a : c = 1 : 1.569) by the Debye-Scherrer technique showed that there is no difference in the lattice structure of the plate- and needle-shaped Be crystals [28]. The Be flakes are considered to be aggregates of Be single crystals with the basal planes in the plane of the flake, though slightly displaced [29]. They are held together by minute metallic "bridges" [30]. The surfaces of flakes produced from 1:1 $BeCl_2$–NaCl melts (by Pechiney) showed projecting pyramidal structures with basal dimensions of 30 to 100 μm and heights of ~3 μm. They consist of up to 20 to 30 hexagonal terraces, forming closed loops and capped by plateaus with minute pits of ~0.1 μm diameter, resembling screw dislocations. Many flakes also contained voids. Foreign material, possibly complex molecules (such as $BeO \cdot BeCl_2$), seems to occupy the pits. Impurity layers about 10 to 20 μm apart are found parallel to the planes of the flakes and are thought to be halide complexes deposited cataphoretically during electrolysis [30, 31]. Surface steps on the powder particles and deposited impurities may influence the behavior of the Be powder during compacting and sintering and hence also the properties of the compacts [30]. In the Milford Haven experimental cell fine, acicular or coarse, dendritic Be flakes were produced, consisting of hexagonal plates with terraced steps of the order of 100 μm. The impurity layers consisted of groups of particles rather than continuous layers and were much more widely spaced than in the Pechiney flakes [30, 31].

The **purity** of the Be deposits was reported to be ~99.95% Be [7, 8], ≧99% [13, 32 to 36]. A spectrochemical analysis of Be flakes is given in [30]. An O content of 1000 to 2000 ppm in Be flakes of 1 to 5 mm diameter was attributed to Be oxide on the surface of the flakes [37, 38]. The Cl content of the Be flakes could not be reduced below a minimum of 200 to 300 ppm by any treatment applied; see [31]. Metallographic and chemical analyses indicated that the Cl content is not due to irreversible chemisorption. Dissolution of the Cl in the Be lattice is improbable on account of the sizes of the Cl ion and the interstices of the Be lattice. Discrete inclusions of Cl containing compounds (average diameter of inclusions < 5 μm) are considered the most likely [31].

Preelectrolysis, i.e., electrolysis at a potential below the decomposition potential of $BeCl_2$, was applied to remove metal impurities which are more electropositive than Be (replacing the

Be anode by a graphite anode, in the case of Be refining); Be of higher purity is then deposited on a fresh cathode. For details, see [6, 8, 9, 34 to 36]. Preelectrolysis does not necessarily improve the purity of the Be deposit [22, 23]. Beryllium of high purity is obtained from pure starting materials and with an apparatus free of leaks.

For data on the analysis of electrolytic Be, see, e.g. [21, 23, 26, 38]. Details on the reduction of impurities in Be refining are given in [6].

In **special procedures** continuous Be electrodeposition can be effected by mobile Ni or Ni alloy cathodes [20], or by graphite-lined tubular Ni cathodes with a movable ring scraper [21].

The purity of Be deposits is improved by a two-stage electrolysis of BeCl$_2$–NaCl melts (1:1 by wt) at 320 to 350°C with exchangeable Ni cathodes (and graphite anodes) in a semi-continuous process: impurity metals more electropositive than Be (e.g., Si, Al, Fe, Cu) are deposited at a lower applied voltage (~1.5 V), before Be scales of purity better than 99.8% are deposited on a fresh cathode [34 to 36]. Beryllium flakes of 99.95 to 99.99% purity are obtained from BeCl$_2$–NaCl melts at 350 ± 70°C according to [8]. Continuous production of Be of ~99.97% purity is possible with an apparatus using several cathodes (BeCl$_2$–NaCl melt with 50 to 60 wt% BeCl$_2$ at 330 to 400°C); see [39]. Only one cathode is used in the following two-step electrolysis: the above impurity metals are deposited along with Be on an Ni cathode from a BeCl$_2$–NaCl melt (wt ratio 5.7 : 4.3) at 350°C, low current density (10 mA/cm²), and low applied voltage (2.4 V, graphite anode). This deposit adheres strongly to the Ni cathode. Coarse and large crystalline Be flakes of high purity (~99.95%) are then deposited onto the impure Be at 80 mA/cm² and an applied voltage of 3.5 V. The Be flakes can be removed easily, leaving the impure Be on the cathode [40].

In the electrolysis of fused equimolar or eutectic BeCl$_2$–NaCl mixtures with **Hg cathodes** the discharged Be forms a "quasi-amalgam", i.e., a dispersion of finely divided Be or of a Be-Hg intermetallic phase in Hg; see [41 to 47], cf. [48 to 50]. This is of interest for the preparation and refining of Be for several reasons: 1) the Hg cathode facilitates the transport of the discharged Be out of the electrolysis cell, e.g., by Hg circulation [41, 49]; 2) the interaction of the Be with the electrolyte is thus prevented [48, 49]; 3) the operating temperature is low [41, 49, 50], viz. between the melting point of the electrolyte mixture (melting point of the BeCl$_2$–NaCl eutectic 210°C [4]) and the boiling point of Hg (~356°C); 4) the Be deposited into the Hg has a fine particle size (0.1 to 10 μm [45, 46]) and can be filtered off or separated from the Hg by hot pressing or low-temperature vacuum distillation (preparation of ultrafine Be powder) [41, 44 to 46]; 5) the Be obtained is not contaminated by the electrolyte [43 to 45].

The sensitivity of the amalgam formation towards traces of oxygen requires rigorous exclusion of air and moisture, viz. the use of anhydrous salt mixtures and an inert atmosphere of pure, dry Ar or N$_2$ (e.g. 99.997% N$_2$). The inert gas flow also serves to remove Cl$_2$ formed at the anode [47]. Graphite anodes were found to disintegrate fairly rapidly [41, 43, 46], but baked C anodes proved satisfactory [41, 43, 44]. The use of a Be anode reduced the C contamination of the deposited Be [47]. Stirring of the Hg cathode is necessary to constantly expose fresh cathode surface and to prevent the amalgam from climbing towards the anode during electrolysis [41 to 47]. The stainless steel electrolysis cell and the stirrer shaft were lined with Pyrex to minimize the contact of the electrolyte with the metal parts; see [41, 43 to 47].

Eutectic BeCl$_2$–NaCl melts (≈ equimolar BeCl$_2$–NaCl) were used in most cases, though the BeCl$_2$ concentrations ranged from 45 to 60 mol% [41, 44]. An advantageous electrolyte may also contain KCl, CaCl$_2$, or MgCl$_2$ [41]. Electrolyses were carried out at 275 to 300°C [45], 308 to 326°C [44], and 300 to 330°C [41 to 43, 46]. Current efficiencies of more than 70% were obtained [41, 44].

The amalgam produced can be diluted with Hg in all proportions and is distributed evenly throughout the Hg cathode by the stirring [41]. It is drawn off periodically or continuously with a Be content of 0.04% [46], 0.25% [47]; 0.25% Be is considered a limit with respect to practical considerations [41]. On standing the amalgam "ages", forming a thick pasty layer on top of the Hg. The agglomerated amalgam has a higher Be content and contains coarser grains (~1 to 2 μm) than the initial amalgam (particle size of 0.1 to 0.2 μm). The agglomeration is accelerated by an increase in temperature [46]. Vacuum filtration (100 to 200 mesh screens) of the drawn-off amalgam yields a semisolid amalgam of a shiny, silvery appearance with 1.1 to 2.3 wt% Be [44, 46, 47]. This corresponds to $Be(Hg)_2$ with a slight excess of Hg; the Cl content was ~0.3 to 2%, the O content less than 0.7% [41, 44]. More Hg could be removed by cold pressing (pressures up to 18 t/in²) until a limiting composition of 24 wt% or 87.5 at% Be was obtained, corresponding to a formula of Be_7Hg [46]. The remaining Hg could be removed completely either by hot pressing or vacuum distillation at 340°C. The hot pressing yielded solid Be compacts (grain sizes < 50 μm) of higher purity than the loose sinter cakes obtained by the vacuum distillation. The O and Cl contents of the Be compacts depended on temperature, pressure, and duration of the hot pressing; but O contents less than 0.8% and Cl contents of 0 to 0.3% were obtained. Other main impurities were Fe and C; for details, see [41, 44, 47]. The content of metallic impurities depends directly on the purity of the $BeCl_2$ feed and the disintegration and/or corrosion of parts of the apparatus [46]. X-ray diffraction analysis of the Be powder obtained after vacuum distillation showed the presence of Be_3N_2 and C [46].

A black powder (particle size ~0.1 to 1.5 μm [46]) appeared on the surface of the amalgam if left standing for a considerable time or on exposure to air [41 to 46]. The composition of this black powder is not unequivocal: a Be content of ~75 wt% was reported by [42]; 90% metallic Be and the remainder BeO (particle size ~1 μm) were indicated by X-ray analysis in [45]; Habashi et al. [46] found that a chemical analysis (15.6% metallic Be, 40% BeO, 10% C, 1.5% Fe, balance: probably residual Hg) did not agree with X-ray analysis results. For spectrographic analyses of the black powder, see [45, 46].

References:

[1] Jaeger, G. (Metall **4** [1950] 183/91).
[2] Bellamy, R. G. (AERE-M-534 [1959] 41 pp., 20/6, 30/3, 35, 41).
[3] Schimmel, F. A. (Y-1380 [1962] 31 pp., 11/2; N.S.A. **16** [1962] No. 15127).
[4] Furby, E.; Wilkinson, K. L. (J. Inorg. Nucl. Chem. **14** [1960] 123/6).
[5] Jaeger, G.; Degussa (Brit. 434338 [1935]).
[6] Silina, G. F.; Grinberg, L. L. (Tsvetn. Met. [Moscow] **33** No. 12 [1960] 47/53; Soviet Non-Ferrous Metals **1** No. 12 [1960] 46/52; C.A. **1961** 19555).
[7] Vil'komirskii, I. E.; Silina, G. F.; Berengard, A. S.; Semakin, V. N. (At. Energiya SSSR **11** [1961] 233/9; Soviet At. Energy **11** [1961] 882/8; C.A. **56** [1962] 5702).
[8] Kida, K.; Abe, M.; Nishigaki, S.; Nakamura, T.; Yokoi, M.; Usami, S.; Ishikawa, Y.; Miyazaki, Y. (Japan. 3802 [1961/63] 6 pp. from C.A. **59** [1963] 14938).
[9] Bellamy, R. G.; Hill, N. A. (International Series of Monographs on Nuclear Energy Division 8: Materials. Vol. 1, Extraction and Metallurgy of U, Th, Be, Pergamon, New York 1963, 198 pp., 76/9, 82/4).
[10] Schmidt, M. J. M. (Bull. Soc. Chim. France [2] **39** [1926] 1686/703).

[11] Tien, J. M. (Trans. Electrochem. Soc. **89** [1946] 237/45).
[12] Sheiko, I. N.; Delimarskii, Yu. K. (Ukr. Khim. Zh. **25** [1959] 295/300; C.A. **1960** 93).
[13] Menzies, I. A.; Hill, D. L.; Owen, L. W. (J. Less-Common Metals **1** [1959] 321/30).
[14] Higgins, J. K.; Bellamy, R. G.; Buddery, J. H. (AERE-M-R-2750 [1958] 18 pp., 16).
[15] Delimarskii, Yu. K.; Skobets, E. M. (Zh. Fiz. Khim. **20** [1946] 1005/10; C.A. **1947** 2341).

[16] Sheiko, I. N.; Delimarskii, Yu. K. (Ukr. Khim. Zh. **23** [1957] 713/20; C.A. **1958** 14384).
[17] Kuroda, T.; Matsumoto, D. (J. Electrochem. Soc. Japan **33** [1965] 29/34).
[18] Delimarskii, Yu. K.; Sheiko, I. N.; Feshchenko, V. G. (Zh. Fiz. Khim. **29** [1955] 1499/507; C.A. **1957** 830).
[19] Smirnov, M. V.; Chukreev, N. Ya. (Zh. Neorg. Khim. **4** [1959] 2536/43; Russ. J. Inorg. Chem. **4** [1959] 1168/72).
[20] Jaeger, G. (Ger. 860281 [1949/52] 1/12; N.S.A. **15** [1961] No. 3009).

[21] Morana, S. J. (Ger. Offen. 1115936 [1956/62] 6 pp.; Patentblatt **81** [1961] 5598).
[22] Morana, S. J. (U.S. 2843544 [1958]; C.A. **1959** 103).
[23] Beryllium Corporation (Brit. 812702 [1959]; N.S.A. **13** [1959] No. 12739).
[24] Delimarskii, Yu. K.; Baranenko, V. M.; Zarubitskii, O. G. (Deposited Doc. VINITI-3083-76 [1976] 1/13; Elektrokhimiya **12** [1976] 1887; Soviet Electrochem. **12** [1976] 1717).
[25] Menzies, I. A.; Hill, D. L.; Owen, L. W. (Nature **183** [1959] 816/7).
[26] Morana, S. J. (Fr. 1153825 [1956/58]; C. **1961** 10736).
[27] Degussa (Fr. 785072 [1935]; C.A. **1936** 394).
[28] Jaeger, G. (Z. Metallk. **41** [1950] 243/6).
[29] Williams, J. (Progr. Nucl. Energy V **1** [1956] 300/4).
[30] Blainey, A.; Johnston, T. L.; Jones, J. W. S. (AERE-M-R-1442 [1954] 35 pp., 1/3, 17/8; Rev. Met. [Paris] **52** [1955] 735/49).

[31] Bellamy, R. G. (M-619 [1960] 14 pp.; C.A. **1960** 18127).
[32] Potvin, R.; Farnham, G. S. (Trans. Can. Inst. Mining Met. **49** [1946] 525/38, 533/5; C.A. **1947** 66).
[33] Potvin, R. (Metal Ind. **1946** 469/70).
[34] Meerson, G. A.; Kaplan, G. E.; Silina, G. F.; Sokolov, D. D. (Issled. Obl. Geol. Khim. Met. Sbornik **1955** 125/31; C.A. **1957** 17484).
[35] Meerson, G. A. (Proc. 1st Intern. Conf. Peaceful Uses At. Energy, Geneva 1956, Vol. 8, pp. 587/9).
[36] Meerson, G. A.; Sokolov, D. D.; Mironov, N. F.; Bogorad, N. M.; Pakhomov, Ya. D.; L'vovskii, D. S.; Ivanov, E. S.; Shmelev, V. M. (At. Energiya SSSR **5** [1958] 624/30; N.S.A. **13** [1959] No. 5589).
[37] Syre, R. (Metaux Corros. Ind. **33** [1958] 406/19; C.A. **1959** 2991).
[38] Syre, R.; Saulnier, A.; Perez, M. (Rev. Met. [Paris] **56** [1959] 359/70; C.A. **1959** 12117).
[39] Nakamura, T. (U.S. 3278402 [1961/66] 5 pp.; C.A. **66** [1967] No. 7911).
[40] Nakamura, T. (U.S. 3296107 [1962/67] 4 pp.; C.A. **66** [1967] No. 61309).

[41] Kells, M. C.; Holden, R. B.; Whitman, C. (SEP-207 [1956] 43 pp.; C.A. **1957** 4174).
[42] Kells, M. C.; Holden, R. B.; Whitman, C. I. (J. Am. Chem. Soc. **79** [1957] 3925).
[43] Holden, R. B.; Kells, M. C.; Whitman, C. I. (Proc. 2nd U.N. Intern. Conf. Peaceful Uses At. Energy, Geneva 1958, Vol. 4, pp. 306/8; C.A. **1960** 7375).
[44] Kopelman, B.; Holden, R. B. (U.S. 2987454 [1961]; C.A. **1961** 21921).
[45] Gale, C. O.; Griffiths, V.; Habashi, F.; Hanson, G. T. (Technical Report NASA Contract NASW 1099 [1965]).
[46] Habashi, F.; Murray, R. J.; Toivonen, R. W.; Griffiths, V.; Hanson, G. T. (Progress Report NASA Contract NASW 1542 [1967]; LN67-26451 [1966] 29 pp.) cited in [47].
[47] Lidman, W. G.; Griffiths, V. (N-72-27577 [1972] 14 pp.; N.S.A. **28** [1973] No. 27799).
[48] Reynolds Metals Co. (Indian 50387 [1953/55]; C. **1960** 17273).
[49] Schmidt, W. (Fr. 1128402 [1953/57]; C. **1961** 8464).
[50] Schmidt, W. (Brit. 781457 [1953/57]; C. **1959** 7626).

16.7.5.4 BeCl₂–KCl

The Be deposition on Pt and W cathodes from $BeCl_2$–KCl melts, containing 0.05 to 3 wt% of $BeCl_2$, at 400 to 850°C was found to be similar to that from $BeCl_2$–LiCl melts [1]. The formation of complexes, e.g., K_2BeCl_4, $(BeCl_4)^{2-}$, is indicated by emf measurements of the cell Be(s)|$BeCl_2$–KCl(l)|Cl_2(g), in the temperature range 400 to 530°C for $BeCl_2$ concentrations of 0.4 to 0.7 mole fraction [2]. For the electrolysis of $BeCl_2$–KCl melts containing 2 to 3 wt% BeS, see [3] and p. 272.

References:

[1] Delimarskii, Yu. K.; Baranenko, V. M.; Zarubitskii, O. G. (Deposited Doc. VINITI-3083-76 [1976] 1/13; Elektrokhimiya **12** [1976] 1887; Soviet Electrochem. **12** [1976] 1717).
[2] Kuroda, T.; Oyamada, R. (J. Electrochem. Soc. Japan **35** [1967] 125/9).
[3] Gibson, A. R. (Brit. 821 091 [1959]; C. A. **1960** 2051).

16.7.5.5 BeCl₂–LiCl–KCl

References for this section are given on pp. 262/4.

General Remarks

All investigations on the Be electrodeposition from melts with the composition given above are based on LiCl–KCl melts of eutectic composition (59 mol% LiCl, melting point 358°C). The relatively high solubility of $BeCl_2$ in this melt in the temperature ranges used for electrolyses (~ 400 to 500°C), low losses in $BeCl_2$ due to vaporization, and the availability of a considerable number of emf measurements with eutectic LiCl–KCl melts are advantages in favor of this melt [1, 2]. An extensive study of the Be electrorefining process was carried out by Wohlfarth [2]; see pp. 260/1.

Decomposition Voltages

For $BeCl_2$ added to eutectic LiCl–KCl melts in various amounts, the following decomposition voltages and mean temperature coefficients for the temperature range studied (760 to 820 K) were determined by emf measurements of the cell Be(s)|$BeCl_2$(l), LiCl–KCl(l)|Cl_2(gas) using a graphite anode and Ar atmosphere (for the cell arrangement, see figure 1 in the paper [3]):

mole fraction of BeCl₂	temperature T in K	Cl₂ pressure in atm	decomposition voltage U in V	dU/dT in V/K
0.0021	761	0.967	2.4726	-3.37×10^{-4}
	789		2.4537	
	817		2.4524	
0.0259	765	0.972	2.3650	-4.00×10^{-4}
	782		2.3594	
	793		2.3547	
	800		2.3509	
	817		2.3434	

mole fraction of $BeCl_2$	temperature T in K	Cl_2 pressure in atm	decomposition voltage U in V	dU/dT in V/K
0.0416	768	0.966	2.3330	-4.66×10^{-4}
	781		2.3280	
	801		2.3185	
	817		2.3109	
0.1514	771	0.975	2.2456	-5.88×10^{-4}
	778		2.2410	
	781		2.2394	
	808		2.2237	
	816		2.2186	
	818		2.2175	

The results indicate a two-electron reaction and the presence of relatively stable Be complexes in the melt [3].

Similarly, the following decomposition voltages were obtained for higher $BeCl_2$ additions (spectrographic-grade carbon anode; Ar atmosphere) [4, 5]:

mole fraction of $BeCl_2$	temperature T in °C	decomposition voltage U in V	dU/dT in V/K[3]
0.188	410	2.165	-8.1×10^{-4}
	450	2.133[1]	
		2.163[2]	
	500	2.092	
0.410	410	2.103	-8.7×10^{-4}
	450	2.068	
	500	2.025	
0.613	410	2.045	-9.2×10^{-4}
	450	2.010	
	500	1.962	
1.000 (extrapolated values)	410	(1.950)	-9.9×10^{-4}
	450	(1.912)	
	500	(1.860)	

[1] From [4]. – [2] From [5]. – [3] Mean values in the temperature range 410 to 500°C.

The formation of complexes such as K_2BeCl_4 was indicated by thermodynamic data calculated from the above results [4 to 6]. For a corresponding discussion of the mechanism of Be deposition from $BeCl_2$–alkali chloride melts, see [5] and pp. 247/8.

Nucleation and Crystallization Overvoltage

Studies of nucleation and crystallization overvoltages (see p. 176) in the electrodeposition of Be from eutectic LiCl–KCl melts containing $BeCl_2$ were carried out with microcathodes of Be, Ta, Mo, W, Ni, and Au by means of galvanostatic curves and simultaneous microscopic observation of the cathode surfaces at temperatures from 400 to 750°C by Chukreev et al. [7 to 14].

Typical overvoltage (η) vs. time curves are shown in **Fig. 16-23** according to [9]. For these studies current impulses of 200 to 1500 ms duration were applied; the dotted lines in the figure indicate the breaking off of the current impulses. The anode and reference electrode were pure Be. The rising branches reflect the processes of charging the cathode electric double layer, the discharge of impurity ions, and the discharge of Be ions up to the maximum supersaturation of the cathode with adsorbed Be atoms at η_{max}, which is the maximum crystallization overvoltage. At $\eta = 0$ the concentration of adsorbed Be atoms on the cathode corresponds to the activity of metallic Be, viz. $a = 1$. When the rising branch is linear, the cathode material probably does not affect nucleation of Be crystals, and the slope is influenced by interactions of the discharged Be with the cathode metal, e.g., alloy formation. The peak maxima η_{max} indicate the appearance of Be crystal nuclei and the descending branches correspond to the growth of the crystal nuclei [9, 14]; cf. [8, 13]. The drop in voltage on switching off the current, as indicated in the figure, is associated with the ohmic drop in the electrolyte and the "tail ends" of the curves are attributed to diffusion overvoltage [9].

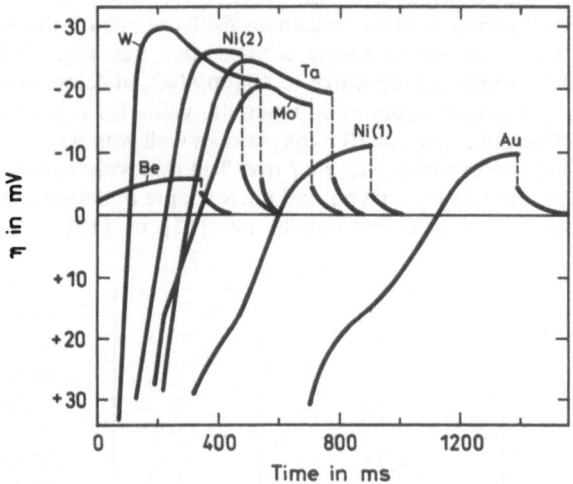

Fig. 16-23. Galvanostatic η vs. time curves for various cathode materials during Be electrodeposition from a eutectic LiCl–KCl melt containing 2.78×10^{-4} mol $BeCl_2/cm^3$; $j = 0.14$ A/cm², T = 500°C for Ta, Mo, W, and Ni(1); $j = 0.4$ A/cm², T = 400°C for Au; $j = 1.0$ A/cm², T = 500°C for Ni(2). The dotted lines indicate the switching off of the current impulses.

The overvoltage maxima in the η vs. time curves for Ta, Mo, and W at 500°C, $j = 0.14$ A/cm² in Fig. 16-23 correspond to three-dimensional Be crystal nuclei; this is confirmed by microscopic observations (see p. 256) [9, 12]. At 400°C the Be nucleation on W required a minimum j of 7.6×10^{-3} A/cm² [13]. For Ni at 500°C with $j = 0.14$ A/cm² no peak is reached on the η vs. time curve, indicating the dissolving of discharged Be atoms in the Ni; but $j = 1.0$ A/cm² provides the Be adatom supersaturation necessary for nucleation. Considerable depolarization due to Be-Au alloy formation was believed responsible for the absence of an η_{max} in the case of the Au cathode. On a clean pure Be substrate nucleation proceeds without difficulties [11, 13].

For the above cathodes η_{max} was $\leqq 30$ mV, and for the Ta, Mo, and W cathodes η_{max} varied inversely with t_{max}, the time taken to reach η_{max}. This was attributed to partial diffusion of Be into the substrate, and hence η_{max} is a qualitative measure of the inertness of the cathode material towards Be [9]. Thus W was inert towards Be at temperatures below 750°C [8]. For comparable experimental conditions, the degree of interaction between Be and the cathode material (and hence the adhesion of the Be crystals to the substrate) increases in the order W < Ta < Mo < Ni < Au [9]; cf. [13]. Surface concentrations of Be adatoms at $\eta = 0$ and η_{max} and the supersaturations, γ, involved in the formation of Be crystal nuclei on Ta, Mo, W ($j = 0.14$ A/cm²) and Ni ($j = 1.0$ A/cm²) cathodes at 500°C were also estimated from the experimental η vs. time curves. These supersaturations correlate with the solubility and the diffusion of Be in the substrate metals [9]. For a W cathode in a melt at 400°C containing 1.2×10^{-3} mol/cm³ $BeCl_2$ for $j = 0.0076$ to 0.94 A/cm² the following relations were obtained: $\eta_{max} = 80.1 + 31.5 \log j$ (± 5.4) mV and $t_{0\,to\,max} = 0.651 \cdot j^{-1.464}$ (± 2) ms, where $t_{0\,to\,max}$ is the time between $\eta = 0$ and η_{max}. At higher current densities, the ohmic drop and diffusion overvoltage contribute increasingly to η_{max} which therefore differs considerably from the true phase formation overvoltage. Values for η_{max}, ohmic resistance of the electrolyte, and phase formation overvoltage are tabulated for $j = 7.6 \times 10^{-3}$ to 1.3 A/cm² and discussed in [14]; cf. [13].

Oxide and impurity layers on the cathode surfaces increase the ohmic component of η, and the cathode potential is shifted to more positive values. When Be microcathodes were covered with Al_2O_3 layers up to 3 μm thick there was no overvoltage peak observed on the η vs. time curves at 400°C in LiCl–KCl melts with 5.85 wt% $BeCl_2$ added; see **Fig. 16-24**. It was suggested that the discharged Be atoms penetrate the porous oxide film or partially dissolve in it. Thicker Al_2O_3 coatings increase the ohmic component and hinder Be phase formation: Be cathodes with a layer thickness $\delta \geqq 8$ μm behave like a "foreign" (here: Al_2O_3) electrode, and the η vs. time curve shows a maximum, see Fig. 16-24c from [11]. The value of η_{max} for a W cathode coated with 8 to 10 μm Al_2O_3 at current densities of ~0.15 A/cm² (other conditions as above) is about three times the value for a pure W cathode (see figure in the paper). The difference, $\Delta\eta_{max} = 70.3$ mV, agrees well with an estimated difference in ohmic resistances for the two systems, viz. 71.7 mV. The relatively narrow width of the overvoltage peak obtained with the Al_2O_3 coating and microscopic observations indicated the formation of only one Be nucleus. For further details, see [11]; cf. [13].

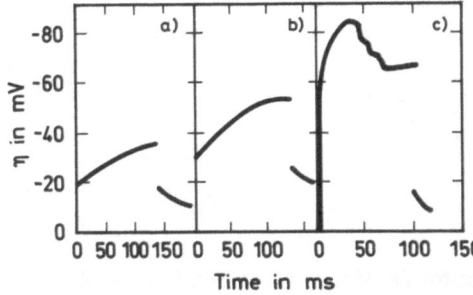

Fig. 16-24. Galvanostatic η vs. time curves for a Be cathode (covered with Al_2O_3 of thickness δ) in molten eutectic LiCl–KCl containing 5.85 wt% $BeCl_2$ at 400°C;
a) $\delta = 2$ to 3 μm, $j = 0.28$ A/cm²;
b) $\delta = 5$ μm, $j = 0.46$ A/cm²;
c) $\delta = 8$ to 10 μm, $j = 0.72$ A/cm².

Optical observation of the nucleation of Be on Be, Mo, and W microcathodes was carried out with a specially designed apparatus; the melts contained 5.85 wt% $BeCl_2$, the temperature ranged from 400 to 750°C, and current densities of 3.3×10^{-2} to 1.9 A/cm² were applied, lasting for up to 180 s. It was found that Be crystal nuclei only formed at the beginning of polarization, at potentials corresponding to η_{max} on the η vs. time curve, and were distributed extremely irregularly on the cathode surface. On continued electrolysis the existing crystal nuclei grew, but new nuclei did not form. The higher current densities in the above range led to a larger number of nuclei owing to the larger supersaturation of the cathode surface and because of the involvement of less active substrate sites in the process [12]. With rather impure melts the appearance of a large number of crystallization centers and subsequent formation of fine Be crystals was observed [12, 13].

The location of the Be nuclei on the cathode surface was found to be reproducible at low current densities, e.g., at 500°C with $j = (3.3 \text{ to } 3.8) \times 10^{-2}$ A/cm² and a W microcathode. So if cathodic deposition and anodic dissolution were alternated continually on the same electrode, the crystal nuclei always appeared at the same positions ("site memory"). This indicates a considerable energetic heterogeneity of the cathode surface [12]; cf. [11, 13].

Effect of Temperature

The η vs. time curves for Mo microcathodes with a macroscopic defect applied in advance and at a $BeCl_2$ concentration of 5.85 wt% in eutectic LiCl–KCl melt showed maxima of η at 400 to 600°C with $j = 6.36 \times 10^{-2}$ to 7.8×10^{-1} A/cm², but not at 650 to 700°C with $j = 1.2$ A/cm². At 750°C a small depolarization wave ($j = 1.9$ A/cm²) indicated alloy formation. The nucleation on Mo increases with rising temperature. This was attributed to an increase in the number of vacancies present on the surface and in the bulk of the cathode metal [12, 13]. The crystallization overvoltage on W microcathodes decreases with increasing temperature, see **Fig. 16-25** from [13], disappearing completely at 750°C.

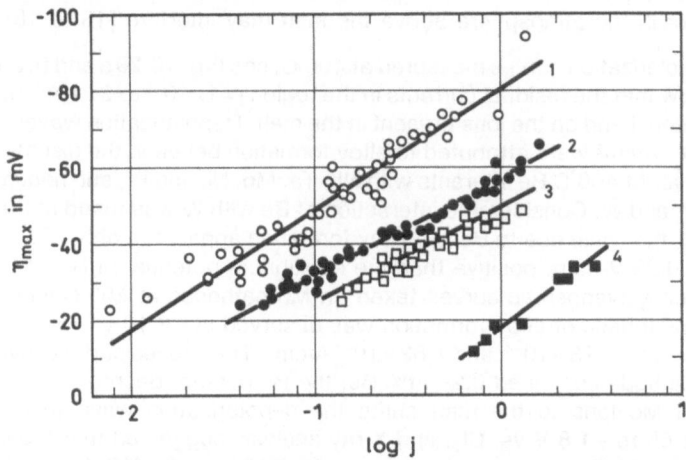

Fig. 16-25. Maximum crystallization overvoltage, η_{max}, of Be on a W electrode as a function of log j (with j measured in A/cm²) at the following temperatures: 1 = 400°C, 2 = 500°C, 3 = 600°C, and 4 = 700°C.

In the temperature range 400 to 700°C, the numerical value of η_{max} for W cathodes is represented by $\eta_{max} = 183.7 + 27.5 \log j - 0.1594\, T\ (\pm 5.7)$ mV, where T = absolute temperature in K and j = current density in A/cm². For expressions for t_{max} and plots of t_{max} vs. log j at 400, 500, 600, and 700°C, see the paper [13]; cf. p. 256. A rise in temperature of 100 K (in the above temperature range and for $j = 0.178$ A/cm²) decreased η_{max} by 4.7 mV and increased t_{max} by 16 ms. Estimated values for the penetration of Be into W at 400 and 850°C indicated that at low temperatures the buildup of the Be supersaturation on the cathode surface is the determining process, and that the diffusion mass transport is determining at the higher temperatures [13]. Ac measurements and the η vs. time curves indicated that ohmic contributions and the diffusion overvoltage of the Be^{2+} ions in the near-cathode layer become considerable at 800 to 850°C. Be-W alloy formation causes depolarization; thus at 850°C the potential of the W cathode is shifted by +10 to +12 mV. For details, see the paper [13].

The results [13] were used for a theoretical estimate of the Be nucleation rate on (110) faces and large-angle grain boundaries (LGB) of a W cathode at 773 K at constant j. It was shown that Be nucleates on W at low supersaturations ($\eta_{max} \leqq 30$ mV) in adsorbed layers of Be atoms on LGB. Active sites exert a decisive effect [15].

Experimental Conditions and Polarization Curves

For conditions in connection with the electrorefining of Be, i.e., simultaneous deposition and anodic dissolution, see pp. 260/2.

The electrodeposition of Be on a number of cathode materials (Zn, Ti, Zr, Nb, Ta, Mo, W, Ni, Pt) from eutectic LiCl–KCl melts with added $BeCl_2$ was studied for $BeCl_2$ concentrations of 1.62 to 15.07 wt% in the temperature range 380 to 800°C [16, 17]. High-purity materials and an Ar atmosphere were used. The melt was purified by preelectrolysis, because impurities more electropositive than Be are deposited prior to Be and may shield and passivate the cathode, i.e., interfere with Be deposition. Similarly, impurities such as oxides present in the melt, on the cathode surface, or in the atmosphere above the melt may interfere [16 to 18].

Potentiostatic polarization curves measured at 400°C, see **Fig. 16-26** a and b, and figures in the paper [17], show that the residual currents in the region $j = 1 \times 10^{-5}$ to 2×10^{-3} A/cm² depend on the cathode material and on the ions present in the melt. Depolarization waves in the range $j = 4 \times 10^{-4}$ to 2×10^{-2} A/cm² were attributed to alloy formation between the discharged Be and the cathode material. At 400°C Be interacts with Nb, Ta, Mo, Ni, and Pt, and negligibly with Ti, but not with Zn, Zr, and W. Considerable interaction of Be with W was found at 800°C. On Mo cathodes at 400°C, the wave due to Be-Mo alloy formation appears at about −2.15 V vs. Cl_2, i.e., at a potential 0.23 V more positive than the equilibrium potential of Be under the same conditions. Also, on galvanostatic curves taken on Mo cathodes at 400°C with 11.82 wt% $BeCl_2$, a step characteristic of alloy formation was observed at −2.10 to −2.15 V vs. Cl_2 for values of j of 4.07×10^{-4}, 7.13×10^{-4}, and 7.63×10^{-4} A/cm². The alloyed surface layer formed at 400°C on an Mo cathode contained 0.24 at% Be; the Be content decreased with increasing depth. Addition of Mo ions to the melt shifts the depolarization wave to more positive potentials, i.e., −1.65 to −1.8 V vs. Cl_2, and X-ray analysis suggested that Mo is deposited along with metastable $MoBe_2$, which is also formed at −2.15 V vs. Cl_2 [16, 17]. Metallic Be was deposited on Mo at potentials corresponding to the Be equilibrium potential. The limiting diffusion current is proportional to the $BeCl_2$ concentration and a two-electron mechanism is indicated. A wave appearing at −2.50 to −2.65 V vs. Cl_2 was attributed to the reaction $Be^{2+} + e \rightarrow Be^+$, and the simultaneous formation of a metallic mirror on the glass and quartz parts of the apparatus was attributed to the reaction $2Be^+ \rightarrow Be^{2+} + Be$ [16].

The polarization of Mo cathodes was measured at the moment of current interruption after 4 s polarization at a fixed j in LiCl–KCl eutectic melts additionally containing 0.1, 1.9, and 7.25 wt% $BeCl_2$ (corresponding mole fractions: 3.5×10^{-4}, 6.6×10^{-3}, and 2.5×10^{-2}, respectively) at 400, 500, and 600°C with j values of $10^{-3} \leqq j \leqq 3$ A/cm²; see figure 3 in the paper [19]. The polarization curves thus obtained indicate that the residual currents are mainly due to the reduction of alkali ions to subions while the reduction of Be^{2+} to Be^+ does not play a significant role. The Be deposition starts at the equilibrium potential in the melt and then proceeds without significant polarization. Concentration polarization was expected on account of the presence of complex ions of the type $BeCl_4^{2-}$ in the melt. But the active cathode surface increases during

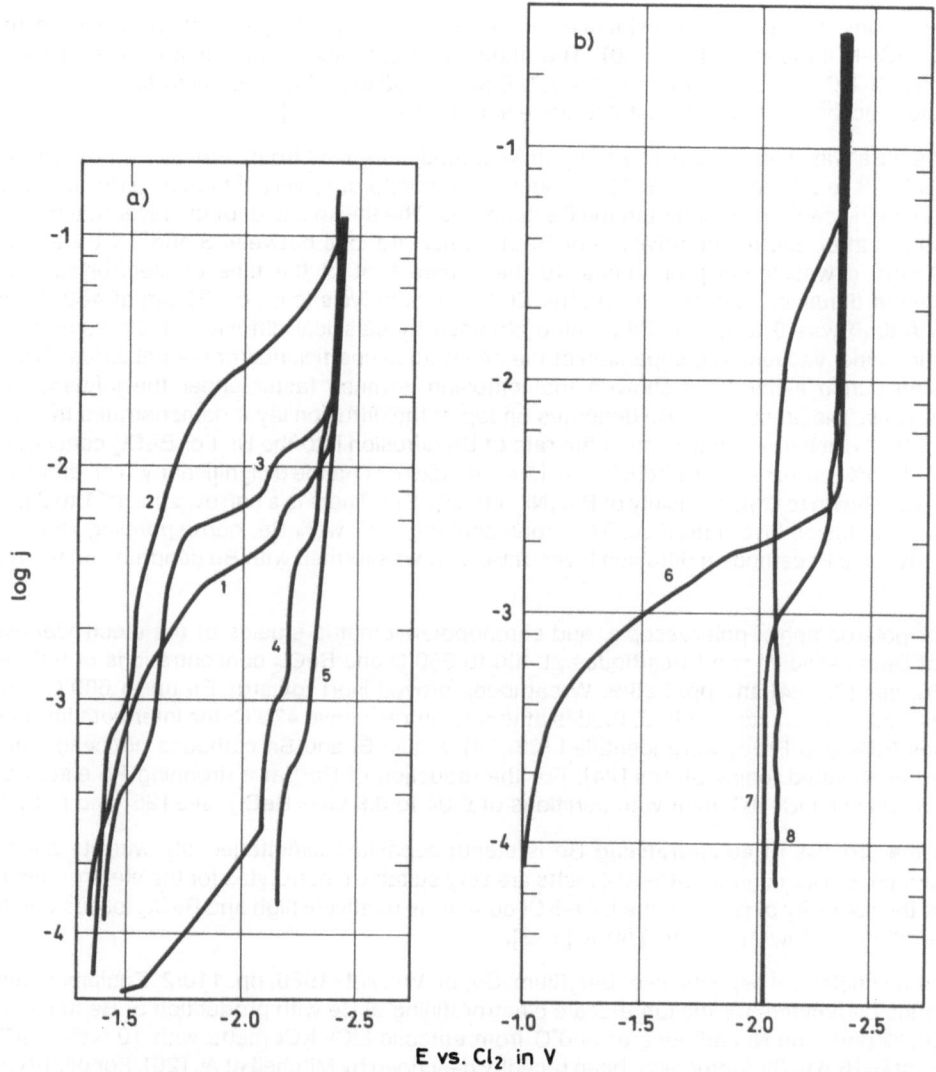

Fig. 16-26 (a and b).   Cathodic polarization curves, taken in a eutectic LiCl–KCl melt with 5.27 wt% $BeCl_2$ added, at 400°C with the cathodes 1 = Nb, 2 = Ta, 3 = Ni, 4 = Pt, 5 = Ti, 6 = W, 8 = Zr, at 380°C with the cathode 7 = Zn (j measured in A/cm²).

the deposition of crystalline Be, and so the true cathodic current density grows more slowly than the nominal current density, especially for higher $BeCl_2$ concentrations. At 500°C the nominal limiting j for the Be^{2+} discharge is not discernible for a 0.1 wt% $BeCl_2$ melt. It is clearly expressed for 1.9 wt% $BeCl_2$, viz. 10^{-3} to 10^{-2} A/cm², and is beyond 2.5 A/cm² for 7.25 wt% $BeCl_2$. The potentials of deposition become more positive with increasing temperature and Be^{2+} concentration. Addition of KF to a eutectic LiCl–KCl melt containing 1.9 wt% $BeCl_2$, such that the molar ratio is Be : F = 1 : 4, shifted the deposition potential of Be at 500°C to more negative values by ~0.3 V and decreased the limiting j for the Be^{2+} discharge significantly.

These results are considered evidence for the process of primary reduction of Be^{2+} in the $BeCl_2$–LiCl–KCl melts [19]; cf. [20]. The anodic and cathodic polarization curves of Be in molten LiCl–KCl eutectic, containing 10 wt% $BeCl_2$ at 450 to 500°C, demonstrate an extremely fast electrochemical system, suitable for electrorefining [20, 22].

When Be was electrodeposited onto an Ni cathode at 480°C from eutectic LiCl–KCl melts containing 3 to 20 wt% $BeCl_2$ [21], a continuous, nonporous, very adherent diffusion layer formed on the cathode underneath the Be dendrites. The thickness, d, of the layer reached up to 100 μm after 650 h electrolysis. For $BeCl_2$ concentrations between 3 and 7 wt% and j = 1.5 A/dm², d was found proportional to the square root of the time of electrolysis, i.e., the time of diffusion. The value of d after 20 h of electrolysis (e.g., d ≈ 32 μm at 480°C and j = 1.5 A/dm²) was 6 to 8 times the value obtained by classical diffusion at the same temperature under vacuum. A comparison of the penetration coefficients for normal diffusion and diffusion during electrolysis showed that diffusion is much faster under the influence of electrolysis. The presence of Be dendrites on top of the diffusion layer demonstrates that the rate of Be^{2+} discharge is higher than the rate of Be diffusion into the Ni. For $BeCl_2$ concentrations > 5 wt%, an increase in j from 0.3 to 15 A/dm² did not change d significantly. The diffusion layer was shown to consist mainly of $Be_{21}Ni_5$ (40 wt% Be). There is a narrow zone (≦ 1 to 2 μm) at the Ni–diffusion layer interface. This zone contains ~13 wt% Be, corresponding to BeNi. Similarly, on a Pt cathode a diffusion layer of $Be_{21}Pt_5$ was formed with Be dendrites on the top [21].

For polarographic, polaroscopic, and chronopotentiometric studies of the electrodeposition of Be on solid W and Pt cathodes at 400 to 850°C and $BeCl_2$ concentrations of 0.04 to 3 wt%, see [23, 24] and pp. 228/9. W cathodes proved inert towards Be up to 600°C. The discharged Be interacts with Pt cathodes at temperatures above 470°C; the intermetallic compounds PtBe and $PtBe_3$ were identified [23, 24]. Liquid Bi and Sn cathodes noticeably dissolved at elevated temperatures [24]. For the reduction of Be^{2+} at a dropping Pb electrode from a eutectic LiCl–KCl melt with additions of 0.04 to 0.9 wt% $BeCl_2$, see [25] and p. 227.

In the process of **electrorefining** Be is electrodeposited simultaneously with its anodic dissolution. $BeCl_2$–eutectic LiCl–KCl melts are very suitable electrolytes for the electrorefining of Be; the solubility of $BeCl_2$ in the LiCl–KCl eutectic is relatively high and $BeCl_2$ losses due to evaporation are low at 720 to 750 K [1, 2].

For technological aspects, see "Beryllium" Suppl. Vol. A 1, 1986, pp. 110/2. Equipment and operating parameters for the large-scale electrorefining of Be with production of Be flakes of ~99.95% purity on Ni cathodes at 500°C from eutectic LiCl–KCl melts with 10 wt% $BeCl_2$ added at j = 15.6 to 39 A/dm² have been recently described by Mitchell et al. [26]. For pilot-type equipment and operation, see [27]; cf. [28].

A careful investigation of the purification of Be by electrorefining in LiCl–KCl–$BeCl_2$ melts was carried out by Wohlfarth [2]. To a eutectic LiCl–KCl melt (59 and 41 mol%, respectively) $BeCl_2$ was added in amounts of about 0.02, 0.09, and 0.12 mole fraction (according to analysis, cf. table 5, p. 31 in [2]). The $BeCl_2$ was prepared in the apparatus from Be and HCl gas. Electrolysis was performed under Ar at 720 K for the lowest and at 750 K for higher $BeCl_2$ concentrations. Cathodic current densities ranged from 2.5 to 200 mA/cm², anodic current densities from 0.8 to 25 mA/cm² (for cathodic and anodic polarization values, see the paper). The deposited Be consisted of flakes or dendrites adherent to the Be cathode; descriptions of the morphology and structure as well as figures are given in the paper. Three kinds of commercially pure Be (see table 15 on p. 53 in [2]) were used as anode material and, in addition, Be metal enriched in Al and Fe (0.3 and 0.16 at%, respectively), and materials enriched in Cu (0.014, 0.14, 0.7, and 4.5 at%). In commercial Be the main impurities Si, Mg, Al,

Cr, Fe, and Cu were present in concentrations of 0.1 to 200 atppm, while oxygen was in the range of 100 to 1000 atppm, and Li, K, Zn were each in the range of 1 atppm. When the melts were purified by preelectrolysis, it was found that with anodic current densities >7 mA/cm^2, independent of the $BeCl_2$ concentration and temperature in the above ranges, nearly the same remaining impurity contents were achieved, viz. (in atppm) for the more electropositive metals Si (2 to 3), Zn (0.1 to 0.2), Al (2 to 3), Cr (0.1 to 0.2), Fe (0.8), Cu (0.1), as well as for the more electronegative metals Li (200 to 300), K (100 to 200), and Mg (0.02 to 0.03). Only at anodic current densities $\leqq 2$ mA/cm^2, i.e., far below that used for technical procedures, an increase of the $BeCl_2$ concentration from about 0.025 to 0.1 mole fraction in the melt led to lower impurity concentrations of Al, Cr, and Fe by a factor of 10, and to higher concentrations in Li and K by a factor of 5. The Li and K contaminations may be diminished by heating the Be flakes in vacuum at 970 K for several hundred hours. The dissolution of Al and Fe, or Cu into the electrolyte melt from the Be anode materials enriched with these metals (viz., Al 0.3 at% and Fe 0.16 at%, or Cu up to 4.5 at%) leads to an increase of their concentration in the cathode deposits (viz., Al to 280 atppm, Fe to 1 atppm, Cu to 7 atppm) [2].

For electrorefining of Be with the aim of preparing high-purity metal and of reclaiming scrap Be, eutectic LiCl–KCl melts, additionally containing 3.5 to 16.5 mol% $BeCl_2$, were electrolyzed at 450 to 600°C with j ranging from 45 to 770 A/ft^2. The $BeCl_2$ component of the bath was produced in situ by reacting Be metal, supplied in a graphite tube, with HCl prior to electrolysis. The electrolysis was carried out in steel chambers lined with graphite under an He atmosphere; see [29 to 31], cf. [32]. In the investigations no significant differences were found for Fe or Mo cathodes. Similarly, metallic impurities (except Ca) in electrodeposited Be were reduced to concentrations below the spectrographic determination limits by a second cycle of electrorefining with 11.4 mol% $BeCl_2$ in eutectic LiCl–KCl at 500°C and j in the range ~150 to 450 A/ft^2 (using an Ni chlorinator tube, an Ni-lined steel chamber, and an Ni cathode) [33]. Also in the temperature range from ~400 to 500°C, the electrorefining of Be was investigated with eutectic LiCl–KCl melts additionally containing ~10 wt% $BeCl_2$. Various Ni cathodes were used with cathodic current densities of 10 to 30 A/dm^2; see [20, 22, 27, 34, 35]. Cathodes of Ni, stainless steel, Inconel, or graphite in melts containing 6 to 20 wt% $BeCl_2$ were also tried in the temperature range 460 to 550°C with current densities of 1075 to 3800 A/m^2. The optimum temperature range proved to be 500 ± 50°C; the adherence of the Be dendrites deposited was equally good on all cathodes used in the above temperature range. The Be obtained was of spectrographic purity. The purity depended on the purity of the melt; the use of a graphite crucible did not increase the C content of Be [1].

High $BeCl_2$ concentrations and low current densities favored the formation of steel gray, flat, platelike Be crystals, whereas low $BeCl_2$ concentrations and high current densities yielded long fibers consisting of platelets. Large lamellae formed in baths with 16.5 mol% $BeCl_2$. The size of the Be crystals decreased with increasing current density. The adherence of the Be deposits on the cathodes was considerably better at the higher temperatures, but sublimation of $BeCl_2$ from a 16.5 mol% $BeCl_2$ bath at 550°C was remarkable. During these electrorefining experiments all impurities in Be were reduced, except for C; this may be due to the graphite used in the apparatus. Thus, the following reductions were achieved: O from 3.27% (in the anode feed) to ~0.04 to 0.06% (in refined Be); Al from 0.5 to 1.0% to $\leqq 0.004$%; Fe from 0.32 to $\leqq 0.08$%; Mg from 2.66 to <0.02%; Ni from 0.05 to <0.002%; Si from 0.1 to 1.0% to $\leqq 0.02$% [29]. The Cl content in the Be crystals appeared to depend on the crystal size. The best results were obtained with the 10.3 mol% $BeCl_2$ bath at 550°C; see [29 to 31].

In experiments with Ni cathodes, e.g., the purity of the refined Be was ~99.8%. The metallic impurities were reduced from 1000 to 4000 ppm in the anode Be to $\leqq 35$ ppm in the refined Be, the O content from 1000 to 5000 ppm to 100 to 300 ppm; traces of the electrolyte were reduced to several ppm after fusion of the refined Be. The purity of the Be deposit did not appear to depend on the size of the Be crystals; however, the larger surface area of fine needles increased the O content, and efficient washing of fine structures was more difficult [20, 22].

Already in early experiments high-purity Be flakes were obtained at 400°C on Cu cathodes from melts of eutectic LiCl–KCl with Be being simultaneously dissolved from a Be anode at anodic current densities of up to 50 A/dm² [36]. The process was also studied at 400 to 550°C with Be, Ni, and stainless steel cathodes. Beryllium of 99.0% purity was obtained adherent to an Ni cathode at 460°C, j = 50 A/dm², whereas with j = 170 A/cm² at 550°C and a stainless steel cathode, the Be deposits consisted of nonadherent flakes of 99.5% purity. With Be cathodes the deposited Be was of low purity (85% at 400°C, j = 70 A/dm² and 95% at 500°C, j = 94 A/dm²). As alkali metal ions are discharged especially in the early stages of electrolysis, part of the Be obtained will result from secondary reduction of Be^{2+}. Flakes are thought to be due to the primary (the electrochemical) reduction process, and powdery, partially adherent Be to the secondary process [37]. For experiments with a diaphragm separating anode and cathode compartments, and also the use of BeO–C anodes, see [38]. Mo and Zr cathodes were investigated for Be electrorefining at 760 to 900°F (404.5 to 482°C) in $BeCl_2$–eutectic LiCl–KCl melts (ratios by wt% 1:2 to 1:3) with best results at 900°F; see [39].

In the electrorefining of Be in eutectic LiCl–KCl melts with additional $BeCl_2$, certain impurities such as Al, Mn, Ni, Fe, Cu (i.e., metals reduced more easily than Be) were found to concentrate in the salt layer surrounding the Be flakes at the cathode. Chemical and spectrographic analyses showed that their concentrations increased rapidly towards the cathode and reached values up to 100 times their concentration in the bulk electrolyte and even 20 times that in the electrodeposited Be. Apparently this phenomenon is independent of the $BeCl_2$ concentration; see [20, 22, 27]; cf. [34, 35]. Anodic current efficiencies indicated that at low current densities Be^+ was also formed at the anode [20]. In some cases the formation of white particles of 95% BeO was observed in the bath. This did not increase the O content of the Be deposit, but reduced the current efficiency. The formation of oxide is attributed to oxidation by moisture, present despite all precautions, and/or to the interaction of the melt with Pyrex and quartz parts of the apparatus [27].

The direct introduction of the very hygroscopic $BeCl_2$ into the melt can be avoided by reacting $SnCl_2$ with Be at 450 to 500°C in the LiCl–KCl melt according to $Be + SnCl_2 \rightarrow BeCl_2 + Sn$. The Sn collects at the bottom of the cell and residual $SnCl_2$ is removed by pre-electrolysis [20, 34]. The Be deposited in the form of needles and two-dimensional dendrites in the case of a contamination with Sn (500 to 2500 ppm) when the residual Sn^{2+} (from the $BeCl_2$ preparation) was not removed by preelectrolysis [27].

References:

[1] Schimmel, F. A. (Y-1380 [1962] 31 pp.; N.S.A. **16** [1962] No. 15127).

[2] Wohlfarth, H. (Diss. Univ. Stuttgart 1981, pp. 1/119, 31, 60, 101).

[3] Yang, L.; Hudson, R. G. (Trans. AIME **215** [1959] 589/601).

[4] Kuroda, T.; Matsumoto, O. (J. Electrochem. Soc. Japan **30** [1962] E 193/E 197).

[5] Kuroda, T.; Matsumoto, O. (Bull. Tokyo Inst. Technol. No. 61 [1964] 29/41; C.A. **64** [1966] 1629).

[6] Kuroda, T; Matsumoto, O. (Denki Kagaku **31** [1963] 688; J. Electrochem. Soc. Japan **31** [1963] 153).

[7] Chukreev, N. Ya.; Polishchuk, V. A.; Shapoval, V. I. (Fiz. Khim. Elektrokhim. Rasplavl. Tverd. Elektrolitov Tezisy Dokl. 7th Vses. Konf. Fiz. Khim. Ionnykh Rasplavov Tverd. Elektrolitov, Sverdlovsk 1979, Vol. 2, pp. 9/11; C.A. **93** [1980] No. 122440).

[8] Shapoval, V. I.; Chukreev, N. Ya.; Polishchuk, V. A. (Fiz. Khim. Elektrokhim. Rasplavl. Tverd. Elektrolitov Tezisy Dokl. 7th Vses. Konf. Fiz. Khim. Ionnykh Rasplavov Tverd. Elektrolitov, Sverdlovsk 1979, Vol. 2, p. 9; C.A. **93** [1980] No. 122381).

[9] Chukreev, N. Ya.; Shapoval, V. I.; Polishchuk, V.A. (Elektrokhimiya **18** [1982] 385/9; Soviet Electrochem. **18** [1982] 341/5).

[10] Chukreev, N. Ya.; Polishchuk, V. A. (Tezisy Dokl. Vses. Konf. Elektrokhim. **6** Vol. 2 [1982] 1/360, 294).

[11] Shapoval, V. I.; Chukreev, N. Ya.; Polishchuk, V. A. (Ukr. Khim. Zh. **49** No. 8 [1983] 841/5; Soviet Progr. Chem. **49** No. 8 [1983] 58/61).

[12] Chukreev, N. Ya.; Shapoval, V. I.; Polishchuk, V. A. (Elektrokhimiya **20** [1984] 520/4; Soviet Electrochem. **20** [1984] 490/4).

[13] Chukreev, N. Ya.; Shapoval, V. I.; Polishchuk, V. A. (Termodin. Elektrokhim. Svoistva Ionnykh Rasplavov **1984** 104/26; C.A. **101** [1984] No. 218544).

[14] Chukreev, N. Ya.; Polishchuk, V. A.; Shapoval, V. I. (Ukr. Khim. Zh. **52** No. 3 [1986] 282/7; Soviet Progr. Chem. **52** No. 3 [1986] 61/6).

[15] Shapoval, V. I.; Polishchuk, V. A.; Chukreev, N. Ya. (Teor. Eksperim. Khim. **23** No. 2 [1987] 245/9; Theor. Exptl. Chem. [USSR] **23** [1987] 229/33).

[16] Chukreev, N. Ya.; Sunegin, G. P. (Fiz. Khim. Elektrokhim. Rasplavl. Solei Tverd. Elektrolitov **1973** No. 2, pp. 10/2; C.A. **82** [1975] No. 9210).

[17] Chukreev, N. Ya.; Sunegin, G. P. (Elektrokhimiya **9** [1973] 842/5; Soviet Electrochem. **9** [1973] 806/9).

[18] Sunegin, G. P.; Chukreev, N. Ya. (Ukr. Khim. Zh. **38** No. 11 [1972] 1091/6; Soviet Progr. Chem. **38** No. 11 [1972] 13/6).

[19] Smirnov, M. V.; Ivanovskii, L. E. (Zh. Fiz. Khim. **32** [1958] 2174/82; C.A. **1959** 6833).

[20] Boisdé, G.; Broc, M.; Chauvin, G.; Coriou, H.; Hardy, L.; Jarny, P. (Bull. Inform. Sci. Tech. [Paris] No. 62 [1962] 29/38; C.A. **58** [1963] 12172).

[21] Broc, M. (CEA-R-4921 [1978] 103 pp., 21, 79/99; C.A. **91** [1979] No. 201108).

[22] Boisdé, G.; Broc, M.; Chauvin, G.; Coriou, H.; Hure, J.; Jarny, P. (J. Nucl. Mater. **6** [1962] 256/64).

[23] Delimarskii, Yu. K.; Baranenko, V. M.; Zarubitskii, O. G. (Deposited Doc. VINITI-3083-76 [1976] 1/13; Elektrokhimiya **12** [1976] 1887; Soviet Electrochem. **12** [1976] 1717).

[24] Delimarskii, Yu. K.; Baranenko, V. M.; Barchuk, V. T.; Shapoval, V. I. (Ukr. Khim. Zh. **45** No. 1 [1979] 3/7; Soviet Progr. Chem. **45** No. 1 [1979] 1/5).

[25] Naryshkin, I.; Minin, N. A. (Zh. Prikl. Khim. **34** [1961] 2353/6; J. Appl. Chem. [USSR] **34** [1961] 2230/2).

[26] Mitchell, D. L.; Nieweg, R. G.; Ledford, J. A.; Richen, M. J.; Burton, D. A.; Harder, R. V.; Watson, L. E.; Thomas, R. L. (RFP-4188 [1989] 26 pp.; C. A. **112** [1990] No. 127793).

[27] Bilard, J.; Boisdé, G.; Broc, M.; Chauvin, G.; Coriou, H.; Hardy, J. (Metaux **42** [1967] 259/69).

[28] Wong, M. M.; Klosterman, J. E. (BM-RI-6489 [1963] 19 pp.; N.S.A. **18** [1964] No. 34031).

[29] Wong, M. M.; Campbell, R. E.; Baker, D. H., Jr. (J. Metals **12** [1960] 786/8; C.A. **1960** 24008).

[30] Wong, M. M.; Cattoir, F. R.; Baker, D. H., Jr. (Rept. Invest. U.S. Bur. Mines No. 5581 [1960] 9 pp.; C.A. **1960** 10591).

[31] Wong, M. M.; Campbell, R. E.; Baker, D. H., Jr. (Rept. Invest. U.S. Bur. Mines No. 5959 [1961]; N.S.A. **16** [1962] No. 15117).

[32] Bellamy, R. G.; Hill, N. A. (International Series of Monographs on Nuclear Energy Division 8: Materials. Vol. 1. Extraction and Metallurgy of U, Th, Be, Pergamon, New York 1963, 198 pp., 81/4).

[33] Wong, M. M.; O'Keefe, D. A. (BM-RI-6570 [1964] 10 pp.; N.S.A. **19** [1965] No. 13870).

[34] Chauvin, G.; Coriou, H.; Hure, J. (Metaux Corrosion Ind. **37** [1962] 112/36; C.A. **57** [1962] 8261).

[35] Coriou, H. (Bull. Inform. Sci. Tech. [Paris] No. 84 [1964] 17/35; C.A. **65** [1966] 16397).

[36] Menzies, I. A.; Hill, D. L.; Owen, L. W. (Nature **183** [1959] 816/7).

[37] Menzies, I. A.; Hill, D. L.; Owen, L. W. (J. Less-Common Metals **1** [1959] 321/30).

[38] Higgins, J. K.; Bellamy, R. G.; Buddery, J. H. (AERE-M-R-2750 [1958] 18 pp.).

[39] Gurklis, J. A.; Beach, J. G.; Faust, C. L. (BMI-781 [1955] 26 pp.; N.S.A. **10** [1956] No. 1367).

16.7.5.6   BeCl$_2$–NaCl–KCl

References for this section are given on p. 268.

General Remarks

The melts investigated were usually based on equimolar NaCl–KCl mixtures (melting at 658°C) with BeCl$_2$ added in varying amounts, mostly in a minor extent. The diffusion coefficients of the Be^{2+} ion in such melts containing 0.108 or 0.18 wt% Be were determined from potential vs. time curves at 680 to 800°C using an Mo cathode [1]. An empirical equation relating the diffusion coefficient to the temperature and ionic moment in the above melt for the temperature range 1000 to 1100 K was formulated [2].

Cathodes of Solid Al, Mo, Ni, and Alloyed Steel

Beryllium deposition on solid Al cathodes from BeCl$_2$–NaCl–KCl melts was investigated in comparison to the deposition with cathodes of liquid Al; see pp. 267/8 and [3].

When a cleaned Mo rod cathode was used in the electrolysis of an equimolar NaCl–KCl melt additionally containing 3.7 wt% BeCl$_2$ at 678°C, the polarization curve, see Fig. 16-28 on p. 267 from [4], indicated interaction between the Mo and the deposited Be because at low current densities the cathode potential is more positive than according to the equilibrium. The cathode potential changed linearly from −2.1 V vs. Cl$_2$ at j = 4×10^{-4} A/cm^2 to −2.4 V at 0.4 A/cm^2. The limiting current was obscured by the growth of the effective cathode surface in the course of the Be deposition [4].

Experiments for the refining of Be with an Ni cathode and a Be rod anode with equimolar (eutectic) NaCl–KCl mixtures melted together with varying amounts of BeCl$_2$ at temperatures between 280°C (50% BeCl$_2$) and 450°C (30% BeCl$_2$) showed that at the higher BeCl$_2$ contents (and lower temperatures) larger Be crystals (needles) are obtained. Small Be crystals and even spongy metal appeared at the cathode in the case of only 30% BeCl$_2$ in the melt (at 450°C); see table in the paper [5]. Also the current efficiency was highest (93%) at j = 5 A/dm^2 when the melt contained 50% BeCl$_2$ at 280 to 300°C [5].

For the deposition of Be on Ni or alloyed steel cathodes from melts based on NaCl and KCl in 1:1 ratio (by wt%) with 13 to 15% BeCl$_2$ added at 760 to 790°C, see [6]. With alloyed steel cathodes, Be flakes were obtained from the above melts at 900 to 925°C [7]. At 750 to 900°C the Be deposition on iron alloy cathodes from melts of equimolar (eutectic) NaCl–KCl with 15 wt% BeCl$_2$ added involves corrosion problems and toxic hazards [8].

If the composition of BeF_2–NaF melts is near that corresponding to Na_2BeF_4, only Na is deposited on electrolysis according to [8]. Sodium was not found to deposit from NaF–BeF_2 melts (2:1 parts by wt) up to temperatures of 1000°C at current densities from several tenths to 10 to 20 A/dm², using a Be anode. Under these conditions, diffusion coatings were formed on cathodes of Y, Ti, U, and Ni with the respective compositions of YBe_{12}, $TiBe_{12}$, UBe_{13}, and Ni_5Be_{21} [9, 10].

References:

[1] Bellamy, R. G.; Hill, N. A. (International Series of Monographs on Nuclear Energy Division 8: Materials Vol. 1. Extraction and Metallurgy of U, Th, Be, Pergamon, New York 1963, 198 pp., 79).

[2] Fischer, H.; Schwan, W. (Metallwirtsch. Metallwiss. Metalltech. 12 [1933] 187/9).

[3] Illig, K.; Fischer, H.; Birett, W. (in: Eger, G.; Die Technische Elektrolyse im Schmelzfluß, 2nd Ed., Leipzig 1955, pp. 571/641, 622/3).

[4] Hubbard, D. O. (Trans. Electrochem. Soc. 89 [1946] 260).

[5] Claflin, H. C. (U.S. 2022404 [1934]; C.A. 1936 687).

[6] Claflin, H. C.; Arnold, J. B. (U.S. 2046148 [1934]; C.A. 1936 5510).

[7] Claflin, H. C.; Arnold, J. B. (Fr. 791704 [1935]; C.A. 1936 4102).

[8] Bellamy, R. G. (AERE-M-534 [1959] 41 pp., 16) cited in [11].

[9] Cook, N. C. (Prot. Corros. Metal Finish. Proc. Intern. Conf., Basel 1966 [1967], pp. 151/5; C.A. 68 [1968] No. 107321).

[10] Cook, N. C.; Fosnocht, B. A.; Evans, J. D. (24th Conf. Natl. Assoc. Corros. Eng. Proc., Cleveland 1968 [1969], pp. 520/4; C.A. 72 [1970] No. 8624).

[11] Chauvin, G.; Coriou, H. (in: Bard, A. J.; Encyclopedia of the Electrochemistry of the Elements, Vol. 5, 1976, pp. 227/60).

16.7.4.4 BeF_2–NaF–KF

By electrolysis of a fluoride melt of the composition NaF–KF–$2BeF_2$ (m.p. ~600°C), fine-crystalline Be was deposited on Ni cathodes using a Be anode. The melt was preelectrolyzed at 1 A/dm² with a graphite anode and the Ni cathode was preberyllized for 1 h at a low current density (0.4 to 0.5 A/dm²) to ensure better cohesion of the Be to the cathode. The process is not suitable for refining Be, as the adhering electrolyte could be removed only with difficulty [1]. Beryllium produced from fluoride melts containing NaF and/or KF besides BeF_2 could not be heated to coalesce the Be, because of fuming, burning of alkali metals, and the disappearance of Be. This was attributed to a reaction between Be and NaF and/or KF to form BeF_2 and alkali metals [2].

References:

[1] Silina, G. F.; Grinberg, L. L. (Tsvetn. Met. [Moscow] 33 No. 12 [1960] 47/53; Soviet J. Non-Ferrous Metals 1 [1960] 46/52; C.A. 1961 19555).

[2] Wong, M. M.; Couch, D. E.; O'Keefe, D. A. (J. Metals 21 No. 1 [1969] 43/5).

16.7.4.5   BeF₂–LiF–NaF–KF

The electrolysis of molten eutectic LiF–NaF–KF (29.2, 11.7, and 59.1% by wt, respectively) with added BeF_2 (15 to 35 wt%) was studied in detail; see [3 to 5]. For this purpose crucibles of high-purity graphite or metallic materials under an Ar atmosphere in a sealed metallic (e.g. Ni-Cr-Mo alloy) apparatus were used. Commercial Be rods or plates served as anodes and Ta, Mo, Ni, Fe, Cu, Ag, and Pt as cathodes. In these experiments precautions were taken to prevent the interference of O_2, H_2O, and other impurities in the Be deposition: use of high-purity salts (purity of BeF_2 should be >99%), degassing of the apparatus at a temperature above that of electrolysis, introduction of pure Ar (O_2 and H_2O content $\leqq$ 5 vpm), preheating of the cathodes and pretreatment of the cathode surfaces, and dehydration of the melt by blowing HF through it for 2 h at temperatures between 600 and 800°C, followed by the passage of Ar. The best temperature proved to be 600°C, though 550 to 750°C was suitable. The current densities ranged from 2 to 10 A/dm²; the applied voltage was ~0.5 V. Mechanical vibration of the cathode (50/s) removed adsorbed gas at the cathode–melt interface and reduced the concentration gradient near the cathode [3]. In general, high-purity Be deposits were obtained; they were compact, coherent, and perfectly crystallized. Sound, continuous, slightly dendritic Be deposits (thickness ~20 μm) formed on Ta cathodes at 600°C and 8 A/dm² after 4.5 h electrolysis. They were nonadherent and detachable (useful for electroforming). Adherent, homogeneous, polycrystalline Be layers (thickness 30 to 60 μm) of excellent purity were obtained on Mo plate cathodes during 5 h electrolysis at 600°C and 9 A/dm². An electron-microscopic analysis showed a weak penetration of Mo into Be (<5 μm) [3 to 5]. According to [5], this diffusion layer corresponds to $MoBe_{13}$. On an Ni cathode, electrolysis for 2 h at 650°C (4 A/dm², BeF_2 concentration 20 wt%) yielded a 30 μm thick diffusion layer composed of $Be_{21}Ni_5$. A thin deposit of dendritic Be formed on the diffusion layer, while a narrow zone ($\leqq$2 μm) corresponding to BeNi formed at the Ni–diffusion layer interface. The thickness of the alloy obtained varied with the square root of time. The electrolysis current promotes diffusion as compared to normal diffusion. A detailed discussion is given in the paper [5]. An undefined phase formed on Fe cathodes at 800°C with the formation of $FeBe_5$ underneath Be dendrites on the top [5]. Beryllium dendrites formed on top of continuous Be-Cu alloy phases (solid α-, β-, δ-, ε-Be-Cu phases) at 600°C [5]; cf. [3, 4]. On Ag cathodes Be dendrites formed directly [5].

According to an earlier patent specification [6], "beryllide" coatings (i.e., coatings of Be alloys with the cathode metals) form on cathodes consisting of, e.g., Y, Ti, Zr, U, Ni, Fe, Cu, Pt, or their alloys, when these are immersed in fused BeF_2–alkali fluoride (10 to 66.6 mol%) baths and connected externally to a Be anode in the same melt. Deposition already proceeds without applied voltage, and very smooth, adherent, tough, and corrosion-resistant beryllides are obtained with applied voltages of <0.5 V, total current densities of <3 A/dm², and temperatures in the range 600 to 900°C. Total current densities >3 A/dm² may be applied for the deposition of nonadherent grains or crystals of Be. Concentrations of BeF_2 of >33.3 mol% prevent the formation of alkali metal at the cathode. At the higher temperatures the melts should contain 50 to 66.6 mol% of the alkali fluorides because of the considerable vapor pressure of BeF_2. Results are given for a melt consisting of 1565 g BeF_2 in 2800 g of an LiF–NaF–KF melt (45, 10, and 45 mol%, respectively). A Y rod was covered at 700°C and 1.2 to 1.5 A/dm² with 0.5×10⁻³ in of a hard beryllide containing YBe_{13}. Gray, hard beryllides formed at 700°C on Ti, Zr, Fe, and Pt at 0.8, 3, 2, and 3 A/dm², respectively. The beryllide coating (thickness 0.5×10⁻³ in) produced on a U rod at 630 to 730°C and 0.09 to 1.5 A/dm² was resistant to oxidation. Nickel beryllide obtained at 700°C and 2.2 to 4.5 A/dm² consisted of Ni_5Be_{21}. A beryllide coating consisting of five Cu-Be alloy layers was formed on Cu strip cathodes at 675 to 700°C and 0.5 to 1.5 A/dm² [6]; cf. [5].

References:

[1] Silina, G. F.; Grinberg, L. L. (Tsvetn. Met. [Moscow] **33** No. 12 [1960] 47/53; Soviet J. Non-Ferrous Metals **1** [1960] 46/52; C.A. **1961** 19555).

[2] Wong, M. M.; Couch, D. E.; O'Keefe, D. A. (J. Metals **21** No. 1 [1969] 43/5).

[3] Binard, M.; Boisdé, G.; Broc, M.; Chauvin, G.; Coriou, H. (Fr. 1521522 [1967/68] 4 pp.; C.A. **71** [1969] No. 18301).

[4] Broc, M.; Chauvin, G.; Coriou, H. (Molten Salt Electrolysis Met. Prod. Intern. Symp., Grenoble 1977, pp. 69/73; C.A. **88** [1978] No. 43101).

[5] Broc, M. (CEA-R-4921 [1978] 103 pp., 21/31, 78/99; C.A. **91** [1979] No. 201108).

[6] Cook, N. C. (U.S. 3024175 [1959/62]; C.A. **56** [1962] 15291).

16.7.4.6 Other Melts Based on Fluorides

Beryllium deposited from $BaBeF_4$ or $MgBeF_4$ melts did not contain any detectable amounts of Ba or Mg [1]. Wong et al. [2, 3] electrolyzed fluoride melts with BeO dissolved in numerous different fluoride mixtures selected from LiF, NaF, KF, BeF_2, MgF_2, CaF_2, SrF_2, and BaF_2. Beryllium could be electrodeposited from these melts only if BeF_2 was also added.

The deposits from melts with BeO–BeF_2 dissolved in MgF_2 or CaF_2 contained intermetallic compounds of Be with Mg or Ca, respectively. The Be deposits from melts based on NaF or KF could not be heated in an open crucible to $\sim1300°C$ in order to coalesce the Be, because of fuming and burning of the alkali metals and the disappearance of the Be [2, 3].

Good deposits consisting of fine Be crystals intermixed with electrolyte were obtained with BeO dissolved in LiF–BeF_2 or LiF–BeF_2–BaF_2 melts (solubility of BeO at 700°C: 0.09 mol BeO/kg melt). The electrolyses were carried out under an inert gas atmosphere in the optimum temperature range 700 to 750°C. At temperatures below 700°C the solubility of BeO and the electrical conductivity of the baths were too low; at temperatures above about 700°C the volatilization of the BeF_2 increased and caused corrosion of the apparatus. A graphite crucible served as anode and closed-end Ni and Fe tubes as cathodes in batch and semicontinuous operations, respectively. The corresponding cathodic current densities were 26 and 18 A/dm². The highest current yields, viz. 51 and 74%, respectively, were obtained with BeO in a 60 mol% LiF–40 mol% BeF_2 melt at 700°C in batch and semicontinuous operations. The purity of the Be obtained after coalescing was 98.8%, the major impurities being Ni, Ca, Al, and Fe. For a cell used in semicontinuous operation and for other details, see [2, 3]. Beryllium of 99.3% purity was obtained from a melt consisting of 30 g BeO, 196 g LiF, and 354 g BeF_2 at 720°C and 721 A/ft². Similarly, a melt consisting of 30 g BeO, 169 g LiF, 196 g BeF_2, and 135 g BaF_2 yielded 99.2% pure Be at 740°C and 784 A/ft²; the same melt without BeO yielded less pure Be. For details of the impurity concentrations in the Be deposit, see [3].

In studies with an LiF–BeF_2–BeO melt (56, 37, and 7 mol%, respectively), Be powder was produced from the BeO at about 790°C. Rods of Mo as cathodes, a porous C anode, and graphite crucibles were used; the range of temperature investigated was 550 to 850°C. The BeO was replenished while the electrolysis was in progress, the feed rate requiring careful regulation. The optimum temperature was 780 to 800°C. For current densities of 50 to 125 A/dm², the current efficiency varied from 47 to 50%, respectively. The size of the Be crystals produced decreased with increasing current density. At a cathodic current density of 5 A/dm², flat Be crystals, 3 mm across, formed. At $j = 75$ A/dm² satisfactory Be powder deposits of $\sim99\%$ purity were obtained, which could be coalesced by heating to 1300°C. Main impurities were Ca, Al, and Fe (0.150, 0.125, and 0.44%, respectively) [4].

Compact Be of up to 99.5% purity can be obtained by electrolysis of BaF_2–Be oxyfluoride melts at 1400°C (Siemens-Halske method), using an H_2O-cooled Fe rod cathode. The current yield has a maximum value of 80% for a 1:1 (by wt) mixture of BaF_2 and Be oxyfluoride. Up to 1% H_2O in the electrolyte can be tolerated [1]. At an NaF content $\geqq$14% the current yield decreases, and pores in the Be deposit and codeposition of Na occur [5]. Sputtering of the melt happens when the BeO content is higher than that corresponding to $5 BeF_2 \cdot 2 BeO$. In this case, the Be deposit contains slags [1].

A comparison of the decomposition voltages of BeO, MgO, and Al_2O_3, as calculated from thermodynamic data, indicated that only a small amount of Be deposits together with Al at 950 to 1000°C from cryolite melts which contain BeO and Al_2O_3 [6].

Beryllium Alloys with the Cathode Metal

A **Be-Cu alloy** with 2.85 wt% Be is formed by electrolysis of molten mixtures of $2 BeO \cdot 5 BeF_2$–NaF–BaF_2 in a graphite crucible at 1150 to 1240°C with a liquid Cu cathode and graphite rod anode. A maximum current efficiency of 48.7% was obtained at 1160°C for a melt containing 50, 34, and 16 parts by weight of Be oxyfluoride, NaF, and BaF_2, respectively. The NaF considerably improved the conductivity of the Be oxyfluoride–BaF_2 mixtures. The current efficiencies were low. This was attributed to anode effects and the formation of Be carbide at the crucible [7].

According to a patent specification, an anode of BeO and C is used in the deposition of Be from fluoride melts on liquid metal cathodes, e.g. Cu, to form alloys. The F_2 discharged at the anode reacts with the BeO and C, mainly according to $2 BeO + C + 2 F_2 = 2 BeF_2 + CO_2$. Thus BeF_2 is replenished at the same rate as it is used up. Further advantages are reported to be the elimination of the anode effect, the increase in current yield, and continuous operation [8].

References:

[1] Illig, K.; Fischer, H.; Birett, W. (in: Eger, G.; Die Technische Elektrolyse im Schmelzfluß, 2nd Ed., Leipzig 1955, pp. 571/641, 596/623).
[2] Wong, M. M.; Couch, D. F.; O'Keefe, D. A. (J. Metals **21** No. 1 [1969] 43/5).
[3] Wong, M. M. (U.S. 3 666 444 [1968/72] 4 pp.; C.A. **77** [1972] No. 82 990).
[4] O'Keefe, D. A.; Couch, D. E. (BM-RI-7347 [1970] 11 pp.; C.A. **73** [1970] No. 122 648).
[5] Songina, O. A. (Redkie Metally, 2nd Ed., Moscow 1955, pp. 1/384, 312/6; C.A. **1956** 13 707).
[6] Belyaev, A. I. (Tsvetn. Met. [Moscow] **18** No. 3 [1945] 57/9).
[7] Fink, C. G.; Shen, T.-N. (Trans. Am. Electrochem. Soc. **72** [1937] 317/25).
[8] Ostroumov, D. D. (U.S.S.R. 49 252 [1935/36] 2 pp.).

16.7.5   Electrolytes Based on Molten Chlorides

16.7.5.1   General Remarks

Low-temperature (330 to 450°C) electrolysis of $BeCl_2$–NaCl and/or –KCl melts containing between 30 and 60 wt% $BeCl_2$ yielded compact Be leaflets and spangles on Ni cathodes at a high current efficiency (up to ~100%). At the low temperatures, heat and energy losses are reduced and reactions with materials of the apparatus and $BeCl_2$ losses due to vaporization are practically avoided [1 to 5]. The chloride melts may hydrolyze on contact with air, yielding oxide particles suspended in the melt. These serve as seeds for the formation of fine Be crystals, but the oxide occlusions are difficult to remove and render the Be brittle. Thus hygroscopic and impure salts must be avoided and the atmosphere above the melt should be inert. A smooth Be plate is obtained on Cu cathodes at current densities up to 4 A/dm², and Be dendrites are formed at current densities >4 A/dm² [6, 7]. Larger Be flakes may form at the later stages of electrolysis [8]; cf. p. 248. A reduction of the $BeCl_2$ content in the chloride melt reduces the size of the Be crystals, and very low $BeCl_2$ concentrations may even produce a spongy Be deposit at the cathode. Small amounts of Be sponge may also form at high current densities, e.g., 16 A/dm². For the influence of the purity of the $BeCl_2$ used and the duration of electrolysis on the cathodic current efficiency, see [9].

The polarization curves of an Mo cathode in alkali chloride melts containing, e.g., 2 wt% $BeCl_2$ at 400, 500, and 600°C show residual currents ($<10^{-2}$ A/cm²) caused by the discharge of alkali ions to subions, which may superimpose on the process of electrolysis. Beryllium deposition does not start before a certain value of the residual current ("threshold value") has been reached [10, 14].

Current yields of Be deposition can be increased by separating the anode and cathode compartments by a diaphragm; for details see [10]. According to a patent specification, the electrodeposition of Be from $BeCl_2$ containing melts can be improved by the addition of 0.2 to 5% (by wt, presumably) $BeCl_2 \cdot n NH_3$ (n = 2, 4, 6, or 12) or $BeCl_2 \cdot 2 N_2H_4$. The additions act as crystal nuclei around which Be monocrystals form [11]. For the use of BeO-carbon anodes for a continuous $BeCl_2$ supply in $BeCl_2$–NaCl and $BeCl_2$ containing eutectic LiCl–KCl melts, see [12]. The formation of anionic Be complexes in melts of $BeCl_2$ in LiCl, KCl, CsCl, and eutectic LiCl–KCl was studied by measuring equilibrium potentials at $BeCl_2$ concentrations of 0.46 to 10.1 mol% in the temperature range 400 to 1050°C [13]; see also pp. 169/70.

References:

[1] Jaeger, G. (Brit. 434338 [1935]).
[2] Jaeger, G. (U.S. 2041131 [1936]).
[3] Kaufmann, A. R.; Kjellgren, B. R. F. (Proc. 1st Intern. Conf. Peaceful Uses At. Energy, Geneva 1956, Vol. 8, pp. 590/9, 593).
[4] Williams, L. R.; Eyre, P. B. (in: McIntosh, A. B.; Heal, T. J.; Materials for Nuclear Engineers, New York 1960, pp. 269/318, 272/5).
[5] Pruvot, E. (Bull. Soc. Chim. France **1961** 172/6; C.A. **1961** 13254).
[6] Kroll, W. J. (Trans. Electrochem. Soc. **87** [1945] 551/69).
[7] Andrieux, J. L. (Rev. Met. [Paris] **45** [1948] 49/59, 49/52).
[8] Menzies, I. A.; Hill, D. L.; Owen, L. W. (J. Less-Common Metals **1** [1959] 321/36).
[9] Silina, G. F.; Grinberg, L. L. (Tsvetn. Met. [Moscow] **33** No. 12 [1960] 47/53; Soviet J. Non-Ferrous Metals **1** No. 12 [1960] 46/52; C.A. **1961** 19555).
[10] Smirnov, M. V.; Yushina, L. D.; Ivanovskii, L. E. (Tr. Inst. Khim. Akad. Nauk SSSR Ural'sk. Filial **1958** No. 2, pp. 161/70; C.A. **1960** 9555).

[11] Berghaus, B.; Staesche, M. (Brit. 970360 [1961/64] 11 pp.; C.A. **62** [1965] 238).
[12] Higgins, J. K.; Bellamy, R. G.; Buddery, J. H. (AERE-M-R-2750 [1958] 18 pp.).
[13] Smirnov, M. V.; Chukreev, N. Ya. (Zh. Neorg. Khim. **6** [1961] 1361/8; Russ. J. Inorg. Chem. **6** [1961] 699/704).
[14] Smirnov, M. V. (Electrochemistry of Molten and Solid Electrolytes, New York 1961, pp. 3/5; N.S.A. **16** [1962] No. 5358).

16.7.5.2 BeCl$_2$–LiCl

When Be is electrodeposited from BeCl$_2$–LiCl melts (0.05 to 3 wt% BeCl$_2$) at 400 to 850°C on Pt cathodes, Be diffuses into Pt while no such diffusion is observed with W cathodes; see [1] and pp. 228/9.

Reference:

[1] Delimarskii, Yu. K.; Baranenko, V. M.; Zarubitskii, O. G. (Deposited Doc. VINITI-3083-76 [1976] 1/13; Elektrokhimiya **12** [1976] 1887; Soviet Electrochem. **12** [1976] 1717).

16.7.5.3 BeCl$_2$–NaCl

References for this section are given on pp. 251/2.

General Remarks

BeCl$_2$–NaCl melts are of greatest importance for the electrolytic preparation of Be; see, e.g. [1] and "Beryllium" Suppl. Vol. A 1, 1986, pp. 79/88. For the electrolytic refining of Be in eutectic BeCl$_2$–NaCl melts, see ibid., pp. 112/3. A review of the literature concerning Be electrodeposition from BeCl$_2$–NaCl melts up to 1959 is given in [2]. The advantages and disadvantages of BeCl$_2$–NaCl melts (in comparison to other chloride melts) for use in Be refining are discussed by Schimmel [3].

The BeCl$_2$–NaCl eutectic temperature (210°C at 55 mol% BeCl$_2$ according to [4], 215°C at 51 mol% BeCl$_2$ according to [10]) allows relatively low operating temperatures. Also, the partial pressure of BeCl$_2$ is low at these temperatures. The cathodes usually consist of Ni. BeCl$_2$–NaCl melts containing between 30 and 70 wt% BeCl$_2$ can be used for electrolyses at temperatures below 450°C; melts with 40 to 60 wt% BeCl$_2$ are preferably electrolyzed at 320 to 380°C [5, 6]. The Be deposits mostly consist of flakes or dendrites, see, e.g. [7, 8]. A fine Be powder is produced at high current densities (>40 A/ft²) [9]. The purity of the deposited Be is ≧99%; lustrous Be dendrites of 99.966% purity were obtained [7], and flakes of 99.99% Be [8]. For further details, see p. 249. At the rather low temperatures of 220 to 245°C, a black Be powder of lower purity (88.4 to 91.5%) was obtained on Ni plate cathodes in fused BeCl$_2$–NaCl (49 mol% BeCl$_2$) with current densities of 1.33 to 1.5 A/cm², using a C rod anode and passing H$_2$ through the melt. Some of the Be powder was dispersed in the electrolyte [11].

Decomposition Potential

In the following first table at the top of the next page decomposition potentials, E, of BeCl$_2$ in eutectic BeCl$_2$–NaCl melts are given for various cathode materials at temperatures, T; they were determined from current vs. potential (vs. Cl$_2$|C) curves with carbon anodes. The table also contains reversible potential values, E$_1$, measured at zero current with Pt electrodes. Preelectrolysis was found to reduce the difference between the values of E and E$_1$ [15].

Sheiko and Delimarskii [16] determined E and E$_1$ in the temperature range 300 to 700°C with BeCl$_2$–NaCl melts containing 0.25 to 0.80 mole fraction of BeCl$_2$ (Pt cathode, calcined graphite anode). Values of E for the following mole fractions of BeCl$_2$ are given in the second table on the next page.

Decomposition potentials were calculated from free energy data for molten BeCl$_2$ at 300, 350, and 400°C to be −1.978, −1.943, and −1.910 abs. V, respectively [2].

T in °C	cath-ode	−E in V	−E₁ in V	Ref.	T in °C	cath-ode	−E in V	−E₁ in V	Ref.
300	Pt	2.12*)	2.153	[12]	500	Pt	1.93*)	1.954	[12]
350	Ni	2.3		[13]	540	Pt	2.02	2.026	[15]
400	Sn	2.00		[2, 14]	550	C	2.01		[15]
400	Pt	2.01*)	2.044	[12]	550	Mo	2.01		[15]
420	Pt	2.08	2.084	[15]	600	C	1.98		[15]
450	Mo	2.06		[15]	600	Pt	1.99	2.008	[15]
500	C	2.03		[15]	640	Pt	1.96	1.957	[15]
500	Mo	2.04		[15]	700	Pt	1.93	1.934	[15]

*) Indicates the use of calcined graphite anodes.

T in °C	−E in V for mole fraction BeCl₂					
	0.25	0.35	0.45	0.51	0.65	0.80
300	—	—	—	2.13	—	—
400	—	2.15	2.11	2.02	2.02	2.02
450	—	—	—	—	1.97	1.97
500	2.14	2.03	2.00	1.93	1.92	1.92
600	2.03	1.96	—	—	—	—
700	1.92	—	—	—	—	—

Corresponding results for E_1 are given in a table in the paper [16].

Mean temperature coefficients of 0.56×10^{-3} and 0.55×10^{-3} V/K were calculated for E and E_1, respectively, in the temperature range 420 to 640°C from results with Pt cathodes [15]. The temperature dependence of the decomposition potential E for C and Mo cathodes corresponds to that for Pt cathodes; see figure in the paper [15]. A mean temperature coefficient of 1.0×10^{-3} V/K was obtained for E in the temperature range 300 to 700°C and for $BeCl_2$ concentrations of 0.25 to 0.80 mole fraction with a Pt cathode. The value of the decomposition potential decreases with increasing $BeCl_2$ concentration in the melt, and a reversible decomposition potential of −1.90 V was obtained for pure $BeCl_2$ at 500°C by extrapolation [16].

The decomposition potentials show considerable deviations from the Nernst equation, and it is suggested that in $BeCl_2$–NaCl melts anionic Be complexes are formed at low Be concentrations and cationic Be complexes at high Be concentrations [16]. The cathodic deposition potential becomes more negative with a lower degree of ionization of the Be halide and with the formation of ionic Be complexes (such as $BeCl_4^{2-}$) as the activity of the Be^{2+} ions is reduced [2, 13].

Deposition and Crystallization Mechanisms

Formation of complexes (such as Na_2BeCl_4) and the presence of a eutectic structure in the homogeneous liquid phase are indicated in fused $BeCl_2$–NaCl systems by emf and conductivity measurements for $BeCl_2$ concentrations of ~30 to 80 mol% at temperatures of 250 to 500°C [17]; cf. [18]. Decomposition, cathode, and anode potentials of $BeCl_2$–NaCl melts with 25 to 80 mol% $BeCl_2$ at 300 to 700°C indicate that processes (such as $BeCl_2 \rightleftharpoons BeCl^+ + Cl^-$,

BeCl$^+ \rightleftharpoons$ Be^{2+} + Cl$^-$, 2 BeCl$_2 \rightleftharpoons$ BeCl$^+$ + BeCl$_3^-$) occur in addition to the formation of Na$_2$BeCl$_4$. It is supposed that low BeCl$_2$ concentrations favor the formation of anionic complexes in the melt, and high BeCl$_2$ concentrations that of cationic complexes [16]; cf. p. 247.

Primary reduction according to Be^{2+} + 2e → Be is expected to yield large flake Be in an uncomplexed halide system. The cathodic deposition potential becomes more negative the lower the degree of ionization of the Be halide. The formation of ionic Be complexes, e.g., BeCl$_4^{2-}$, reduces the activity of the Be^{2+} ions and causes an even more negative deposition potential. Thus the alkali metal ion may be discharged along with or even before the Be^{2+} ion, and subsequently a secondary reduction of Be^{2+} can take place according to 2 Na + Be^{2+} → 2 Na$^+$ + Be, yielding powdery, partially adherent Be deposits. An increase in j will increase the fraction of secondary Be [2, 13]; cf. p. 249. As the growing Be deposit increases the cathode area, thus decreasing the "real" value of j, the fraction of primary Be^{2+} reduction may increase in the course of electrolysis; this is indicated by the deposition of larger Be crystals during the later stages of electrolysis [13]. Some of the chlorine in Be deposits may possibly be due to the reduction of BeCl$^+$ cations [2].

Smirnov and Chukreev [19] showed that in the electrolysis of a BeCl$_2$–NaCl melt (50% BeCl$_2$) at 350°C not more than 0.5% of the total Be is in the monovalent state in the catholyte in a closed electrolytic cell with a diaphragm. For this purpose they determined the equilibrium constant for Be$^{2+}_{melt}$ + Be $\rightleftharpoons$ 2 Be$^+_{melt}$ in alkali chloride melts with a Be anode at 350 to 600°C by measuring the redox potential at an Mo electrode. In an open cell without separation of anode and cathode compartments, the concentration of Be$^+$ is found to be even smaller than the above value. The fraction of Be$^+$ increases with increasing temperature and decreasing BeCl$_2$ concentration. The addition of F$^-$ ions displaces the above equilibrium to the left, thus reducing the Be$^+$ concentration [19].

The Be atoms discharged on an Ni cathode tend to diffuse into the cathode surface. At 300 to 400°C the diffusion layer thickness is small, though the absence of an oxide layer facilitates diffusion. Two-dimensional nuclei of the hcp Be lattice form on the surface. The Be lattice may tend to arrange itself epitaxially to the underlying Ni lattice, the growth rates varying according to the crystallographic faces. Flat platelets are formed as the prismatic faces (10$\bar{1}$0) of Be have the highest, and the basal planes (0001) have the smallest growth rates. The formation of dendrites and needles may result from lower deposition potentials on advancing crystal faces. Impurities can favor the growth of particular crystal faces and thus influence particle size and crystal habit [2].

Experimental Conditions and Properties of Depositions

The **cathode material** is usually Ni; thus Ni or Ni alloy crucibles, rods, plates, or cylinders may serve as cathodes for continuous operation cells [20]. Graphite linings or "sleeves" on the Ni cathodes reduce the contamination of the Be deposit by Ni; in addition, the Be dendrites formed can be removed more easily, see [21 to 23]. The cohesion between the cathode and the Be deposit is improved by a preliminary plating of the Ni or graphite cathodes with Be at low current densities [6, 7]. The Be discharge on Pt and W cathodes was studied by a polaroscopic method in the temperature range 400 to 850°C with BeCl$_2$–NaCl melts containing 0.05 to 3 wt% BeCl$_2$. At the higher temperatures (e.g., 850°C), reproducible polarograms (at 400 mV/min) could be obtained only after up to 15 polarization cycles. For the interaction of Be and Pt, see [24] and pp. 228/9. High-purity Be flakes were obtained on Cu cathodes from the eutectic BeCl$_2$–NaCl melt at 400°C [25]. Melts containing more than 60 wt% BeCl$_2$ were found to attack carbon electrodes; see [21 to 23, 26].

According to patent specifications, graphite, W lamellae, or Be (in the case of Be refining) are used as **anodes** [5, 27]. For a detailed study of the factors influencing the refining of Be, as well as for an anode design and an optimum operating system, see [6]; cf. "Beryllium" Suppl. Vol. A 1, 1986, pp. 112/3.

The **forms of deposited Be**, i.e., the sizes and crystalline shapes of the Be deposits from $BeCl_2$–NaCl melts range from powder to needles, dendrites, and flakes. The dimensions of the $BeCl_2$ crystals were reduced considerably by decreasing the $BeCl_2$ concentration in the $BeCl_2$–NaCl melt from 50 to 30 wt%, and even spongy Be metal appeared at a cathode consisting of preberyllized Ni. Low $BeCl_2$ concentration also leads to an increase in the impurity content of the Be and a decrease in current yield, but good results were obtained with 40 to 50 wt% $BeCl_2$. Well-shaped, bright Be crystals were obtained at an optimum temperature of 340°C with cathodic current densities of 4 to 12 A/dm², while at 16 A/dm² also small amounts of Be sponge formed; see [6]. High current densities yield fine Be powder; however, if the electrolyte is very pure, large Be crystals may be obtained even at high current densities [2, 9]. An increase in the size of the Be crystals in the course of the electrolysis may be due to a decrease in the real current density, as the cathode surface area grows with the Be deposit [13]. As powdery, partially adherent Be deposits are expected to form by a secondary reduction ($2 Na + Be^{2+} \rightarrow 2 Na^+ + Be$, cf. p. 248), larger Be flakes produced during the later stages of an electrolysis may indicate that a secondary reduction is diminished [2, 13]. Jaeger [28] maintains that the $BeCl_2$ concentration and the impurities present in the electrolyte are the main factors determining the crystal shape of the Be (see photos in the paper). The leaflike, dendritic crystals are mostly found to be twins; the platelike crystals have a layered structure with terraces and grow oriented on the substrate; the Be needles show prismatic or obelisk-like structures. The determination of the ratio of the lattice parameters (a : c = 1 : 1.569) by the Debye-Scherrer technique showed that there is no difference in the lattice structure of the plate- and needle-shaped Be crystals [28]. The Be flakes are considered to be aggregates of Be single crystals with the basal planes in the plane of the flake, though slightly displaced [29]. They are held together by minute metallic "bridges" [30]. The surfaces of flakes produced from 1:1 $BeCl_2$–NaCl melts (by Pechiney) showed projecting pyramidal structures with basal dimensions of 30 to 100 μm and heights of ~3 μm. They consist of up to 20 to 30 hexagonal terraces, forming closed loops and capped by plateaus with minute pits of ~0.1 μm diameter, resembling screw dislocations. Many flakes also contained voids. Foreign material, possibly complex molecules (such as $BeO \cdot BeCl_2$), seems to occupy the pits. Impurity layers about 10 to 20 μm apart are found parallel to the planes of the flakes and are thought to be halide complexes deposited cataphoretically during electrolysis [30, 31]. Surface steps on the powder particles and deposited impurities may influence the behavior of the Be powder during compacting and sintering and hence also the properties of the compacts [30]. In the Milford Haven experimental cell fine, acicular or coarse, dendritic Be flakes were produced, consisting of hexagonal plates with terraced steps of the order of 100 μm. The impurity layers consisted of groups of particles rather than continuous layers and were much more widely spaced than in the Pechiney flakes [30, 31].

The **purity** of the Be deposits was reported to be ~99.95% Be [7, 8], ≧99% [13, 32 to 36]. A spectrochemical analysis of Be flakes is given in [30]. An O content of 1000 to 2000 ppm in Be flakes of 1 to 5 mm diameter was attributed to Be oxide on the surface of the flakes [37, 38]. The Cl content of the Be flakes could not be reduced below a minimum of 200 to 300 ppm by any treatment applied; see [31]. Metallographic and chemical analyses indicated that the Cl content is not due to irreversible chemisorption. Dissolution of the Cl in the Be lattice is improbable on account of the sizes of the Cl ion and the interstices of the Be lattice. Discrete inclusions of Cl containing compounds (average diameter of inclusions < 5 μm) are considered the most likely [31].

Preelectrolysis, i.e., electrolysis at a potential below the decomposition potential of $BeCl_2$, was applied to remove metal impurities which are more electropositive than Be (replacing the

Be anode by a graphite anode, in the case of Be refining); Be of higher purity is then deposited on a fresh cathode. For details, see [6, 8, 9, 34 to 36]. Preelectrolysis does not necessarily improve the purity of the Be deposit [22, 23]. Beryllium of high purity is obtained from pure starting materials and with an apparatus free of leaks.

For data on the analysis of electrolytic Be, see, e.g. [21, 23, 26, 38]. Details on the reduction of impurities in Be refining are given in [6].

In **special procedures** continuous Be electrodeposition can be effected by mobile Ni or Ni alloy cathodes [20], or by graphite-lined tubular Ni cathodes with a movable ring scraper [21].

The purity of Be deposits is improved by a two-stage electrolysis of BeCl$_2$–NaCl melts (1:1 by wt) at 320 to 350°C with exchangeable Ni cathodes (and graphite anodes) in a semi-continuous process: impurity metals more electropositive than Be (e.g., Si, Al, Fe, Cu) are deposited at a lower applied voltage (~1.5 V), before Be scales of purity better than 99.8% are deposited on a fresh cathode [34 to 36]. Beryllium flakes of 99.95 to 99.99% purity are obtained from BeCl$_2$–NaCl melts at 350 ± 70°C according to [8]. Continuous production of Be of ~99.97% purity is possible with an apparatus using several cathodes (BeCl$_2$–NaCl melt with 50 to 60 wt% BeCl$_2$ at 330 to 400°C); see [39]. Only one cathode is used in the following two-step electrolysis: the above impurity metals are deposited along with Be on an Ni cathode from a BeCl$_2$–NaCl melt (wt ratio 5.7:4.3) at 350°C, low current density (10 mA/cm²), and low applied voltage (2.4 V, graphite anode). This deposit adheres strongly to the Ni cathode. Coarse and large crystalline Be flakes of high purity (~99.95%) are then deposited onto the impure Be at 80 mA/cm² and an applied voltage of 3.5 V. The Be flakes can be removed easily, leaving the impure Be on the cathode [40].

In the electrolysis of fused equimolar or eutectic BeCl$_2$–NaCl mixtures with **Hg cathodes** the discharged Be forms a "quasi-amalgam", i.e., a dispersion of finely divided Be or of a Be-Hg intermetallic phase in Hg; see [41 to 47], cf. [48 to 50]. This is of interest for the preparation and refining of Be for several reasons: 1) the Hg cathode facilitates the transport of the discharged Be out of the electrolysis cell, e.g., by Hg circulation [41, 49]; 2) the interaction of the Be with the electrolyte is thus prevented [48, 49]; 3) the operating temperature is low [41, 49, 50], viz. between the melting point of the electrolyte mixture (melting point of the BeCl$_2$–NaCl eutectic 210°C [4]) and the boiling point of Hg (~356°C); 4) the Be deposited into the Hg has a fine particle size (0.1 to 10 μm [45, 46]) and can be filtered off or separated from the Hg by hot pressing or low-temperature vacuum distillation (preparation of ultrafine Be powder) [41, 44 to 46]; 5) the Be obtained is not contaminated by the electrolyte [43 to 45].

The sensitivity of the amalgam formation towards traces of oxygen requires rigorous exclusion of air and moisture, viz. the use of anhydrous salt mixtures and an inert atmosphere of pure, dry Ar or N$_2$ (e.g. 99.997% N$_2$). The inert gas flow also serves to remove Cl$_2$ formed at the anode [47]. Graphite anodes were found to disintegrate fairly rapidly [41, 43, 46], but baked C anodes proved satisfactory [41, 43, 44]. The use of a Be anode reduced the C contamination of the deposited Be [47]. Stirring of the Hg cathode is necessary to constantly expose fresh cathode surface and to prevent the amalgam from climbing towards the anode during electrolysis [41 to 47]. The stainless steel electrolysis cell and the stirrer shaft were lined with Pyrex to minimize the contact of the electrolyte with the metal parts; see [41, 43 to 47].

Eutectic BeCl$_2$–NaCl melts (≈ equimolar BeCl$_2$–NaCl) were used in most cases, though the BeCl$_2$ concentrations ranged from 45 to 60 mol% [41, 44]. An advantageous electrolyte may also contain KCl, CaCl$_2$, or MgCl$_2$ [41]. Electrolyses were carried out at 275 to 300°C [45], 308 to 326°C [44], and 300 to 330°C [41 to 43, 46]. Current efficiencies of more than 70% were obtained [41, 44].

The amalgam produced can be diluted with Hg in all proportions and is distributed evenly throughout the Hg cathode by the stirring [41]. It is drawn off periodically or continuously with a Be content of 0.04% [46], 0.25% [47]; 0.25% Be is considered a limit with respect to practical considerations [41]. On standing the amalgam "ages", forming a thick pasty layer on top of the Hg. The agglomerated amalgam has a higher Be content and contains coarser grains (~1 to 2 μm) than the initial amalgam (particle size of 0.1 to 0.2 μm). The agglomeration is accelerated by an increase in temperature [46]. Vacuum filtration (100 to 200 mesh screens) of the drawn-off amalgam yields a semisolid amalgam of a shiny, silvery appearance with 1.1 to 2.3 wt% Be [44, 46, 47]. This corresponds to $Be(Hg)_2$ with a slight excess of Hg; the Cl content was ~0.3 to 2%, the O content less than 0.7% [41, 44]. More Hg could be removed by cold pressing (pressures up to 18 t/in²) until a limiting composition of 24 wt% or 87.5 at% Be was obtained, corresponding to a formula of Be_7Hg [46]. The remaining Hg could be removed completely either by hot pressing or vacuum distillation at 340°C. The hot pressing yielded solid Be compacts (grain sizes < 50 μm) of higher purity than the loose sinter cakes obtained by the vacuum distillation. The O and Cl contents of the Be compacts depended on temperature, pressure, and duration of the hot pressing; but O contents less than 0.8% and Cl contents of 0 to 0.3% were obtained. Other main impurities were Fe and C; for details, see [41, 44, 47]. The content of metallic impurities depends directly on the purity of the $BeCl_2$ feed and the disintegration and/or corrosion of parts of the apparatus [46]. X-ray diffraction analysis of the Be powder obtained after vacuum distillation showed the presence of Be_3N_2 and C [46].

A black powder (particle size ~0.1 to 1.5 μm [46]) appeared on the surface of the amalgam if left standing for a considerable time or on exposure to air [41 to 46]. The composition of this black powder is not unequivocal: a Be content of ~75 wt% was reported by [42]; 90% metallic Be and the remainder BeO (particle size ~1 μm) were indicated by X-ray analysis in [45]; Habashi et al. [46] found that a chemical analysis (15.6% metallic Be, 40% BeO, 10% C, 1.5% Fe, balance: probably residual Hg) did not agree with X-ray analysis results. For spectrographic analyses of the black powder, see [45, 46].

References:

[1] Jaeger, G. (Metall **4** [1950] 183/91).

[2] Bellamy, R. G. (AERE-M-534 [1959] 41 pp., 20/6, 30/3, 35, 41).

[3] Schimmel, F. A. (Y-1380 [1962] 31 pp., 11/2; N.S.A. **16** [1962] No. 15127).

[4] Furby, E.; Wilkinson, K. L. (J. Inorg. Nucl. Chem. **14** [1960] 123/6).

[5] Jaeger, G.; Degussa (Brit. 434338 [1935]).

[6] Silina, G. F.; Grinberg, L. L. (Tsvetn. Met. [Moscow] **33** No. 12 [1960] 47/53; Soviet Non-Ferrous Metals **1** No. 12 [1960] 46/52; C.A. **1961** 19555).

[7] Vil'komirskii, I. E.; Silina, G. F.; Berengard, A. S.; Semakin, V. N. (At. Energiya SSSR **11** [1961] 233/9; Soviet At. Energy **11** [1961] 882/8; C.A. **56** [1962] 5702).

[8] Kida, K.; Abe, M.; Nishigaki, S.; Nakamura, T.; Yokoi, M.; Usami, S.; Ishikawa, Y.; Miyazaki, Y. (Japan. 3802 [1961/63] 6 pp. from C.A. **59** [1963] 14938).

[9] Bellamy, R. G.; Hill, N. A. (International Series of Monographs on Nuclear Energy Division 8: Materials. Vol. 1, Extraction and Metallurgy of U, Th, Be, Pergamon, New York 1963, 198 pp., 76/9, 82/4).

[10] Schmidt, M. J. M. (Bull. Soc. Chim. France [2] **39** [1926] 1686/703).

[11] Tien, J. M. (Trans. Electrochem. Soc. **89** [1946] 237/45).

[12] Sheiko, I. N.; Delimarskii, Yu. K. (Ukr. Khim. Zh. **25** [1959] 295/300; C.A. **1960** 93).

[13] Menzies, I. A.; Hill, D. L.; Owen, L. W. (J. Less-Common Metals **1** [1959] 321/30).

[14] Higgins, J. K.; Bellamy, R. G.; Buddery, J. H. (AERE-M-R-2750 [1958] 18 pp., 16).

[15] Delimarskii, Yu. K.; Skobets, E. M. (Zh. Fiz. Khim. **20** [1946] 1005/10; C.A. **1947** 2341).

[16] Sheiko, I. N.; Delimarskii, Yu. K. (Ukr. Khim. Zh. **23** [1957] 713/20; C.A. **1958** 14384).
[17] Kuroda, T.; Matsumoto, D. (J. Electrochem. Soc. Japan **33** [1965] 29/34).
[18] Delimarskii, Yu. K.; Sheiko, I. N.; Feshchenko, V. G. (Zh. Fiz. Khim. **29** [1955] 1499/507; C.A. **1957** 830).
[19] Smirnov, M. V.; Chukreev, N. Ya. (Zh. Neorg. Khim. **4** [1959] 2536/43; Russ. J. Inorg. Chem. **4** [1959] 1168/72).
[20] Jaeger, G. (Ger. 860281 [1949/52] 1/12; N.S.A. **15** [1961] No. 3009).

[21] Morana, S. J. (Ger. Offen. 1115936 [1956/62] 6 pp.; Patentblatt **81** [1961] 5598).
[22] Morana, S. J. (U.S. 2843544 [1958]; C.A. **1959** 103).
[23] Beryllium Corporation (Brit. 812702 [1959]; N.S.A. **13** [1959] No. 12739).
[24] Delimarskii, Yu. K.; Baranenko, V. M.; Zarubitskii, O. G. (Deposited Doc. VINITI-3083-76 [1976] 1/13; Elektrokhimiya **12** [1976] 1887; Soviet Electrochem. **12** [1976] 1717).
[25] Menzies, I. A.; Hill, D. L.; Owen, L. W. (Nature **183** [1959] 816/7).
[26] Morana, S. J. (Fr. 1153825 [1956/58]; C. **1961** 10736).
[27] Degussa (Fr. 785072 [1935]; C.A. **1936** 394).
[28] Jaeger, G. (Z. Metallk. **41** [1950] 243/6).
[29] Williams, J. (Progr. Nucl. Energy V **1** [1956] 300/4).
[30] Blainey, A.; Johnston, T. L.; Jones, J. W. S. (AERE-M-R-1442 [1954] 35 pp., 1/3, 17/8; Rev. Met. [Paris] **52** [1955] 735/49).

[31] Bellamy, R. G. (M-619 [1960] 14 pp.; C.A. **1960** 18127).
[32] Potvin, R.; Farnham, G. S. (Trans. Can. Inst. Mining Met. **49** [1946] 525/38, 533/5; C.A. **1947** 66).
[33] Potvin, R. (Metal Ind. **1946** 469/70).
[34] Meerson, G. A.; Kaplan, G. E.; Silina, G. F.; Sokolov, D. D. (Issled. Obl. Geol. Khim. Met. Sbornik **1955** 125/31; C.A. **1957** 17484).
[35] Meerson, G. A. (Proc. 1st Intern. Conf. Peaceful Uses At. Energy, Geneva 1956, Vol. 8, pp. 587/9).
[36] Meerson, G. A.; Sokolov, D. D.; Mironov, N. F.; Bogorad, N. M.; Pakhomov, Ya. D.; L'vovskii, D. S.; Ivanov, E. S.; Shmelev, V. M. (At. Energiya SSSR **5** [1958] 624/30; N.S.A. **13** [1959] No. 5589).
[37] Syre, R. (Metaux Corros. Ind. **33** [1958] 406/19; C.A. **1959** 2991).
[38] Syre, R.; Saulnier, A.; Perez, M. (Rev. Met. [Paris] **56** [1959] 359/70; C.A. **1959** 12117).
[39] Nakamura, T. (U.S. 3278402 [1961/66] 5 pp.; C.A. **66** [1967] No. 7911).
[40] Nakamura, T. (U.S. 3296107 [1962/67] 4 pp.; C.A. **66** [1967] No. 61309).

[41] Kells, M. C.; Holden, R. B.; Whitman, C. (SEP-207 [1956] 43 pp.; C.A. **1957** 4174).
[42] Kells, M. C.; Holden, R. B.; Whitman, C. I. (J. Am. Chem. Soc. **79** [1957] 3925).
[43] Holden, R. B.; Kells, M. C.; Whitman, C. I. (Proc. 2nd U.N. Intern. Conf. Peaceful Uses At. Energy, Geneva 1958, Vol. 4, pp. 306/8; C.A. **1960** 7375).
[44] Kopelman, B.; Holden, R. B. (U.S. 2987454 [1961]; C.A. **1961** 21921).
[45] Gale, C. O.; Griffiths, V.; Habashi, F.; Hanson, G. T. (Technical Report NASA Contract NASW 1099 [1965]).
[46] Habashi, F.; Murray, R. J.; Toivonen, R. W.; Griffiths, V.; Hanson, G. T. (Progress Report NASA Contract NASW 1542 [1967]; LN67-26451 [1966] 29 pp.) cited in [47].
[47] Lidman, W. G.; Griffiths, V. (N-72-27577 [1972] 14 pp.; N.S.A. **28** [1973] No. 27799).
[48] Reynolds Metals Co. (Indian 50387 [1953/55]; C. **1960** 17273).
[49] Schmidt, W. (Fr. 1128402 [1953/57]; C. **1961** 8464).
[50] Schmidt, W. (Brit. 781457 [1953/57]; C. **1959** 7626).

16.7.5.4 BeCl$_2$–KCl

The Be deposition on Pt and W cathodes from BeCl$_2$–KCl melts, containing 0.05 to 3 wt% of BeCl$_2$, at 400 to 850°C was found to be similar to that from BeCl$_2$–LiCl melts [1]. The formation of complexes, e.g., K$_2$BeCl$_4$, (BeCl$_4$)$^{2-}$, is indicated by emf measurements of the cell Be(s)|BeCl$_2$–KCl(l)|Cl$_2$(g), in the temperature range 400 to 530°C for BeCl$_2$ concentrations of 0.4 to 0.7 mole fraction [2]. For the electrolysis of BeCl$_2$–KCl melts containing 2 to 3 wt% BeS, see [3] and p. 272.

References:

[1] Delimarskii, Yu. K.; Baranenko, V. M.; Zarubitskii, O. G. (Deposited Doc. VINITI-3083-76 [1976] 1/13; Elektrokhimiya **12** [1976] 1887; Soviet Electrochem. **12** [1976] 1717).
[2] Kuroda, T.; Oyamada, R. (J. Electrochem. Soc. Japan **35** [1967] 125/9).
[3] Gibson, A. R. (Brit. 821091 [1959]; C.A. **1960** 2051).

16.7.5.5 BeCl$_2$–LiCl–KCl

References for this section are given on pp. 262/4.

General Remarks

All investigations on the Be electrodeposition from melts with the composition given above are based on LiCl–KCl melts of eutectic composition (59 mol% LiCl, melting point 358°C). The relatively high solubility of BeCl$_2$ in this melt in the temperature ranges used for electrolyses (~400 to 500°C), low losses in BeCl$_2$ due to vaporization, and the availability of a considerable number of emf measurements with eutectic LiCl–KCl melts are advantages in favor of this melt [1, 2]. An extensive study of the Be electrorefining process was carried out by Wohlfarth [2]; see pp. 260/1.

Decomposition Voltages

For BeCl$_2$ added to eutectic LiCl–KCl melts in various amounts, the following decomposition voltages and mean temperature coefficients for the temperature range studied (760 to 820 K) were determined by emf measurements of the cell Be(s)|BeCl$_2$(l), LiCl–KCl(l)|Cl$_2$(gas) using a graphite anode and Ar atmosphere (for the cell arrangement, see figure 1 in the paper [3]):

mole fraction of BeCl$_2$	temperature T in K	Cl$_2$ pressure in atm	decomposition voltage U in V	dU/dT in V/K
0.0021	761	0.967	2.4726	-3.37×10^{-4}
	789		2.4537	
	817		2.4524	
0.0259	765	0.972	2.3650	-4.00×10^{-4}
	782		2.3594	
	793		2.3547	
	800		2.3509	
	817		2.3434	

mole fraction of $BeCl_2$	temperature T in K	Cl_2 pressure in atm	decomposition voltage U in V	dU/dT in V/K
0.0416	768	0.966	2.3330	-4.66×10^{-4}
	781		2.3280	
	801		2.3185	
	817		2.3109	
0.1514	771	0.975	2.2456	-5.88×10^{-4}
	778		2.2410	
	781		2.2394	
	808		2.2237	
	816		2.2186	
	818		2.2175	

The results indicate a two-electron reaction and the presence of relatively stable Be complexes in the melt [3].

Similarly, the following decomposition voltages were obtained for higher $BeCl_2$ additions (spectrographic-grade carbon anode; Ar atmosphere) [4, 5]:

mole fraction of $BeCl_2$	temperature T in °C	decomposition voltage U in V	dU/dT in V/K[3]
0.188	410	2.165	-8.1×10^{-4}
	450	2.133[1]	
		2.163[2]	
	500	2.092	
0.410	410	2.103	-8.7×10^{-4}
	450	2.068	
	500	2.025	
0.613	410	2.045	-9.2×10^{-4}
	450	2.010	
	500	1.962	
1.000 (extrapolated values)	410	(1.950)	-9.9×10^{-4}
	450	(1.912)	
	500	(1.860)	

[1] From [4]. – [2] From [5]. – [3] Mean values in the temperature range 410 to 500°C.

The formation of complexes such as K_2BeCl_4 was indicated by thermodynamic data calculated from the above results [4 to 6]. For a corresponding discussion of the mechanism of Be deposition from $BeCl_2$–alkali chloride melts, see [5] and pp. 247/8.

Nucleation and Crystallization Overvoltage

Studies of nucleation and crystallization overvoltages (see p. 176) in the electrodeposition of Be from eutectic LiCl–KCl melts containing $BeCl_2$ were carried out with microcathodes of Be, Ta, Mo, W, Ni, and Au by means of galvanostatic curves and simultaneous microscopic observation of the cathode surfaces at temperatures from 400 to 750°C by Chukreev et al. [7 to 14].

Typical overvoltage (η) vs. time curves are shown in **Fig. 16-23** according to [9]. For these studies current impulses of 200 to 1500 ms duration were applied; the dotted lines in the figure indicate the breaking off of the current impulses. The anode and reference electrode were pure Be. The rising branches reflect the processes of charging the cathode electric double layer, the discharge of impurity ions, and the discharge of Be ions up to the maximum supersaturation of the cathode with adsorbed Be atoms at η_{max}, which is the maximum crystallization overvoltage. At $\eta = 0$ the concentration of adsorbed Be atoms on the cathode corresponds to the activity of metallic Be, viz. $a = 1$. When the rising branch is linear, the cathode material probably does not affect nucleation of Be crystals, and the slope is influenced by interactions of the discharged Be with the cathode metal, e.g., alloy formation. The peak maxima η_{max} indicate the appearance of Be crystal nuclei and the descending branches correspond to the growth of the crystal nuclei [9, 14]; cf. [8, 13]. The drop in voltage on switching off the current, as indicated in the figure, is associated with the ohmic drop in the electrolyte and the "tail ends" of the curves are attributed to diffusion overvoltage [9].

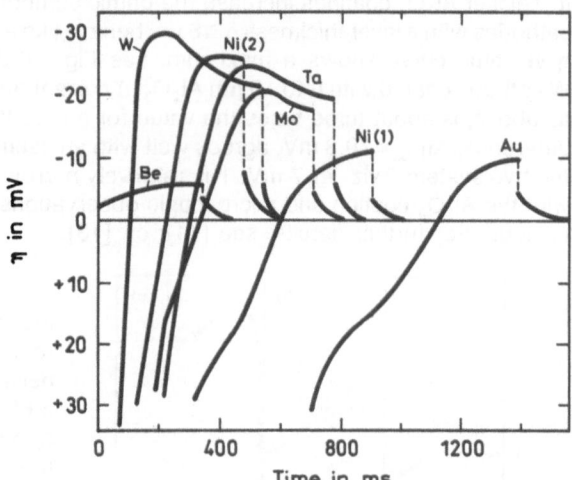

Fig. 16-23. Galvanostatic η vs. time curves for various cathode materials during Be electrodeposition from a eutectic LiCl–KCl melt containing 2.78×10^{-4} mol $BeCl_2/cm^3$; $j = 0.14$ A/cm², T = 500°C for Ta, Mo, W, and Ni(1); $j = 0.4$ A/cm², T = 400°C for Au; $j = 1.0$ A/cm², T = 500°C for Ni(2). The dotted lines indicate the switching off of the current impulses.

The overvoltage maxima in the η vs. time curves for Ta, Mo, and W at 500°C, $j = 0.14$ A/cm² in Fig. 16-23 correspond to three-dimensional Be crystal nuclei; this is confirmed by microscopic observations (see p. 256) [9, 12]. At 400°C the Be nucleation on W required a minimum j of 7.6×10^{-3} A/cm² [13]. For Ni at 500°C with $j = 0.14$ A/cm² no peak is reached on the η vs. time curve, indicating the dissolving of discharged Be atoms in the Ni; but $j = 1.0$ A/cm² provides the Be adatom supersaturation necessary for nucleation. Considerable depolarization due to Be-Au alloy formation was believed responsible for the absence of an η_{max} in the case of the Au cathode. On a clean pure Be substrate nucleation proceeds without difficulties [11, 13].

For the above cathodes η_{max} was ≤ 30 mV, and for the Ta, Mo, and W cathodes η_{max} varied inversely with t_{max}, the time taken to reach η_{max}. This was attributed to partial diffusion of Be into the substrate, and hence η_{max} is a qualitative measure of the inertness of the cathode material towards Be [9]. Thus W was inert towards Be at temperatures below 750°C [8]. For comparable experimental conditions, the degree of interaction between Be and the cathode material (and hence the adhesion of the Be crystals to the substrate) increases in the order W < Ta < Mo < Ni < Au [9]; cf. [13]. Surface concentrations of Be adatoms at $\eta = 0$ and η_{max} and the supersaturations, γ, involved in the formation of Be crystal nuclei on Ta, Mo, W ($j = 0.14$ A/cm²) and Ni ($j = 1.0$ A/cm²) cathodes at 500°C were also estimated from the experimental η vs. time curves. These supersaturations correlate with the solubility and the diffusion of Be in the substrate metals [9]. For a W cathode in a melt at 400°C containing 1.2×10^{-3} mol/cm³ BeCl$_2$ for $j = 0.0076$ to 0.94 A/cm² the following relations were obtained: $\eta_{max} = 80.1 + 31.5 \log j$ (± 5.4) mV and $t_{0 \text{ to max}} = 0.651 \cdot j^{-1.464}$ (± 2) ms, where $t_{0 \text{ to max}}$ is the time between $\eta = 0$ and η_{max}. At higher current densities, the ohmic drop and diffusion overvoltage contribute increasingly to η_{max} which therefore differs considerably from the true phase formation overvoltage. Values for η_{max}, ohmic resistance of the electrolyte, and phase formation overvoltage are tabulated for $j = 7.6 \times 10^{-3}$ to 1.3 A/cm² and discussed in [14]; cf. [13].

Oxide and impurity layers on the cathode surfaces increase the ohmic component of η, and the cathode potential is shifted to more positive values. When Be microcathodes were covered with Al$_2$O$_3$ layers up to 3 μm thick there was no overvoltage peak observed on the η vs. time curves at 400°C in LiCl–KCl melts with 5.85 wt% BeCl$_2$ added; see **Fig. 16-24**. It was suggested that the discharged Be atoms penetrate the porous oxide film or partially dissolve in it. Thicker Al$_2$O$_3$ coatings increase the ohmic component and hinder Be phase formation: Be cathodes with a layer thickness $\delta \geqq 8$ μm behave like a "foreign" (here: Al$_2$O$_3$) electrode, and the η vs. time curve shows a maximum, see Fig. 16-24c from [11]. The value of η_{max} for a W cathode coated with 8 to 10 μm Al$_2$O$_3$ at current densities of ~0.15 A/cm² (other conditions as above) is about three times the value for a pure W cathode (see figure in the paper). The difference, $\Delta\eta_{max} = 70.3$ mV, agrees well with an estimated difference in ohmic resistances for the two systems, viz. 71.7 mV. The relatively narrow width of the overvoltage peak obtained with the Al$_2$O$_3$ coating and microscopic observations indicated the formation of only one Be nucleus. For further details, see [11]; cf. [13].

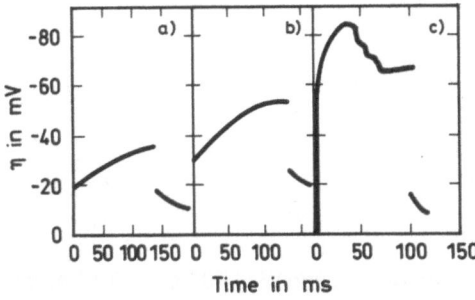

Fig. 16-24. Galvanostatic η vs. time curves for a Be cathode (covered with Al$_2$O$_3$ of thickness δ) in molten eutectic LiCl–KCl containing 5.85 wt% BeCl$_2$ at 400°C;
a) $\delta = 2$ to 3 μm, $j = 0.28$ A/cm²;
b) $\delta = 5$ μm, $j = 0.46$ A/cm²;
c) $\delta = 8$ to 10 μm, $j = 0.72$ A/cm².

Optical observation of the nucleation of Be on Be, Mo, and W microcathodes was carried out with a specially designed apparatus; the melts contained 5.85 wt% BeCl$_2$, the temperature ranged from 400 to 750°C, and current densities of 3.3×10^{-2} to 1.9 A/cm² were applied, lasting for up to 180 s. It was found that Be crystal nuclei only formed at the beginning of polarization, at potentials corresponding to η_{max} on the η vs. time curve, and were distributed extremely irregularly on the cathode surface. On continued electrolysis the existing crystal nuclei grew, but new nuclei did not form. The higher current densities in the above range led to a larger number of nuclei owing to the larger supersaturation of the cathode surface and because of the involvement of less active substrate sites in the process [12]. With rather impure melts the appearance of a large number of crystallization centers and subsequent formation of fine Be crystals was observed [12, 13].

The location of the Be nuclei on the cathode surface was found to be reproducible at low current densities, e.g., at 500°C with $j = (3.3$ to $3.8) \times 10^{-2}$ A/cm² and a W microcathode. So if cathodic deposition and anodic dissolution were alternated continually on the same electrode, the crystal nuclei always appeared at the same positions ("site memory"). This indicates a considerable energetic heterogeneity of the cathode surface [12]; cf. [11, 13].

Effect of Temperature

The η vs. time curves for Mo microcathodes with a macroscopic defect applied in advance and at a BeCl₂ concentration of 5.85 wt% in eutectic LiCl–KCl melt showed maxima of η at 400 to 600°C with $j = 6.36 \times 10^{-2}$ to 7.8×10^{-1} A/cm², but not at 650 to 700°C with $j = 1.2$ A/cm². At 750°C a small depolarization wave ($j = 1.9$ A/cm²) indicated alloy formation. The nucleation on Mo increases with rising temperature. This was attributed to an increase in the number of vacancies present on the surface and in the bulk of the cathode metal [12, 13]. The crystallization overvoltage on W microcathodes decreases with increasing temperature, see **Fig. 16-25** from [13], disappearing completely at 750°C.

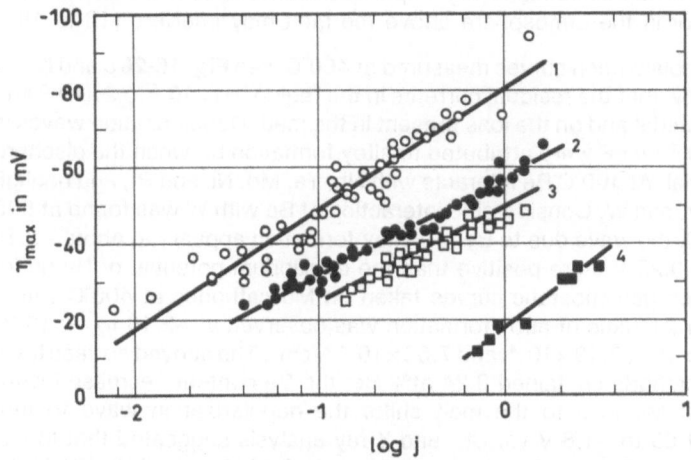

Fig. 16-25. Maximum crystallization overvoltage, η_{max}, of Be on a W electrode as a function of log j (with j measured in A/cm²) at the following temperatures: 1=400°C, 2=500°C, 3=600°C, and 4=700°C.

In the temperature range 400 to 700°C, the numerical value of η_{max} for W cathodes is represented by $\eta_{max} = 183.7 + 27.5 \log j - 0.1594\,T\ (\pm 5.7)$ mV, where T = absolute temperature in K and j = current density in A/cm². For expressions for t_{max} and plots of t_{max} vs. log j at 400, 500, 600, and 700°C, see the paper [13]; cf. p. 256. A rise in temperature of 100 K (in the above temperature range and for $j = 0.178$ A/cm²) decreased η_{max} by 4.7 mV and increased t_{max} by 16 ms. Estimated values for the penetration of Be into W at 400 and 850°C indicated that at low temperatures the buildup of the Be supersaturation on the cathode surface is the determining process, and that the diffusion mass transport is determining at the higher temperatures [13]. Ac measurements and the η vs. time curves indicated that ohmic contributions and the diffusion overvoltage of the Be^{2+} ions in the near-cathode layer become considerable at 800 to 850°C. Be-W alloy formation causes depolarization; thus at 850°C the potential of the W cathode is shifted by +10 to +12 mV. For details, see the paper [13].

The results [13] were used for a theoretical estimate of the Be nucleation rate on (110) faces and large-angle grain boundaries (LGB) of a W cathode at 773 K at constant j. It was shown that Be nucleates on W at low supersaturations ($\eta_{max} \leqq 30$ mV) in adsorbed layers of Be atoms on LGB. Active sites exert a decisive effect [15].

Experimental Conditions and Polarization Curves

For conditions in connection with the electrorefining of Be, i.e., simultaneous deposition and anodic dissolution, see pp. 260/2.

The electrodeposition of Be on a number of cathode materials (Zn, Ti, Zr, Nb, Ta, Mo, W, Ni, Pt) from eutectic LiCl–KCl melts with added $BeCl_2$ was studied for $BeCl_2$ concentrations of 1.62 to 15.07 wt% in the temperature range 380 to 800°C [16, 17]. High-purity materials and an Ar atmosphere were used. The melt was purified by preelectrolysis, because impurities more electropositive than Be are deposited prior to Be and may shield and passivate the cathode, i.e., interfere with Be deposition. Similarly, impurities such as oxides present in the melt, on the cathode surface, or in the atmosphere above the melt may interfere [16 to 18].

Potentiostatic polarization curves measured at 400°C, see **Fig. 16-26** a and b, and figures in the paper [17], show that the residual currents in the region $j = 1 \times 10^{-5}$ to 2×10^{-3} A/cm² depend on the cathode material and on the ions present in the melt. Depolarization waves in the range $j = 4 \times 10^{-4}$ to 2×10^{-2} A/cm² were attributed to alloy formation between the discharged Be and the cathode material. At 400°C Be interacts with Nb, Ta, Mo, Ni, and Pt, and negligibly with Ti, but not with Zn, Zr, and W. Considerable interaction of Be with W was found at 800°C. On Mo cathodes at 400°C, the wave due to Be-Mo alloy formation appears at about −2.15 V vs. Cl_2, i.e., at a potential 0.23 V more positive than the equilibrium potential of Be under the same conditions. Also, on galvanostatic curves taken on Mo cathodes at 400°C with 11.82 wt% $BeCl_2$, a step characteristic of alloy formation was observed at −2.10 to −2.15 V vs. Cl_2 for values of j of 4.07×10^{-4}, 7.13×10^{-4}, and 7.63×10^{-4} A/cm². The alloyed surface layer formed at 400°C on an Mo cathode contained 0.24 at% Be; the Be content decreased with increasing depth. Addition of Mo ions to the melt shifts the depolarization wave to more positive potentials, i.e., −1.65 to −1.8 V vs. Cl_2, and X-ray analysis suggested that Mo is deposited along with metastable $MoBe_2$, which is also formed at −2.15 V vs. Cl_2 [16, 17]. Metallic Be was deposited on Mo at potentials corresponding to the Be equilibrium potential. The limiting diffusion current is proportional to the $BeCl_2$ concentration and a two-electron mechanism is indicated. A wave appearing at −2.50 to −2.65 V vs. Cl_2 was attributed to the reaction $Be^{2+} + e \rightarrow Be^+$, and the simultaneous formation of a metallic mirror on the glass and quartz parts of the apparatus was attributed to the reaction $2\,Be^+ \rightarrow Be^{2+} + Be$ [16].

The polarization of Mo cathodes was measured at the moment of current interruption after 4 s polarization at a fixed j in LiCl–KCl eutectic melts additionally containing 0.1, 1.9, and 7.25 wt% $BeCl_2$ (corresponding mole fractions: 3.5×10^{-4}, 6.6×10^{-3}, and 2.5×10^{-2}, respectively) at 400, 500, and 600°C with j values of $10^{-3} \leqq j \leqq 3$ A/cm²; see figure 3 in the paper [19]. The polarization curves thus obtained indicate that the residual currents are mainly due to the reduction of alkali ions to subions while the reduction of Be^{2+} to Be^+ does not play a significant role. The Be deposition starts at the equilibrium potential in the melt and then proceeds without significant polarization. Concentration polarization was expected on account of the presence of complex ions of the type $BeCl_4^{2-}$ in the melt. But the active cathode surface increases during

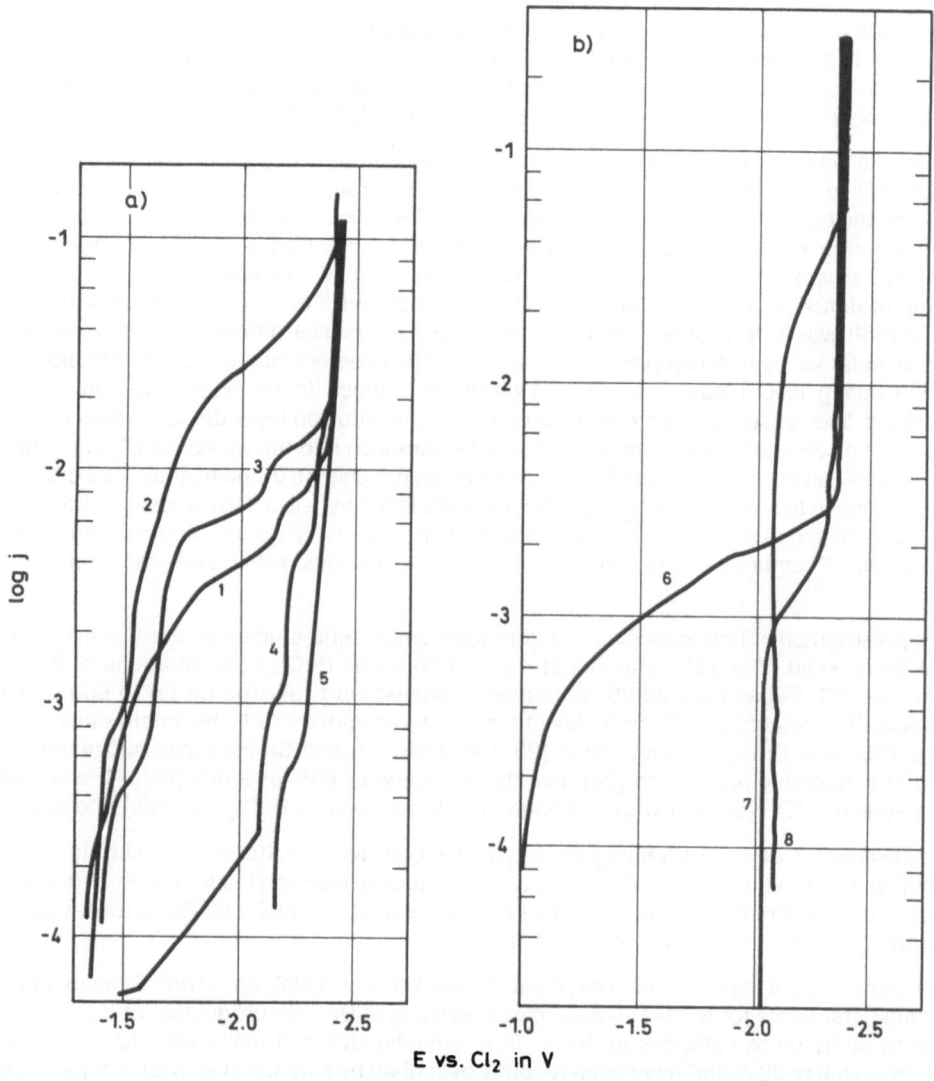

Fig. 16-26 (a and b).   Cathodic polarization curves, taken in a eutectic LiCl–KCl
melt with 5.27 wt% $BeCl_2$ added, at 400°C with the cathodes 1=Nb, 2=Ta,
3=Ni, 4=Pt, 5=Ti, 6=W, 8=Zr, at 380°C with the cathode 7=Zn (j measured
in A/cm²).

the deposition of crystalline Be, and so the true cathodic current density grows more slowly
than the nominal current density, especially for higher $BeCl_2$ concentrations. At 500°C the
nominal limiting j for the Be^{2+} discharge is not discernible for a 0.1 wt% $BeCl_2$ melt. It is clearly
expressed for 1.9 wt% $BeCl_2$, viz. 10^{-3} to 10^{-2} A/cm², and is beyond 2.5 A/cm² for 7.25 wt%
$BeCl_2$. The potentials of deposition become more positive with increasing temperature and
Be^{2+} concentration. Addition of KF to a eutectic LiCl–KCl melt containing 1.9 wt% $BeCl_2$, such
that the molar ratio is Be:F=1:4, shifted the deposition potential of Be at 500°C to more
negative values by ~0.3 V and decreased the limiting j for the Be^{2+} discharge significantly.

These results are considered evidence for the process of primary reduction of Be^{2+} in the $BeCl_2$–LiCl–KCl melts [19]; cf. [20]. The anodic and cathodic polarization curves of Be in molten LiCl–KCl eutectic, containing 10 wt% $BeCl_2$ at 450 to 500°C, demonstrate an extremely fast electrochemical system, suitable for electrorefining [20, 22].

When Be was electrodeposited onto an Ni cathode at 480°C from eutectic LiCl–KCl melts containing 3 to 20 wt% $BeCl_2$ [21], a continuous, nonporous, very adherent diffusion layer formed on the cathode underneath the Be dendrites. The thickness, d, of the layer reached up to 100 μm after 650 h electrolysis. For $BeCl_2$ concentrations between 3 and 7 wt% and j = 1.5 A/dm², d was found proportional to the square root of the time of electrolysis, i.e., the time of diffusion. The value of d after 20 h of electrolysis (e.g., d ≈ 32 μm at 480°C and j = 1.5 A/dm²) was 6 to 8 times the value obtained by classical diffusion at the same temperature under vacuum. A comparison of the penetration coefficients for normal diffusion and diffusion during electrolysis showed that diffusion is much faster under the influence of electrolysis. The presence of Be dendrites on top of the diffusion layer demonstrates that the rate of Be^{2+} discharge is higher than the rate of Be diffusion into the Ni. For $BeCl_2$ concentrations > 5 wt%, an increase in j from 0.3 to 15 A/dm² did not change d significantly. The diffusion layer was shown to consist mainly of $Be_{21}Ni_5$ (40 wt% Be). There is a narrow zone (≦ 1 to 2 μm) at the Ni–diffusion layer interface. This zone contains ~13 wt% Be, corresponding to BeNi. Similarly, on a Pt cathode a diffusion layer of $Be_{21}Pt_5$ was formed with Be dendrites on the top [21].

For polarographic, polaroscopic, and chronopotentiometric studies of the electrodeposition of Be on solid W and Pt cathodes at 400 to 850°C and $BeCl_2$ concentrations of 0.04 to 3 wt%, see [23, 24] and pp. 228/9. W cathodes proved inert towards Be up to 600°C. The discharged Be interacts with Pt cathodes at temperatures above 470°C; the intermetallic compounds PtBe and $PtBe_3$ were identified [23, 24]. Liquid Bi and Sn cathodes noticeably dissolved at elevated temperatures [24]. For the reduction of Be^{2+} at a dropping Pb electrode from a eutectic LiCl–KCl melt with additions of 0.04 to 0.9 wt% $BeCl_2$, see [25] and p. 227.

In the process of **electrorefining** Be is electrodeposited simultaneously with its anodic dissolution. $BeCl_2$–eutectic LiCl–KCl melts are very suitable electrolytes for the electrorefining of Be; the solubility of $BeCl_2$ in the LiCl–KCl eutectic is relatively high and $BeCl_2$ losses due to evaporation are low at 720 to 750 K [1, 2].

For technological aspects, see "Beryllium" Suppl. Vol. A 1, 1986, pp. 110/2. Equipment and operating parameters for the large-scale electrorefining of Be with production of Be flakes of ~99.95% purity on Ni cathodes at 500°C from eutectic LiCl–KCl melts with 10 wt% $BeCl_2$ added at j = 15.6 to 39 A/dm² have been recently described by Mitchell et al. [26]. For pilot-type equipment and operation, see [27]; cf. [28].

A careful investigation of the purification of Be by electrorefining in LiCl–KCl–$BeCl_2$ melts was carried out by Wohlfarth [2]. To a eutectic LiCl–KCl melt (59 and 41 mol%, respectively) $BeCl_2$ was added in amounts of about 0.02, 0.09, and 0.12 mole fraction (according to analysis, cf. table 5, p. 31 in [2]). The $BeCl_2$ was prepared in the apparatus from Be and HCl gas. Electrolysis was performed under Ar at 720 K for the lowest and at 750 K for higher $BeCl_2$ concentrations. Cathodic current densities ranged from 2.5 to 200 mA/cm², anodic current densities from 0.8 to 25 mA/cm² (for cathodic and anodic polarization values, see the paper). The deposited Be consisted of flakes or dendrites adherent to the Be cathode; descriptions of the morphology and structure as well as figures are given in the paper. Three kinds of commercially pure Be (see table 15 on p. 53 in [2]) were used as anode material and, in addition, Be metal enriched in Al and Fe (0.3 and 0.16 at%, respectively), and materials enriched in Cu (0.014, 0.14, 0.7, and 4.5 at%). In commercial Be the main impurities Si, Mg, Al,

Cr, Fe, and Cu were present in concentrations of 0.1 to 200 atppm, while oxygen was in the range of 100 to 1000 atppm, and Li, K, Zn were each in the range of 1 atppm. When the melts were purified by preelectrolysis, it was found that with anodic current densities >7 mA/cm², independent of the $BeCl_2$ concentration and temperature in the above ranges, nearly the same remaining impurity contents were achieved, viz. (in atppm) for the more electropositive metals Si (2 to 3), Zn (0.1 to 0.2), Al (2 to 3), Cr (0.1 to 0.2), Fe (0.8), Cu (0.1), as well as for the more electronegative metals Li (200 to 300), K (100 to 200), and Mg (0.02 to 0.03). Only at anodic current densities ≤ 2 mA/cm², i.e., far below that used for technical procedures, an increase of the $BeCl_2$ concentration from about 0.025 to 0.1 mole fraction in the melt led to lower impurity concentrations of Al, Cr, and Fe by a factor of 10, and to higher concentrations in Li and K by a factor of 5. The Li and K contaminations may be diminished by heating the Be flakes in vacuum at 970 K for several hundred hours. The dissolution of Al and Fe, or Cu into the electrolyte melt from the Be anode materials enriched with these metals (viz., Al 0.3 at% and Fe 0.16 at%, or Cu up to 4.5 at%) leads to an increase of their concentration in the cathode deposits (viz., Al to 280 atppm, Fe to 1 atppm, Cu to 7 atppm) [2].

For electrorefining of Be with the aim of preparing high-purity metal and of reclaiming scrap Be, eutectic LiCl–KCl melts, additionally containing 3.5 to 16.5 mol% $BeCl_2$, were electrolyzed at 450 to 600°C with j ranging from 45 to 770 A/ft². The $BeCl_2$ component of the bath was produced in situ by reacting Be metal, supplied in a graphite tube, with HCl prior to electrolysis. The electrolysis was carried out in steel chambers lined with graphite under an He atmosphere; see [29 to 31], cf. [32]. In the investigations no significant differences were found for Fe or Mo cathodes. Similarly, metallic impurities (except Ca) in electrodeposited Be were reduced to concentrations below the spectrographic determination limits by a second cycle of electrorefining with 11.4 mol% $BeCl_2$ in eutectic LiCl–KCl at 500°C and j in the range ~150 to 450 A/ft² (using an Ni chlorinator tube, an Ni-lined steel chamber, and an Ni cathode) [33]. Also in the temperature range from ~400 to 500°C, the electrorefining of Be was investigated with eutectic LiCl–KCl melts additionally containing ~10 wt% $BeCl_2$. Various Ni cathodes were used with cathodic current densities of 10 to 30 A/dm²; see [20, 22, 27, 34, 35]. Cathodes of Ni, stainless steel, Inconel, or graphite in melts containing 6 to 20 wt% $BeCl_2$ were also tried in the temperature range 460 to 550°C with current densities of 1075 to 3800 A/m². The optimum temperature range proved to be 500 ± 50°C; the adherence of the Be dendrites deposited was equally good on all cathodes used in the above temperature range. The Be obtained was of spectrographic purity. The purity depended on the purity of the melt; the use of a graphite crucible did not increase the C content of Be [1].

High $BeCl_2$ concentrations and low current densities favored the formation of steel gray, flat, platelike Be crystals, whereas low $BeCl_2$ concentrations and high current densities yielded long fibers consisting of platelets. Large lamellae formed in baths with 16.5 mol% $BeCl_2$. The size of the Be crystals decreased with increasing current density. The adherence of the Be deposits on the cathodes was considerably better at the higher temperatures, but sublimation of $BeCl_2$ from a 16.5 mol% $BeCl_2$ bath at 550°C was remarkable. During these electrorefining experiments all impurities in Be were reduced, except for C; this may be due to the graphite used in the apparatus. Thus, the following reductions were achieved: O from 3.27% (in the anode feed) to ~0.04 to 0.06% (in refined Be); Al from 0.5 to 1.0% to $\leq$0.004%; Fe from 0.32 to $\leq$0.08%; Mg from 2.66 to <0.02%; Ni from 0.05 to <0.002%; Si from 0.1 to 1.0% to $\leq$0.02% [29]. The Cl content in the Be crystals appeared to depend on the crystal size. The best results were obtained with the 10.3 mol% $BeCl_2$ bath at 550°C; see [29 to 31].

In experiments with Ni cathodes, e.g., the purity of the refined Be was ~99.8%. The metallic impurities were reduced from 1000 to 4000 ppm in the anode Be to $\leq$35 ppm in the refined Be, the O content from 1000 to 5000 ppm to 100 to 300 ppm; traces of the electrolyte were reduced to several ppm after fusion of the refined Be. The purity of the Be deposit did not appear to depend on the size of the Be crystals; however, the larger surface area of fine needles increased the O content, and efficient washing of fine structures was more difficult [20, 22].

Already in early experiments high-purity Be flakes were obtained at 400°C on Cu cathodes from melts of eutectic LiCl–KCl with Be being simultaneously dissolved from a Be anode at anodic current densities of up to 50 A/dm² [36]. The process was also studied at 400 to 550°C with Be, Ni, and stainless steel cathodes. Beryllium of 99.0% purity was obtained adherent to an Ni cathode at 460°C, j = 50 A/dm², whereas with j = 170 A/cm² at 550°C and a stainless steel cathode, the Be deposits consisted of nonadherent flakes of 99.5% purity. With Be cathodes the deposited Be was of low purity (85% at 400°C, j = 70 A/dm² and 95% at 500°C, j = 94 A/dm²). As alkali metal ions are discharged especially in the early stages of electrolysis, part of the Be obtained will result from secondary reduction of Be²⁺. Flakes are thought to be due to the primary (the electrochemical) reduction process, and powdery, partially adherent Be to the secondary process [37]. For experiments with a diaphragm separating anode and cathode compartments, and also the use of BeO–C anodes, see [38]. Mo and Zr cathodes were investigated for Be electrorefining at 760 to 900°F (404.5 to 482°C) in BeCl$_2$–eutectic LiCl–KCl melts (ratios by wt% 1:2 to 1:3) with best results at 900°F; see [39].

In the electrorefining of Be in eutectic LiCl–KCl melts with additional BeCl$_2$, certain impurities such as Al, Mn, Ni, Fe, Cu (i.e., metals reduced more easily than Be) were found to concentrate in the salt layer surrounding the Be flakes at the cathode. Chemical and spectrographic analyses showed that their concentrations increased rapidly towards the cathode and reached values up to 100 times their concentration in the bulk electrolyte and even 20 times that in the electrodeposited Be. Apparently this phenomenon is independent of the BeCl$_2$ concentration; see [20, 22, 27]; cf. [34, 35]. Anodic current efficiencies indicated that at low current densities Be⁺ was also formed at the anode [20]. In some cases the formation of white particles of 95% BeO was observed in the bath. This did not increase the O content of the Be deposit, but reduced the current efficiency. The formation of oxide is attributed to oxidation by moisture, present despite all precautions, and/or to the interaction of the melt with Pyrex and quartz parts of the apparatus [27].

The direct introduction of the very hygroscopic BeCl$_2$ into the melt can be avoided by reacting SnCl$_2$ with Be at 450 to 500°C in the LiCl–KCl melt according to Be + SnCl$_2$ → BeCl$_2$ + Sn. The Sn collects at the bottom of the cell and residual SnCl$_2$ is removed by pre-electrolysis [20, 34]. The Be deposited in the form of needles and two-dimensional dendrites in the case of a contamination with Sn (500 to 2500 ppm) when the residual Sn²⁺ (from the BeCl$_2$ preparation) was not removed by preelectrolysis [27].

References:

[1] Schimmel, F. A. (Y-1380 [1962] 31 pp.; N.S.A. **16** [1962] No. 15 127).

[2] Wohlfarth, H. (Diss. Univ. Stuttgart 1981, pp. 1/119, 31, 60, 101).

[3] Yang, L.; Hudson, R. G. (Trans. AIME **215** [1959] 589/601).

[4] Kuroda, T.; Matsumoto, O. (J. Electrochem. Soc. Japan **30** [1962] E 193/E 197).

[5] Kuroda, T.; Matsumoto, O. (Bull. Tokyo Inst. Technol. No. 61 [1964] 29/41; C.A. **64** [1966] 1629).

[6] Kuroda, T; Matsumoto, O. (Denki Kagaku **31** [1963] 688; J. Electrochem. Soc. Japan **31** [1963] 153).

[7] Chukreev, N. Ya.; Polishchuk, V. A.; Shapoval, V. I. (Fiz. Khim. Elektrokhim. Rasplavl. Tverd. Elektrolitov Tezisy Dokl. 7th Vses. Konf. Fiz. Khim. Ionnykh Rasplavov Tverd. Elektrolitov, Sverdlovsk 1979, Vol. 2, pp. 9/11; C.A. **93** [1980] No. 122440).

[8] Shapoval, V. I.; Chukreev, N. Ya.; Polishchuk, V. A. (Fiz. Khim. Elektrokhim. Rasplavl. Tverd. Elektrolitov Tezisy Dokl. 7th Vses. Konf. Fiz. Khim. Ionnykh Rasplavov Tverd. Elektrolitov, Sverdlovsk 1979, Vol. 2, p. 9; C.A. **93** [1980] No. 122381).

[9] Chukreev, N. Ya.; Shapoval, V. I.; Polishchuk, V.A. (Elektrokhimiya **18** [1982] 385/9; Soviet Electrochem. **18** [1982] 341/5).

[10] Chukreev, N. Ya.; Polishchuk, V. A. (Tezisy Dokl. Vses. Konf. Elektrokhim. **6** Vol. 2 [1982] 1/360, 294).

[11] Shapoval, V. I.; Chukreev, N. Ya.; Polishchuk, V. A. (Ukr. Khim. Zh. **49** No. 8 [1983] 841/5; Soviet Progr. Chem. **49** No. 8 [1983] 58/61).

[12] Chukreev, N. Ya.; Shapoval, V. I.; Polishchuk, V. A. (Elektrokhimiya **20** [1984] 520/4; Soviet Electrochem. **20** [1984] 490/4).

[13] Chukreev, N. Ya.; Shapoval, V. I.; Polishchuk, V. A. (Termodin. Elektrokhim. Svoistva Ionnykh Rasplavov **1984** 104/26; C.A. **101** [1984] No. 218544).

[14] Chukreev, N. Ya.; Polishchuk, V. A.; Shapoval, V. I. (Ukr. Khim. Zh. **52** No. 3 [1986] 282/7; Soviet Progr. Chem. **52** No. 3 [1986] 61/6).

[15] Shapoval, V. I.; Polishchuk, V. A.; Chukreev, N. Ya. (Teor. Eksperim. Khim. **23** No. 2 [1987] 245/9; Theor. Exptl. Chem. [USSR] **23** [1987] 229/33).

[16] Chukreev, N. Ya.; Sunegin, G. P. (Fiz. Khim. Elektrokhim. Rasplavl. Solei Tverd. Elektrolitov **1973** No. 2, pp. 10/2; C.A. **82** [1975] No. 9210).

[17] Chukreev, N. Ya.; Sunegin, G. P. (Elektrokhimiya **9** [1973] 842/5; Soviet Electrochem. **9** [1973] 806/9).

[18] Sunegin, G. P.; Chukreev, N. Ya. (Ukr. Khim. Zh. **38** No. 11 [1972] 1091/6; Soviet Progr. Chem. **38** No. 11 [1972] 13/6).

[19] Smirnov, M. V.; Ivanovskii, L. E. (Zh. Fiz. Khim. **32** [1958] 2174/82; C.A. **1959** 6833).

[20] Boisdé, G.; Broc, M.; Chauvin, G.; Coriou, H.; Hardy, L.; Jarny, P. (Bull. Inform. Sci. Tech. [Paris] No. 62 [1962] 29/38; C.A. **58** [1963] 12172).

[21] Broc, M. (CEA-R-4921 [1978] 103 pp., 21, 79/99; C.A. **91** [1979] No. 201108).

[22] Boisdé, G.; Broc, M.; Chauvin, G.; Coriou, H.; Hure, J.; Jarny, P. (J. Nucl. Mater. **6** [1962] 256/64).

[23] Delimarskii, Yu. K.; Baranenko, V. M.; Zarubitskii, O. G. (Deposited Doc. VINITI-3083-76 [1976] 1/13; Elektrokhimiya **12** [1976] 1887; Soviet Electrochem. **12** [1976] 1717).

[24] Delimarskii, Yu. K.; Baranenko, V. M.; Barchuk, V. T.; Shapoval, V. I. (Ukr. Khim. Zh. **45** No. 1 [1979] 3/7; Soviet Progr. Chem. **45** No. 1 [1979] 1/5).

[25] Naryshkin, I.; Minin, N. A. (Zh. Prikl. Khim. **34** [1961] 2353/6; J. Appl. Chem. [USSR] **34** [1961] 2230/2).

[26] Mitchell, D. L.; Nieweg, R. G.; Ledford, J. A.; Richen, M. J.; Burton, D. A.; Harder, R. V.; Watson, L. E.; Thomas, R. L. (RFP-4188 [1989] 26 pp.; C. A. **112** [1990] No. 127793).

[27] Bilard, J.; Boisdé, G.; Broc, M.; Chauvin, G.; Coriou, H.; Hardy, J. (Metaux **42** [1967] 259/69).

[28] Wong, M. M.; Klosterman, J. E. (BM-RI-6489 [1963] 19 pp.; N.S.A. **18** [1964] No. 34031).

[29] Wong, M. M.; Campbell, R. E.; Baker, D. H., Jr. (J. Metals **12** [1960] 786/8; C.A. **1960** 24008).

[30] Wong, M. M.; Cattoir, F. R.; Baker, D. H., Jr. (Rept. Invest. U.S. Bur. Mines No. 5581 [1960] 9 pp.; C.A. **1960** 10591).

[31] Wong, M. M.; Campbell, R. E.; Baker, D. H., Jr. (Rept. Invest. U.S. Bur. Mines No. 5959 [1961]; N.S.A. **16** [1962] No. 15117).

[32] Bellamy, R. G.; Hill, N. A. (International Series of Monographs on Nuclear Energy Division 8: Materials. Vol. 1. Extraction and Metallurgy of U, Th, Be, Pergamon, New York 1963, 198 pp., 81/4).

[33] Wong, M. M.; O'Keefe, D. A. (BM-RI-6570 [1964] 10 pp.; N.S.A. **19** [1965] No. 13870).

[34] Chauvin, G.; Coriou, H.; Hure, J. (Metaux Corrosion Ind. **37** [1962] 112/36; C.A. **57** [1962] 8261).

[35] Coriou, H. (Bull. Inform. Sci. Tech. [Paris] No. 84 [1964] 17/35; C.A. **65** [1966] 16397).

[36] Menzies, I. A.; Hill, D. L.; Owen, L. W. (Nature **183** [1959] 816/7).

[37] Menzies, I. A.; Hill, D. L.; Owen, L. W. (J. Less-Common Metals **1** [1959] 321/30).

[38] Higgins, J. K.; Bellamy, R. G.; Buddery, J. H. (AERE-M-R-2750 [1958] 18 pp.).

[39] Gurklis, J. A.; Beach, J. G.; Faust, C. L. (BMI-781 [1955] 26 pp.; N.S.A. **10** [1956] No. 1367).

16.7.5.6 $BeCl_2$–NaCl–KCl

References for this section are given on p. 268.

General Remarks

The melts investigated were usually based on equimolar NaCl–KCl mixtures (melting at 658°C) with $BeCl_2$ added in varying amounts, mostly in a minor extent. The diffusion coefficients of the Be^{2+} ion in such melts containing 0.108 or 0.18 wt% Be were determined from potential vs. time curves at 680 to 800°C using an Mo cathode [1]. An empirical equation relating the diffusion coefficient to the temperature and ionic moment in the above melt for the temperature range 1000 to 1100 K was formulated [2].

Cathodes of Solid Al, Mo, Ni, and Alloyed Steel

Beryllium deposition on solid Al cathodes from $BeCl_2$–NaCl–KCl melts was investigated in comparison to the deposition with cathodes of liquid Al; see pp. 267/8 and [3].

When a cleaned Mo rod cathode was used in the electrolysis of an equimolar NaCl–KCl melt additionally containing 3.7 wt% $BeCl_2$ at 678°C, the polarization curve, see Fig. 16-28 on p. 267 from [4], indicated interaction between the Mo and the deposited Be because at low current densities the cathode potential is more positive than according to the equilibrium. The cathode potential changed linearly from −2.1 V vs. Cl_2 at $j = 4 \times 10^{-4}$ A/cm² to −2.4 V at 0.4 A/cm². The limiting current was obscured by the growth of the effective cathode surface in the course of the Be deposition [4].

Experiments for the refining of Be with an Ni cathode and a Be rod anode with equimolar (eutectic) NaCl–KCl mixtures melted together with varying amounts of $BeCl_2$ at temperatures between 280°C (50% $BeCl_2$) and 450°C (30% $BeCl_2$) showed that at the higher $BeCl_2$ contents (and lower temperatures) larger Be crystals (needles) are obtained. Small Be crystals and even spongy metal appeared at the cathode in the case of only 30% $BeCl_2$ in the melt (at 450°C); see table in the paper [5]. Also the current efficiency was highest (93%) at $j = 5$ A/dm² when the melt contained 50% $BeCl_2$ at 280 to 300°C [5].

For the deposition of Be on Ni or alloyed steel cathodes from melts based on NaCl and KCl in 1:1 ratio (by wt%) with 13 to 15% $BeCl_2$ added at 760 to 790°C, see [6]. With alloyed steel cathodes, Be flakes were obtained from the above melts at 900 to 925°C [7]. At 750 to 900°C the Be deposition on iron alloy cathodes from melts of equimolar (eutectic) NaCl–KCl with 15 wt% $BeCl_2$ added involves corrosion problems and toxic hazards [8].

Alloy Formation on Cu and Pt Cathodes

Anfinogenov et al. [9, 10] studied the deposition of Be onto Cu cathodes from an equimolar NaCl–KCl melt containing 16 wt% $BeCl_2$. The cathode potential at constant current densities was measured at 710, 750, 800, and 835°C under an He atmosphere. **Fig. 16-27** shows a potential vs. time diagram obtained at 710°C for various current densities. The broken horizontal lines in the figure indicate the potential boundaries of the various Be-Cu alloy phases formed in the course of electrolysis [10]. (For the relationship of the equilibrium electrode potentials of Be-Cu alloys to their composition at the above temperatures, see [11].) The α-phase has the lowest, the δ-phase the highest Be content. Two competing processes determine the composition and the structure of the surface layer of the cathode: the discharge of Be^{2+} ions (depending on j) and the diffusion of Be into the cathode (depending on the temperature). The potential vs. time curves for the various temperatures investigated (see figures in the papers [9, 10]) show that the lower the temperature and the higher the current density, the more rapidly the potential changes with time. Metallographic, chemical, X-ray, and spectral analyses as well as microhardness measurements were used to study the effect of changes in current density, temperature, and duration of electrolysis on the structure and phase composition of the surface layers. Phase compositions and layer thicknesses are tabulated for j = 0.004, 0.01, 0.02, and 0.04 A/cm², temperatures of 710, 750, 800, and 835°C, and 1, 2, 4, 6, and 8 h of electrolysis in [9]. Thus electrolysis for 8 h at 710°C yielded surface layers of 0.095 and 0.265 mm thickness with current densities of 0.004 and 0.04 A/cm², respectively. The corresponding values at 835°C were 0.100 and 0.835 mm, respectively. One or several defined layers corresponding to the Be-Cu alloy phases are formed, depending on the conditions of electrolysis. The maximum Be content of the surface layer changed from 6.09% at 710°C to 3.62% at 835°C for j = 0.01 A/cm², and from 7.09% at 710°C to 4.75% at 835°C for j = 0.02 A/cm². The thickness depended linearly on the time of electrolysis as long as no metallic Be was formed on the surface. When this was the case, the relationship became parabolic. The deposited Be was found to spread by frontal diffusion from the cathode surface into the bulk. The cathode potential is related to the phase in the surface layer only under conditions for which the Be diffuses rapidly into the cathode, i.e., at high temperatures and high current

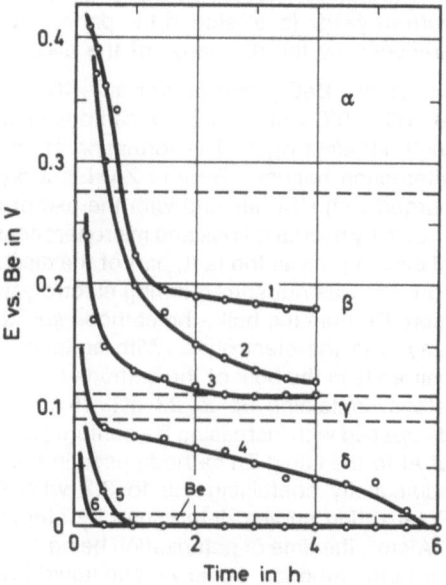

Fig. 16-27.   Change in the potential of a Cu cathode with time during Be deposition at 710°C from molten equimolar NaCl–KCl containing 16 wt% $BeCl_2$. Current densities (in A/ cm²): 1 = 0.001, 2 = 0.002, 3 = 0.004, 4 = 0.01, 5 = 0.02, and 6 = 0.04.

densities with short electrolysis times or at low current densities with long times of electrolysis [9]. For a theoretical investigation on the kinetics of diffusion, such as that of Be electrodeposited from salt melts onto a Cu cathode with alloy formation, see [12].

The formation of Be-Pt-alloy phases in Be electrodeposition on Pt cathodes was indicated by the polarization curves obtained in equimolar NaCl–KCl melts containing 0.75 wt% BeCl$_2$ at 710°C [13], and 0.12, 0.17, and 0.30 wt% BeCl$_2$ at 710°C [14, 15]. For low-frequency polarographic studies at 400 to 850°C with 0.05 to 3 wt% BeCl$_2$, see also [15]; cf. pp. 228/9 and 260. Prolonged potentiostatic electrolysis was carried out at −1.60, −1.78, and −2.35 V vs. Cl$_2$ at 710°C with BeCl$_2$ concentrations of 0.12, 0.17, and 0.30 wt%. The surface layers formed on the Pt cathode were analyzed chemically, metallographically, and by X-rays. Three intermetallic compounds may form when Be is electrodeposited on Pt. At −1.60 V the gray brown PtBe (bcc, 2.7909 Å) was formed. After electrolysis at −1.78 V the Pt cathode showed a three-layered structure consisting of a partially ordered, light gray solid solution, PtBe$_3$ (fcc, 3.870 Å) on the outside, and a PtBe layer intermediate between the PtBe$_3$ and the Pt. At −2.35 V the deposition of some Be plates and the conversion of the Pt to PtBe$_5$ (fcc, 5.974 Å) occurred. The formation potential of PtBe$_5$ (and possibly of PtBe$_{12}$) is difficult to determine experimentally, as it is close to the deposition potential of Be. On Pt cathodes Be is electrodeposited at more positive potentials (by ∼0.2 V at 600°C) than on W cathodes, because the intermetallic Pt-Be compounds form [14, 15]; cf. p. 230.

Liquid Cathodes of Bi, Zn, Al

Pyatkov et al. [16] found that at 700°C the Be deposition onto a liquid **Bi** electrode from equimolar NaCl–KCl melts additionally containing 2.85 wt% BeCl$_2$ proceeded at the same potential as on an indifferent electrode of Mo (cf. [17]), viz. at about −2.4 V vs. Cl$_2$. The extremal character with a flat maximum of the E vs. time curves of the Be deposition onto Bi was attributed to the appearance of dendritic Be on the surface of the cathode, thus decreasing the true cathodic current density. Additions of NaF to the above melt with [F$^-$]:[Be^{2+}]>0.9 shift the Be deposition potential to more negative values; for [F$^-$]:[Be^{2+}]≅1.74 the E vs. time curves indicate that the discharge of alkali metal ions is the cathodic process. Furthermore, the current yields (see table in the paper) show that the Be deposition on a liquid Bi cathode is preceded by the discharge of the alkali metal ions for [F$^-$]:[Be^{2+}]≅2 [16]; cf. p. 272.

When a BeCl$_2$–equimolar NaCl–KCl melt (BeCl$_2$ concentration not given) was electrolyzed at 700±10°C with a liquid **Zn** cathode in an open electrolyzer, the discharged Be dispersed in the bulk electrolyte. The formation of an oxide layer on the cathode surface prevented the interaction between Be and Zn. Electrolysis in a "closed" electrolyzer with a purified melt purged with HCl gas and with the use of distilled Zn yielded a two-phase Be-Zn alloy, as is shown by microsections and microhardness measurements of the solidified cathode. If the rate of discharge was too fast, part of the discharged Be stayed in the bulk electrolyte. Also, if the cathode was not stirred during electrolysis, the upper part of the cathode product contained more Be than the bulk, the cathode surface was nonuniform, and most of the discharged Be stayed in the electrolyte. With agitation of the liquid Zn cathode, the Be was distributed uniformly in the bulk of the cathode and its surface became more uniform. Nonviscous Be-Zn alloys containing up to 23 at% (5 wt%) Be could be obtained. The fluidity of the alloy decreased with increasing Be content [18]. Nichkov et al. [4] studied the influence of additions of Al to the liquid Zn cathode used in the electrolysis at 700°C of equimolar NaCl–KCl melts additionally containing 1.5 to 3.7 wt% BeCl$_2$ (Ar atmosphere, closed quartz apparatus). Polarization curves determined by stepwise increasing the current density from 1×10^{-4} to 4 A/cm^2, the time of polarization being 15 s at each j value, are given in **Fig. 16-28**. The curves for both cathodes, liquid Zn and liquid Zn-Al with 9% Al, each show three regions for 1) the deposition of Zn dissolved in the electrolyte ($j \approx 6 \times 10^{-3}$ A/cm^2), 2) the deposition of Be, and 3) that of the alkali metals. The Be deposition first proceeds at potentials more positive than the equilibrium value because of the formation of Zn-Be alloys. The potential is then shifted in the negative direction, as the surface layer of the cathode becomes saturated with Be. For melts containing 1.5 wt% BeCl$_2$, the limiting current for the discharge of the Be^{2+} ions at the liquid Zn cathode is 0.1 to 0.15 A/cm^2. The Be deposition on the Zn-Al cathode proceeds nearly at the same potentials as on the Zn cathode. Differences in the polarization curves only occur in the regions of low and of high current densities. At $j = 1 \times 10^{-4}$ to 1×10^{-2} A/cm^2 the potential of the Zn-Al cathode is more negative than that of the Zn cathode by ~ 0.15 V, probably due to the deposition of dissolved Al. The difference at the high current densities was attributed to an influence of Al on the solubility of the alkali metals in Zn [4].

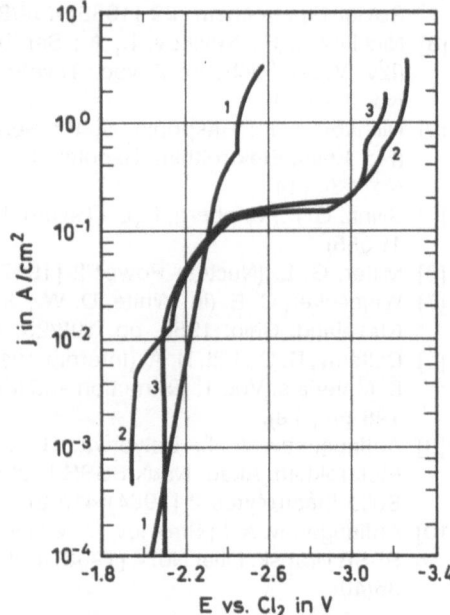

Fig. 16-28.   Polarization curves for BeCl$_2$–equimolar NaCl–KCl melts.
1 = Mo cathode, 678°C, 3.7 wt% BeCl$_2$;
2 = liquid Zn cathode, 700°C, 1.5 wt% BeCl$_2$;
3 = liquid Zn-Al cathode, 700°C, 2.3 wt% BeCl$_2$.

Polarization curves (as shown in the figures 1 and 2 in the paper [3]) were obtained with liquid and solid **Al** cathodes in the electrolysis of fused equimolar NaCl–KCl additionally containing up to 30 wt% BeCl$_2$ at 645 to 750°C. Liquid Al cathodes give linear regions for 10^{-2} A/cm$^2 < j < 0.3$ A/cm^2, corresponding to the discharge of Be ions with alloy formation, in the temperature range 700 to 750°C and 1.7 to 8.4 wt% BeCl$_2$ in the melt. With 6.5 wt% BeCl$_2$ in the melt a linear section in the interval 0.4 A/cm$^2 < j < 0.9$ A/cm^2 indicates the formation of a pure Be phase on the surface of its alloy with Al. With increasing BeCl$_2$ content and rising temperature, the limiting diffusion currents and the corrosion currents increase; the corresponding cathode potentials become more positive. In the case of a melt containing 8.40 wt% BeCl$_2$, the limiting diffusion current for the Be^{2+} discharge was not reached. For Be deposition at 700°C on the liquid Al cathode, the current yields were comparatively high (85.3 to 97.4%) except in the limiting current region, where coprecipitation of alkali metals led to a secondary reduction process. For low BeCl$_2$ concentrations, considerable initial polarization of the liquid Al cathode was observed. Metallographic analyses and microhardness measurements of the liquid Al cathodes after solidification showed that if the obtained Al-Be contained > 2 wt% Be, it consisted of two phases, viz. possibly a solid solution of Be in Al and crystals of metallic Be.

The polarization curves with solid Al cathodes and 12.7 and 13.6 wt% $BeCl_2$ in the melts at 650 and 645°C, respectively, indicate alloy formation for $j > 10^{-2}$ A/cm². The alloy formation shifts the potential for the Be^{2+} discharge to more positive potentials, compared to indifferent cathodes. Then, at $j > 0.3$ A/cm², the potential of the solid Al cathode shifts to more negative values and metallic Be appears on the cathode. Above $j = 1$ A/cm² the Be and alkali metal ions are discharged, jointly forming alloys [3].

References:

[1] Nichkov, I. F.; Novikov, E. A.; Raspopin, S. P.; Butorov, V. P. (Zh. Prikl. Khim. **42** [1969] 2828/9; J. Appl. Chem. [USSR] **42** [1969] 2674/5).

[2] Komarov, V. E.; Borodina, N. P.; Pakhnutov, I. A. (Elektrokhimiya **22** [1986] 478/82; Soviet Electrochem. **22** [1986] 444/9).

[3] Nichkov, I. F.; Novikov, E. A.; Serebryakov, G. A.; Kanashin, Yu. P.; Sardyko, G. N. (Izv. Vyssh. Uchebn. Zaved. Tsvetn. Metall. **23** No. 2 [1980] 58/62; C. A. **93** [1980] No. 338059).

[4] Nichkov, I. F.; Raspopin, S. P.; Serebryakov, G. A.; Butorov, V. P.; Novikov, E. A. (Fiz. Khim. Elektrokhim. Rasplav. Solei Shlakov **1969** No. 2, pp. 146/9; C. A. **74** [1971] No. 134114).

[5] Silina, G. F.; Grinberg, L. L. (Tsvetn. Met. [Moscow] **33** No. 12 [1960] 47/53; C. A. **1961** 19555).

[6] Miller, G. L. (Nuclear Power **2** [1957] 362/5; Metall. Abstr. [2] **25** [1957/58] 724).

[7] Windecker, C. E. (in: White, D. W., Jr.; Burke, J. E.; The Metal Beryllium, Vol. 1, ASM, Cleveland, Ohio, 1955, pp. 102/23, 114/8, 123).

[8] Bellamy, R. G.; Hill, N. A. (International Series of Monographs on Nuclear Energy Division 8: Materials. Vol. 1. Extraction and Metallurgy of U, Th, Be, Pergamon, New York 1963, 198 pp., 79).

[9] Anfinogenov, A. I.; Belyaeva, G. I.; Smirnov, M. V.; Ilyushchenko, N. G. (Tr. Inst. Elektrokhim. Akad. Nauk SSSR Ural'sk. Filial No. 4 [1963] 55/66; Electrochem. Molten Solid Electrolytes **2** [1964] 41/52).

[10] Anfinogenov, A. I.; Smirnov, M. V.; Ilyushchenko, N. G. (Tr. Inst. Elektrokhim. Akad. Nauk SSSR Ural'sk. Filial No. 4 [1963] 47/53; Electrochem. Molten Solid Electrolytes **2** [1964] 35/40).

[11] Anfinogenov, A. I.; Smirnov, M. V.; Ilyushchenko, N. G.; Belyaeva, G. I. (Tr. Inst. Elektrokhim. Akad. Nauk SSSR Ural'sk. Filial No. 3 [1962] 83/100; C. A. **59** [1963] 8187).

[12] Andreev, Yu. Ya. (Elektrokhimiya **15** [1979] 49/54; Soviet Electrochem. **15** [1979] 39/43).

[13] Delimarskii, Yu. K.; Baranenko, V. M.; Barchuk, V. T.; Shapoval, V. I. (Ukr. Khim. Zh. **45** No. 1 [1979] 3/7; Soviet Progr. Chem. **45** No. 1 [1979] 1/5).

[14] Delimarskii, Yu. K.; Khandros, E. L.; Baranenko, V. M. (Ukr. Khim. Zh. **42** No. 1 [1976] 3/6; Soviet Progr. Chem. **42** No. 1 [1976] 1/4).

[15] Delimarskii, Yu. K.; Baranenko, V. M.; Zarubitskii, O. G. (Deposited Doc. VINITI-3083-76 [1976] 1/13; Elektrokhimiya **12** [1976] 1887; Soviet Electrochem. **12** [1976] 1717).

[16] Pyatkov, V. I.; Lebedev, V. A.; Nichkov, I. F.; Raspopin, S. P. (Izv. Vyssh. Uchebn. Zaved. Tsvetn. Metall. **17** No. 2 [1974] 111/2; C. A. **81** [1974] No. 9043).

[17] Smirnov, M. V.; Ivanovskii, L. E. (Zh. Fiz. Khim. **32** [1958] 2174/81; C. A. **1959** 6833).

[18] Nichkov, I. F.; Smirnov, M. V. (Izv. Vyssh. Uchebn. Zaved. Tsvetn. Metall. **4** No. 3 [1961] 105/7; C. A. **56** [1962] 1248).

16.7.5.7 BeCl₂–LiCl–NaCl–KCl

Only melts based on the eutectic LiCl–NaCl–KCl composition (in wt%: 40.4 LiCl, 6.4 NaCl, 53.2 KCl, m.p. 362°C) were investigated. For a graph of the melting points of its mixtures with $BeCl_2$, see the paper [1]. The electrodeposition of Be from such $BeCl_2$–LiCl–NaCl–KCl melts is described in early patent specifications for the purpose of obtaining pure Be in a coherent form which may be easily stripped from the cathode, using relatively low temperatures and an apparatus allowing a nearly continuous operation of the electrolysis [1, 2]. The melts preferably contain 40 to 50 wt% $BeCl_2$. A temperature range of 260 to 326°C can thus be used; Al, Ni, and Cu sheet cathodes are considered to be suitable [1]. Fe, Cr-Fe, Cu, and stainless steel cathodes were found to corrode severely in the melt; Mg cathodes reacted rapidly with the melt and the Cl_2 fumes. A suitable cathode material was W except for the poor adherence of Be to the W [2]. With Al cathodes and graphite anodes a current efficiency of 94.8% was attained, the decomposition potential being about –2.2 V [1].

Beryllium deposits with $\geqq$99.8% purity may be produced. At deposition temperatures below ~290°C the Be is spongy and noncoherent, wheras dense, strong, and coherent metal was obtained above 290°C, especially in the region 325 to 500°C. In this temperature range, current densities of ~0.1 to 0.25 A/dm² produced very fine adherent Be crystals on clean cathodes of Al or Be. A subsequent increase of j to $\geqq$4 A/dm² did not reduce the density and strength of the Be deposit; this is attributed to a "seeding" effect of the initial crystals and to an extensive increase in the effective cathode area due to the deposited crystals. Melts containing 30 to even 90 wt% $BeCl_2$ yield satisfactory Be deposits, especially at the higher temperatures of the above temperature range [2]. For obtaining large crystals of dendritic Be the optimum conditions were found to be: $BeCl_2$ concentrations >10 wt% and temperatures between 400 and 800°C (Mo cathode, Ar atmosphere). The decomposition voltage, U (in V), of $BeCl_2$ in these melts is given by $U(T)=1.91-0.6\times10^{-3}(T-410)$, where T is the temperature in °C. Current efficiencies up to 99.1% were obtained, see [3, 4].

The Be deposition from eutectic LiCl–NaCl–KCl melts, additionally containing $\sim3\times10^{-3}$ to $\sim2\times10^{-2}$ mole fraction of $BeCl_2$, was studied in detail by polarization measurements at 450°C. Two reduction waves observed at –2.3 and –2.6 V vs. Cl_2 on the current-potential curves (determined with Ni cathodes) were attributed to $Be^{2+}+2e\to Be$ and $Be^{2+}+e\to Be^+$, respectively. Additional waves appearing at the higher $BeCl_2$ concentrations ($>10^{-2}$ mole fraction) were believed to be due to the formation of complexes, such as Na_2BeCl_4. Plots of polarization potential vs. time of electrolysis at fixed current values were determined also at 450°C with a mole fraction of 4.8×10^{-2} $BeCl_2$; see figure in the paper [5]. Electrolysis at cathode potentials more positive than the decomposition potentials of the alkali metal chlorides, i.e. $\geqq$–3.0 V vs. Cl_2, yielded Be dendrites. Electrolysis of the above melt at potentials more negative than the decomposition potentials of the alkali metal chlorides produced a mixture of powdery and dendritic Be. The dendritic Be is presumably formed by primary reduction, i.e., $Be^{2+}+2e\to Be$ (current efficiency: 75 to 85%), while the formation of Be powder was attributed to the secondary reduction of Be^{2+} by discharged alkali metal [5]; cf. [6].

For polarograms of eutectic LiCl–NaCl–KCl melts additionally containing up to 2.09 wt% $BeCl_2$ at 460°C, see [7] and p. 230.

References:

[1] Kjellgren, B. R. F.; Sawyer, C. B. (U.S. 2188904 [1940]; C.A. **1940** 3601).
[2] Sawyer, C. B.; Kjellgren, B. R. F. (U.S. 2311257 [1943]; C.A. **1943** 4309).
[3] Kuroda, T.; Matsumoto, O.; Suzuki, T.; Hasegawa, S. (Bull. Tokyo Inst. Technol. No. 49 [1962] 195/203; C.A. **1961** 5201).

[4] Kuroda, T.; Matsumoto, O. (Denki Kagaku **31** [1963] 338/41; J. Electrochem. Soc. Japan **31** No. 1/2 [1963] 75).

[5] Kuroda, T.; Matsumoto, O. (Bull. Tokyo Inst. Technol. No. 61 [1964] 29/41; C. A. **64** [1966] 1629).

[6] Kuroda, T.; Matsumoto, O. (Denki Kagaku **32** No. 2 [1964] 105/8; J. Electrochem. Soc. Japan **32** No. 2 [1964] 53).

[7] Ohmae, K.; Kuroda, T. (J. Electrochem. Soc. Japan **36** No. 3 [1968] 163/9).

16.7.5.8   Other Chloride Melts

$BeCl_2$–LiCl–KCl–CsCl

The effect of cathode vibrations on the structure of Be electrodeposited from eutectic LiCl–KCl–CsCl melts containing up to 10 mol% $BeCl_2$ was examined by Kozhanov et al. [1] at temperatures of 300 to 600°C using current densities of 0.001 to 0.2 A/cm². Under normal deposition conditions (i.e., "natural" convection of the electrolyte), electrocrystallization of Be supposedly takes place homogeneously over the whole cathode surface and proceeds along the directions $\langle 11\bar{2}0 \rangle$ and $\langle 10\bar{1}0 \rangle$ of the hcp Be lattice with the formation of spiral dislocations. When ultrasonic vibrations (22 kHz, amplitude up to 75 μm) were applied perpendicular to the end face of a cylindrical cathode, the Be deposits formed a thin layer with craters and fine cracks at the center of the end face. The crystals changed from fine to coarse as the distance from the center increased. A cubic crystal structure in the deposits is indicated at temperatures above the Be recrystallization temperature, while the structure is clearly hexagonal at lower temperatures. Low-frequency vibrations (50 Hz, amplitude 1 to 2 mm) produced a compact Be deposit at the end face, and columnar, dendritic, and bushy structures on the side surfaces of the cylindrical cathode, resulting from different "hydrodynamic" conditions at the end and side surfaces of the cathode [1]; see also "Beryllium" A 2, 1991, Section 9.1.4, p. 16.

$BeCl_2$–NaCl–KCl–$MgCl_2$

The polarization curves of Zn-Al alloy cathodes (9 wt% Al) in $BeCl_2$–$MgCl_2$–equimolar NaCl–KCl melts containing 2.2 to 9.9 wt% $MgCl_2$ and 2.2 to 9.8 wt% $BeCl_2$ were measured at 700 and 800°C under an Ar atmosphere. The polarization curve at 800°C is similar to that at 700°C, except that the initial cathode potential is more positive by ~80 mV. The curves show that the discharge of Al^{3+} and impurity ions takes place at an approximately constant cathode potential for j up to ~0.01 A/cm², but the cathode potential shifts to the negative side by ~70 mV when the Be^{2+} ions are discharged in the current density region ~0.01 to ~0.05 A/cm². A Be-Zn-Al alloy is formed at the cathode. For j = 0.07 to 0.10 A/cm², the rate of discharge of the Be ions is higher than the rate of diffusion of discharged Be into the Zn-Al alloy. Hence a metallic Be phase forms on the cathode surface and determines its potential. At j > 0.1 A/cm², first Mg^{2+} and then also alkali ions are discharged, and the potential values are more negative. A higher $MgCl_2$ content of the melt shifts the Mg deposition potential to more positive values, thus reducing the selectivity of the discharge of Be and Mg ions. Joint discharge of Be and Mg ions at Zn-Al cathodes is possible at current densities higher than the limiting diffusion current for the discharge of the Be ions [2].

$BeCl_2$–NaCl–KCl–$SrCl_2$

A decomposition potential of −2.06 V vs. Cl_2 (in the same melt) and a temperature coefficient of 0.00075 V/K were obtained at 700°C for $BeCl_2$ in molten 50% $SrCl_2$ – 28% NaCl – 22%

KCl (m.p. 540°C). The $BeCl_2$ concentration was 10 mol% with regard to $SrCl_2$ and the graphite electrodes were saturated with the products of electrolysis by preelectrolysis [3].

$BeCl_2$–NaCl–$CaCl_2$; $BeCl_2$–NaCl–KCl–$CaCl_2$, –$SrCl_2$, –$BaCl_2$

Eutectic melts of NaCl–$CaCl_2$ (m.p. 508°C), NaCl–KCl–$CaCl_2$ (m.p. 504°C), NaCl–KCl–$SrCl_2$ (m.p. 504°C), and NaCl–KCl–$BaCl_2$ (m.p. 522°C) were found suitable solvents for $BeCl_2$ with regard to Be electrodeposition; see [4].

References:

[1] Kozhanov, V. N.; Vasil'ev, A. V.; Novikov, E. A.; Nichkov, I. F.; Semavin, Yu. N. (Rasplavy 2 No. 6 [1988] 111/3; C.A. **110** [1989] No. 201510).

[2] Nichkov, I. F.; Novikov, E. A.; Serebryakov, G. A.; Kanashin, Yu. P.; Snigirev, A. A. (Izv. Vyssh. Uchebn. Zaved. Tsvetn. Metall. **12** No. 4 [1979] 94/5; C.A. **91** [1979] No. 148455).

[3] Delimarskii, Yu. K.; Ryabokon, V. D.; Kolotti, A. A. (Ukr. Khim. Zh. **15** [1949] 149/58).

[4] Kjellgren, B. R. F.; Sawyer, C. B. (U.S. 2188904 [1940]; C.A. **1940** 3601).

16.7.6 Electrolytes Based on Other Melts

Fluoride–Chloride Melts

Fused $2BeF_2$–NaCl–KCl was found unsuitable for a useful electrodeposition of Be on preliminarily beryllized Ni cathodes because the current yield was low and a complete removal of electrolyte adhering to the Be crystals proved difficult [1]. Beryllium crystals of up to 99.58% purity could be obtained on Ni or steel cathodes by electrolysis at 800°C of molten 25 wt% $BeF_2 \cdot NaF$ – 75 wt% NaCl or BeF_2–NaF–NaCl–KCl (10, 10, 40, and 40 wt%, respectively) using Be_2C or BeO-coke anodes; see [2].

Beryllium ions in equimolar NaCl–KCl melts containing 0.028 to 0.132 mole fraction of BeF_2 at 700°C were discharged at a liquid Zn cathode at potentials of −2.230 to −2.420 V (vs. Cl_2) and current densities from ∼0.01 to ∼0.1 A/cm² respectively. A Be-Zn alloy is formed. The rate of diffusion of metallic Be into the liquid Zn seems to be comparable to the rate of discharge of the Be ions. But for $j > 0.1$ A/cm² the polarization curve reflects the discharge of Be ions on the surface of a solid phase of metallic Be. If YCl_3 is added to BeF_2–equimolar NaCl–KCl melts (mole fraction of BeF_2: 0.059 to 0.185, mole fraction of YCl_3: 0.0055 to 0.0222; $[Be^{2+}]:[Y^{3+}] = 2.9$ to 11.9), the polarization curves determined at 700°C in the current density range 0.03 to ∼0.3 A/cm² indicate the joint discharge of Be and Y ions and alloy formation at the liquid Zn cathode. The cathode potential is shifted to more positive values by an increase in the BeF_2 concentration at constant YCl_3 concentration or by an increase in the YCl_3 concentration at constant BeF_2 concentration. This is attributed to increases in the activities of the Y^{3+} and Be^{2+} ions by addition of BeF_2 or YCl_3, respectively. At $j > 0.25$ A/cm² the cathode potential becomes more negative and the solid phase YBe_{13} is formed on the cathode. From melts additionally containing NaF (BeF_2 mole fraction 0.044 to 0.142; NaF 5.7 to 20.9 wt%; YCl_3 mole fraction 0 to 0.065; $[F^-]:[Be^{2+}] = 4$) the Be ions are discharged at $j = 2.0 \times 10^{-3}$ to 1×10^{-1} A/cm². At $j = 0.1$ to 0.4 A/cm² the cathode potential corresponds to the discharge of Be ions on Be with concentration polarization, possibly accompanied by the discharge of alkali ions. Codeposition of Be and Y is possible in the current density range 6.0×10^{-3} to 1.8×10^{-1} A/cm², again with significant depolarization. At $j > 1.0$ A/cm², as in the case without NaF, the solid intermetallic compound YBe_{13} is formed on the cathode surface from a melt containing 3.7 wt% BeF_2, 17.8 wt% YCl_3, and 6.6 wt% NaF in equimolar NaCl–KCl; for details, see [3].

Small additions of NaF to equimolar NaCl–KCl melts containing 2.85 wt% of $BeCl_2$ showed no significant effect on the Be deposition potential at a liquid Bi cathode at 700°C. For higher F^- ion concentrations ($[F^-]:[Be^{2+}]\geqq 2$) the Be deposition potential is shifted considerably to more negative values, which indicates the discharge of alkali metals; this is supported by results on the current yields [4].

Iodide Melts

The decomposition potential of BeI_2 (5 mol%) in NaI was found to be 1.48, 1.38, and 1.28 V (between graphite electrodes) at 600, 700, and 800°C, respectively [8].

Melts Containing BeS

Electrolysis of anhydrous BeS dissolved in a fused fluoride or in fused mixtures of anhydrous fluorides, e.g., BeF_2, cryolite, BaF_2, or BeF_2–BaF_2, yields Be at the cathode [5]; see also "Beryllium" Suppl. Vol. A 1, 1986, p. 88.

The electrodeposition of Be from alkali metal chlorides, e.g., $BeCl_2$–KCl melts (60 : 40 mol%), can be improved by the addition of at least 2 to 3 wt% BeS to the melt. The electrolysis is carried out at 450°C in an Fe or stainless steel vessel with an Ni cathode and graphite anode at j = 240 A/ft²; the applied voltage is 3.2 V. By the BeS addition metallic impurities are kept in their lowest oxidation states, the energy required for the electrodeposition is reduced, and the evolution of Cl_2 at the anode is largely or wholly avoided [6].

Other Melts

Electrolysis of a fused mixture of ethyl pyridinium bromide and $BeCl_2$ (quantities not specified) at 135 ± 5°C yielded a brownish coating on the Pt cathode (applied voltage 6 to 12 V, C rod anode) [7]. Beryllium was electrodeposited from a molten mixture of a Be halide and an alkyl-substituted imidazolium halide [9].

References:

[1] Silina, G. F.; Grinberg, L. L. (Tsvetn. Met. [Moscow] **33** No. 12 [1960] 47/53; C.A. **1961** 19555).

[2] Pruvot, E.; Moutach, M. (Fr. 1217839 [1958/60]; C.A. **1961** 17312).

[3] Butorov, V. P.; Nichkov, I. F.; Novikov, E. A. (Izv. Vyssh. Uchebn. Zaved. Tsvetn. Metall. **19** No. 6 [1976] 50/6; C.A. **86** [1977] No. 147726).

[4] Pyatkov, V. I.; Lebedev, V. A.; Nichkov, I. F.; Raspopin, S. P. (Izv. Vyssh. Uchebn. Zaved. Tsvetn. Metall. **17** No. 2 [1974] 111/2; C.A. **81** [1974] No. 9043).

[5] Gardner, D. (Brit. 482468 [1938]; C.A. **1938** 7009).

[6] Gibson, A. R. (Brit. 821091 [1959]; C.A. **1960** 2051).

[7] Hurley, F. H.; Wier, T. P., Jr. (J. Electrochem. Soc. **98** [1951] 203/6).

[8] Delimarksii, Yu. K.; Kolotti, A. A. (Zh. Fiz. Khim. **23** [1949] 97/100; C.A. **1949** 4581).

[9] Takahashi, S.; Saeki, I.; Mori, S.; Sakura, Y. (Jpn. Kokai Tokkyo Koho 02-240289 [90-240289] [1989/90] 5 pp. from C.A. **115** [1991] No. 80929).

16.7.7   Electrolytes Based on Molten Organic Complex Beryllium Salts

According to patent specifications very pure Be can be electrodeposited at about 65 to 95°C from melts of the salt-like organometallic complexes of the general formula $(MeX)_m(BeRR')_n$, where Me is an alkali metal or substituted ammonia (e.g., $(C_2H_5)_4N$, (iso-butyl)$_4$N), X is a halogen (F, Cl) or CN^-, R and R' are saturated aliphatic or aromatic radicals, or R' may be an H atom; m, n are whole numbers, and the ratio n : m is preferably 1:1 or 2:1. These compounds melt between room temperature and ~100°C. The electrolyte may contain additions of alkyl or aryl Be compounds or mixtures of these [1 to 3]; cf. [9].

According to Hans [1], a deposit containing 99.995 (±0.004)% Be (Na, Ca, Al, Mn, Fe: each <0.001%; Si, Mg: each 0.001%) can be obtained by electrolysis of $KF \cdot 2Be(C_2H_5)_2$ in an excess of $Be(C_2H_5)_2$: decomposition voltage 1 V, current density 0.2 A/dm², temperature range 65 to 95°C [1]. But the Be thus prepared was not analyzed for Be_2C, of which there could be up to 20 to 30% [4]. Intense agitation of the electrolyte, e.g., by using rotating cathodes, and atmospheres of pure, dried N_2 or Ar were found necessary. The applied voltage should not exceed a certain value, which depends on the electrolyte; e.g., at 75°C $KF \cdot 2Be(C_2H_5)_2$–$Be(C_2H_5)_2$ (1 and 3 parts by wt, respectively) was electrolyzed preferably at 1.5 to 1.8 V and ~0.4 A/dm² [4], cf. [2, 3, 5 to 7].

Deposition of Be was also studied from various Be diethyl complexes, viz. from $KF \cdot 2Be$-$(C_2H_5)_2$, $KF \cdot Be(C_2H_5)_2$, $CsF \cdot 2Be(C_2H_5)_2$, $RbF \cdot 2Be(C_2H_5)_2$, $R_4NX \cdot 2Be(C_2H_5)_2$ with R = CH_3 or C_2H_5 and X = Cl or F, and $[(CH_3)_3(CH_2C_6H_5)N]F \cdot 2Be(C_2H_5)_2$; see [2, 4 to 7]; also from $[(CH_3)_3(CH_2C_6H_5)N]Cl \cdot 2Be(i-C_3H_7)_2$ [3, 7], and $KF \cdot 2Be(i-C_3H_7)_2$ and $[(CH_3)_4N]F \cdot 2Be(i-C_3H_7)_2$ [7]. The electrolysis was carried out with the pure as well as with the diluted complex salts, e.g., $KF \cdot 2Be(C_2H_5)_2$ or $CsF \cdot 2Be(C_2H_5)_2$ were diluted with 2 [2] and 3 [4] parts by wt of $Be(C_2H_5)_2$; see below. Cu, Pt, and Ag cathodes were used, and the temperature ranged from 60 to 100°C; voltages from 0.6 to 3.0 V (preferably $\leq$ 2.0 V) were applied with current densities of 0.5 to 0.8 A/dm².

The Be deposits on Cu and Ag cathodes were adherent, rough, and light gray to gray brown, while those on Pt cathodes were loose, gray shining scales or powder, which could be removed easily. The difference in adherence may result because Be forms intermetallic compounds with Cu and Ag, but not with Pt. The deposits contained only ~43 to ~84% Be. This was attributed to be mostly due to occluded electrolyte, but in the case of the Cs complex, metallic Cs seems to have been electrodeposited [4, 5].

The current–voltage curves for $KF \cdot 2Be(C_2H_5)_2$ and $CsF \cdot 2Be(C_2H_5)_2$, dissolved in excess $Be(C_2H_5)_2$ at 63 to 85°C (see figures in [4, 5]), depended considerably on temperature. They show that only one cation transports the charge and is discharged, i.e., K^+ is not discharged simultaneously with Be^{2+} [4, 5]. It is suggested that K^+ is discharged and Be is deposited by a secondary reaction: in the melt the complex dissociates according to $KF \cdot 2Be(C_2H_5)_2 \underset{T \gg 80°C}{\overset{melt}{\rightleftharpoons}}$ $KF \cdot Be(C_2H_5)_2 + Be(C_2H_5)_2$ and $KF \cdot Be(C_2H_5)_2 \rightleftharpoons K^+ + [Be(C_2H_5)_2 \cdot F]^-$. The K^+ is discharged at the cathode and then reacts according to $K + Be(C_2H_5)_2 \rightarrow KC_2H_5 + \cdot Be(C_2H_5)$ and $K + \cdot Be(C_2H_5) \rightarrow$ $KC_2H_5 + Be$. The Be deposition by a secondary reaction is supported by further observations: Be deposits with identical properties are obtained on both sides of a cylindrical Cu cathode surrounding a cylindrical anode. Electrolysis of molten $KF \cdot 2Be(C_2H_5)_2$ at 80°C yielded a Be deposit on a Cu cathode, but only K amalgam on an Hg cathode. Be_2C (present in the Be deposits in amounts up to 25%) is formed when $Be(C_2H_5)_2$ is reacted with K or Na suspended in toluene [5, 6]. The formation of the $\cdot Be(C_2H_5)$ radical was proved by the electrolysis at 80°C of a solution of 3.5 g (0.018 mol) of $KF \cdot 2Be(C_2H_5)_2$ in 11 g (0.12 mol) of toluene: Be was deposited at the cathode. The addition of 3.2 g pyridine to the above solution prevented the Be deposition because of the formation of the complex radical $[C_5H_5N:Be(C_2H_5)]$; see [6].

Strohmeier and Popp [7] studied the effect of the composition of the electrolyte, the current density, and the temperature on the composition of the deposits obtained by electrolysis of Be dialkyl complexes. The addition of $Be(C_2H_5)_2$ to fused $KF \cdot 2Be(C_2H_5)_2$ (up to a molar ratio of 3:1, respectively) hardly affected the Be_2C content (10 to 12%) of the deposits produced at 80°C on a Cu cathode with 0.06 A/dm². The K content of the deposits decreased from 18% for the pure $KF \cdot 2Be(C_2H_5)_2$ electrolyte to 5% for molten $KF \cdot 2Be(C_2H_5)_2 - Be(C_2H_5)_2$ (mole ratio 1:3). Electrolysis of molten $KF \cdot 2Be(C_2H_5)_2 - (C_4H_9)_2O$ yielded deposits containing 7% Be_2C and 24% K for a mole ratio of 1:0.12, but 3% Be_2C and 13% K for a mole ratio of 1:0.45, respectively, at 80°C and 0.06 A/dm². Electrolysis of the molten complexes $A \cdot 2Be(C_2H_5)_2$, where $A = [(CH_3)_4N]F$, $[(CH_3)_4N]Cl$, $[(C_2H_5)_4N]Cl$, or $[(CH_3)_3(CH_2C_6H_5)N]Cl$, at 80°C and 0.05 to 0.15 A/dm² (Cu cathode, Be anode), yielded Be deposits containing 2 to 13% Be_2C and 4 to 24% of the discharged ammonium radical. A considerably lower Be_2C content was achieved if $KF \cdot 2Be(i-C_3H_7)_2$ was electrolyzed instead of $KF \cdot 2Be(C_2H_5)_2$, e.g., 6 instead of 17 mol% Be_2C at 80°C, 0.12 to 0.15 A/dm², using Be anodes. Electrolysis of $[(CH_3)_3(CH_2C_6H_5)N]F \cdot 2Be(C_2H_5)_2$ at 80°C and 0.04 A/dm² (anode: Be) yielded no Be deposit at the Cu cathode. An increase in the current density from 0.06 to 0.92 A/dm² in the electrolysis at 80°C of fused $KF \cdot 2Be(C_2H_5)_2 - Be(C_2H_5)_2$ (mole ratio 1:3) increased the Be_2C content of the deposit from 12 to 18% and that of K from 5 to 13%, respectively. An increase in temperature increased the Be_2C content. Thus the electrolysis of $[(CH_3)_4N]F \cdot 2Be(C_2H_5)_2$ (liquid at room temperature) diluted with $Be(C_2H_5)_2$ in the mole ratio 1:3, respectively, yielded Be deposits containing 1% each of Be_2C and $(CH_3)_4N$ at 50°C, but 6% Be_2C and 9% $(CH_3)_4N$ at 80°C and 0.11 to 0.12 A/dm² (Cu cathode, Cu anode). Be_2C could not be detected in Be deposits obtained at 20°C from $[(CH_3)_4N]F \cdot 2Be(i-C_3H_7)_2$, with a slight excess of $Be(i-C_3H_7)_2$ and diluted with toluene in the mole ratio 1:4, respectively, at 0.1 A/dm², but the Be_2C content increased to 10% at 80°C. However, the current efficiencies were very low; for details, see [7].

The gray, adherent layer formed on a Cu cathode by electrolysis at 80°C of molten $KF \cdot 2Be(C_2H_5)_2$ (anode: Be) consisted of crystalline Be with occluded crystalline Be_2C and polycrystalline K. Electronmicroscopic studies showed the lattice parameters for the K and the Be_2C to agree with values from literature. The lattice parameters for the crystalline Be were twice the values for the normal α-Be modification. This was suggested to be due to the occluded Be_2C, or possibly a new Be modification may be involved [6, 7].

Electrolysis of $KF \cdot 2Be(C_2H_5)_2$ in $Be(C_2H_5)_2$ at 85°C and 0.5 to 1.0 A/dm² produced a shining Be coating of 0.01 to 50 mm thickness on a steel cathode after covering with a 0.01 to 0.1 mm Be layer by thermal decomposition of an organic Be compound ("dual plating") [8].

Dötzer et al. [3] electrolyzed $[(CH_3)_3(C_6H_5CH_2)N]Cl \cdot 2Be(i-C_3H_7)_2$, containing ~0.2 mol $Be(i-C_3H_7)_2$ in excess and diluted with ~2 mol toluene per mol of complex, at 80°C using an Ag plate cathode rotating at 80 rpm and a Be pipe anode. The cell was rinsed with dry Ar or N_2, the applied voltage was 1.5 V, and the initial cathodic current density 0.6 A/dm². Pure, bright, dendritic Be formed at the cathode with contaminations $< 1 \times 10^{-4}$% (in ppm: < 5 C; < 3 Si; 0.6 Al; < 0.2 Cr; < 0.002 W; 0.02 Mn; < 2 Ni; 0.05 Fe; 0.05 Cu) [3]. Strohmeier and Popp [7] failed to obtain a Be deposit at an Ag cathode, when they electrolyzed a bath containing $[(CH_3)_3(C_6H_5CH_2)N]Cl \cdot 2Be(i-C_3H_7)_2$ with a slight excess of $Be(i-C_3H_7)_2$ and 0.60 mol of toluene per mol of complex with 0.015 A/dm² at 20°C or with 0.13 A/dm² at 80°C.

References:

[1] Hans, G. (Ger. Offen. 1 162 576 [1959/64]; C. **1964** No. 51-2313).
[2] Strohmeier, W. (Ger. Offen. 1 179 074 [1961/64]; Patentblatt **84** [1964] 5129, **85** [1965] 2084).

[3] Dötzer, R.; Engelbrecht, F.; Todt, E. (Ger. Offen. 1 236 208 [1960/67] 6 pp.).
[4] Strohmeier, W.; Gernert, F. (Z. Naturforsch. **20 b** [1965] 829/31).
[5] Gernert, F. (Diss. Univ. Würzburg 1962) cited in [7].
[6] Strohmeier, W.; Popp, G. (Z. Naturforsch. **23 b** [1968] 38/41).
[7] Strohmeier, W.; Popp, G. (Z. Naturforsch. **23 b** [1968] 870/1).
[8] Norman, V.; Whaley, T. P.; Prestridge, H. B. (U.S. 3 294 654 [1966] 6 pp.; C. A. **66** [1967] No. 43 209).
[9] Dötzer, R. (Chem. Ing. Tech. **36** [1964] 616/37, 623, 635/6).

Physical Constants and Conversion Factors

Avogadro constant N_A (or L) = 6.02214×10^{23} mol⁻¹   Planck constant $h = 6.62608 \times 10^{-34}$ J·s

Faraday constant $F = 9.64853 \times 10^{4}$ C/mol   elementary charge $e = 1.60218 \times 10^{-19}$ C

molar gas constant $R = 8.31451$ J·mol⁻¹·K⁻¹   electron mass $m_e = 9.10939 \times 10^{-31}$ kg

molar volume (ideal gas) $V_m = 2.24141 \times 10^{1}$ L/mol   proton mass $m_p = 1.67262 \times 10^{-27}$ kg

(273.15 K, 101 325 Pa)

1 kg = 2.205 pounds

1 m = 3.937×10^{1} inches = 3.281 feet

1 m³ = 2.642×10^{2} gallons (U.S.)

1 m³ = 2.200×10^{2} gallons (Imperial)

Force	N	dyn	kp
1 N	1	10^{5}	1.019716×10^{-1}
1 dyn	10^{-5}	1	1.019716×10^{-6}
1 kp	9.80665	9.80665×10^{5}	1

Pressure	Pa	bar	kp/m²	at	atm	Torr	lb/in²
1 Pa = 1 N/m²	1	10^{-5}	1.019716×10^{-1}	1.019716×10^{-5}	9.86923×10^{-6}	7.50062×10^{-3}	1.450378×10^{-4}
1 bar = 10^{6} dyn/cm²	10^{5}	1	1.019716×10^{4}	1.019716	9.86923×10^{-1}	7.50062×10^{2}	1.450378×10^{1}
1 kp/m² = 1 mm H_2O	9.80665	9.80665×10^{-5}	1	10^{-4}	9.67841×10^{-5}	7.35559×10^{-2}	1.422335×10^{-3}
1 at (technical)	9.80665×10^{4}	9.80665×10^{-1}	10^{4}	1	9.67841×10^{-1}	7.35559×10^{2}	1.422335×10^{1}
1 atm = 760 Torr	1.01325×10^{5}	1.01325	1.033227×10^{4}	1.033227	1	7.60×10^{2}	1.469595×10^{1}
1 Torr = 1 mm Hg	1.333224×10^{2}	1.333224×10^{-3}	1.359510×10^{1}	1.359510×10^{-3}	1.315789×10^{-3}	1	1.933678×10^{-2}
1 lb/in² = 1 psi	6.89476×10^{3}	6.89476×10^{-2}	7.03069×10^{2}	7.03069×10^{-2}	6.80460×10^{-2}	5.17149×10^{1}	1

Work, Energy, Heat	J	kW·h	kcal	Btu	eV
1 J = 1 W·s = 1 N·m = 10^7 erg	1	2.778×10^{-7}	2.39006×10^{-4}	9.4781×10^{-4}	6.242×10^{18}
1 kW·h	3.6×10^6	1	8.604×10^2	3.41214×10^3	2.247×10^{25}
1 kcal	4.1840×10^3	1.1622×10^{-3}	1	3.96566	2.6117×10^{22}
1 Btu (British thermal unit)	1.05506×10^3	2.93071×10^{-4}	2.5164×10^{-1}	1	6.5858×10^{21}
1 eV	1.602×10^{-19}	4.450×10^{-26}	3.8289×10^{-23}	1.51840×10^{-22}	1

$1 \text{ cm}^{-1} \triangleq 1.239842 \times 10^{-4} \text{ eV}$

$2 \text{ rydberg} = 1 \text{ hartree} = 27.2114 \text{ eV}$

$1 \text{ Hz} \triangleq 4.135669 \times 10^{-15} \text{ eV}$

$1 \text{ eV} \triangleq 96.485 \text{ kJ/mol}$

Power	kW	hp	$kp \cdot m \cdot s^{-1}$	kcal/s
1 kW = 10^3 J/s	1	1.35962	1.01972×10^2	2.39006×10^{-1}
1 hp (horsepower, metric)	7.3550×10^{-1}	1	7.5×10^1	1.7579×10^{-1}
1 $kp \cdot m \cdot s^{-1}$	9.80665×10^{-3}	1.333×10^{-2}	1	2.34384×10^{-3}
1 kcal/s	4.1840	5.6886	4.26650×10^2	1

References:

Mills, I. (Ed.), International Union of Pure and Applied Chemistry, Quantities, Units and Symbols in Physical Chemistry, Blackwell Scientific Publications, Oxford 1988.

The International System of Units (SI), National Bureau of Standards Spec. Publ. 330 [1972].

Landolt-Börnstein, 6th Ed., Vol. II, Pt. 1, 1971, pp. 1/14.

ISO Standards Handbook 2, Units of Measurement, 2nd Ed., Geneva 1982.

Cohen, E. R., Taylor, B. N., Codata Bulletin No. 63, Pergamon, Oxford 1986.

Key to the Gmelin System
of Elements and Compounds

System Number	Symbol	Element
1		Noble Gases
2	H	Hydrogen
3	O	Oxygen
4	N	Nitrogen
5	F	Fluorine
6	**Cl**	**Chlorine**
7	Br	Bromine
8	I	Iodine
8a	At	Astatine
9	S	Sulfur
10	Se	Selenium
11	Te	Tellurium
12	Po	Polonium
13	B	Boron
14	C	Carbon
15	Si	Silicon
16	P	Phosphorus
17	As	Arsenic
18	Sb	Antimony
19	Bi	Bismuth
20	Li	Lithium
21	Na	Sodium
22	K	Potassium
23	NH_4	Ammonium
24	Rb	Rubidium
25	Cs	Caesium
25a	Fr	Francium
26	Be	Beryllium
27	Mg	Magnesium
28	Ca	Calcium
29	Sr	Strontium
30	Ba	Barium
31	Ra	Radium
32	**Zn**	**Zinc**
33	Cd	Cadmium
34	Hg	Mercury
35	Al	Aluminium
36	Ga	Gallium

System Number	Symbol	Element
37	In	Indium
38	Tl	Thallium
39	Sc, Y La—Lu	Rare Earth Elements
40	Ac	Actinium
41	Ti	Titanium
42	Zr	Zirconium
43	Hf	Hafnium
44	Th	Thorium
45	Ge	Germanium
46	Sn	Tin
47	Pb	Lead
48	V	Vanadium
49	Nb	Niobium
50	Ta	Tantalum
51	Pa	Protactinium
52	**Cr**	**Chromium**
53	Mo	Molybdenum
54	W	Tungsten
55	U	Uranium
56	Mn	Manganese
57	Ni	Nickel
58	Co	Cobalt
59	Fe	Iron
60	Cu	Copper
61	Ag	Silver
62	Au	Gold
63	Ru	Ruthenium
64	Rh	Rhodium
65	Pd	Palladium
66	Os	Osmium
67	Ir	Iridium
68	Pt	Platinum
69	Tc	Technetium[1]
70	Re	Rhenium
71	Np, Pu . . .	Transuranium Elements

Flow diagram labels: HCl, $ZnCl_2$, $CrCl_2$, $ZnCrO_4$

Material presented under each Gmelin System Number includes all information concerning the element(s) listed for that number plus the compounds with elements of lower System Number.

For example, zinc (System Number 32) as well as all zinc compounds with elements numbered from 1 to 31 are classified under number 32.

[1] A Gmelin volume titled "Masurium" was published with this System Number in 1941.

A Periodic Table of the Elements with the Gmelin System Numbers is given on the Inside Front Cover